Particle Technology Forum 2016

Core Programming Area at the 2016 AIChE Annual Meeting

San Francisco, California, USA
13 - 18 November 2016

ISBN: 978-1-5108-3444-6

Printed from e-media with permission by:

Curran Associates, Inc.
57 Morehouse Lane
Red Hook, NY 12571

Some format issues inherent in the e-media version may also appear in this print version.

Printed by Curran Associates, Inc. (2017)

AIChE
120 Wall Street, FL 23
New York, NY 10005-4020

Phone: (800) 242-4363
Fax: (203) 775-5177

www.aiche.org

Additional copies of this publication are available from:

Curran Associates, Inc.
57 Morehouse Lane
Red Hook, NY 12571 USA
Phone: 845-758-0400
Fax: 845-758-2633
Email: curran@proceedings.com
Web: www.proceedings.com

TABLE OF CONTENTS

468962 Experimental Techniques to Study Non-Spherical Particles in Granular Flows

Abstract:

Modelling of gas-solid flows has long been restricted to spherical particles. As a result the translational velocity has gained most interest as this determines the overall solids flow. Recently a lot of effort is being put into the simulation of non-spherical particles and as such the particle orientation and rotation becomes more and more important. This in turn sets challenges to develop novel experimental techniques as well. In this paper an introduction is given to two new techniques that are expected to have a large impact in the study of non-spherical particulate flows.

Most experimental techniques are not so well suited to study the orientation of particles. Magnetic Particle Tracking (MPT) however is a technique that tracks a single magnetic marker that acts as a dipole introducing a small magnetic field that can be tracked with a set of Anisotropic MagnetoResistive (AMR) sensors. Because the dipole/marker has a position and the magnetic field a direction, both position and orientation of the markers can be tracked over time. Experiments in a 3D cylindrical fluidized bed have been performed with MPT for three different types of particles with similar mass and volume but different aspect ratios: spheres and two cylinders with aspect ratio of 2.25 and 4.5. Experiments show distinctly different flow patterns for the cylinders with respect to the spheres. The orientation of the cylinders is largely determined by their interaction with the walls., whereas the spheres have no preferred orientation as expected.

Because MPT can determine the orientation as an intrinsic property of the technique, it is a very powerful tool in studying particulate flows involving non-spherical particles. It can however not determine position and orientation of multiple particles at the same time and as such can also not determine the mutual alignment of particles. Image analysis techniques can do this but are restricted to visually accessible systems, i.e.; pseudo 2D fluidized beds. A Digital Image Analysis (DIA) technique has been developed that can determine position and orientation of cylinders of

particles in plane with the wall of the pseudo 2D fluidized bed. This technique was used to study the segregation and mixing of a mixture of spheres and cylinders of aspect ratio 10 with equal mass and volume. Orientation and position distributions as well as mutual alignment, determined with a nematic order parameter, are used to show that segregation occurs through formation of semi-aligned clusters of particles. These clusters are seen by the flow as larger particles, segregating from the bulk of the system.

These two techniques, MPT and DIA, are capable of determining two very important parameters to characterize non-spherical particles in granular flow, that previously could not be studied; orientation and mutual alignment.
Acknowledgements

This research was funded by the European Research Council, under the Advanced Investigator Grant Scheme, contract no. 247298 (Multiscale flows), and the 3TU Centre of Excellence â€“ multiscale phenomena

462498 Experimental and Numerical Investigation on Agglomerates in Fully-Developed Riser Flows

Abstract:

The behavior of cohesive particles is studied in the riser section of a circulating fluidized bed, where gas-solid flows are fully-developed. The effects of cohesion are quantified by the distribution of agglomerates experimentally measured by shadowgraphy. In additional to relative humidity, which induces cohesion via small capillary bridges that condense between particles, agglomerate properties are affected by superficial gas velocity and solid concentration. To predict agglomerates in risers, DEM-CFD simulations were conducted with a smaller domain than the experiments and used cyclic boundary conditions for both gas and solid phases, in an attempt to mimic the fully-developed flow in experiments. By incorporating a capillary force model that considers particle surface roughness and applying the same agglomeration detection method in simulations, direct comparison between numerical and experimental results are performed. Based on the validated simulation conditions, the response of agglomerate properties to relative humidity, superficial gas velocity and solid concentration was systematically studied, which helps aid the understanding of particle entrainment in riser flows.

463710 Detailed Modelling of Fluid Catalytic Cracking

Abstract:

The design and performance of the fluid catalytic cracking process (FCC) is studied by modelling the riser in detail. The selectivity and conversion of the cracking kinetics need to be coupled with the hydrodynamics to undertake reliable simulations of the process. Modelling the catalytic cracking of heavy oils is not an easy task, since it involves a very large number of components and reactions. The lumping kinetic approach, although unable to handle detailed reaction kinetics, is extremely powerful to simulate industrial FCC risers, for its ability to treat a large number of reactive networks. In this work, a novel approach which implements the continuum lumping kinetics in conjunction with CFD. The reactive systems is coupled with the complex hydrodynamics of the riser and experiments are implemented to validate the modelling framework. The results show non-uniformity of the gas-solid flow which affects the reaction temperature and residence time distribution. The methodology presented shows good ability to predict the product yields in an industrial FCC unit.

468546 Computational Study of the Influence of Injection Nozzles in FCC Industrial Risers

Abstract:

Petroleum is considered one of the most important resources nowadays because, in addition to dominating the world energy matrix, its derivatives are present in almost all human activities. Among its derivatives, gasoline is the most important and its production depends largely on the Fluid Catalytic Cracking process, also known as FCC. However, the study of the FCC process is complex as it involves several simultaneous phenomena. In this sense, modeling and simulation, in this case by computational fluid dynamics, are tools that enable the development and optimization of these equipments. Among the process variables that were highlighted in the literature in recent years, we can found the injection nozzles, however, simulations of industrial risers with these structures are still scarce. Therefore, this study validated a specific modeling able to simulate these structures, highlighting the weaknesses of the models currently used. Finally, a numerical study was carried out, through statistical tools, to evaluate the influence of the nozzles, varying some of the geometric parameters that define this equipment. Thus, it was possible to evaluate the parameters which have more influence on the sense of the gasoline yield. The results showed the existence of a statistical correlation between the yield of gasoline and the angle of the nozzle relative to the riser.

Numerical simulation of coal gasification in a counterflow fluidized bed reactor

Ronak Ghandriz and M. Reza H. Sheikhi*
Northeastern University, Boston, MA

Abstract

This study focuses on modeling and simulation of coal gasification in a bench-scale counterflow fluidized bed. The mathematical model is based on Eulerian-Eulerian description of multi-phase flows. The flow governing equations are solved by a finite volume method in which a time-splitting scheme is used to integrate the chemical reaction source term in a separate step using a stiff ordinary differential equation solver. The solid phase consists of Kansas bituminous coal, which contains four pseudo-species: ash, moisture, volatile matter and fixed carbon. The gas phase includes six species: CO, CO_2, CH_4, H_2, H_2O and N_2. The bed is a counterflow gasifier in which coal is fed to the reactor from the top by gravity. Steam serves as the fluidizing media which enters form the bottom. Gasification occurs following devolatilization and cracking processes as incoming coal particles heated rapidly to the gasification temperature. The char gasification and water-gas shift reactions are described by a four-step global mechanism. Simulations are conducted in a three-dimensional domain and capture the transient behavior of the reactor. Results show good agreement with experimental data.

1 Introduction

Fluidized beds provide suitable reactors for gasification due to thorough mixing of solid particles with gaseous reactants which leads to high rate of heat transfer and chemical reaction [1]. There are two types for fluidized bed gasifiers: circulating beds, which use Geldart A particles at very high velocities, and bubbling fluidized bed with lower gas velocity and negligible carryover of solids. Gasification of coal in fluidized beds has been the subject of many studies. Depending on the gasifying medium direction of flow into the bed of solids, gasifiers can be of three types: coflow, counterflow and crossflow. In the present study, coal gasification is considered in a counterflow fluidized bed which has been studied experimentally by Neogi [2]. Counterflow gasifiers provide several advantages such as simplicity in design and construction, high conversion efficiency and the possibility of operation with

*Current address: Dena Scientific, Wellesley, MA.

many types of feedstock. In these gasifiers gas stream enters the reactor from the bottom while the solid fuel is introduced from the top. In addition to laboratory measurements, Neogi [2] also used a modified version of an existing coal gasification kinetics for smaller size coal particles to simulate the gasification using a one-dimensional mathematical model. In this model, empirical equations for hydrodynamics of the bubbling regime was employed without taking account of momentum conservation equations. The objective of the present work is to provide more detailed study of this gasifier by considering gas and solid momentum balances along with mass and species conservation equations. Multiphase flow modeling is done using Eulerian-Eulerian approach (also known as two-fluids model). Simulations are conducted on a three-dimensional (3D) domain. The char gasification kinetics model is that provided by Neogi [2]. The model does not consider the initial devolatilization. Therefore, the devolatilization product distribution proportion is derived from pyrolysis results of the experiment by Yeboah *et al.* [3].

2 Formulation

In the Eulerian-Eulerian multiphase flow modeling, different phases are considered to be interpenetrating continuous media. A volume fraction is associated with each phase to describe the amount of space occupied by each phase in an infinitesimal volume within the computational domain. Following this assumption, conservation laws are expressed for each phase to account for transport of mass, momentum and energy. In this study, these equations include the conservation of mass and momentum for individual phases as well as transport of chemical species for each phase. Conservation of mass for each phase includes generation or destruction of bulk phase mass due to chemical reaction. Conservation of momentum includes forces resulting from interactions between different phases as well as solid stress and granular pressure which need to be modeled. The fluid-solid momentum transfer is represented by Syamlal-O'Brien model [4]. The gas phase is considered to be a Newtonian fluid. The solid stress tensor is described by Johnson *et al.* [5] model and the granular pressure is modeled by Lun *et al.* [6]. The species transport equation is expressed for each species present in each phase and includes the effect of heterogeneous chemical reaction. These equations are solved using a time splitting method in which the species transport equations are first solved without the chemical reaction source term. Subsequently, this source term is directly integrated using a stiff ordinary differential equation solver.

The coal considered in this study is Kansas bituminous coal. The gasification kinetics parameters for this coal is a four step reaction as presented in Table 1. The first three reactions [7] are heterogeneous, transforming char into CO, H_2 and CH_4. Rate of these reactions are expressed as $R_i = A_p k_i C_i$, $(i = 1, 2, 3)$ where C_i $[mol/cm^3]$ is the concentration of the reactant gas in each reaction and A_p $[cm^2/cm^3]$ is the total particle surface area per unit volume. In this equation, the rate coefficient $k_i = A_i \exp(\frac{-E_i}{R_u T})$ where A_i, E_i,, R_u and T are the pre-exponential factor, activation energy, universal gas constant and temperature, respectively. The pre-exponential factor and activation energy for these reaction (Table 1) are the adjusted values provided by Neogi [2]. The fourth reaction is the water-gas shift reaction whose rate is expressed as $R_4 = k_4 \epsilon_g (C_{\mathrm{CO}} C_{\mathrm{H_2O}} - \frac{1}{K_{eq}} C_{\mathrm{CO_2}} C_{\mathrm{H_2}})$, where C_i and K_{eq} are concentration of chemical species and the equilibrium constant, respectively. The kinetic parameters for this reaction is reported by Biba *et al.* [7] and Yoon [8].

Table 1: Kinetic parameters for the reactions.

No.	Reaction	E_i $[kJ/kmol]$	A_i		K_{eq}
1	$C + H_2O \longrightarrow CO + H_2$	121,417	13.9	$[cm/s]$	-
2	$C + CO_2 \longrightarrow 2\,CO$	360,065	5.56×10^6	$[cm/s]$	-
3	$C + 2\,H_2 \longrightarrow CH_4$	230,274	20.8	$[cm/s]$	-
4	$CO + H_2O \rightleftharpoons CO_2 + H_2$	12,560	2.78×10^6	$[cm^3/mol.s]$	$0.0265\exp(3955.7/T)$

3 Flow configuration and numerical specifications

The counterflow fluidized bed configuration is based on the experimental work of Neogi [2]. Figure 1 shows a schematic of the reactor. The lower part of the reactor has an internal diameter and height of $10\,cm$ and $30\,cm$, respectively. To reduce slugging, an expansion section is considered on top with $14\,cm$ diameter and $20\,cm$ height. The bed is initially filled with inert matrix consisting of 25% limestone and 75% silica (by weight), which is initially fluidized by N_2 stream from the bottom. When fluidized, this stream is replaced with steam at temperature of $1073\,[K]$ to start the gasification process. The coal is fed into the reactor from top by gravity through a vertical feed pipe with $3\,cm$ diameter. This feed is discharged at a location about $8\,cm$ above the static bed, which has an initial height of $10\,cm$. A purge flow of nitrogen assists flow of coal through the feed pipe and prevents gas back-flow. The gasification products exit the reactor from the top through an outlet pipe with $3\,cm$ diameter, as shown in Fig. 1. The outlet stream subsequently passes through a cyclone to remove fine solid particles. The experimental study is performed with uniform temperature in the bed and freeboard [2]. This is done by supplying heat for gasification endothermic reactions using electrical heaters. The value of parameters related to gasifier operating condition are provided in Table 2.

The governing equations are discretized using finite volume method with first and second order accuracy in time and space, respectively. To approximate fluxes across control volume faces the Superbee scheme along with Downwind Weighting Factor (DWF) normalized function are used. In momentum equations, the interphase source terms include velocity components of various phases which cause a strongly coupled set of equations leading to a nonstandard solution matrix structure. For this reason, the partial elimination of interphase coupling is employed for decoupling the equations [9]. The Semi-Implicit Method for Pressure-Linked Equations (SIMPLE), modified for multiphase flow calculations, is used for solving the momentum equations. Similar to that in gas phase, SIMPLE algorithm is applied for dispersed phases with solid pressure correction [10]. To integrate the chemical reaction source term in the reaction sub-step, the stiff ODE solver LSODA [11] is used with the relative and absolute error tolerance values of 10^{-6} and 10^{-8}, respectively. Simulations are carried out on a three-dimensional (3D) cylindrical mesh with grid resolutions of 20, 15 and 12 coal particle diameters (d_p), where $20\,d_p$ corresponds to $\Delta r = 0.5\,cm$, $\Delta z = 1.0\,cm$ and $\Delta\theta = 0.52\,rad$; $15\,d_p$ corresponds to $\Delta r = 0.39\,cm$, $\Delta z = 0.39\,cm$ and $\Delta\theta = 0.44\,rad$;

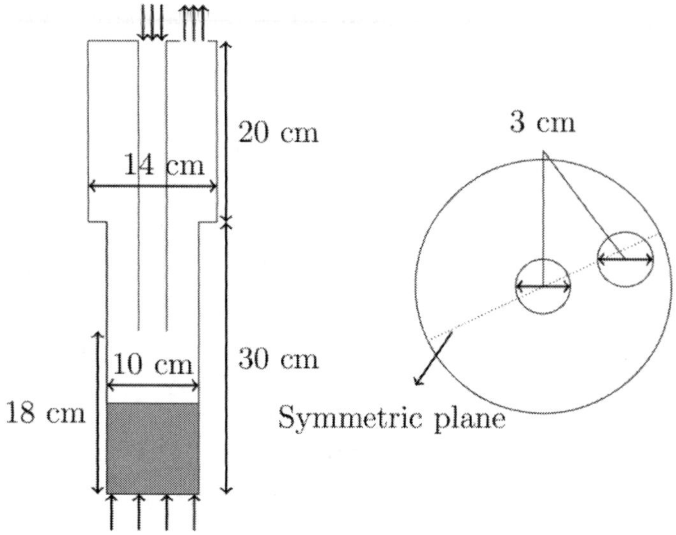

Figure 1: Schematics of the axial cross section (along the symmetric plane) and top view of the gasifier [2].

Table 2: Operating condition [2]

Feed	
Coal feed rate$[g/s]$	0.085
Steam supply rate$[g/s]$	0.22
Operating pressure $[atm]$	1.0
Entrance temperature $[K]$	1073
Char surface area $[cm^2/g]$	13130
Coal particle diameter $[cm]$	0.0297
Coal density $[g/cm^3]$	1.58
Bed	
Operating temperature $[K]$	1073
Minimum fluidization void fraction	0.43
Minimum fluidization velocity $[cm/s]$	12.0
Mean particle diameter $[cm]$	0.03
Superficial velocity $[cm/s]$	16.5
Static bed height $[cm]$	10.0
Bed density $[g/cm^3]$	2.608

and $12\,d_p$ corresponds to $\Delta r = 0.35\,cm$, $\Delta z = 0.36\,cm$ and $\Delta\theta = 0.34\,rad$. Δr, Δz, $\Delta\theta$ denote grid spacings in radial, axial and circumferential directions, respectively. The initial condition of the reactor is considered to be the onset of the gasification when the bed is already at the fluidized state and steam begins to flow from the bottom. At the initial time, the bed is assumed to include a single solid phase representing the inert matrix. The inlet condition is specified according to gasifier operating conditions (Table 2). The inlet stream entering the reactor from the top consists of solid and gas phases resulting from coal devolatilization accompanied by a nitrogen purge flow. The solid phase includes 81% char and 19% tar (by weight). The gas phase is composed of 0.008% CO, 0.006% CO_2, 99.958% N_2, 0.01% H_2, 0.01% H_2O and 0.008% CH_4 (by volume). The inlet stream from the bottom consists of steam. The outlet boundary condition is set to constant pressure of $1\,bar$. The Johnson-Jackson [5] and no-slip boundary conditions are used for solid and gas phases near the walls, respectively.

4 Results

The Multiphase Flow with Interphase eXchanges (MFiX) code [10], coupled with the time splitting method, is employed for computational fluid dynamics study of the fluidized bed gasifier. Simulations are carried out for $t = 40\,s$ of the flow time and mean (time-averaged) quantities are calculated after the flow reaches statistically stationary state. Grid resolution study is preformed by considering $12\,d_p$, $15\,d_p$ and $20\,d_p$ mesh sizes specified in §3. Figures 2 shows the mean void fraction profiles across radial direction for the three grid resolutions at $z = 5\,cm$. In this figure, the peak values are caused by the rise of pockets of gas (bubbles) on both sides of the inlet pipe in the middle. Near the centerline where the solid particles are introduced void fraction show lower values. With $20\,d_p$ mesh, profiles look overly smooth due to excessive numerical diffusion. Simulations with $15\,d_p$ and $12\,d_p$ mesh sizes give more clear distinction between gas and solid regions. It is noticed that there is no significant change in the results when the grid is refined from $15\,d_p$ to $12\,d_p$ and the solution becomes less sensitive to mesh resolution. Figure 3 shows the mean void fraction along the vertical coordinate for

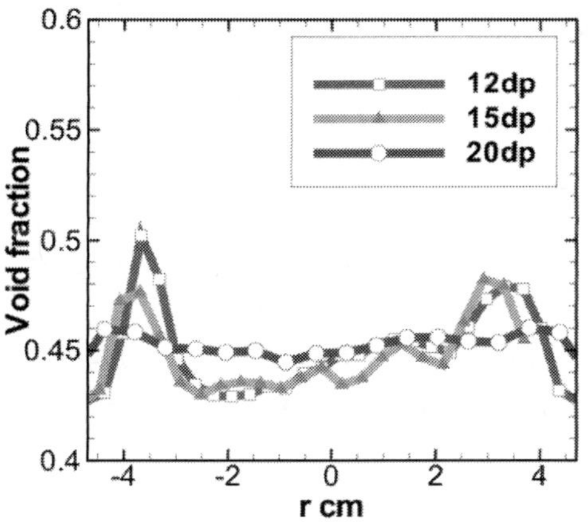

Figure 2: Grid resolution study for mean void fraction along radial coordinate at $z = 5\,cm$ on the symmetric plane.

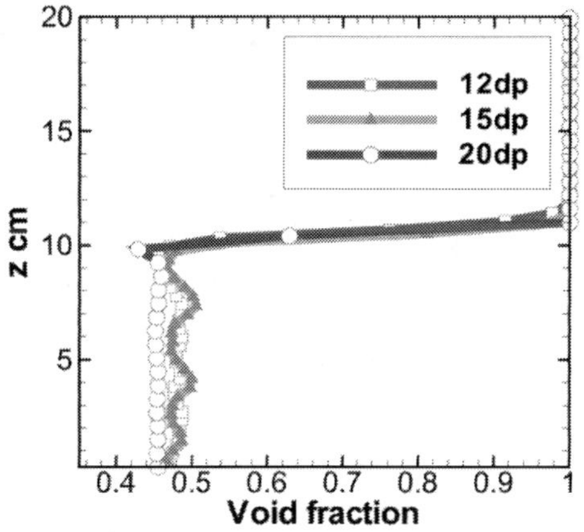

Figure 3: Grid resolution study for mean void fraction along vertical coordinate on the symmetric plane at $r = 3.5\,cm$ (towards the outlet pipe).

different grid sizes. Despite the differences in radial profiles shown above, the bed height is predicted almost similarly using the three mesh resolutions. It is clear that refining the mesh requires larger simulation time which can be prohibitive in reacting flow simulations. Considering the similarity of the results obtained with $12\,d_p$ and $15\,d_p$ mesh sizes, a uniform grid of $15\,d_p$ is used in the remainder of this study as a compromise between accuracy and computational cost.

Figure 4 illustrates the mean void fraction contours on the axial cross section (symmetric plane) of the reactor. The bed is fluidized with the formation and convection of bubbles from the bottom on both sides of the centerline. As the bubbles move, they transport the solid particles to the wake region underneath them which enhances the mixing of gas with coal particles. Near the center of the reactor, the higher solid volume fraction values indicate the location where coal particles enter the bed. The freeboard region is filled with gas devoid of solid particles. The mean velocity vectors show that near the centerline both gas and solid flow downward due to dominant inflow of coal and nitrogen from the middle pipe. The particles move upward farther away from the centerline due to the steam injection from the bottom of the reactor which causes raising of bubbles and solid particles. Two strong circulation regions near the wall help accelerate mixing between particles and gas flow regimes. The H_2 and CO mass fraction contours are shown in Fig. 5. These species show higher concentrations near the coal feed pipe where the rate of gasification is high

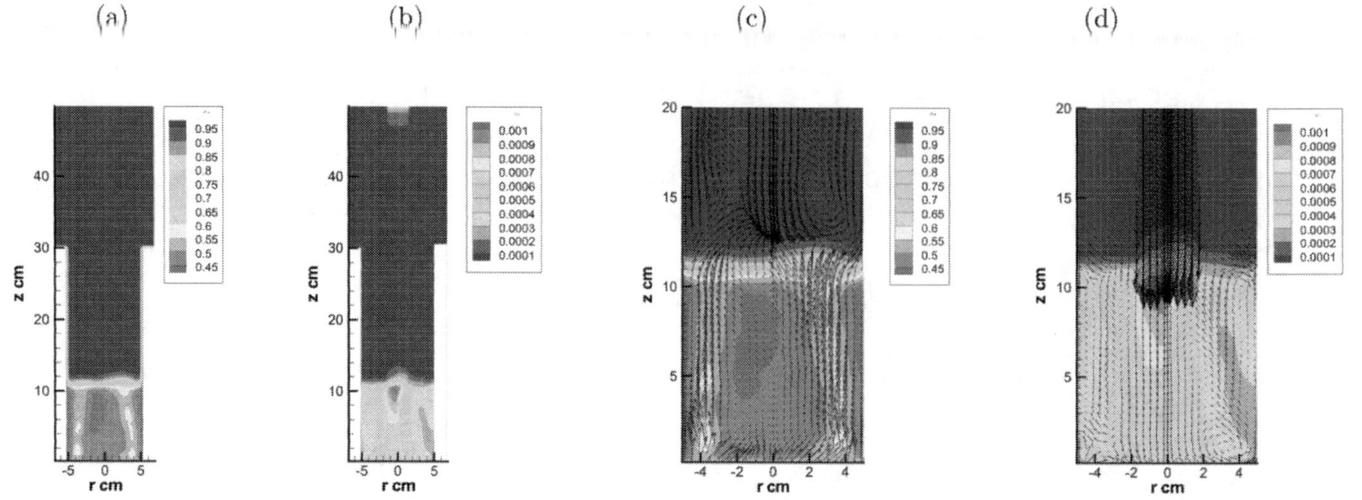

Figure 4: Mean volume fraction and velocity vectors on the symmetric plane. a) Void fraction, b) coal volume fraction, c) gas velocity vectors superimposed on the void fraction contours, and d) coal velocity vectors superimposed on the coal fraction contours.

resulting from strong local mixing of coal and steam. Small concentration of these products is observed in the bottom of the bed near the steam inlet where the coal concentration is too low for the gasification reactions to proceed. Radial profiles of gasification gaseous product mole fractions at $z = 5\,cm$ are shown in Fig. 6. Considering that char participates in the the first three reaction steps, the gaseous product concentrations depends strongly on the solid volume fraction entering the reactor. The lowest gaseous concentration belongs to CH_4, due to the slow methanation rate in the gasification mechanism. CO and CO_2 show nearly similar values at this location. CO shows higher concentration than CO_2 as gasification reactions become more dominant throughout the bed. At the outlet (Fig. 7) CO_2 has higher molar fraction than CO as the forward water-gas shift reaction (fourth reaction in Table 1) becomes more dominant in the absence of char. This is consistent with Wen *et al.* [12] observations. H_2 has slightly smooth gradient along the lateral direction with the highest molar concentration inside the bed and at the outlet. Figure 7 shows a comparison of the simulation results with the experimental data. In general, results are in reasonable agreement with the data. It is noticed that CH_4 has relatively small concentration at the outlet. Simulations yield almost no methane at the outlet whereas in the experiment larger concentration is captured. This slight discrepancy is currently under investigation. Simulation results are further compared with the experimental data by considering the gas mass yield. The gas mass yield (kg gas/kg feed, dry and ash free (DAF)) is defined as $\sum_{i=1}^{N_g} Y_i W_i$ where Y_i and W_i denote mole fraction and molecular weight for N_g number of gaseous species, respectively. Simulations and experiment values of gas yield are 1.7 and 1.64 kg gas/kg feed DAF, respectively, which are in reasonable agreement. The lower value obtained in the simulation is due to neglecting heavier C_2-C_4 hydrocarbons in the chemical mechanism.

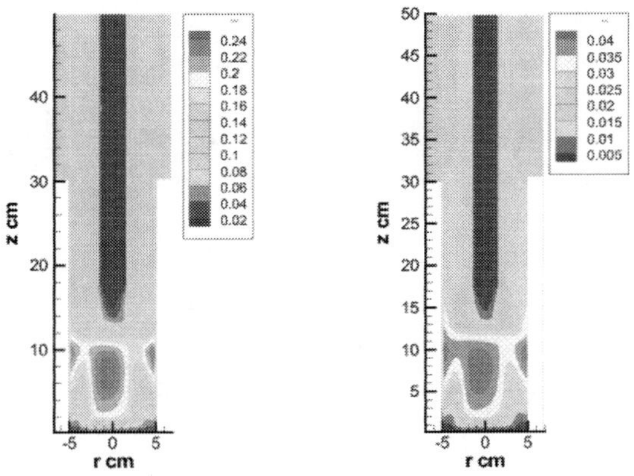

Figure 5: Mean species mass fraction contours on the symmetric plane: a) CO, and b) H_2.

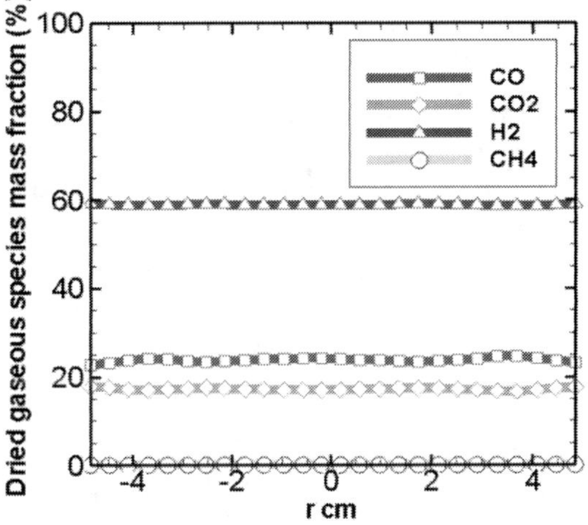

Figure 6: Mean species molar fraction profiles vs. radial coordinate at $z = 5\,cm$ on the symmetric plane.

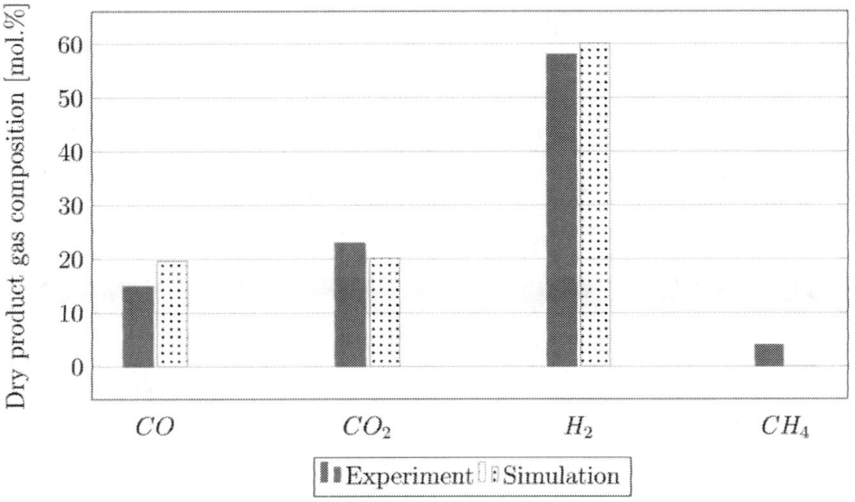

Figure 7: Predicted mean dried gas phase species molar fractions compared with the experimental data.

References

[1] Froment, G. F., Bischoff, K. B., and De Wilde, J., *Chemical Reactor Analysis and Design*, John Wiley and Sons Inc., Hoboken, NJ, 2011.

[2] Neogi, D., Coal Gasification in an Experimental Fluidized-Bed Reactor, M.Sc. Thesis, Department of Mechanical Engineering, Kansas State University, Manhattan, KS, 1984.

[3] Yeboah, Y. D., Longwell, J. P., Howard, J. B., and Peters, W. A., Effect of Calcined Dolomite on the Fluidized Bed Pyrolysis of Coal, *Ind. Eng. Chem. Process Des. Dev.*, **19**:646–653 (1980).

[4] Syamlal, M. and O'Brien, T. J., Hydrodynamics of Fluidization: Prediction of Wall to Bed Heat Transfer coefficients, *AIChE J.*, **31**:127–135 (1987).

[5] Johnson, P. C. and Jackson, R., Frictional-Collision Constitutive Relations for Granular Materials with Application to Plane Shearing, *Russ. J. Math. Phys.*, **176**:67–93 (1987).

[6] Lun, C. K. K., Savage, S. B., Jeffrey, D. J., and Chepurniy, N., Kinetic Theories for Granular Flow: Inelastic Particles in Couette Flow and Slightly Inelastic Particles in a General Flow Field, *J. Fluid Mech.*, **140**:223–256 (1984).

[7] Biba, V., Macak, J., Klose, E., and Malecha, J., Mathematical Model for the Gasification of Coal under Pressure, *Ind. Eng. Chem. Process Des. Dev.*, **17**:92–98 (1978).

[8] Yoon, H., Modeling and Analysis of Moving Bed Coal Gasifier, Ph.D Thesis, Department of Chemical Engineering, The University of Delaware, Newark, DE, 1978.

[9] Spalding, D. B., Numerical Computation of Multi-Phase Fluid Flow and Heat Transfer, Pineridge Press., Whiting, NJ, 1980.

[10] Symlal, M., Rogers, W., and O' Brien, T., MFiX Open Source Documentation, National Energy Technology Laboratory, Morgantown, WV, 1993.

[11] Hindmarsh, A. C., ODEPACK, A Systemized Collection of ODE Solvers, Lawrence Livermore National Labratory, Livermore, CA, 1982.

[12] Wen, C. Y. and Dutta, S., Rates of Coal Pyrolysis and Gasification Reactions, Addison-Wesley Publishing Co., Boston, MA, 1979.

464821 Cold Flow Validation of Catalyst-Biomass Hydrodynamics in Catalytic Fast Pyrolysis

Abstract:

Catalytic Fast Pyrolysis (CFP) is a promising technology for producing renewable transportation fuels and chemicals from lignocellulosic biomass. In the CFP process, circulating fluidized beds (CFBs) of catalyst particles (ranging from dilute-phase (risers) to dense-phase bubbling beds) are used to simultaneously pyrolyze biomass and catalytically upgrade the vapors to a thermally stable, partially deoxygenated bio-oil. The bio-oil is subsequently upgraded and fractionated to produce valuable fuels and chemicals.

Inaeris Technologies (formerly KiOR) is a world leader in development and commercialization of the CFP process. There are many commercial applications of CFBs, but because of the special requirements of biomass pyrolysis, the applicability of the large and well-established body of CFB learnings to CFP is quite limited. For this reason, Inaeris has been executing an aggressive research program into the relevant physics and chemistry of the CFP process, with parallel development of a CFD model capable of reactor design, scale-up, troubleshooting and optimization. To achieve the model development objectives at an accelerated pace, Inaeris has been collaborating with CPFD Software, utilizing the Barracuda Virtual Reactor® software.

The Inaeris-CPFD collaboration has investigated, modeled and reported on catalyst fluidization and catalyst-biomass mixing in fixed fluidized beds [1], followed by catalyst circulation in CFBs[2]. The current stage of the program is investigating and modeling catalyst-biomass hydrodynamics in CFBs, specifically addressing:

(1) Pneumatic transport, agglomeration and deagglomeration of biomass, and

(2) Mixing of catalyst and biomass under dynamic conditions

Both these objectives involve generating experimental validation data under cold flow conditions against which CFD predictions are compared and verified. A lab-scale cold-flow CFB unit has been constructed such that catalyst and biomass can be independently pneumatically conveyed from separate, positive displacement feeders through transport lines to mock reactors under differential pressure conditions. This unit is flexible enough to test different geometries for mock reactors, different configurations for catalyst-biomass contacting and mixing, and wide ranges of catalyst, biomass and gas flows. Existing measurement capabilities include mass holdups of catalyst and biomass in the mock reactor, pressure profiles, particle size distribution (PSD) analysis, quality of mixing, bed-building kinetics, residence time distributions, and visual observations.

Both experimental and CFD results for catalyst only circulation confirm transition from dense regime to dilute regime as a function of transport gas flowrate. This transition is evident in solids holdup in the mock reactor and the corresponding pressure drop across the unit. Furthermore, such flow regime shifts are determined to be sensitive to catalyst PSD and solids feed rate. We will present the CFD models for catalyst circulation and detail model applicability for a range of configurations, feed rates and flow ranges. Next, we will report on biomass incorporation and co-feeding configurations in the experimental setup. Experimental methods to ascertain catalyst-biomass mixing in the mock reactor will be discussed and corresponding CFD models will be presented.

References

[1] B. Adkins, N. Kapur, J. Pendergrass, J. Parker, P. Blaser, KiOR Update: Incorporating Barracuda in our CFP Development Process, Proceedings of the Inaugural Barracuda Virtual Reactor Usersâ€™ Conference, Santa Ana Pueblo, New Mexico, 2015.

[2] B. Adkins, N. Kapur, T. Dudley, S. Webb, P. Blaser, Experimental validation of CFD hydrodynamic models for catalytic fast pyrolysis (CFP), Fluidization XV, May 22-27, 2016, Quebec, Canada.

471165 Three-Dimensional CFD Simulation of a Circulating Fluidized Bed Reactor for CO_2 Capture Processes Using Solid Sorbents

Abstract:

Circulating fluidized beds (CFB) have been widely used in the fluid catalytic cracking (FCC) units, gasification of coal and biomass, synthesis of Olefins from Methanol, and chemical looping and will continue to be used in forthcoming processes such as CO_2 capture units using solid sorbents. The simulation of individual parts of the system was recently performed and can provide qualitative, but not quantitative, estimation of the system behavior. However, to quantitatively simulate the entire loop, CFD simulation of the full-loop of the circulating fluidized bed (NETLâ€™s C2U unit) was performed. For this system, our calculations showed that the circulating fluidized bed behavior through the transient period, and reaching to a statistical steady state after about 10 seconds. This time strongly depends on the initial condition of the simulations; however, the order of magnitude of this time depends mainly on the geometry and scale of the system and should be in vicinity of 10 seconds for different initial conditions that was also reported by NETL for the bench-scale C2U unit. The calculated fluidization regime, pressure drop, and solid circulating rate using our simulation were in agreement with NETL experimental observation. The results of our simulation showed the excellent capability of our model in describing the flow behavior in a circulating fluidized bed system which includes the whole range of solid volume fraction from very dilute parts (i.e., riser) to very dense parts (i.e., standpipe).

449928 Situating the High-Density Circulating Fluidized BED with the EMMS MODEL

Abstract:

The dense suspension upflow (DSU) regime is characterized with high suspension densities, high solids flux and no downward flux near the wall. It was defined to explain the different flow behavior in high-density circulating fluidized bed (CFB) from those in fast fluidization and pneumatic conveying. Experimental correlations have been proposed in literature to define the boundary of this regime but we are still far from clear understanding of the underlying mechanisms. In this work, we will apply the energy-minimization multi-scale (EMMS) model to predict the flow regime transitions, in particular, the DSU regime and the choking transition in CFBs. Comparison between model prediction and available experimental data is provided.

462053 Maximizing Poorly Water-Soluble API Nanoparticle Loading in Polymer Strip Films

Abstract:

One of the most prevalent misconceptions of the polymer strip film format for pharmaceutical applications is its limited active pharmaceutical ingredient (API) loading. While this is partially due to the size restriction of a typical orodispersible dosage (~6 cm^2 and ~40 μm thick), this limitation also stems from API particle stability issues in traditional pharmaceutical film manufacturing processes. For instance, solvent casting imposes an inherent limit on the API loading in the resulting film, as exceeding a certain API loading leads to uncontrollable precipitation of API particles upon removal of organic solvent during drying. The objective of this work was to challenge the previously conceived limits of API loading in pharmaceutical strip films using a combination of API nanomilling and slurry casting, as well as evaluate the practical limitations of strip films with high loadings of API nanoparticles. Hydroxypropyl methylcellulose (HPMC) was used as the film-forming polymer and griseofulvin (GF) was used as a model BCS class II API. HPMC-E15 films were prepared with API loadings between 9â€“49 wt% GF and HPMC-E4M films were prepared with API loadings between 30â€“73 wt% GF. All HPMC-E15 films exhibited good content uniformity and API nanoparticle redispersibility, whereas HPMC-E4M films had slightly worse content uniformity and poor API nanoparticle redispersibility above 50 wt% GF loading. Although film strength remained strong with increasing GF loading, a significant decrease in percent elongation was observed, resulting in film brittleness. Results demonstrate that, with proper particle engineering and stabilization protocols, high loading of poorly water-soluble API nanoparticles is indeed achievable in polymer films.

451652 Computational Approaches for Modeling Fluid Bed Granulation Process

Abstract:

Understanding the fundamental multi-physics behind the fluid bed granulation process and its projections in macroscopic scale provide the essentials to fabricate granulate with desired material properties. A proper prediction of physicochemical properties and morphological characteristics of a continuous wet granulation process product, particularly in response to operation conditions, is contingent upon the knowledge of the contact pattern evolution and the microstructure formation under the effect of the fluidizing medium. In this regard, current study looks at unveiling correlation between the process conditions and the main mechanisms that determine the material properties of granulate during fluid bed granulation process.

A multi-component modeling approach was explored that enables us to understand the fundamentals of fluid bed granulation process through both statistical modeling and first-principle modeling tools. A supervised machine learning algorithm that can simulate the evolution of process variables during fluid bed granulation process was explored, including: particle diameter, particle size distribution and loss on drying. Model predictions agree well with experimental analysis. Currently the focus is on a numerical framework that integrates design of experiment and the findings of statistical models with multi-scale mechanistic models such as discrete element model and computational fluid dynamics models. Using such models in early and late stage development phases will enhance the development timelines and increase the process product understanding.

461274 Comparison of the Electrostatic Properties of Amorphous and Crystalline Lactose

Abstract:

Background

Conversion from a crystalline to amorphous state in solid materials may be observed as a result of various processes applied to pharmaceutical materials during manufacturing. Examples include rapid crystallisation from solution to mechanical activation of the crystalline phase upon milling.[1-2] The resulting products usually display different thermodynamic properties to pure crystalline forms, hence affecting their subsequent performance.[2-3] Furthermore, powders are susceptible to a process called tribocharging whenever movement of particles across the surface is involved. If surface amorphization is observed during processes involving the movement of powder, such as during milling, generation and accumulation of electrostatic charge could be affected significantly, hence affecting the overall powder flow and sticking.

Aims

The aim of this work is to determine how the presence of an amorphous state influences the electrostatic performance of lactose. The main objectives involve determining whether the material electrostatic characteristics such as the maximum charge accumulated and charge decay behaviour are directly dictated by the materialâ€™s state.

Methods

Amorphous lactose was prepared by spray drying from water and storing at dry conditions to limit its re-crystallisation. The electrostatic properties of both pure crystalline and pure amorphous lactose with known particle size were studied using JCI 155v6 Charge Decay Analyser[4] under controlled environmental conditions (25 Â°C and 25% RH). Subsequently, four blends containing 15%, 30%, 50% and 75% amorphous content by mass were prepared and characterised using the same method. All samples studied were further characterised using PXRD, RST and SEM. The analysis for further applied to anhydrous lactose prepared by dehydration of crystalline lactose by exposing to elevated temperatures.

Results

Electrostatic characterisation of pure amorphous and crystalline lactose revealed that the materials showed significantly different behaviour in terms of maximum charge accumulated as a result of corona charging and decay characteristics. The maximum charge of -684.65 ± 45.04 V/g acquired by the pure amorphous lactose was twice as large as the charge accumulated by the pure crystalline sample (-337.64 ± 5.56 V/g). Although the smaller particle size of amorphous lactose could potentially account for its greater overall charge accumulation, the decay profiles are significantly different for the two forms with the amorphous lactose showing significantly slower charge decay. The time of 113.19 ± 8.30 min is required for the charge on the amorphous lactose to decay to 10% of its initial value, whereas only 10.03 ± 0.16 min is needed for the crystalline sample to reach the same level. Results obtained for four blends with varied amorphous content show that the material performance is directly dictated by its composition and a linear correlation was obtained between charge accumulated and amorphous content of the blends. The same correlation was observed for the charge decay profiles.

Conclusion

The results obtained show that amorphous lactose is able to accumulate a larger amount of electrostatic charge, which is retained for a prolonged time compared to its crystalline form. This behaviour is then directly translated into lactose blends having varied amorphous content, where both components exist in a form of physical mixture. Even though amorphous lactose is unstable and tends to convert to crystalline form quite rapidly, its generation during processing could lead to significant charge generation and accumulation that otherwise would not be expected for a crystalline material.

Acknowledgments

The authors acknowledge Dr Jin Wang Kwek for his contribution to this work.

References:

(1) Descamps, M.; Willart, J. F.; Dudognon, E.; Caron, V., Transformation of pharmaceutical compounds upon milling and comilling: the role of T(g). *J. Pharm. Sci.* **2007,** *96* (5), 1398-1407.

(2) Hancock, B. C.; Zografi, G., Characteristics and significance of amorphous states in pharmaceutical systems. *J. Pharm. Sci.* **1997,** *86* (1), 1-12.

(3) Fitzpatrick, J. J.; Hodnett, M.; Twomey, M.; Cerqueira, P. S. M.; O'Flynn, J.; Roos, Y. H., Glass transition and the flowability and caking of powders containing amorphous lactose. *Powder Technol.* **2007,** *178* (2), 119-128.

(4) Rowland, M. G. Electrostatic Properties of Particles for Inhalation. University of Bath, 2014.

458202 Particle Engineering Research from the NSF Erc Research Experience for Teachers (RET) Program at Njit

Abstract:

We aim to give highlights of particle engineering research from the New Jersey Institute of Technology (NJIT)'s Research Experiences for Teachers (RET) Program, which is a collaboration of the NSF Engineering Research Center for Structured Organic Particulate Systems (ERC-SOPS) and the Center for Pre-College Programs. This NSF-funded program engaged pairs of 14 high school teachers in 2015 summer research program, where they performed research on various particulate materials and processes relevant to pharmaceutical products. In addition, various faculty and educational experts delivered workshops on technical writing, best research practice, effective collaboration in a team, pharmaceutical industry and process–formulation development, educational module development, etc. The teachers acquired the skills and knowledge of research and the subject of particle engineering and pharmaceutical engineering that they incorporated into their teaching practice. In the research component of the program, the teachers collaborated with graduate student mentors under the guidance of professors and worked on various particle–pharmaceutical engineering projects. In the educational component, the teachers came up with a professional development plan and prepared "educational modules", which were delivered to high school students. While this presentation will expose few educational highlights including sample modules developed by the teachers, it will mainly focus on two summer research projects about the bioavailability enhancement of poorly water-soluble drugs via drug nanoparticle composites and drug nanoparticle-laden polymer strip films. Through application of particle engineering approaches such as nanoparticle formation–stabilization and drug encapsulation within hydrophilic polymeric matrices, we have ensured fast redispersibility of drug nanoparticles from solid dosages and demonstrated significant improvement of the dissolution rate of poorly water-soluble drugs.

463026 Study of DPI Carriers Design Parameters for Dry Powder Inhalation through Artificial Neural Network Modeling

Abstract:

Carriers are commonly used in dry powder inhalation to enhance the flowability, aerosolization and drug delivery efficiency of fine drug particles. The properties of carriers such as, size, shape, density, surface roughness, as well as carrier-to-drug ratio, inhalation flow rate, type of dry powder inhalers have important impacts on the drug delivery efficiency. However, there are contradictory reports in the literature on the effects of individual parameters on drug delivery efficiency. In this study, the effect of carrier properties, carrier-to-drug ratio, and inhalation flow rate on drug delivery efficiency in terms of emitted dose (ED) and fine particle fraction (FPF) are examined using artificial neural network (ANN) modeling. Image analysis is performed on scanning electron microscopy (SEM) images of carriers to determine the surface properties. Sensitivity analysis is carried out to determine important parameters that have the most impact on ED and FPF. Genetic algorithm (GA) is applied to determine the ANN modeling parameters. The effects and interactions of various factors on ED and FPF are revealed.

453857 The Surface Properties of Organic Crystalline Solids and Their Interactions with Polymeric Excipients and Binders

Abstract:

The physiochemical properties of solid state pharmaceuticals are crucial in understanding drug manufacturing, processing, and biological absorbtion. They govern the flowability, dispersibility, bulk density, dustiness, compressibility, and particle aggregation of active pharmaceutical ingredients (APIs)[1]. Binders and excipients are added to the formulation so as to achieve greater uniformity in the final product, and to enlarge the drug particles by agglomeration. Thus, particle design and characterozation are crucial for the understanding of the product being delivered to the consumer[2,3]. Many advances have been made in this research but given there are many mechanisms that affect the final product, itâ€™s difficult to model and predict what will occur accurately, such as adhesion, attrition, coalescence etc[4].

In pharmaceutical processes micron and submicron sized particles are used. In this scale, surface properties such as surface free energy heavily influence particle behaviour[5]. Surface free energy, or more conveniently surface energy, is defined as the energy required to form a surface of a particular material per unit area, where various components make up the total surface energy for one specific facet of a crystal. By having a better understanding of these components and the surface energies, researchers can better predict the properties of drugs. This includes how particles form, grow, break, stick, flow, and more. This then offers more data for researchers to understand how the APIs interact with polymeric excipients and binders.

Current research and models have too often made assumptions disregarding the surface heterogeneity of the particles[6]. Due to the anisotropy of the crystal, the properties are vastly more complex than the simple assumptions of spherical homogeneous particles[6]. Measuring these differences is crucial in the further development of this field and drug formulation as a whole. The most appropriate way to measure the facet specific surface energies is by sessile drop contact angle measurements or inverse gas chromatography[5]. This can be expanded further to compare the energetic interactions of pharmaceuticals with solvated excipients and see how we may better predict the physiochemical properties of the resulting particles[4].

Our study chooses to focus on the formulation of pharmaceuticals alongside polymers so as to offer a better understanding of their properties during such processes. This will offer better control in their manufacturing and later ties in to the drugs bio-availability. Our investigation is divided into two parts, the first involves property analyses of the crystals and powders of the API, and the second the interactions between the drugs and excipients. Here we chose to use both three polymorphs of anhydrous Carbamazepine and Copovidone as our model API and excipient respectively. Together, we offer further data to facilitate the modelling in the particle design of pharmaceuticals.

We performed studies on the polymorphs of anhydrous Carbamazepine using: sessile drop contact angle measurements (CA), finite dilution inverse gas chromatography analysis (FD-iGC), and X-ray photoelectron spectroscopy (XPS). The interactions between Copovidone and the main facets of P-Monoclinic Carbamazepine were analysed with two experiments. In the first, we measured the contact angle of Copovidone dissolved in water at various amounts, on the crystal facets of Carbamazepine. In the other we measured the contact angle of water on the crystals facets with spun cast Copovidone on them. The combination of these instruments helps elucidate and predict some of the surface properties that may affect drug formulation, offering more comprehensive data on their anisotropy.

P-Monoclinic seeds (with dimensions about 0.5 x 0.5 x 0.1 cm) were prepared by two step cooling crystallization of a supersaturated methanol solution. These seeds were suspended to grow in a, slowly evaporating, supersaturated methanol solution6[1]. The crystals grew to sizes up to 5x3x3 cm, with three main facets exposed; (100), (002) and (010). CA measurements, using diiodomethane, water, formamide and ethylene glycol, were performed on these facets to determine their surface energy, both the Lifshitz-van der Waals and the acid-base components (using VOCG approach), and their affinity to water. The dispersive surface energy ranges from about 48 mJ/m^2 to 40 mJ/m^2 depending on the facet. The acid-base component has a smaller contribution, with its value going up to 10 mJ/m^2. However, there is large variation between the acid and the base component. The acidic component can be more than one order of magnitude larger than the basic one.

XPS analysis was conducted on the facets showing that strong amine presence on the surface of two of them. Contact angle measurements suggests that these two facets are the more hydrophilic. From the other hand the carbonyl component of the facets, less abundant than the amine one, does not seem to correlate with the water affinity.

P-Monoclinic Carbamazepine was recrystallized by stirring a supersaturated ethanol solution and the Triclinic polymorph was prepared by heating the P-Monoclinic polymorph at 160°C overnight[7]. FD-IGC measurements were performed on the two polymorphs to determine their surface energy heterogeneity. It was found that the most stable Monoclinic polymorph is lower energetically than the Triclinic. Computational analysis of the surface energy map of the Monoclinic polymorph shows some agreement with the measured surface energies[8].

We measured the contact angles of 0.1%, 0.25%, 0.5%, 1%, 1.5%, 2%, 2.5%, 3%, 4%, 5%, 10%, and 20% polymeric solutions by weight on three facets of P-Monoclinic Carbamazepine: (002), (010) and (100), to better understand how Copovidone interacts with facets of Carbamazepine. The amount of polymer in the solution was positively correlated to the solutions wettability on all three facets; the contact angle became smaller. Between the three facets the more apolar they were according to the FD-iGC, the more wettable they became as polymeric content increased. These increases were not identical, but were more pronounced as the dispersive component on the contact angle solution increased, i.e. as the polymeric content increased.

The data above was corroborated with water contact angle measurements of crystalline facets spun cast with Copovidone on top of them. This led to a mark increase in wettability, therefore the surface became more hydrophilic, in agreement with the data supplied in the previous paragraph. Della Volpe proposes that the acidic component of the surface energy of water is 6.5 times stronger than the basic one, from the other hand FD-IGC measurements on Copovidone reveal that it has a very strong basic component and a Lifsitz-Van der Waals component similar to that of Carbamazepine. Thus, it can be concluded that the increase in hydrophilicity is attributed to the acid-base interactions.

Previous studies have established the concept of anisotropy in pharmaceutical crystals. These anisotropic properties were proved to affect dry properties of pharmaceutical processes such as cohesion. However, this study goes beyond dry interactions revealing the importance of crystal anisotropy in solid-liquid interactions, particularly important in a wide range of pharmaceutical processes such as wet granulation, dissolution etc. It highlights the need for in depth understanding of crystal anisotropy in order to develop robust model predictive tools describing solid-liquid processes. Furthermore, it highlights the need for the application of complementary techniques to accurately describe crystal anisotropy and it also indicates the importance of computational models in these studies

References:

1. Ho, Raimundo, Majid Naderi, Jerry Y. Y. Heng, Daryl R. Williams, Frank Thielmann, Peter Bouza, Adam R. Keith, Greg Thiele, and Daniel J. Burnett. 2012. "Effect Of Milling On Particle Shape And Surface Energy Heterogeneity Of Needle-Shaped Crystals". *Pharm Res* 29 (10): 2806-2816.
2. Krycer, Ian, David G. Pope, and John A. Hersey. 1983. "An Evaluation Of Tablet Binding Agents Part I. Solution Binders". *Powder Technology* 34 (1): 39-51.
3. Simons, S.J.R., D. Rossetti, P. Pagliai, R. Ward, and S. Fitzpatrick. 2005. "The Relationship Between Surface Properties And Binder Performance In Granulation". *Chemical Engineering Science* 60 (14): 4055-4060.
4. Ho, Raimundo, Sarah E. Dilworth, Daryl R. Williams, and Jerry Y. Y. Heng. 2011. "Role Of Surface Chemistry And Energetics In High Shear Wet Granulation". *Industrial & Engineering Chemistry Research* 50 (16): 9642-9649.
5. Ho, Raimundo, Steven J. Hinder, John F. Watts, Sarah E. Dilworth, Daryl R. Williams, and Jerry Y.Y. Heng. 2010. "Determination Of Surface Heterogeneity Of D-Mannitol By Sessile Drop Contact Angle And Finite Concentration Inverse Gas Chromatography". *International Journal Of Pharmaceutics* 387 (1-2): 79-86.
6. Heng JYY, Bismarck A, Lee AF, Wilson K, Williams DR. Anisotropic surface energetics and wettability of macroscopic form I paracetamol crystals. *Langmuir*. 2006;22:2760-2769.
7. Grzesiak AL, Lang M, Kibum K, Matzger AJ. Comparison of the four anhydrous polymorphs of carbamazepine and the crystal structure of form I. *Journal of Pharmaceutical Sciences*. 2003;92(11):2260-2271.
8. Smith RR, Williams DR, Burnett DJ, Heng JYY. A new method to determine dispersive surface energy site distributions by inverse gas chromatography. *Lagmuir*. 2014;30(27):8029-8035.

NARROW TUBE SPRAY DRYING WITH PRECISE DRYING HISTORY CONTROL FOR PARTICLE ENGINEERING

May Ginn (Jaclyn) Lee[1], Shahnaz Mansouri[1], Karen Hapgood[1] and Meng Wai Woo[1],
(1) Chemical Engineering, Monash University, Clayton, Melbourne, Australia
Email: meng.woo@monash.edu

Introduction

The use of spray drying technology has become more prevalent and popular over the years in the drying industry. It is a one step process, using hot air to dry atomised sample solution, resulting dry powdered product in a matter of seconds. Spray drying is particularly suitable to be used in the pharmaceutical industry as the spray dried products offers consistency and meet the standards on particle size, density and morphology.[1] Amorphous and crystalline structures are most commonly discovered in spray dried particles. Amorphous structure provides high dissolution rate which is an advantage.[2] However, in the amorphous state, products are generally unstable and undergo phase transition during shelf life. This alters the functionality and properties of product which could lead to degradation.[3] In addition, amorphous particles are prone to develop into sticky particles causing wall deposition and reduce the overall efficiency.[4] In contrast, crystalline particle exhibit properties that are opposite to that of amorphous. Partial or complete crystallisation is capable of reducing stickiness and prevents wall deposition. They are flowable thereby making them simple and easy in handling. Most importantly, crystalline particle are stable and does not undergo phase transition upon storage, unlike amorphous particles which is only stable in low humidity.[5] Thus, many researchers are focusing on increasing the degree of particle crystallinity in spray dried products in order to overcome the issue cause by amorphous particles.

Many factors and parameters are involved in determining the degree of particle crystallinity with temperature being the highest impact followed by atomiser rate, pump rate, feed concentration and aspirator rate.[6] Previously, Islam and Langrish, 2009 used lactose as the study material and reported with a higher inlet drying temperature from 134°C to 210°C, the degree of crystallinity increases from 55% to 76% using a Buchi-290 laboratory scale spray dryer.[4] On the contrary, when it is scaled-up to a pilot plant, results shows the opposite as an increase in inlet drying temperature from 170°c to 210°C yields a decrease in particle crystallinity from 72% to 22%.[7] These contradicting results were due to inlet temperature being used as the parameter to describe the condition suitable to encourage particle crystallinity. Experiments carried out thus far were focusing on controlling the inlet or outlet drying temperature without any in-situ study on the effect of particle drying history along the chamber on particle crystallinity. Hence, this paper would be addressing this issue by creating various drying history using temperature as the primary factor in controlling particle crystallinity.

Current laboratory spray dryers have limitations in which only the inlets or outlet temperature is explicitly controlled. Thus, a new spray dryer capable of controlling the drying history is required for this study. Mansouri et al, 2015 proposed using a narrow tube spray dryer with internal diameter of 1.27-4.8 cm by introducing hot air at the inlet similar to current spray dryers.[8] With such narrow tube, heat loss through the walls was significant, therefore insulation was essential to meet the targeted drying temperature. A new concept was discovered based on the narrow tube dryer by utilising the rapid heat transfer through the tube walls as a mean to control the particle drying history. Instead of providing heat at the inlet of the drying chamber, heat can be provided through the walls of the tubes. The amount of heat transferred to the drying chamber can be easily manipulated by varying the wall tube temperature. As a result, in depth study on the effect of various drying history on particle crystallinity can be undertaken.

Modified Narrow Tube Dryer

In order to address the current limitations of spray dryers in controlling the drying history of particles throughout the drying chamber, a new modified spray dryer called *'narrow tube dryer'* was designed. The narrow tube dryer was designed based on the idea of controlling the temperature profile of air in the drying chamber through the heated tube walls. This is of paramount importance as the air would ultimately act as a drying medium for the particles through convective heat transfer. Thus, by controlling the temperature profile of air in the drying chamber, the effect of drying history on particles can be studied.

The narrow tube dryer consists of twelve 0.5 m length of copper tubes with inner and outer diameters of 1.7 and 1.9 cm respectively. Resistors connected in series are used as heating elements and are directly mounted on the tube walls which are controlled by a PID controller. Two thermocouples are placed on each tube to ensure uniform heating is achieved for each tube. In addition, a thermocouple is placed in between two connecting copper tubes at 0.5 m interval to measure the bulk temperature in the drying chamber. Since the wall temperature of each tube needs to be controlled independently, all tube are isolated from one another by placing gasket between them. By doing this, different wall temperature can be applied and varied accordingly for each independent tube. Particles are collected at the end of the drying chamber using a high efficiency particulate arrestance (HEPA) filter with the aid of a vacuum suction.

Material and Method

Lactose solution with 10 wt% solute was used as the study material in this experiment. Alpha lactose monohydrate was purchased from Sigma-Aldrich while milli-Q water was used to prepare the solution. The lactose solution was left to be stirred for 4 hours before it is spray dried in the narrow tube dryer. The average duration of spray drying for each run is approximately 1 minute. Final samples were collected on the HEPA filter paper before they are transferred to a Scanning Electron Microscope (SEM) stubs for viewing. All image analysis on particle morphologies were carried out using FEI Nova Nano SEM 450 FEG and a bench top SEM (Phenom World). Prior to viewing on SEM, samples are prepared on aluminium stub using double sided carbon tape and sputter coated with approximately 1nm Iridium. The accelerating voltage used was 3 kV with magnification ranging from 1000 to 40000 x. All images were taken using secondary electron imaging.

Manipulating the Particle Drying History

As the objective of this paper is to study the effect of drying history on particle morphology, 1-Dimension (1D) modelling of the temperature profile along the tube is paramount as it is capable of predicting important parameters of both the drying medium and the particle itself such as bulk temperature, particle temperature and moisture content. With 1D modelling, it helps in experimental planning since the results can be predicted before the actual run.

Heat, mass and momentum balance equations were used to determine the bulk temperature profile along the tube. Equation 1, 2 and 3 below capture the temperature profile of a particle, the change in absolute humidity during the drying process and the velocity of particle travelling horizontally respectively. These three equations were previously used by Patel for a vertical plug flow spray dryer. With the narrow tube dryer, equations for heat and mass balance will still be valid while some modifications were done for momentum balance since particles are travelling horizontally. [9]

$$\frac{dT_p}{dt} = \frac{hA_p(T_b - T_p) + \Delta H_L m_s \frac{d\bar{X}}{dt}}{m_w C_{p,w} + m_s C_{p,s}} \qquad \text{Equation 1}$$

$$\frac{dY}{dl} = -\frac{\theta}{\dot{V}\rho_b v_p}\frac{dm_w}{dt} \qquad \text{Equation 2}$$

$$\frac{dv_p}{dt} = \frac{-0.75C_D\rho_b(v_p - v_b)|v_p - v_b|}{d_p\rho_p} \qquad \text{Equation 3}$$

Another important parameter to be considered is the rate of evaporation of the droplet which accounts for the changes in the particle moisture content. Equation 4 was used to predict the rate of evaporation.[10]

$$-\frac{dm_w}{dt} = -m_s\frac{d\bar{X}}{dt} = h_m A_p\left[\rho_{v,sat}exp\left(-\frac{\Delta E_v}{R_g T_p}\right) - \rho_{v,b}\right] \qquad \text{Equation 4}$$

The apparent activation energy, ΔE_v in Equation 4 is unique for each type of material and its initial concentration. Though Equation 5 below is used for lactose solution with initial concentration of 20%, it is concluded that Equation 5 is suitable to be used for initial concentration of 10% as well since the difference is negligible. [11]

$$\frac{\Delta E_v}{\Delta E_{v,b}} \approx 1.017exp[-1.678(\bar{X} - X_b)^{1.018}] \qquad \text{Equation 5}$$

By incorporating Equation 1-5, the bulk temperature profile along the tube can be predicted using Equation 6 and this was used to compare with the experimental results.

$$\dot{V}\rho_b C_{p,b}\frac{dT_b}{dl} = \frac{\theta}{v_p}\left[\frac{dm_w}{dt}\{\Delta H_L + C_{p,v}(T_b - T_p)\} - hA_p(T_b - T_p)\right] - \dot{V}\rho_b(\Delta H_L + C_{p,v}T_b)\frac{dY}{dl} + [U(\pi D_e)(T_w - T_b)] \qquad \text{Equation 6}$$

The overall heat transfer coefficient, U in equation 6 was calculated by taking into consideration the thermal conductivity of the copper tube, k which is 400 W/(m.K) and the internal convective heat transfer coefficient. The external heat transfer coefficient can be eliminated since a constant wall temperature was maintained for all tubes. The overall heat transfer coefficient, U is calculated using equation 7.

$$\frac{1}{U_o} = \frac{d_o ln\left(\frac{d_o}{d_i}\right)}{2k} + \frac{d_o}{d_i} \times \frac{1}{h_i} \qquad \text{Equation 7}$$

All the parameters in Equation 6 were calculated using the equations mentioned above. The only manipulated parameter was the wall temperature of the tube, T_w . Thus, by varying the wall temperatures of each tube, the bulk temperature, T_b was calculated along the tube creating unique drying history specific to the wall temperatures. Equation 6 was used as the basis in comparing and validating the experimental data. With this verification, other crucial parameters such as T_p and X can be predicted and aid in analysing particle morphology. Overall, 1D modelling was used as a tool to predict particle drying history and study its effect on particle morphology.

Results and Discussion

Control Model

Prior to using 1D model as a tool in predicting crucial parameters throughout the drying process, it was important to first validate its applicability on the modified narrow tube dryer. In the first section, two runs were carried out and for each run, theoretical bulk temperatures were calculated with and without the presence of atomised droplets by equating θ which refers to the mass flow rate of reactant to 1and 0 respectively in Equation 6.

Table 1. Wall temperatures for Run 1 and 2

Tube No.	1	2	3	4	5	6	7	8	9	10
Run 1 (°C)	29	42	49	61	34	79	84	91	92	99
Run 2 (°C)	68	76	82	74	34	60	69	67	67	86

Table 1 shows for both Run 1 and 2, in which the wall temperatures were set in an increasing manner with a sudden decrease in the middle section. Run 1 was set for a gradual increase while Run 2 was set for a rapid increase in bulk temperature. Based on Figure 1 and 2, experimental results follow closely to the theoretical results, displaying similar trends. Theoretical results shows that with the presence of atomised droplets, the bulk temperature was approximately 2-3°C lower than the run with air alone. This agrees with the experimental results whereby bulk temperature with atomised droplet is lower than that of air alone. The small difference is contributed by the low mass loading of atomised droplets, leading to a negligible effect on bulk temperature. [12]

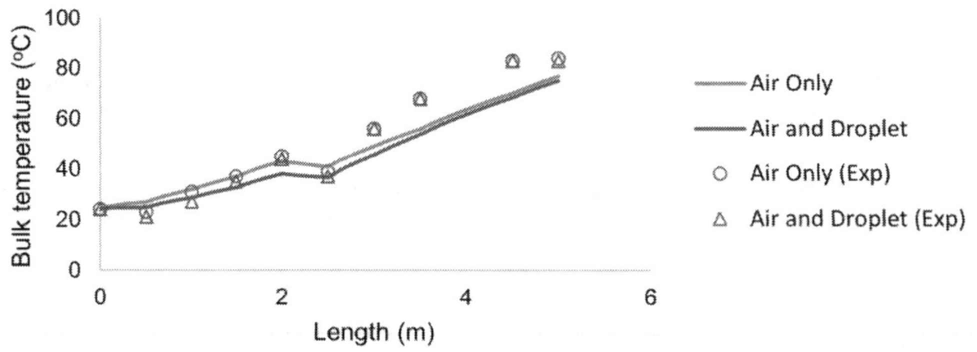

Figure 1. Run 1-Bulk temperature with and without the presence of atomised droplet

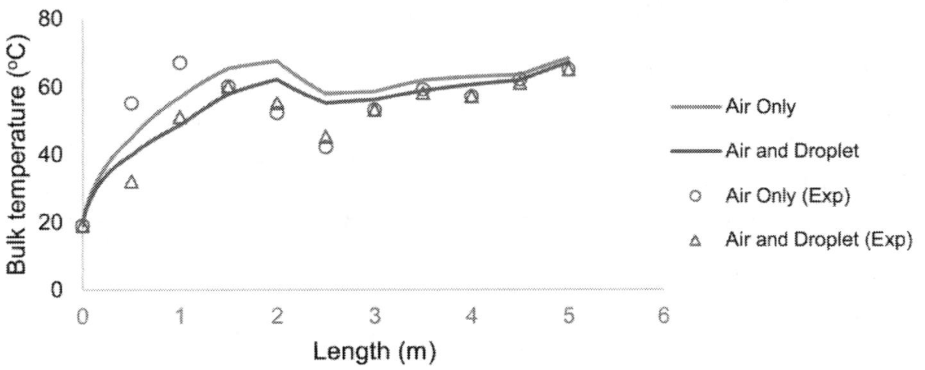

Figure 2. Run 2-Bulk temperature with and without the presence of atomised droplet

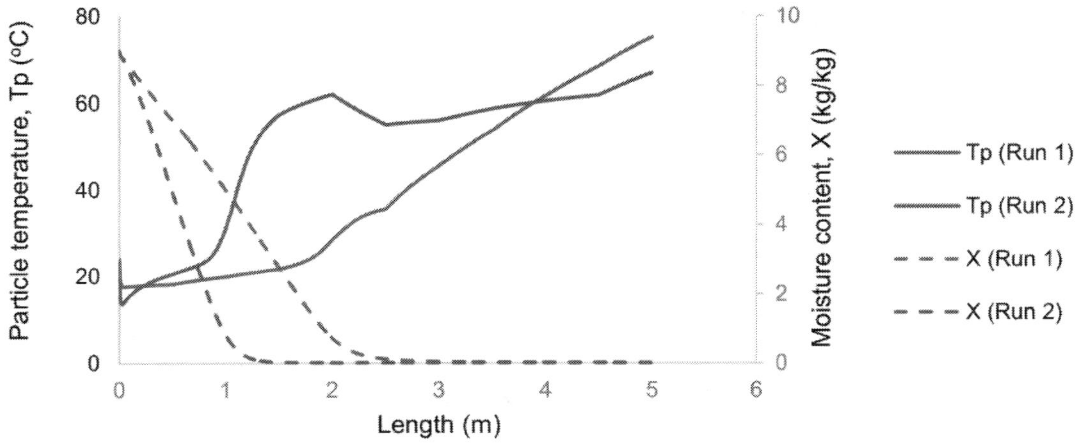

Figure 3. Particle temperature and moisture content of Run 1 and 2

The validation of control model is pivotal for the prediction of particle temperature and moisture content. Figure 3 shows the different drying rates are for Run 1 and 2 as the bulk temperature for both runs are notably different. Both runs exhibit a kink in the middle section due to the sudden drop in wall temperature before it increases again. This demonstrates the sensitivity of bulk temperature to wall temperature and substantiates the idea of precisely controlling the drying history of particle using the narrow tube dryer.

Effect of Drying History on Particle Morphology

In this section, the effect of drying history on particle morphology will be analysed with the aid of the predicted particle temperature and moisture content as shown in Figure 4. Three runs with different wall temperatures applied were carried out by mainly manipulating the first six tubes while maintaining the remaining tubes at a constant temperature as shown in Table 2. This was aimed at investigating the significance of initial drying rate on spray dried particles. The last two tubes were intentionally left at room temperature to induce a cooling effect at the end right before particle collection to maintain a T_p-T_g<0, avoiding any sticky particle.

Table 2. Wall temperatures for Run 3,4, and 5

Tube No.	**1**	**2**	**3**	**4**	**5**	**6**	**7**	**8**	**9**	**10**
Run 3 (°C)	90	90	60	80	90	90	90	90	90	90
Run 4 (°C)	70	30	30	30	90	75	90	90	90	90
Run 5 (°C)	30	35	44	47	57	80	90	90	90	90

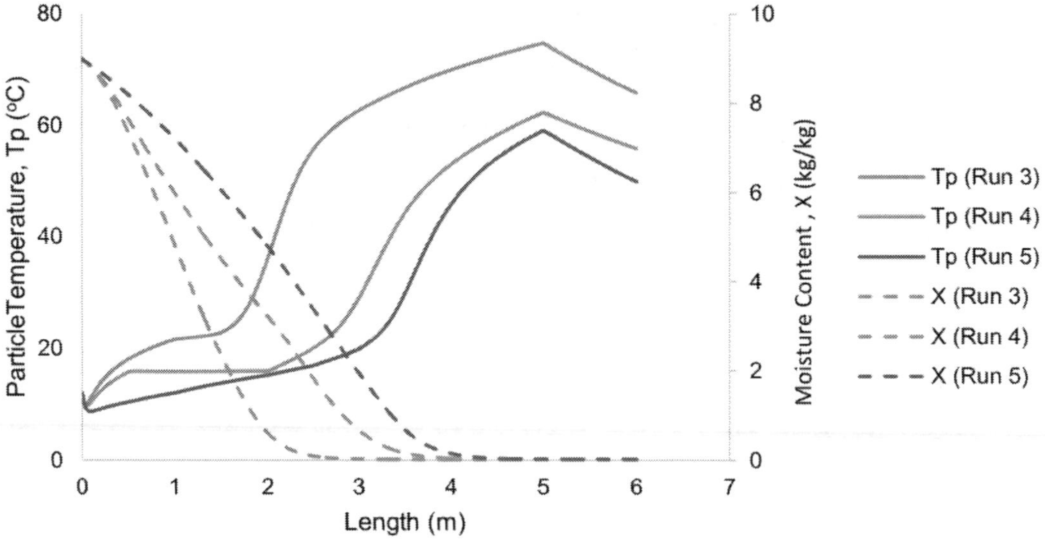

Figure 4. Particle temperature and moisture content of Run 3,4 and 5

Wall temperature in Run 3 was set to be the highest followed by Run 4 and Run 5. Based on the predicted moisture curve in Figure 4, this agrees with the wall condition as Run 3 and 5 display the highest and lowest gradient signifying the evaporation rate as well. These 3 different runs resulted in various particle morphology as shown in the SEM images tabulated in Table 3. It is notable that successful spray drying was achieved for all three runs despite the low drying temperature in comparison to typical spray drying conditions which are above 100°C. Mixture of amorphous and crystalline structures were observed for all runs.

Run 3 with the highest particle evaporation rate produces bigger crystalline particles, approximately 2-3 µm while Run 5 with the lowest particle evaporation rate yields finer crystalline structure, less than 1 µm. This could be due to a lower particle evaporation rate allowing sufficient time for molecules to rearrange themselves as the drying process continues. For the condition in Run 3 with high particle evaporation rate, limited time was provided for molecules rearrangement during the evaporation stage. Once evaporation slows down, a sudden increase in particle temperature can be observed. Based on Figure 2, this situation happened at length 1.8, 2.8 and 3.4 m for Run 3, 4 and 5 respectively which is in accordance to the time frame for particle evaporation. In addition, flat piece shaped crystals were found in both Run 3 and 4. This suggests that particle morphology is not subjected to a specific drying history but an evaporation time frame may exist which results in a unique particle morphology.

Table 3. SEM results for Run 3, 4 and 5

Run 3		
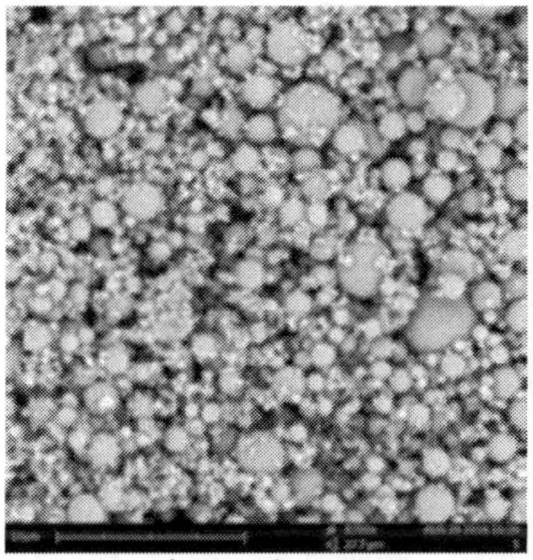 Amorphous	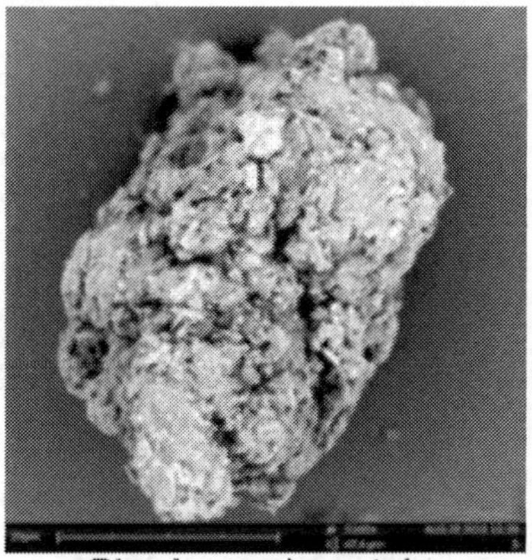 Big clumped crystals	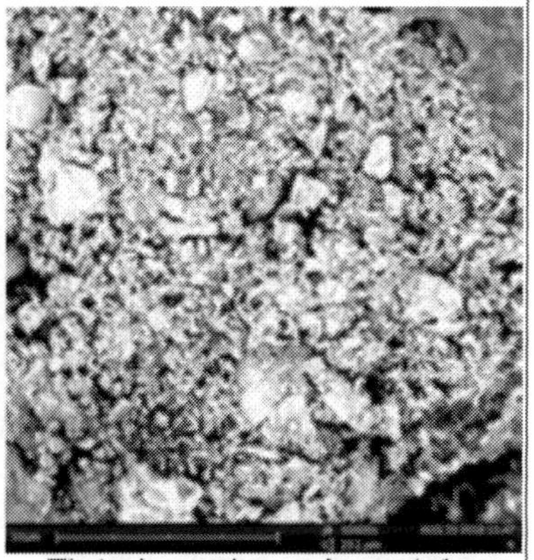 Flat piece shaped crystals
Run 4		
 Amorphous	 Mixture of amorphous and big crystals	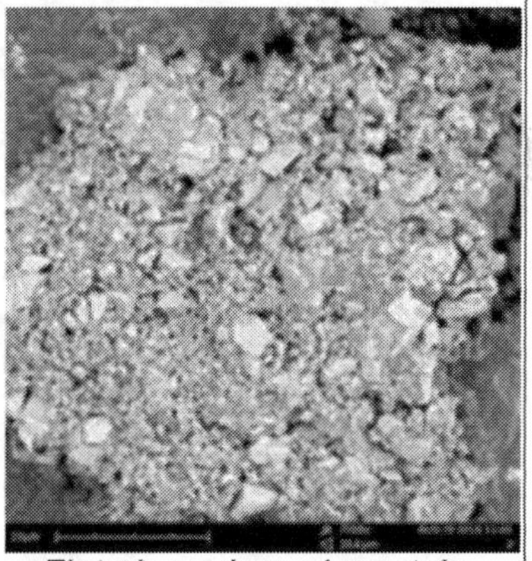 Flat piece shaped crystals
Run 5		
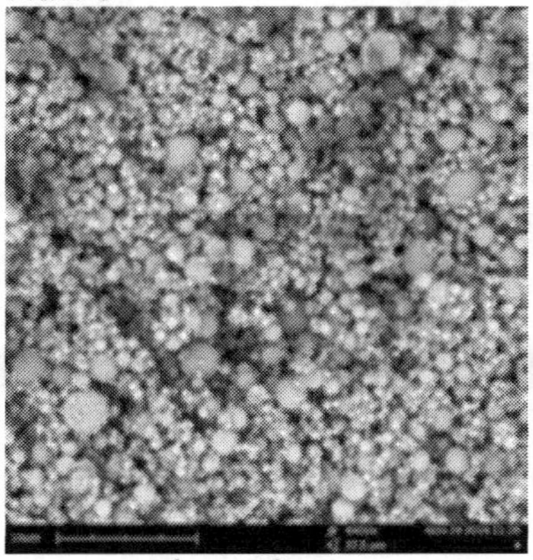 Amorphous	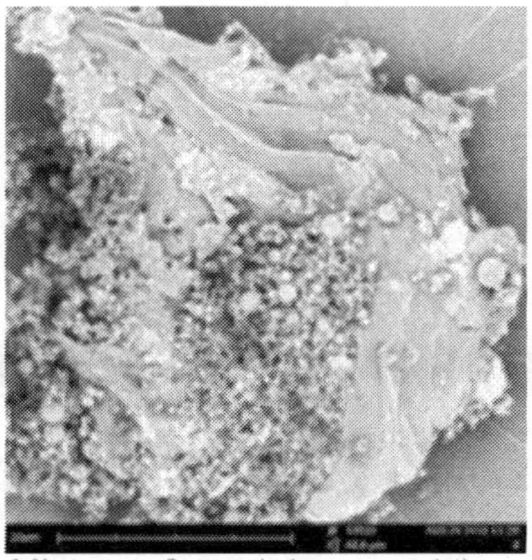 Mixture of amorphous and tiny crystals	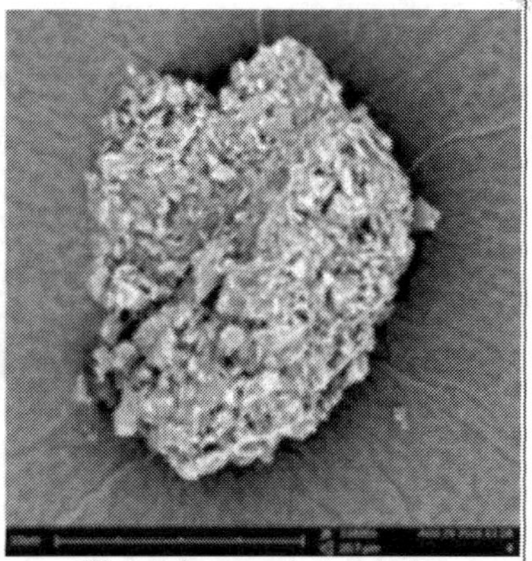 Fine clumped crystals

Conclusion

The newly modified narrow tube dryer is capable of precisely manipulating the drying temperature along the tube through convective heat transfer from the heated tube walls. 1D modelling has proven to be a useful tool in predicting temperature profile and providing valuable insights along the dryer. By incorporating 1D modelling, the drying history can be controlled to study its effect on particle morphology. Results shows a low drying temperature, below 100°C is sufficient to yield successful spray dried product which is in fact beneficial for thermal labile materials. Mixture of amorphous and crystalline structures were produced for all runs. However, different initial drying rate affects the size and shape of crystalline particles. Further research is currently underway to unveil the underlying phenomena on crystallinity.

References

1. Sollohub, K. and K. Cal, *Spray drying technique: II. Current applications in pharmaceutical technology.* Journal of Pharmaceutical Sciences, 2010. **99**(2): p. 587-597.
2. Chen, J., B. Sarma, J.M.B. Evans, and A.S. Myerson, *Pharmaceutical Crystallization.* Crystal Growth & Design, 2011. **11**(4): p. 887-895.
3. Shekunov, B.Y. and P. York, *Crystallization processes in pharmaceutical technology and drug delivery design.* Journal of Crystal Growth, 2000. **211**(1–4): p. 122-136.
4. Chiou, D., T.A.G. Langrish, and R. Braham, *The effect of temperature on the crystallinity of lactose powders produced by spray drying.* Journal of Food Engineering, 2008. **86**(2): p. 288-293.
5. Woo, M.W., N. Fu, F.T. Moo, and X.D. Chen, *Unveiling the Mechanism of in situ Crystallization in the Spray Drying of Sugars.* Industrial & Engineering Chemistry Research, 2012. **51**(36): p. 11791-11802.
6. Chiou, D. and T.A.G. Langrish, *A comparison of crystallisation approaches in spray drying.* Journal of Food Engineering, 2008. **88**(2): p. 177-185.
7. Das, D. and T.A.G. Langrish, *An activated-state model for the prediction of solid-phase crystallization growth kinetics in dried lactose particles.* Journal of Food Engineering, 2012. **109**(4): p. 691-700.
8. Mansouri, S., V.S. Hena, and M.W. Woo, *Narrow Tube Spray Drying.* Drying Technology, 2015.
9. Patel, K., X. Chen, R. Jeantet, and P. Schuck, *One-dimensional simulation of co-current, dairy spray drying systems — pros and cons.* Dairy Science & Technology, 2010. **90**(2-3): p. 181-210.
10. Chen, X.D. and S.X.Q. Lin, *Air drying of milk droplet under constant and time-dependent conditions.* AIChE Journal, 2005. **51**(6): p. 1790-1799.
11. Lin, S.X.Q. and X.D. Chen, *A Model for Drying of an Aqueous Lactose Droplet Using the Reaction Engineering Approach.* Drying Technology, 2006. **24**(11): p. 1329-1334.
12. Mostafa, A.A. and H.C. Mongia, *On the interaction of particles and turbulent fluid flow.* International Journal of Heat and Mass Transfer, 1988. **31**(10): p. 13.

465150 Soluble Tag-Assisted Liquid-Phase Peptide Synthesis Using Hydrophobic Benzyl Alcohols

Abstract:

[Introduction]
Peptides have attracted much attention as promising medicinal candidates. As compared with traditional small molecules and antibody medicines, peptides have medium-sized molecular structures. Therefore, they are expected to have both advantages of small molecules and antibody medicines, for example, chemical synthetic accessibility and highly selective pharmacological activity. Current chemical synthesis of peptides is mainly carried out through solid-phase methods, which use insoluble resins as supports to realize easy separation processes and also enable fully automated synthesis. However, soluble tags assisted liquid-phase approach has also been studied intensively since large scale preparations are still challenging in solid-phase method. In this context, polyethylene glycols(PEGs) are the first option as soluble tag and they are central to many applications in the field. Since PEGs are soluble in polar solvents and insoluble in less-polar solvents, a reaction of polymer-attached reactant or catalyst can be performed under homogeneous conditions in a polar solvent. After the reaction completion, the polymer is precipitated by the addition of a less-polar solvent. However, as for peptide synthesis, because both amino acids and coupling reagents are polar, the opposite solubility is preferable for soluble tag to rinse away excess chemicals by washing with polar solvents.
We have been working on the use of hydrophobic benzyl alcohols as soluble tags in liquid-phase method. In this method, reactions take place in homogeneous conditions in less-polar solvents and after the reaction completion, polar solvents are added to precipitate the tag-attached product. Since excess chemicals can be rinsed away efficiently by washing with polar solvents, this approach allows a production of peptide with high quality in relatively larger scale. For this strategy to be successful, high control of the tag-attached product solubility and the physical property must be addressed. In this study, the liquid-phase chemical synthesis of several therapeutic peptides and the investigation on poor-solvents-addition-induced precipitation mechanism will be presented.

[Experimental]
The hydrophobic benzyl alcohols synthesized from the corresponding hydroxyl benzaldehyde in 2 steps, including alkylation by respective bromide and reduction by sodium borohydride. At each step, the desired products were washed with polar solvents repeatedly to give pure forms, which were characterized by NMR. The chemical synthesis of peptides was carried out in less-polar solvents using N-protected amino acids and COMU as a coupling reagent. After the reaction completion, polar solvents were used as poor solvents to induce precipitations of the tag-attached products. The physical property of the recovered precipitates were characterized by scanning electron microscopy, particle size distribution measurement, and X-ray diffraction.

[Results and discussion]
High control of the solubility property was achieved for the hydrophobic benzyl alcohols. The tag-attached products were able to be precipitated quantitatively at most steps. Although dichloromethane was a good solvents for all chemicals used, including the tags, N-protected amino acids, and a coupling reagent, we found that tetrahydrofuran was also effectives as a reaction solvent. This is preferable for the large scale production of peptides from the viewpoint of environmental aspect. As for the poor solvent, methanol and acetonitrile were found to be effective to induce precipitation of the tag-attached products. By using this methodology, both couplings and deprotections were carried out in homogeneous reaction conditions to give several therapeutic peptides in gram-scales. In addition, the physical property of the precipitation was found to be dramatically different when cyclohexane was used as a reaction solvent instead of tetrahydrofuran. Especially, the biphasic system composed of cyclohexane and acetonitrile provided an ideal reaction condition, which solubilize the all chemicals used for the synthesis. The precipitations obtained from this reaction condition were fine powders that were able to be recovered readily through paper filtration.

[Reference]

(a) Okada, Y. et. al, Org. Lett. 2015, 17, 4264-4267. (b) Okada, Y. et. al. J. Org. Chem. 2013, 78, 320-327. (c) Tana, G. et. al. Chem. Commun. 2010, 46, 8219-8221

460254 Measurements of Time-Varying Liquid Spatial Distributions in a Shear Cell Using X-Ray Computed Micro-Tomography

Abstract:

Contact spreading is the transfer of liquid via the formation and breakage of liquid bridges that form between particles. There have been few experimental studies accurately quantifying contact spreading as a function of system properties due to the difficulty of non-intrusively imaging dense granular systems. X-ray computed micro tomography (XRCT) is used in the present work to measure the time-varying spatial distribution of liquid in a shear cell geometry. We discuss the experimental system and procedures in addition to experimental results.

461194 DEM Simulation of the Mixture of Dry and Wet Particles inside Rotary Mixer

Abstract:

Industrial particle coating process often involves spraying the coating fluid on the top layer of moving particles inside mixer. It takes time on the order of several minutes or more for all particles to pass through the "spray zone". During this transient time the bulk of the granular material inside the mixer contains the mixture of dry particles and sticky wet particles with amount of wet particles continuously increasing while the amount of dry particles is continuously decreasing. The surface frictional properties of dry and wet particles can be very different. We developed a numerical model based on the Discrete Element Method (DEM) to simulate this dry-to-wet transition during coating. In this model every dry particle is replaced by wet particle if it is inside the cone-shape spray zone. The model is implemented within commercial software STAR-CCM+ version 11.04 developed by CD-adapco. We tested the model in several mixer geometries and analyzed the bulk properties of the solid phase as a function of time. For the simplest geometry case, a rotating drum, we obtained the expected change in dynamic angle of repose in the cases of all-dry and all-wet particles and, in terms of the flow pattern, repose angle, and the frequency of avalanching, we obtained results for all-wet particles comparable with published experimental results [1]. Figure below shows the side-view of dry (grey color) and wet (blue color) particles inside the rotating drum for increasing values of time.

Time = 5.5 s | Time = 19.2 s | Time = 24.5 s | Time = 68 s

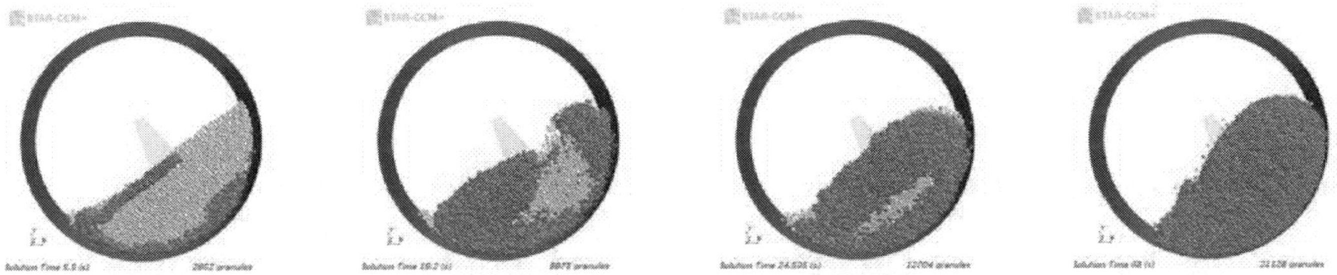

We believe that our model is useful for both better understanding of the bulk flow during dry-to-wet transition and its ability to differentiate different mixer designs in industrial processes where the difference in frictional properties between wet and dry particle cannot be neglected.

[1] P. Y. Liu, R. Y. Yang and A. B. Yu. Dynamics of wet particles in rotating drums: Effect of liquid surface tension. Physics of Fluids, 23, 013304, (2011)

461241 An Investigation into Growth of a Non-Wetting System in High Shear Granulation

Abstract:

An Investigation into growth of a non-wetting system in high shear granulation

D. Al-aaraj, R. Smith, J. Litster, A. Dunbar[1]

[1]Chemical and Biological Engineering, The University of Sheffield, United Kingdom

Email address: DKAl-aaraj1@sheffield.ac.uk

Pharmaceutical manufacturing involves different powder ingredients, each with different wetting characteristics. Granulation of non-wetting powders is a common problem which affects granule and tablet uniformity. The discovery of liquid marbles is a promising way to solve the problem of hydrophobic powder wetting behaviour.

The influence of different variables on granule growth behaviour and a qualitative understanding of the granule growth mechanism have been proposed. However the granulation behaviour of a new formulation is still not able to be predicted from knowledge of its basic properties and process parameters.

The aim of this work is to conduct a systematic study of the effect of formulation and operational variables on size distribution and morphology of granules of hydrophobic powder mixtures in high shear granulation. Salicylic acid and microcrystalline cellulose were granulated with Polyethylene glycol 200 in the high shear granulator. Different liquid to solid (L/S) and impeller speed were applied. Granule size distribution and morphology were identified.

The agglomeration behaviour of granules at a low L/S ratio 60/200 was largely unaffected by the impeller speed. As a level of liquid increases, L/S=80/200 - 120/200, there is an increase in granule size as the impeller speed increases from 350 rpm to 450 rpm. However, at 550 rpm impeller speed, there is a small change in the granules size. The granules are spherical at 350 and 450 rpm, and elongated and irregular in shapes at high impeller speed.

At a 60/200 L/S, the particle size distribution is similar to the raw material. However, larger granules appeared at a L/S ratio of 80/200. At a 100/200 L/S ratio, the granule size distribution moved to the right as the granules size increased. Moreover, the amount of un-granulated powder decreased sharply at a L/S 120 with a dramatic increase in granule size to more than 2.5 mm. The granules appeared to be elongated and collapsed with an increase in liquid content.

The effect of impeller speed was found to depend on L/S ratio, although it had little effect on granule size. However, there was much more of effect on the morphology of the granules, especially at high L/S ratio. The effect of L/S was more significant on granule size and shape. As the L/S ratio increases, the amount of un-granulated fine powder of raw materials decreases and the final granule size increases in the size range from 0.3mm to 3mm.

463844 Effect of Particle Wettability on Particle-Particle Adhesion of Colliding Particles through Droplet

Abstract:

In an actual wet granulation processes, a liquid bridge between particles is not static but dynamic due to continuous particle motion. Therefore, to fully understand the micro-level particle agglomeration phenomena, elucidation of the particle-particle adhesion by the dynamic liquid bridge has been an important issue.

Numerous studies of the liquid bridge have been conducted. However, the particle-particle adhesion by the dynamic liquid bridge cannot be explained by these conventional theories and analytical models, because deformation of the liquid bridge is not taken into account. Thus, we have developed a direct numerical simulation model for the particle-particle adhesion by the dynamic liquid bridge. By using this simulation model, effects of many critical parameters such as particle colliding velocity and physical properties of particle and liquid can be directly analyzed.

We here conducted a numerical simulation of the particle-particle adhesion by a dynamic liquid bridge. The particle-particle adhesion of two colliding particles through a droplet on a particle surface was simulated. In particular, effect of particle wettability on a critical velocity for the particle adhesion (i.e., adhesiveness of the two colliding particles) was investigated. It was found that the critical velocity for the particle adhesion non-linearly changed with the particle wettability. The critical velocity exhibited a local maximum with an increase in the contact angle, while the static liquid bridge force monotonically decreases with an increase in the contact angle. We revealed that a combined effect of the liquid bridge deformation and instantaneous liquid bridge force results in the non-linear dependence on the particle wettability.

465420 Direct Numerical Simulation of Viscous Forces Due to Liquid Bridge Formed Between Two Moving Particles

Abstract:

In many particulate processes in chemical engineering fields, liquid is often added to dry powder to improve powder characteristics and ease powder handling. Examples of such processes include wet agglomeration, particle coating and catalytic reaction. When relatively small amount of liquid is dispersed in powder, liquid bridges are formed between primary particles and bonding forces are exerted on the particles, i.e. surface tension and viscous forces. In particular, viscous forces can significantly influence the powder behaviour in a process since liquid viscosity can vary many orders of magnitude depending on the applications. Therefore, it is of paramount importance to accurately estimate the viscous forces in order to understand the process.

Several viscous force models are proposed in literature based on (a) the solutions of the Reynolds lubrication equation and (b) asymptotic solutions of the Stokes equations. It is shown that these models can estimate viscous forces comparable to experimental measurements when the inter-particle separation distance is sufficiently small. However, the accuracy decreases drastically when the separation distance increases. Therefore, there is a need to derive more accurate viscous force models.

In the present work, liquid bridges between particles are directly simulated using the Volume of Fluid (VOF) method with dynamic mesh as shown in Figure 1. It can be observed that the pressure distribution obtained from simulation is in good agreement with that obtained from the solution of the Reynolds lubrication equation when the separation distance is small (Figure 1a), whilst the difference becomes significant when the separation distance becomes large (Figure 1b). The forces exerted on the particles obtained from the simulations are then used to derive an improved viscous force model.

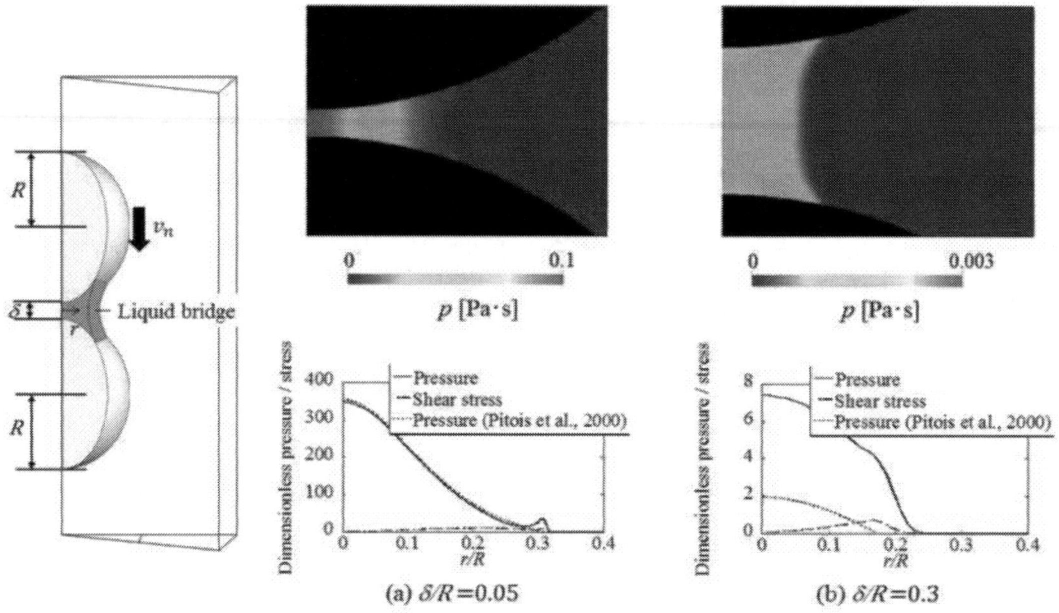

Figure 1 Pressure distribution between particles; (a) small separation distance case and (b) large separation distance case. Dashed line presents the solution of the Reynolds lubrication equation (J. Colloid Interface Sci., **231**, pp 26-31, 2000).

Keywords

Viscous force; Liquid bridge; Direct Numerical Simulation; Volume of Fluid (VOF) method

457430 High Shear Granulation: An Investigation into the Granule Consolidation Mechanism

Abstract:

High-shear granulation, the agglomeration of small particles in heavy mixers to produce agglomerates called granules, is extensively used by many industries. The key mechanisms taking place during the granulation process, wetting and nucleation; consolidation and growth; and breakage and attrition, however, are not yet fully understood.

The aim of this work is to compare a model for consolidation and layered growth to experimental data to allow for the prediction of the kinetics of the consolidation process in high-shear mixers. The model is based on deformation-driven diffusive growth, with small deformations causing a diffusion-like migration of binder liquid and powder, resulting in growth. The growth rate decays exponentially until the granule has reached a final size, which depends on the initial amount of liquid and the powder properties.

In order to properly monitor the growth of particles without the interference of other mechanisms such as breakage and attrition, a Consolidation-Only Granulator (COG) has been designed. Product granules were weighed, and the density of the granules was determined. In this way, the growth and density of the granules could be monitored over time.

Four different powder-binder systems were considered: lactose-silicone oil, lactose-hydroxypropyl methylcellulose (HPMC), glass beads-silicone oil and glass beads-HPMC. Binder liquids were used at varying viscosities to examine the effect of varying binder viscosity on the kinetics.

Here we present the results of granule growth behaviour and compare experimental findings to the proposed model. Growth behaviour is discussed in terms of the critical-packing liquid volume fraction, which is defined as the final liquid volume fraction of a granule.

Future work will include the incorporation of the kinetic model into population balance models, and the performance of DEM simulations to further increase the physical relevance of the model. In this way, consolidation and layered growth kernels for population balances can become more reliable.

465926 A Bonded Sphero-Cylinder Model for Computational Simulation of Large Aspect Ratio, Flexible Fibers

Abstract:

A bonded sphero-cylinder model is developed for computational simulation of large aspect ratio, flexible fibers. In this model, a single fiber is formed by connecting a number of sphero-cylinder segments at their spherical ends using elastic bonds. The model is verified by examining the mechanical behavior of a single fiber subject to external loads. The simulation results of bending, twisting, and stretching deformations of the single fiber are compared with the analytical solutions. Bond damping model is proposed to dissipate the kinetic energy during fiber vibration process. The correlation is established between the macroscopic damping coefficient, which characterizes energy dissipation rate for the whole fiber during vibration, and the damping coefficient for a single bond, which is the input parameter for the simulation.

To illustrate its capability, the developed model is applied to computationally investigate the compressibility of assemblies of fibrous particles. The compression of fibrous materials is frequently encountered in industrial operations, such as briquetting of agricultural materials and processing of wool, cotton and textiles. A better understanding of fiber compression can greatly improve the design of these operating procedures. From the fiber compression simulations, the macroscopic compression properties of bulk material (e.g., bulk compression stiffness) can be measured. The effects of initial particle bed configuration and fiber properties (e.g., aspect ratio, bending stiffness, and friction coefficient) will be explored.

460464 Influence of Normal Contact Force Model on Simulations of Sphero-Cylindrical Particles

Abstract:

This study investigates the impact of the contact force model on various aspects of assemblies of rod-shaped particles. DEM normal force models for cohesionless, rod-shaped particles are compared under static and dynamic conditions. Comparisons of the force, overlap, and contact duration frequency distributions are presented. Differences are observed in the overlap and contact duration distributions, but the force distributions are nearly independent of the choice of force model.

465890 Computational and Experimental Studies of Stress-Controlled Shear Flows of Fibrous Particles

Abstract:

The Discrete Element Method (DEM) has been utilized to simulate stress-controlled shear flows of fibrous particles, in which a shear flow of the material is driven by the relative movement between two parallel plates and a constant normal stress is applied to the material by one of the plates. A typical stress-controlled shear flow is granular material flow in a Schulze shear cell test. The DEM simulations of stress-controlled shear flows of fibrous particles are calibrated with the Schulze shear cell tests of cut fishing wires. The shear flow simulations and experiments are performed with various normal stresses in order to explore the dependence of the yield shear stress on the pressure applied to compress the material. Using the DEM simulations, parametric studies are conducted to examine the effects of particle-particle friction, fiber bending stiffness, and particle aspect ratio on the yield shear stress. The influence of moisture is also investigated by performing DEM simulations and Schulze shear cell tests using wet fibrous particles (i.e. the wetted cut fishing wires). It is found that in dense flows, the effect of cohesion, which is caused by the moisture, is minimal compared to the effect of particle-particle friction.

The stress-controlled shear flows are then compared with concentration-fixed shear flows, in which the solid volume fraction remains constant during the flow process. The average particle velocity profiles are very different between these two types of shear flows: a linear velocity profile is observed in the concentration-fixed flows while the velocity profile is non-linear in the stress-controlled flows, for which much a larger velocity gradient is observed close to the boundary plate that can move perpendicularly to the direction of material flow. With the same normal stress, the average shear stress and solid volume fraction (at steady state) obtained in the stress-controlled shear flow are the same as those in the concentration-fixed shear flow.

463967 Segregating Rods and Spheres: Simulation and Modeling

Abstract:

Many studies of segregation of flowing granular materials focus on spherical particles, even though in practice particles can have non-spherical shapes. Here we focus on mixtures of rod-like (cylindrical) particles of the same diameter but different lengths and mixtures of rod-like and spherical particles by implementing particle shapes described by super-ellipsoids in DEM simulations. To validate the effectiveness of super-ellipsoids, we compare to previous "true geometry" models of cylindrical particles. Then we use super-ellipsoids to study particle size segregation in bounded heap flow for a variety of particle diameters and lengths, comparing results with supporting experiments. In the flowing layer of a bidisperse rod mixture in a bounded heap flow, shorter rods percolate toward the lower portion of the flowing layer while longer rods rise toward the upper portion of the flowing layer. The rods tend to deposit on the underlying bed of particles in the heap such that they are aligned with the flow with the smaller rods deposited upstream of the larger rods due to segregation. Likewise, spherical particles segregate from rods having the same diameter. The percolation velocities related to the segregation depend on the local shear rate and the concentration of the other particle species, just as is the case for size or density bidisperse spherical particles. Using this percolation velocity and an appropriate value for collisional diffusion, the advection-diffusion-segregation continuum model for segregation successfully predicts the segregation of rod-like particles of different lengths as well as rod-like and spherical particles.

459959 Development of Collisional Dissipation Rate Models for Non-Spherical Particles Using the Discrete Element Method

Abstract:

Two-fluid simulations based on Computational Fluid Dynamics (CFD) have the ability to simulate large domains of granular flows. Unfortunately, the accuracy of these granular flow simulations is hampered due to the models available, which tend to assume that the particles are spherical. Introducing models that take particle shape into account offers the ability to improve the accuracy, making two-fluid simulations more viable for process design. This work presents a method that utilizes the Discrete Element Method (DEM) simulations to create a collisional dissipation rate model for non-spherical particle systems.

Homogeneous Cooling System (HCS) simulations of frictionless particles at a wide range of solid volume fractions (from 0.025 to 0.6, nearly the packing limit) are performed to measure the collisional dissipation rate, i.e. the rate of granular temperature decrease with time. Two types of particle models are used in the simulations: the glued-sphere particle, which is formed by rigidly connecting a string of spheres in a straight line, and the true cylindrical particle, which has the shape of true cylinder. Unlike the frictionless spheres, the frictionless, non-spherical particles can rotate after collisions, inducing the rotational granular temperature. Equal partition of rotational and translational granular temperatures is observed for the elongated particles with the particle aspect ratio equal or greater than 2. Larger dissipation rates are obtained for the non-spherical particles compared to the spherical particles. The dissipation rate increases as the particle aspect ratio increases. Most importantly, it is found that the collisional dissipation rate for non-spherical particles can be expressed as a function of the dissipation rate for spherical particles. This finding may allow a simple modification of the granular kinetic theory to account for the effect of particle shape. Overall, the HCS simulation using the DEM provides an effective approach for the development of the dissipation rate models for the particle systems of complex shapes.

467359 Coagulation of Agglomerates with Polydisperse Primary Particles

Abstract:

Agglomeration of nanoparticles is encountered in both atmospheric and industrial processes as in volcanic plumes and aerosol manufacture of carbon black or fumed silica. Even though the dynamics of coagulating spherical particles, such as self-preserving size distribution (SPSD) and coagulation rate are reasonably well-understood, there is significant uncertainty for fractal-like agglomerates (Friedlander, 2000). For the latter, coagulation rates have been proposed (Mountain et al., 1986; Thajudeen et al., 2012), their mobility (Sorensen, 2011) and SPSDs have been determined (Vemury and Pratsinis, 1995) and even the time needed to reach their asymptotic structure has been estimated (Goudeli et al., 2015). All these have been confined to agglomerates with monodisperse PPs. Realistic agglomerates, however, consist of polydisperse primary particles (PPs). So little is known for the effect of constituent PP polydispersity on agglomerate structure and, most importantly, coagulation dynamics.

Here coagulation of nanoparticles of varying PP polydispersity ($s_{g,PP}$ = 1 â€" 3) in the absence of coalescence, sintering or surface growth is investigated by a discrete element method (DEM) in the free molecular regime (Goudeli et al., 2015). This is an early stage of particle formation especially at high temperatures followed typically by rapid quenching to facilitate collection of particles and retention of their ramified structure as in manufacture of fumed SiO_2, Ni or TiO_2. As a result, particle dynamics at this stage frequently dominate the end product characteristics. The effect of PP polydispersity on agglomerate size (radius of gyration, mobility radius and volume-equivalent radius), morphology (fractal dimension, D_f, mass mobility exponent, D_{fm}, and their prefactors) as well as on the attainment of the well-known asymptotic fractal-like structure (D_f = 1.91 and D_{fm} = 2.15) and SPSD is investigated. Increasing the polydispersity of the constituent PPs from $s_{g,PP}$ = 1 to 3 does not affect but only delays the attainment of the asymptotic D_f, D_{fm} and SPSD of the resulting agglomerates. The crossover agglomerate size and critical number of PPs per agglomerate that mark the transition between D_f= 3 and 1.91 scaling are obtained by various ways and increase with PP polydispersity. Only clusters larger than the crossover size should be considered for drawing conclusions on the formation mechanism as well as their structure. Furthermore, the effect of PP polydispersity on agglomerate dynamics (coagulation rate and polydispersity) is elucidated quantitatively also.

Such characteristics affect the environmental impact of agglomerates (climate forcing or visibility impairment by soot) or performance of gas sensors, catalysts and even biomaterials and nutritional products.

Keywords:coagulation rate, self-preserving distribution, kinetic gas theory, crossover to fractal regime, event-driven algorithm, log-normal primary particle size distribution.

References:

Friedlander, S. K., *Smoke, Dust, and Haze: Fundamentals of Aerosol Dynamics*. 2nd ed.; Oxford University Press: New York, 2000.

Goudeli, E., Eggersdorfer, M. L., & Pratsinis, S. E. (2015). Coagulation â€" Agglomeration of Fractal-like Particles: Structure and Self-Preserving Size Distribution. *Langmuir*, 31: 1320-1327.

Mountain, R. D., Mulholland, G. W. and Baum, H. (1986). Simulation of aerosol agglomeration in the free molecular and continuum flow regimes. *J. Colloid Interface Sci.* 114, (1), 67-81.

Sorensen, C. M. (2011). The mobility of fractal aggregates: a review. *Aerosol Science and Technology*45: 765-779.

Thajudeen, T., Gopalakrishnan, R., Hogan, C. J., Jr. (2012). The collision rate of non-spherical particles and aggregates for all diffusive Knudsen numbers. *Aerosol Sci. Technol.*, 46: 1174-1186.

Vemury, S., Pratsinis, S.E. (1995). Self-preserving size distributions of agglomerates. *J. Aerosol Sci.*, 26: 175-185.

460278 Breakage of Fractal Agglomerates

Abstract:

Agglomerates are formed by small primary particles. Breakage of agglomerates has received considerable research attention due to its association with various industrial processes such as the production of pharmaceutical tablets and handling, transportation, or compaction of powders. Discrete element method (DEM) simulations have been widely used to investigate the breakage of agglomerates with regular shapes in previous research. In this study, breakage behavior of fractal agglomerates as a function of impact velocity and surface energy was investigated using DEM simulations. Agglomerates with fractal dimensions (D_f) ranging from about 2.0 to 2.8 were produced in this research by letting primary particles with random initial velocities agglomerate under a centripetal force. The simulation results show that the damage ratio (D_m), defined as the ratio of broken interparticle contacts to the initial total contacts, decreases with increasing fractal dimension in the range between 2.0 to 2.6, which can be explained by the increase of mechanical strength with increasing fractal dimension. Previously published Weber number (W_e) scale of damage ratio has been modified by incorporating the fractal dimension: $D_m=B(W_e/D_f^2)^{Î±}$, where B and Î± depend exponentially and linearly on Weber number, respectively. Visualization of impact process illustrates the feature of ductile fracture of fractal agglomerates irrespective of fractal dimensions.

478079 PTF Young Professional SABIC Award Lecture

Abstract:

Flame Synthesis and Coating of Nanostructured Particles in One Step

Surface properties of inorganic or organic nanostructured particles often have to be modified to meet specific needs in final applications. This is accomplished by hermetically depositing shells and/or attaching functional groups onto the core particles. Here, an aerosol reactor has been developed that enables in-situ coating of flame-made particles with nanothin SiO_2 films. This process is continuous and scalable and serves as a tool to produce particles of increased functionality at limited additional cost. For example, when a SiO_2 coating is applied on UV-absorbing flame-made TiO_2 nanoparticles, their photocatalytic activity is minimized rendering such particles suitable for sunscreens or as nanofiller materials. Silica encapsulation of core nanoparticles offers several advantages, for example it can reduce toxicity of plasmonic Ag nanoparticles, minimize magnetic nanoparticle interactions and reduce particle agglomeration in aqueous solutions. Furthermore, this gas-phase process is not only limited to silica coatings, but is also further extended for the synthesis of nanoparticles that are surface functionalized with organic groups rendering them hydrophobic. This one-step SiO_2 coating process is also utilized for engineering magnetic nanostructured particles for biomedical applications. Superparamagnetic Fe_2O_3 nanoparticles or even hybrid Janus like plasmonic
play a decisive role on their final performance. Finally, results of such stimuli
a multiscale drug delivery system will be shown where the triggered drug release is achieved by the application of an alternate magnetic field.

461920 Modeling the Effect of Inhomogeneities on the Fluid-Particle Drag Force

Abstract:

In a fluidized bed, the fluid-particle drag force balances the force of gravity, and therefore provides the main mechanism for fluidization. Thus, the ability to accurately simulate large-scale fluidized systems depends on the implementation of a quantitatively-accurate constitutive drag model. Despite the fact that inhomogeneities, such as bubble-like voids and particle clusters, are often observed in fluidized beds, commonly-employed drag relations, such as those of Beetstra et al[1] and Wen & Yu[2], are derived under the assumption that the distribution of particles is homogeneous. Prior studies focusing on particle-scale dynamics, such as those of Kriebitzsch et al[3] and Zhou et al[4], have found that inhomogeneities in the particle distribution have a significant effect on the computed fluid-particle drag force. Therefore, the goal of the current study is to quantify the effect of inhomogeneities on the drag in the form of a new constitutive model.

In this work, the dynamics of fluid-particle systems are simulated using a fully-resolved lattice Boltzmann method (LBM), as was previously done by Derksen & Sundaresan[5]. In order to ascertain the effect of particle structures at different length scales, the fluid-particle drag results are computed over a range of filter, or averaging, sizes. Inspired by the work of Rubinstein et al[6], we find that, in the low Reynolds number (Re) regime, the drag force within a fluidized bed with a particular extent of inhomogeneities can be quantified relative to the drag in the high and low Stokes number (St) limit beds, with identical extents of inhomogeneities, using only the local St and particle volume fraction. With this insight into the relative fluidized bed drag behavior, we are able to simplify our analysis of the effect of inhomogeneities on the drag force by focusing solely on the high and low St limit beds, rather than having to separately analyze the effect of the particle configurations on each of the different types of fluidized beds.

By analyzing the drag results over a variety of particle configurations, we find that in the low Re regime, where the effects of fluctuations in the particle velocity on the drag force are negligible, changes in the drag as compared to the random, homogeneous case are due to structures that are unresolved at the scale of the filter. The extent of these sub-filter scale inhomogeneities can be quantified using the squared fluctuations in the particle volume fraction. In both the high and low St limits, there is a clear reduction in the drag due to increases in the extent of sub-filter scale inhomogeneities that are present at the length scale of just a few particle diameters. Thus, we have formulated models for the high and low St drag forces, respectively, as functions of the inhomogeneities in the particle configuration.

Since the extent of sub-filter scale inhomogeneities cannot be directly computed in a larger-scale fluidized bed simulation, we have employed a test filter approach to estimate this sub-filter quantity. This test filter approach is based on a dynamic sub-grid scale modeling technique that was developed by Germano et al[7] to estimate sub-filter scale stress in single-phase turbulence, and adapted by Parmentier et al[8] and Ozel et al[9] to coarse-grained drag in multiphase flows. In this technique, the extent of inhomogeneities is computed at the base filter size by demanding scale similarity between the base and test filters. In this manner, we have developed a method for incorporating the effects of inhomogeneities into the constitutive relation for the drag force, which can be used in coarser simulations based on both two-fluid model and Euler-Lagrange approaches, such as the Computational Fluid Dynamics-Discrete Element Method simulations[10,11]. It is hoped that this improved drag model will lead to more accurate and grid-size independent predictions of fluidized bed behavior; however, these improvements remain to be established.

1. Beetstra, R., van der Hoef, M. A., & Kuipers, J. A. M. (2007). Drag Force of Intermediate Reynolds Number Flow Past Mono- and Bidisperse Arrays of Spheres. *AIChE J.* **53**(2), 489–501.

2. Wen, C. Y., & Yu, Y. H. (1966). Mechanics of fluidization, *Chem. Eng. Prog. Symp. Ser.* **62**, 100.

3. Kriebitzsch, S. H. L., van der Hoef, M. A., & Kuipers, J. A. M. (2013). Drag Force in Discrete Particle Models-Continuum Scale or Single Particle Scale? *AIChE J.* **59**(1), 316-324.

4. Zhou, G., Xiong, Q., Wang, L., Wang, X., Ren, X., & Ge, W. (2014). Structure-dependent drag in gas-solid flows studied with direct numerical simulation. *Chem. Engng. Sci.* **116**, 9-22.

5. Derksen, J. J., & Sundaresan, S. (2007). Direct numerical simulations of dense suspensions: wave instabilities in liquid-fluidized beds. *J. Fluid Mech.* **587**, 303–336.

6. Rubinstein, G. J., Derksen, J. J., & Sundaresan, S. (2016). Lattice Boltzmann simulations of low-Reynolds number flow past fluidized spheres: Effect of Stokes number on drag force. *JFM.* **788**, 576-601

7. Germano, M., Piomelli, U., Moin, P., & Cabot, W. H. (1991). A dynamic subgrid-scale eddy viscosity model. *Phys. Fluids A.* **3**, 1760.

8. Parmentier, J.-F., Simonin, O., & Delsart, O. (2012). A functional subgrid drift velocity model for filtered drag prediction in dense fluidized bed. *AIChE J.* **58**(4), 1084-1098.

9. Ozel, A., Fede, P., & Simonin, O. (2013). Development of filtered Euler-Euler two-phase model for circulating fluidised bed: High resolution simulation, formulation and a priori analyses. *Int. J. Multiph. Flow.* **55**, 43-63.

10. Pepoit, P. & Desjardins, O. (2012). Numerical analysis of the dynamics of two- and three-dimensional fluidized bed reactors using an Euler–Lagrange approach. *Powder Technol.* **220**, 104–121.

11. Radl, S. & Sundaresan, S. (2014). A drag model for filtered Euler–Lagrange simulations of clustered gas–particle suspensions. *Chem. Engng. Sci.* **117**, 416-425.

462637 Linking Defluidization Behavior of Solid Particles in Humid Environments to Micro-Scale Forces

Abstract:

Humidity results in small capillary bridges that condense between particles and induce cohesive behavior. To better understand the effects of humidity-induced cohesion, defluidization experiments are investigated under increasing relative humidity (*RH*) levels, the results of which are explained via particle-level predictions for the resulting capillary force. The sensitivity of the defluidization curve is captured by a new defluidization velocity U_{df}, which is intended to be similar to the complete fluidization velocity used to characterize defluidization of polydisperse beds. In particular, U_{df} characterizes the curvature of the defluidization plot (pressure drop vs. velocity) arising from humidity and leading to a partially fluidized state between the packed bed and fully-fluidized regions. The partially fluidized region, as characterized by U_{df}, is more sensitive to increasing levels of cohesion than the packed-bed region. An increase in *RH* results in a U_{df} that gradually increases approximately linearly, followed by an abrupt (non-linear) increase at *RH* ~55%. Following this abrupt increase in U_{df},, the bed transitions from Group A to Group C behavior over a narrow range of increasing *RH* levels. To explain these macro-scale (many-particle) results, micro-scale (two-particle) theories for capillary forces are employed, in which measured values (i.e. not assumed or fitted) of particle surface roughness are implemented. From this physical description, it is found that the cohesive energy, rather than the maximum cohesive force or Bond number, predicts the same qualitative trend with increasing *RH* as U_{df} (for *RH* < 55%). Moreover, a hypothesis is presented to explain the transition form Group A to Group C behavior that occurs over *RH* ~60-65%; namely, the previously observed adsorption of water onto glass surfaces with liquid-like layers rather than crystal/ice-like layers.

462912 The Dual Liquid-Solid Nature of Fluidized Granular Matter and Its Influence on Pattern Formation

Abstract:

The dual liquid-solid nature of fluidized granular matter and its influence on pattern formation

Lilian de Martin (l.demartin@ucl.ac.uk), Kaiqiao Wu (kaiqiao.wu.13@ucl.ac.uk), and Marc-Olivier Coppens (m.coppens@ucl.ac.uk)

All from:

University College London

Department of Chemical Engineering

Torrington Place

London WC1E 7JE

United Kingdom

T + 44 20 7679 7369

F + 44 20 7383 2348

Granular beds fluidized with an oscillating gas flow can form regular patterns under certain experimental conditions.

In a vertically quasi-2D bed, a row of bubbles with alternating positions forms at every gas pulse, resulting in a triangular bubble tessellation when viewed from the side of the bed. In a 3D bed, patterns manifest themselves as stripe and square configurations on the surface of the bed, similar to those observed in vertically vibrated granular layers and pure liquids (Faraday waves). Patterns excel as a method to impose structure in fluidized bed reactors, control the bubble size, and facilitate bed scale up, but their formation mechanism is poorly understood.

Patterns in very shallow¾few mm¾fluidized beds can appear with the entire bed operating in the kinetic regime, i.e., with the solids packing below the critical packing limit, whereas patterns in deeper beds typically appear with the bed operating in the plastic regime, i.e., with the solids packing above the critical packing limit. These two limit situations involve completely different physics and have deep consequences for the computational modelling and the ability of different approaches to capture this phenomenon [1]. In addition, the fact that vibrated fluids and purely granular systems also form patterns under sinusoidal excitations provides a unique framework to study the physics of granular matter.

In this contribution, we discuss how pattern formation can be used to gain insight into the liquid-solid dual nature of fluidized matter and provide some experimental and computational examples of this duality.

[1] K. Wu, L. de Martin, L. Mazzei, and M.-O. Coppens. Pattern formation in fluidized beds as a tool for model validation: a two fluid model based study. Powder Technology 295, 35 (2016).

465668 Impact of Particle and Column Electrical Properties on Entrainment from Gas-Solid Fluidized Beds

Abstract:

Impact of particle and column electrical properties on entrainment from gas-solid fluidized beds

Farzam Fotovat, John R. Grace, Xiaotao T. Bi

Department of Chemical and Biological Engineering, University of British Columbia, Vancouver, Canada V6T 1Z3

1. Introduction

Entrainment of fine particles is a common pitfall of fluidized beds working at high gas velocities resulting in loss of bed inventory such as precious catalysts, and potential release of fine particulates into the atmosphere. To reduce the amount of the elutriated particles, gas-solid separators like cyclones and filters are installed downstream of the fluidized beds. However, optimal design and operation of these devices greatly depends on estimating the extent of entrainment in the freeboard. Despite the numerous studies conducted on the entrainment process [1–3], it has not been well understood, due to the complexity of gas-solid flows, particularly with respect to inter-particle cohesion and hydrodynamic clustering effects.[4] A handful of correlations have been developed to predict the entrainment rate of fluidized beds involving a variety of particles entrained at different operating conditions. In the absence of a thorough understanding of the underlying mechanisms of entrainment, there are huge discrepancies in the elutriation rates predicted by the existing empirical correlations and semi-empirical relationships.

The nature of gas-solid fluidization processes produces continuous motion and rubbing among bed particles such that the generation of electrostatic charges is inevitable. The electrostatic charges in fluidized systems can interfere with the normal hydrodynamics of the bed, resulting in particle-wall adhesion, inter-particle cohesion, electrostatic discharges, wall sheeting and even explosions, all of which can affect plant safety and economics.[5,6] The particles entrained from the bed are also subject to the electrostatic forces, which are likely to affect the entrainment process by influencing whether particles travel individually or as aggregates/clusters. Although the significant impact of the electrostatics on the particle entrainment from gas-solid fluidized beds has been observed by a number of researchers[7–9], its contribution to the correlations predicting the entrainment rate has been overlooked; an oversight that may well be responsible for the huge discrepancies in predicted entrainment rates.

To allow for the effect of the electrostatic interactions on entrainment, it is imperative to determine the parameters influencing the electrostatic forces exerted on the particles in the freeboard. To this end, focus of this work is exploring the impact of electrical properties of particles and the column wall on entrainment. Outcomes of this study may clarify how choosing appropriate materials can be helpful in controlling electrostatics and entrainment in gas-solid fluidized beds. Furthermore, this study may shed light on the discrepancies observed in the entrainment rates that cannot be explained by hydrodynamics.

2. Experimental

A number of dielectric fines (such as glass, alumina, cork and, porcelain) as well as, conductive fines (such as silver-coated glass, copper, and stainless steel) were elutriated from a stainless steel column with an inner diameter of 0.15 m and a height of 2.0 m. To triboelectrically charge fines in the column, a binary fines and coarse glass beads were mixed before each test and the column was then loaded with the mixture to a total depth of 0.4 m. Each mixture contained 90 wt. % of the coarse material, with the fine species making up the balance. The entrainment rate of fines and their mass charge density were simultaneously measured by means of a sampling vessel into which the elutriated fines were diverted after being recycled by an external cyclone. The average entrained mass divided by the sampling time with at least five repetitions was taken as the entrainment rate for each run. The charge density of entrained particles was measured by a sampling device incorporating the Faraday cup principle developed and described by Alsmari et al.[10] The relative humidity of the bed was set at ~10% for all experiments using a refrigerating unit and a vapor removal filter. Pressure and temperature were maintained constant at 205 kPa and 20°C, respectively. The superficial gas velocity was varied from 0.2 to 0.8 m/s. For the purpose of exploring the impact of the electrical properties of the column wall on entrainment of fines, the stainless steel column will be replaced with a non-conductive acrylic column and all the above-mentioned tests will be carried out in this column.

3. Preliminary Results

Fig. 1 shows the entrainment flux (W_s) vs. gas velocity in excess of the terminal velocity, (U_g-U_t) for each fine material elutriated from the stainless steel column. In general, the elutriation flux of the conductive fine particles (silver-coated glass, copper and stainless steel) was larger than for the non-conductive fine materials (glass, alumina, cork and porcelain), especially for (U_g-U_t)>0.3 m/s. As (U_g-U_t) approached 0.6 m/s, the entrainment rate of the silver-coated fine glass beads became more than six times larger than that of the (uncoated) fine glass beads at the corresponding (U_g-U_t), despite the close similarity of the physical properties (density, size, shape) of these two types of particles.

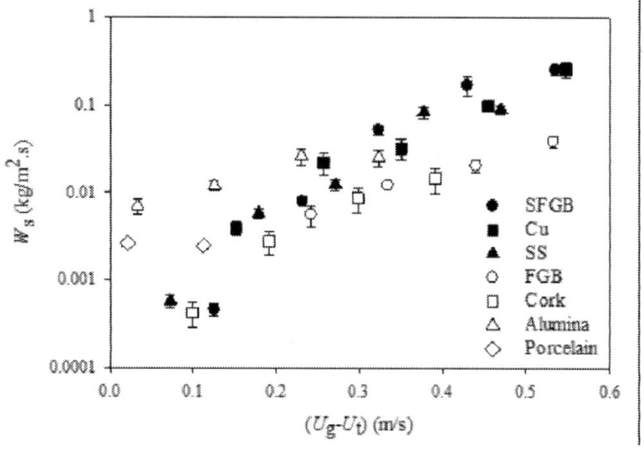

Fig. 1. Entrainment flux of particles as a function of (U_g-U_t). Open symbols: dielectric particles; filled symbols: conductive particles. Error bars show the standard deviations of the experimental measurements.

Assessment of the hydrodynamic forces such as gravity and drag exerted on the fine particles in the freeboard illustrates that these forces are not responsible for the higher entrainment rate of the conductive particles relative to the dielectric fines. On the other hand, Fig. 2 shows that the F_e/F_g (electrostatic to gravity force) ratios are significantly smaller for conductive than for dielectric species. By comparing regular and silver-coated fine glass beads (open and filled circle symbols), Fig. 2 indicates that changing the conductivity of particles can profoundly influence the magnitude of the entrainment flux.

This observation can be attributed to the intensification of electrostatic inter-particle forces for non-conductive particles due to the non-uniform electrical charge distribution over their surfaces. Moreover, dominance of the attractive forces among the dielectric particles is likely in the freeboard region, promoting formation of aggregates or clusters which reduces the entrainment rate. On the other hand, repulsive electrostatic forces between pairs of touching conductive particles can cause these particles to act independently, augmenting their entrainment. This study suggests that decreased electrical conductivity of particles can assist significantly in reducing the entrainment of fine particles from gas-solid fluidized beds.

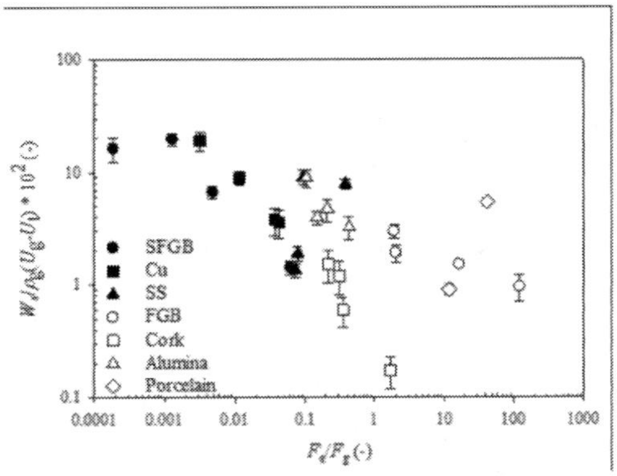

Fig. 2. Normalized entrainment flux vs. electrostatic-to-gravity-force ratio of entrained particles. Open symbols: dielectric particles, filled symbols: conductive particles. Error bars show the standard deviations of the experimental measurements.

The plausible influence of the electrical properties of the column wall on entrainment of the fines will be discussed in the final version of this study.

4. References

(1) George, S. E.; Grace, J. R. Entrainment of particles from a pilot scale fluidized bed. *Can. J. Chem. Eng.* **1981**, 59, 279–284.

(2) Ma, X.; Kato, K. Effect of interparticle adhesion forces on elutriation of fine powders from a fluidized bed of a binary particle mixture. *Powder Technol.* **1998**, 95, 93–101.

(3) Tasirin, S. M.; Geldart, D. The elutriation of fine and cohesive particles from gas fluidized beds. *Chem. Eng. Commun.* **1999**, 173, 175–195.

(4) Chew, J. W.; Cahyadi, A.; Hrenya, C. M.; Karri, R.; Cocco, R. A. Review of entrainment correlations in gas–solid fluidization. *Chem. Eng. J.* **2015**, 260, 152–171.

(5) Cross, J. Electrostatics: principles, problems and applications, **1987**, Adam Hilger Bristol, Bristol.

(6) Hendrickson, G. Electrostatics and gas phase fluidized bed polymerization reactor wall sheeting. *Chem. Eng. Sci.* **2006**, 61, 1041–1064.

(7) Baron, T.; Briens, C. L.; Bergougnou, M. A.; Hazlett, J. D. Electrostatic effects on entrainment from a fluidized bed. *Powder Technol.* **1987**, 53, 55–67.

(8) Briens, C. L.; Baron, T.; Bergougnou, M. A.; Inculet, I. I.; Hazlett, J. D. Size distribution of particles entrained electrostatic effects from fluidized beds: electrostatic effects. *Powder Technol.* **1992**, 70, 57–62.

(9) Alsmari, T. A.; Grace, J. R.; Bi, X. T. Effects of particle properties on entrainment and electrostatics in gas–

solid fluidized beds. *Powder Technol.* **2015**

(10) Alsmari, T. A.; Grace, J. R., Bi, X. T. Effects of superficial gas velocity and temperature on entrainment and electrostatics in gas–solid fluidized beds. *Chem. Eng. Sci.* **2015**, 123, 49–56.

470029 Study of Solid-Liquid Fluidized Bed Using Electrical Impedance Tomography

Abstract:

Introduction:

Liquid fluidization has many applications in industrial processes, such as particle classification, adsorption and ion exchange, and in wastewater treatment[1]. Additionally, a solid-liquid fluidized bed can operate over a wide range of particle and fluid properties (e.g. solids concentration, particle size, density, and fluid viscosity) and therefore provides ideal conditions to study the dynamic behaviour of solid-liquid systems. Specifically particle-particle, particle-fluid, and particle-wall interactions can be investigated by studying liquid fluidization[2].

The turbulent nature of fluidization results in the fluctuations of the velocities and volume fractions of both the solid and liquid phases[3,4]. Experimental investigations with the aim of understanding the nature and magnitude of these fluctuations are necessary since: 1) these fluctuations are directly responsible for the high level of mixing taking place in the fluidization process, which in turn produces effective heat and mass transfer[3]; 2) such investigations are beneficial for the characterization of the dynamic behaviour of other solid-liquid flow systems such as two-phase flows in pipelines; and 3) measurements of the solid concentration fluctuations can be used to validate and improve available numerical simulations.

Solid concentration fluctuations can be classified in two categories: Large global fluctuations resulting from the bulk behaviour of the bed (i.e. circulation); and small-scale fluctuations mainly produced by the local hydrodynamics, and by particle-particle and particle-fluid interactions[3]. Buyevich and Kapbasov[3] were the first to provide a mathematical model for prediction of the magnitude of small-scale fluctuations as a function of bulk solid concentration. Zenit and

Hunt[4] then measured the magnitude of small-scale fluctuations at different fluidization conditions and for different particle properties. Their results showed that in addition to the bulk solid concentration, particle size and density are contributing factors[4]. In this study, for the case of inertial particles (particles with the largest Stokes number tested), the measured values of the Root Mean Square (RMS) fluctuations were close to the values predicted by the Buyevich and Kapbasov model[4]. However, for the other particles tested (lower Stokes numbers) the magnitude of fluctuations measured by Zenit and Hunt where 3 to 10 times lower than the values predicted by the model[3], depending on the particle terminal Stokes number[4]. The particle terminal Stokes number can be expressed as $St_t = (\rho_s/\rho_f)(Re_t/9)$, where ρ_s and ρ_f are solid and fluid density, respectively, and Re_t is the particle Reynolds number at the particle terminal settling velocity.

In 2010, Gevrin et al.[5] conducted a numerical study of the dynamic behaviour of a liquid fluidized bed and showed nonuniform distributions of instantaneous local solid concentration in both the vertical and radial directions. However, since the experimental data available from Zenit and Hunt's study was limited to values averaged over the cross-section, Gevrin et al. were only able to validate their numerical solutions for cross-sectional averaged concentration fluctuations. This indicates that the magnitude of the local instantaneous distribution of particles and local concentration fluctuations must still be measured. In 2013, Hashemi[6] measured the local solid concentration fluctuations for selected particles using high-speed EIT. Results from this work which were limited to particles with moderate Stokes numbers (glass beads) showed that the RMS fluctuations was not uniformly distributed over the bed cross-section (being higher in the near-wall region)[6].

Based on the results of previous studies of the small-scale fluctuations, there are multiple mechanisms involved in their production, including particle-fluid interactions[4], direct particle collisions[3,4] and particle-wall interactions[6]. More studies are required to investigate the relative magnitude of these mechanisms at different fluidization conditions and for different particle properties. The focus of this study is to measure the local instantaneous concentration distribution of particles in a liquid fluidized bed and calculate the RMS fluctuations based on those measurements. These localized measurements are conducted for a wide range of particle properties (i.e. Stokes numbers) at different bulk concentrations. Here we only present the results for particles having relatively low Stokes numbers (St = 122).

Methods:

The solid-liquid fluidized bed used in this study had an internal diameter of 10.16 cm. Mono-sized spherical Nylon particles (d = 6 mm, ρ =1150 kg/m^3) were used as the solid phase and a dilute salt solution (C_{NaCl} = 0.85 g/L) was the liquid phase. The fluid was circulated through the bed at different superficial velocities using a centrifugal pump. Fluid flow rate and temperature were measured upstream of the entrance to the bed. A high-speed EIT instrument (Z8000, Industrial Tomography Systems) was used to obtain the instantaneous solid concentration distribution recorded at a sampling rate of 1160 frames per second. Electrical tomography is a robust noninvasive technique with the unique ability of providing local concentration fluctuations even for opaque concentrated mixtures[7,8]. This technique utilizes an array of sensors to measure the electrical impedance of the medium which will then be used to deduce solid concentration using Maxwell's equations[7]. The measurements conducted by EIT are used to produce a solid concentration map of the bed cross-section with 316 pixels.

Results:

At each bulk solid concentration, C, 8000 instantaneous solid concentration maps were obtained. From these maps, the magnitude of the RMS of the solid concentration fluctuations, C , was then calculated for each pixel. Time-averaged solid concentration maps, such as those shown in Figures 1a, 1c, and 1e, indicate that, on average, the particles are uniformly distributed over the bed cross-section for all the bulk concentrations tested here. The fluctuations, on the other hand, are not uniform. Figures 1b, 1d, and 1f show that C in the near-wall region are 2 to 4 times higher than that measured in the center of the bed. Greater values of C in the near-wall region were also observed for spherical glass particles with moderate Stokes numbers ($153 \leq St \leq 480$)[6], and spherical Delrin particles with low Stokes numbers (St = 94)[9].

At the lowest bulk solid concentration tested ($C = 0.05$), the magnitude of the RMS of fluctuations falls within the range of C = 0.005 to 0.022. By increasing the bulk solid concentration to 0.27, the range of the fluctuations increases to C = 0.01 to 0.038. When the bulk solid concentration is increased from 0.27 to 0.38, the magnitude of the fluctuations does not change appreciably. Further increase in the bulk solid concentration (up to $C = 0.52$) decreases the range of the fluctuations to C = 0.01 to 0.028. Similar trends for the changes in the magnitude of C with the change in the bulk solid concentration were obtained for particles with lower Stokes numbers (Delrin, St = 94)[9].

By increasing the Stokes number from 94 (Delrin) to 122 (Nylon), the magnitude of the RMS of fluctuations increases. The magnitude of fluctuations for both of these particles is smaller than that of particles with moderate Stokes number (Glass, St = 480)[6]. This finding is in agreement with the experimental results of Zenit and Hunt[4], which showed that by increasing the Stokes number the magnitude of the fluctuations increases and approaches the values predicted by the Buyevich and Kapbasov model[3].

Conclusions and future work:

The results of this study demonstrate the non-uniform distribution of the local solid concentration fluctuations over the bed cross-section for Nylon particles (St = 122). The magnitude of the fluctuations in the near-wall region was found to be 2 to 4 times greater than that of the center of the bed. These results, along with the previous work for Delrin particles (St = 94) confirm the importance of local instantaneous concentration measurements. Future experiments will focus on particles with higher Stokes numbers ($800 < St < 3600$). Collectively, these results can be used to create more accurate models for prediction of the magnitude of local concentration fluctuations. They can also be used to validate and improve numerical models of fluidization.

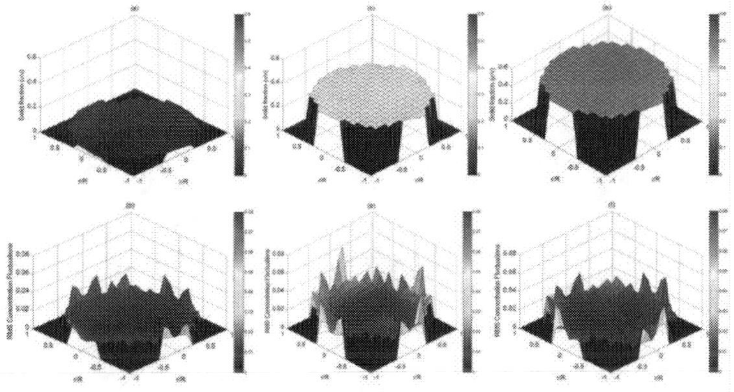

Figure 1. Time-averaged solid concentration maps (first row) and solid concentration fluctuations maps (second row): (a) and (b) $\bar{C} = 0.05$; (c) and (d) $\bar{C} = 0.33$; (e) and (f) $\bar{C} = 0.52$.

References:

[1] Epstein, N., Applications of Liquid-Solid Fluidization, *Int. J. Chem. React. Eng.*, vol. 1, no.1, 2002.

[2] Gevrin, F., Masbernat, O., and Simonin, O., Granular pressure and particle velocity fluctuations prediction in liquid fluidized beds, *Chem. Eng. Sci.*, vol.63, no.9, pp.2450–2464, May 2008.

[3] Buyevich, Y.A., and Kapbasov, S.K., Random fluctuations in a fluidized bed, *Chem. Eng. Sci.*, vol.49, no.8, pp.1229–1243, Apr. 1994.

[4] Zenit, R., and Hunt, M.L., Solid fraction fluctuations in solid–liquid flows, *Int. J. Multiph. Flow*, vol.26, no.5, pp.763–781, May 2000.

[5] Gevrin, F., Masbernat, O., and Simonin, O., Numerical study of solid–liquid fluidization dynamics, *AIChE J.*, vol.56, no.11, pp.2781–2794, Nov. 2010.

[6] Hashemi, S.A., *Velocity and concentration fluctuations in concentrated solids-liquid flows*, PhD thesis, University of Alberta, Edmonton, Canada, 2013.

[7] Dyakowski, T., Jeanmeure, L.F.C. and Jaworski, A.J., Applications of electrical tomography for gas–solids and liquid–solids flows —a review, *Powder Technol.*, vol.112, no.3, pp.174–192, Oct. 2000.

[8] Norman, J.T. and Bonnecaze, R.T., Measurement of Solids Distribution in Suspension Flows using Electrical

Resistance Tomography, *Can. J. Chem. Eng.*, vol.83, no.1, pp.24–36, 2005.

[9] Marefatallah, M. and Sanders, R.S., Solid Volume Fraction Fluctuations in a Liquid Fluidized Bed, *Proceedings of the 9th International Conference on Multiphase Flow*, Firenze, Italy, 2016.

469799 Simulating Fluid-Particle Flows Across All Flow Regimes

Abstract:

In technological applications and nature, fluid-particle flows often exhibit large variations in the local particle volume fraction. For example, in circulating fluidized beds there are regions where the particles are close-packed as well as very dilute regions where particle-particle collisions are infrequent. Thus, in order to simulate such fluid-particle systems, it is necessary to design a numerical algorithm that can accurately treat all flow regimes that can occur in any particular application. Here, a novel algorithm, based on splitting the kinetic flux dynamically and locally in the flow, is implemented in in an open-source CFD package for particle velocity moments up to second order. In close-packed regions, a traditional hydrodynamic solver is employed, while in dilute to very dilute regions a kinetic-based finite-volume solver based on quadrature-based moment methods is used. The accuracy and robustness of the proposed flux-splitting algorithm is demonstrated using gravity-driven, gas-particle flows exhibiting cluster-induced turbulence. By varying the average particle volume fraction in the flow domain, it is shown that the flux-splitting algorithm handles seamlessly all flow regimes present in circulating fluidized beds.

470859 Dynamic Forces Induced By Geldart Group a and Group B Particles in Fluidized Bed Systems

Abstract:

Fluid bed internals such as heat exchanger tube bundles and baffles are subject to random buffeting forces induced by rising bubbles and particle motion in the bed. To understand and quantify these forces, a piezoelectric force sensor was mounted in a 2-D Plexiglas column, 30 cm wide and 2.5 cm deep for a visual depiction of the phenomenon. The magnitude of these forces not only depended on momentum and residence time of bubbles when they were in contact with the sensor, but also on the material that was being tested. Tests were conducted with fluid catalytic cracking catalyst particles (Group A) and glass beads (Group B). The average forces induced by these materials were compared at different fluidization regimes, ranging from a single bubble that was injected at minimum fluidization, as well as in a bubbling and turbulent bed. Tests were also done to quantify the forces during the startup of the bed, when there was no aeration. It was found that glass beads on average generated greater forces during the startup of the bed, as well as during subsequent fluidization. The larger bubbles sizes of Group B material moving with a greater momentum generated higher forces when they came into contact with the sensor. This is an important criterion in the design and fabrication of internals to mitigate structural damages.

469311 Nanocrystal Doping Stabilizes the Perovskite Phase of Cesium Lead Iodide

Abstract:

All-inorganic, perovskite-phase cesium lead iodide is a promising material for photovoltaics. However, under ambient conditions it rapidly transforms to a non-functional, so-called yellow phase. Here we demonstrate more than two orders of magnitude improvement in the stability of nanostructured perovskite-phase cesium lead iodide thin films through a nanocrystal doping approach. The doped, functional perovskite phase is synthesized by co-assembling cesium lead iodide nanocrystals with cesium lead chloride dopant nanocrystals. The resulting doped nanocrystal solid is subsequently fused into a polycrystalline thin film by chemically-induced, room-temperature sintering. This process ensures nanometer-scale mixing even at compositions that exceed the bulk miscibility of the two constituents. Spectroscopy and X-ray diffraction indicate that some of this chloride is further dispersed during sintering and a polycrystalline mixed phase is formed. By comparison to DFT-calculated values, the relative change in band gap and the lattice contraction are shown to be consistent with a Cl:I ratio of a few percent in the mixed phase.

460501 Metal Oxide Electron-Selective Layers for Inverted Perovskite Solar Cells By Atomic Layer Deposition

Abstract:

Perovskites are promising candidates for low cost, high efficiency light absorbers with a large, tunable band gap. Recent advances in photostability and thermal stability from the introduction of mixed-cation lead mixed-halide perovskites in the form of cesium formamidinium (CsFA) perovskites has lead to significant interest in the development of perovskite-silicon tandem solar cells. Improvements in the deposition of transparent contacts, stability of organics within the device structure and blocking of moisture to the perovskite absorber must be made to realize highly efficient and stable perovskite-silicon devices. The improved thermal stability of the CsFA perovskite has opened up new processing options through atomic layer deposition (ALD) for the conformal coating of metal oxide materials on top of perovskite absorbers. In this work, we explore TiO_x and SnO_x ALD thin films deposited on top of CsFA perovskite devices in an inverted structure as dual-purpose layers for electron-selectivity and degradation protection. We will discuss the effects of ALD processing conditions on TiO_x and SnO_x electronic properties as well as on CsFA perovskite stability. We achieve stable >15% power conversion efficiency with 1.6 eV perovskite devices through maximum power point tracking and demonstrate the effectiveness of ALD metal oxide layers as buffer layers for the sputtering of indium tin oxide contacts for semi-transparent devices with >14% efficiency.

467509 Transparent Conductive Oxide Nanocrystals Coated with Insulators By Atomic Layer Deposition

Abstract:

Thin films comprised of transparent conducting oxide (TCO) nanocrystals are attractive for a number of optoelectronic applications. However, it is often observed that the conductivity of such films is very low when in contact with air. It has recently been demonstrated, somewhat surprisingly, that filling in films comprised of initially insulating TCO nanocrystals with another insulator by atomic layer deposition (ALD) dramatically increases the conductivity by many orders of magnitude. This work aims to elucidate the mechanism by which the ALD coating increases the conductivity. We used an *in situ* electrical probe to examine the effect on sheet resistance of removing two adsorbed oxygen species (physisorbed molecular water and chemisorbed hydroxide) and compared this result to thin films comprised of ZnO nanocrystals coated with Al_2O_3 and also HfO_2 by ALD. Although both insulating infills decrease the sheet resistance and increase the stability of the films, there is a stark discrepancy between the two. From the *in situ* measurements, it was found that coating with Al_2O_3 removes both physisorbed water and chemisorbed hydroxide resulting in a net reduction of the ZnO nanocrystals. Coating with HfO_2 only removes physisorbed water, which was confirmed by Fourier transform infrared spectroscopy. A similar phenomenon was observed when coating thin films comprised of Sn-doped In_2O_3 nanocrystals with Al_2O_3 by ALD; suggesting Al_2O_3 can be used to reduce and stabilize metal-oxide nanocrystals in general.

464634 Microstructure Development in $Cu_2znsn(S_xSe_{1-x})_4$ Thin Films during Annealing of Colloidal Nanocrystal Coatings

Abstract:

Copper zinc tin sulfoselenide is a promising material for thin film solar cells because it comprises inexpensive and abundant elements. Thin polycrystalline films of $Cu_2ZnSn(S_xSe_{1-x})_4$, or CZTSSe, may be formed by annealing nanocrystal coatings cast from colloidal dispersions in a selenium and/or sulfur atmosphere. This is a promising low-cost and high-throughput route for fabrication of CZTSSe-based thin film solar cells. During annealing, the nanocrystal films can transform into polycrystalline thin films with micrometer size grains. While benchmark efficiencies of CZTSSe solar cells have risen sharply in the last two decades, continued improvements in the solar cell performance and cost reduction require a fundamental understanding of the factors that influence the film microstructure. We have been investigating the microstructure evolution of polycrystalline CZTSSe thin films through systematic control of parameters such as chalcogen (sulfur or selenium) vapor pressure, annealing temperature, and heating rate in a closed system. In this talk, we will contrast microstructure development in CZTSSe films formed by annealing CZTS nanocrystal coatings in selenium vapor with microstructure development in CZTS films formed by annealing CZTS nanocrystal coatings in sulfur vapor. There are significant differences in the mechanisms of microstructure development despite the similar approaches. These differences are due to differences in thermodynamic phase equilibria of sulfur and selenium as well as due to differences in chemical reactions of sulfur and selenium with the carbon containing ligands that surround the nancrystals: these ligands are used to stabilize the nanocrystal dispersions and often end up in the coatings and films even after annealing. For example, annealing CZTS nanocrystal coatings at high selenium pressure (*e.g.*, 450 Torr) leads to the formation of a bilayer microstructure containing a microcrystalline CZTSSe layer on top of a nanocrystalline layer that is rich in carbon. In contrast, films annealed with sulfur do not show a distinct carbon-rich layer and most of the carbon volatilizes from the film during annealing. We suggest that carbon remains in the film due to the formation of a non-volatile carbon selenide polymer, the presence of which inhibits nanocrystal sintering and leads to the formation of a large-crystal top layer. The microstructure of films annealed in the presence of selenium liquid is influenced by liquid phase sintering. These films exhibit monomodal coarsening and densification into a compact film without the formation of a bilayer structure.

Multiscale study of the Self-Organized Vertical Concentration Profile of PEDOT:PSS for Work Function Optimization

Min Huang*
School of Chemical Science and Engineering, Tongji University
1239 Siping Road, Shanghai, China 200092

ABSTRACT

Conventional conducting polymers, poly(3,4-ethylenedioxythiophene):poly(styrenesulfonate) (PEDOT:PSS) has been widely used as the hole injection layer (HIL). It would be preferable to design a gradient profile of work function near the interface in vertical for individual applications. In the literature, tetrafluoroethylene-perfluoro-3,6-dioxa-4-methyl-7-octene-sulfonic acid copolyme (PFI) and nano Nickel Oxide were experimentally added into the PEDOT:PSS layer to create a self-organized vertical concentration profile. In this contribution, the mechanisms of the self-organized gradient effect were studied using multiscale analysis and dissipative particle dynamics (DPD) simulation. The Flory-Huggins parameters and intramolecular bonding parameters were obtained through reverse mapping of molecular dynamics simulations. DPD simulations with different additive molecular structure and concentration were performed, their influences on the self-organized concentration profile were discussed.

Keywords: multiscale, dissipative particle dynamics, PEDOT:PSS : PFI, NiOx nano particle, work function

1. INTRODUCTION

Transparent conductive polymer pss:pedot is very widely used in many applications such as organic solar cells[1,2], OLED[3], and recently perovskite solar cells[4]. it is often inserted in between indium tin oxide (ITO) and the active layer to serve as the hole injection layer (HIL). The work function of ITO is usually at about 4.8eV. Work function of the active layers according to their applications, maybe 5.8 eV. The work function of PSS:PEDOT can vary strongly according to its composition, different PSS:PEDOT ratio[5]. Therefore, in practice, optimized PSS:PEDOT ratio may be chosen to form a cascaded step-changed profile in work function. However, in some applications, such as OLED, this commonly used PEDOT:PSS has been found to have serious problems due to its inefficient hole-injection, inefficient electron-blocking, and substantial quenching of excitons close to the PEDOT:PSS. It was proposed that a smooth gradient profile can solve or at least lessen these problems. Many experimental studies aiming towards this goal, solvent surface treatment[6], NiOx nano particles[7], PFI diblock copolymer[8]

Modeling of the interface and additives interact with PEDOT:PSS is very important for mechanisms understanding, future device architecture designing and structure optimization. Classical molecular dynamics is a very powerful tool for it can provide atomistic detail of the interface, but it also has an equally great disadvantage, it can only apply to very small size dimension and short time duration, far from the length and time scales in real sense. This disadvantage maybe overcome through recently matured multiscale simulations. In a multiscale simulation mesoscale simulations were

* minontongji@tongji.edu.cn; phone 86 021 65982232; fax 86 021 65982287

performed to provide a bridge to relate microscopic view from molecular dynamics to the dimension and time in real sense. There are two popular mesoscale simulation tools, Coarse grain dynamics which uses parameters from direct mapping of the molecular dynamics, therefore, retained most features of molecular dynamics. The other method is the so called Dissipative particle dynamics (DPD). DPD was developed in the early 90's[9] and reformulated by Espanol[10], which combines coarse graining with a branch of thermodynamics, the Flory-Huggins theory for polymer[11,12,13]. Flory-Huggins theory for polymer mix states that the free energy of a polymer mix including three terms:

$$\left(\frac{\Delta G_{mix}}{RT}\right)\left(\frac{V_r}{V}\right)=\frac{\phi_A}{N_A}\ln\phi_A+\frac{\phi_B}{N_B}\ln\phi_B+\chi_{AB}\phi_A\phi_B \quad (1)$$

the first two terms is entropy of mix from ideal solution, the third term led by the interaction parameter χ_{AB} represents enthalpic interaction between monomer segments and/or solvents which can be calculated form cohesive energy density. The non-ideality of entropy of mix can be incorporated through chain flexibility using intramolecular bonding parameters. According to second law of thermodynamics, the energy will be at its minimum when the system reaches its equilibrium. Therefore, in this way, the DPD simulation has a thermodynamic constrain built in, the simulation process will go towards nothing but thermodynamic final state.

2. MODELS AND METHODS

Dissipative particle dynamics[14] (DPD) is a mesoscale dynamic simulation tool, therefore, atomistic details are not considered. This allowing it to give access to longer time and length scales than classical molecular dynamics (MD) simulations. In DPD simulation different beads with similar volume size and mass must be used to represent polymer monomers in a wormlike structure.

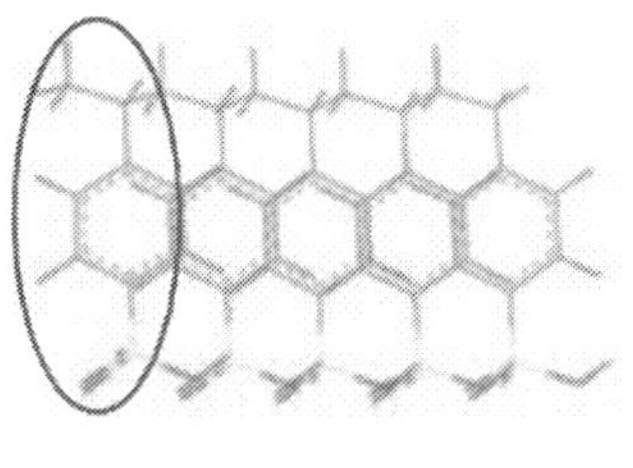

PSS

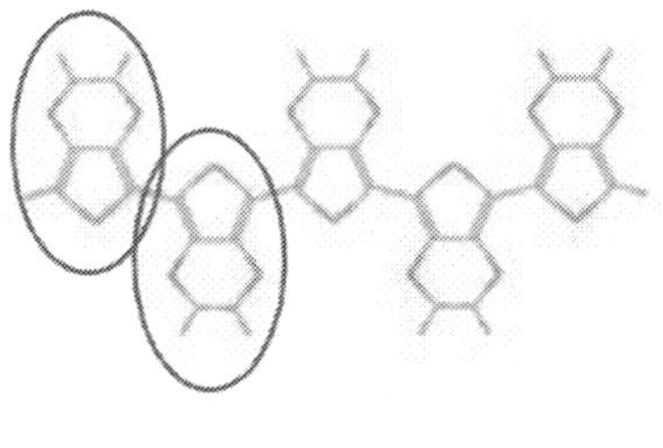

PEDOT

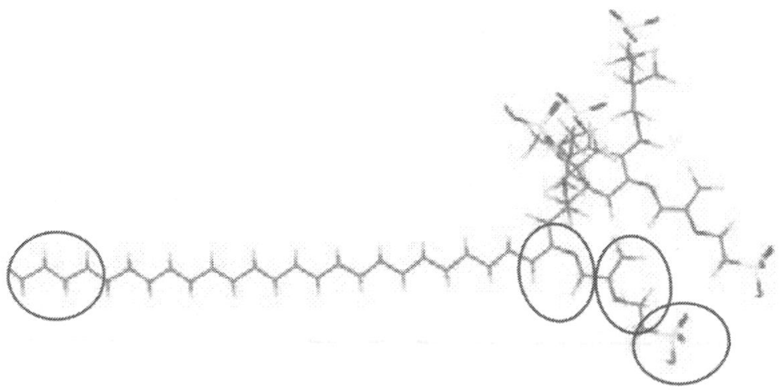

PFI

Figure 1. DPD beads representation of PEDOT, PSS and PFI

All monomers in this study were optimized using DFT, B3LYP[15,16,17] functional and 6-311G++2d,2p basis set[18]. The definition of PSS:PEDOT and PFI beads are shown in Figure 1. The polymer chain length is also very important parameter. In a condensed phase, a polymer chain may fold along its chain length, therefore forming many contact points with self and/or with other chains. As a result, much short chain model is needed. Previous studies, through solubility simulations, found that 15 or more monomers will be suffice. In this study, 25 were used. When block copolymer was involved, the molecular configuration also needs to be considered—there may be differences in degree of polymerization in each branch or alternating structure may exist.

2.1 Flory-Huggins Repulsion parameter

Enthalpic contribution χ_{ij} is obtained through a series of single and bi-component simulation boxes assembled according to bulk average concentration. Molecular Dynamics simulations were then performed. CFF force field was used for PSS:PEDOT:PFI system while UFF force field was used in PSS:PEDOT:NiOx system. Average cohesive energy density were extracted according to equation. Flory-Huggins interaction parameters χ_{ij} and DPD repulsion parameters were calculated accordingly.

$$\chi_{ij} = \left(\frac{\Delta E_{ij}}{RT}\right) V_{brad} = \left(\frac{\Delta E_{ij}}{8.314 \times 298} 6.022 \times 10^{23} \times 10^{-30}\right) V_{brad} = \left(2.43 \times 10^{-10}\right) \Delta E_{ij} V_{brad} \tag{2}$$

where V_{bead} is the volume of a bead.

$$\Delta E_{ij} = \phi_i (E_{coh})_i + \phi_j (E_{coh})_j - (E_{coh})_{ij} \tag{3}$$

where E_{coh} represents cohesive energy.

$$\chi_{ij} = \left(\frac{\Delta E_{ij}}{RT}\right) V_{bead} \tag{4}$$

Once the Flory-Huggins parameters were decided, the DPD repulsion parameters of different beads can then be obtained according to Groot[19]

$$\alpha_{ij} = \frac{\chi_{ij}}{0.306} + \alpha_{ii} \tag{5}$$

2.2 Intramolecular bonding parameter

Simulation box contains all related components was assembled according to again the bulk average concentration. The generated trajectory was converted to obtain intra-molecular distances and angles distributions, the first peak was used to extract bond stretching and bending distribution function for each component, then convert to spring constants.

2.3 DPD simulation

The DPD simulation box were 10x10x20 DPD units for PEDOT:PSS:NiOx and 20x20x30 DPD units for PEDOT:PSS:PFI. Chain length of 25 repeated monomer beads were used, of which PEDOT(monomer):PSS equals to 1:2.5 weight ratio according to the literature[20]. Different PFI concentration were used in experiments in the same literature, low PFI concentrations have been published elsewhere[21], in this contribution, concentration of weight ratio from 5 and 8.5 of PFI will be presented. The concentration of NiOx nano particle is not as clearly defined in the literature, therefore, 0.1 and less were used.

In the experiment, PEDOT:PSS and PFI were in appearance, well mixed in the solution according to the standard device fabrication procedures. However, monomers confined to polymer chains do not have the same freedom as small molecules, total random mix condition at microscopic level is seldom met. The mixture will always be in some level of semi-ordered form. Therefore, DPD simulation boxes with stacking and random PEDOT:PSS and PFI initial conditions were generated side by side to mimic the processes change from ordered to semi-ordered and form random to semi-ordered form; and verify the results for each other. For PEDOT:PSS:NiOx, only random initial condition was used.

3. RESULTS AND DISCUSSES

3.1 DPD parameters

DPD parameters of PSS:PEDOT:PFI system were published in somewhere else[12]. Obtained PSS:PEDOT:NiOx DPD repulsion parameters are listed in Table 1, PEDOT and NiOx nano particle have strong attraction, while PSS and NiOx nano particles repulse mildly. Water molecule attracts to PEDOT and repulse to PSS mildly. Water attracts to NiOx nano particle very strongly. Their intramolecular bond stretching and bending results are listed in Table 2.

Table 1. PSS:PEDOT:NiOx DPD repulsion parameters α_{ij}

ij	EDOT	SS	NiOx	H_2O
EDOT	25			
SS	24.12	25		
NiOx	17.0	28.24	25	
H_2O	23.3	27.8	13.58	25

From Table 2, adding water to the PSS:PEDOT chain flexibility of both polymer have only little change. When NiOx is added, PEDOT chain become twice as rigid. This rigidness will contribute negatively to mixing and tend to form phase segregation.

Table 2. PSS:PEDOT Intramolecular bonding parameters

System	Interactions	AAMD Simu	
	Bond length	ks (kcal/mol/A^2)	r(A)
PEDOT:PSS	EDOT-EDOT	0.8158	4.4201
	SS-SS	0.0436	5.8071
PEDOT:PSS:H2O	EDOT-EDOT	0.6085	4. 3002
	SS-SS	0.0566	6.8432
PEDOT:PSS:NiOx	EDOT-EDOT	1. 6998	4.5438
	SS-SS	0.0179	9.1388
	Bend angle	k_θ (kcal/mol/rad^2)	θ_0 (degree)
PSS:PEDOT	EDOT-EDOT-EDOT	4.0538	128.7380
	SS-SS-SS	0.3159	31.1367
PSS:PEDOT:H2O	EDOT-EDOT-EDOT	4.4385	140.1322
	SS-SS-SS	0.1548	27.7342
PSS:PEDOT:NiOx	EDOT-EDOT-EDOT	2.1044	125.7708
	SS-SS-SS	0.3983	45.8840

3.2 DPD simulation results

DPD simulation from both initial conditions reached similar final state in all the PFI concentrations in study. The effect of concentration of different kinds on the vertical profiles from random initial condition are shown in Figure 2. PSS:PEDOT:PFI forms phase separation with PFI concentrated on the top surface, leading to self-organized gradient profiles for PSS and PEDOT.

PEDOT:PSS:PFI 1:5:5 1:2.5:5 1:2.5:8.5

Figure 2. DPD simulation results of the self-organized gradient effect of ITO-PEDOT:PSS:PFI, random initio condition, where grey, blue, cyan and fulvous beads represent ITO, PEDOT, PSS and PFI respectively.

The same mechanism causing PEDOT to have steeper profile than PSS towards the surface, therefore, large eV value can be expected on PFI rich surface.

The effect of NiOx concentration on the vertical profiles are shown in Figure 3. From Table1 and 2, NiOx nano particles cling to PEDOT polymer chains, causing PEDOT chain to be more stretched and rigid, thus easier to form phase separation. At the same time, the strong attraction between NiOx nano particles and PEDOT pulls PEDOT together to form more condensed PEDOT phase, thus more condensed PSS phase on the opposite direction which lead to the self-organized PEDOT and PSS vertical concentration profile, which, through optimization, can be used to achieve gradient work function in vertical direction.

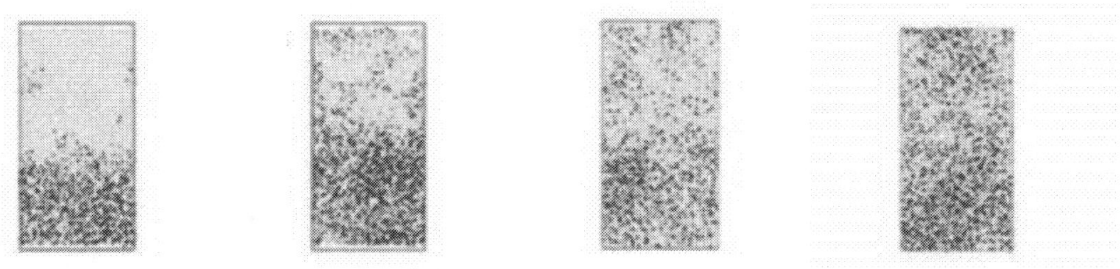

PEDOT:PSS:NiOx 1:5:0.1 1:2.5:0.05 1:5:0.04 1:2.5:0.025

Figure 3. DPD simulation result of the self-organized gradient effect of PEDOT:PSS:NiOx. Where blue, cyan and yellow beads represent PEDOT, PSS and NiOx respectively.

4. CONCLUSIONS

PSS:PEDOT:PFI forms phase separation in a wide PFI concentration region, with PFI concentrated on the top surface, leading to self-organized gradient profiles for PSS and PEDOT, The same mechanism causing PEDOT to have steeper profile than PSS towards the surface, therefore, large eV value can be expected on PFI rich surface. PSS:PEDOT:NiOx takes a different mechanisms. NiOx nano particles cling to PEDOT polymer chains, causing PEDOT chain to be more stretched and rigid, thus easier to form phase separation. At the same time, the strong attraction between NiOx nano particles and PEDOT pulls PEDOT together to form more condensed PEDOT phase, thus more condensed PSS phase on the opposite direction.

The model can be used for future numerical experimenting on effects of composition, structure configuration and additives for potentially tailored design and structure optimization.

REFERENCES

[1] Sun, K., Zhang, S., Li, P., Xia, Y., Zhang, X., Du, D., Isikgor, F. H. and Ouyang, J., "Review on application of PEDOTs and PEDOT:PSS in energy conversion and storage devices," J Mater. Sci: Mater Electron. 26, 4438–4462 (2015)

[2] Nie W., Tsai, H., Asadpour, R., Blancon, J.-C., Neukirch, A.J., Gupta, G., Crochet, J.J., Chhowalla, M., Tretiak, S., Alam, M.A., Wang, H.-L. and Mohite, A.D., "High-efficiency solution-processed perovskite solar cells with millimeter-scale grains," Sci. 367, 522-525 (2015)

[3] Wang, G.-F., Tao, X.-M. and Wang, R.-X., "Fabrication and characterization of OLEDs using PEDOT:PSS and

MWCNT nanocomposites," Compos. Sci. Technol. 68, 2837–2841 (2008)
[4] You J, Hong Z, Yang YM, Chen Q, Cai M, Song TB, Chen CC, Lu S, Liu Y, Zhou H, Yang Y, "Low-temperature solution-processed perovskite solar cells with high efficiency and flexibility", ACS Nano, 8(2):1674-80 (2014)
[5] Petrosino, M., Vacca, P., Miscioscia, R., Nenna, G., Minarini, C. and Rubino, A., "Effect of PEDOT:PSS ratio on the electrical and optical properties of OLEDs," Proc. SPIE 6593, 659310 (2007)
[6] I. Cruz-Cruza, M. Reyes-Reyesa, M.A. Aguilar-Frutisb, A.G. Rodrigueza, c, R. López-Sandoval, "Study of the effect of DMSO concentration on the thickness of the PSS insulating barrier in PEDOT:PSS thin films", Synthetic Metals, 160:1501–1506 (2010)
[7] Anil Kanwat, William Milne, Jin Jang, "Vertical phase separation of PSS in organic photovoltaics with a nickel oxide doped PEDOT: PSS interlayer", Solar Energy Materials and Solar Cells, 132:623–631 (2015)
[8] Chin, B.D., Choi, Y., Baek, H.-I. and Lee, C., "High efficiency phosphorescent OLEDs based on the heterostructured light emission and charge injection layers," Proc. SPIE 7415, 741510 (2009)
[9] P. J. Hoogerbrugge and J. M. V. A. Koelman. Simulating microscopic hydrodynamic phenomena with dissipative particle dynamics. Europhysics Letters, 19(3):155–160, JUN 1 1992
[10] P. Español and P. B. Warren. Statistical-mechanics of dissipative particle dynamics. Europhysics Letters, 30(4):191–196, MAY 1 1995
[11] Flory, P.J., "Thermodynamics of High Polymer Solutions," J. Chem. Phys. 10, 51-61 (1942)
[12] Huggins, M.L., "Some Properties of Solutions of Long-chain Compounds," J. Phys. Chem. 46, 151-158 (1942)
[13] Huggins, M.L., "Theory of Solutions of High Polymers," J. Am. Chem. Soc. 64, 1712-1719 (1942)
[14] Groot, R.D., and Warren, P.B., "dissipative particle dynamics: bridging the gap between atomistic and mesoscopic simulation," J. Chem Phys. 107, 4423-4435, (1997)
[15] Becke, A.D., "Density-functional exchange-energy approximation with correct asymptotic behavior," Phys. Rev. A 38, 3098-3100 (1988)
[16] Chengteh, L., Weitao, Y. and Robert, G.P., "Development of the Colle-Salvetti correlation-energy formula into a functional of the electron density," Phys. Rev. B 37 (2), 785-789 (1988)
[17] Vosko, S. H., Wilk, L. and Nusair, M., "Accurate spin-dependent electron liquid correlation energies for local spin density calculations: a critical analysis," Can. J. Phys. 58, 1200-1211 (1980)
[18] Ditchfield, R., Hehre, W.J., Pople, J.A., "Self-consistent molecular-orbital methods. IX. an Extended Gaussian-type basis for molecular-orbital studies of organic molecules," J. Chem Phys. 54, 724-728 (1971)
[19] Groot, R.D. and Madden, T.J., "Dynamic Simulation of Diblock Copolymer Microphase Separation," J. Chem. Phys. 108(20), 8713-8724, (1998)
[20] Lee, T.-W., Chung, Y., Kwon, O., and Park, J.-J., "Self-Organized Gradient Hole Injection to Improve the Performance of Polymer Electroluminescent Devices," Adv. Funct. Mater. 17, 390–396 (2007)
[21] Min Huang, " Multiscale study of the self-organized gradient effect of novel hole injection material PEDOT:PSS:PFI" , Proc. of SPIE Vol.9941:994116, Organic Light Emitting Materials and Devices XX, 2016 SPIE Optics & Photonics, San Diego, California, USA, Aug. 29, 2016

472603 Study of Charge Transfer Dynamics in Spray Deposited Cu2ZnSnS4 (CZTS) Photoelectrodes for Performance Improvement

Abstract:

Thin film solar cell based on Cu_2ZnSnS_4 (CZTS) has the potential for large scale electricity production using solar radiation while all of the elements are earth abundant, inexpensive and environment friendly. The state of art CZTS solar cell has reached 12.6% efficiency while the commercial production will require at least 15% for economic feasibility. Here, we approach to understand the charge transfer dynamics in CZTS using ultrafast laser spectroscopy and focus on improving the efficiency of CZTS photovoltaics through the incorporation of a mesoscopic hole acceptor, nickel oxide. CZTS photoelectrode is prepared using spray deposition of precursor solution, and characterized using SEM, TEM, XRD, XPS and Raman Spectroscopy. The performance of the material is tested in three electrode photo electrochemical cell.

463080 Highly Efficient Solar Cells Made with $Cu_{1-X}K_XInSe_2$ Alloys: A Foundation for Engineering K in $Cu(In,Ga)Se_2$

Abstract:

For the first time, Cu-KF-In-Se co-deposition was used to form $Cu_{1-x}K_xInSe_2$ (CKIS) films over the full composition range ($0 < x < 1$). Compared to K-free $CuInSe_2$, the CKIS films exhibited increased majority carrier concentrations and wider band gaps, while $0 < x < 0.22$ compositions showed increased minority carrier lifetimes. Thin film solar cells were fabricated using soda-lime glass/Mo/CKIS/CdS/intrinsic-ZnO/Al:ZnO/Ni/Al stacks, where a Ni/Al finger grid with partial shading was used for the top contact. Photovoltaic (PV) performance was optimal at $x \sim 0.07$, which resulted in an officially-measured 15.0%-efficient deviceâ€"equal to the world record $CuInSe_2$ efficiency (grown at 575 Â°C), despite being grown at 500 Â°C, without the three stage process, and without device stack optimization. The devices were characterized with quantum efficiency, temperature- and light-intensity-dependent current-voltage, capacitance-voltage, and admittance spectroscopy. These data were then used to calculate charge carrier recombination rates in the bulk absorber, the depletion region, and at the interface. Bulk recombination was reduced for $x > 0$. Interface recombination was reduced at $x \sim 0.07$, while x â‰¥ 0.22 compositions had poorer performance due to interface recombination, relative to $x \sim 0$. These trends are consistent with both surface and bulk recombination affecting apparent lifetimes. The results showed that large K compositions deteriorated PV performance in CKIS alloys, while nevertheless, $Cu(In,Ga)Se_2$ solar cells with record efficiencies have exhibited large K compositions at the absorber/buffer interface. Processing techniques were employed to address this apparent discrepancy in material quality at high K content. Increasing the substrate's Na composition was found to increase the extent of CKIS alloy formation, relative to $CuInSe_2$ + $KInSe_2$ formation. Similarly, decreased substrate temperature was found to favor CKIS formation, relative to $CuInSe_2$ + $KInSe_2$. Substrate Na and temperature can therefore be used to engineer K bonding in $Cu(In,Ga)Se_2$ absorbers to enhance both initial and long-term PV power generation. Material growth

methods such as these may also be used to investigate the differences between CKIS alloys and Cu-K-In-Se films with very low $KInSe_2$ content.

468083 Study of Electron Transport Mechanism in Dye-Sensitized Solar Cell with the Effect of Morphology, Crystalline Structure and Electron Mobility

Abstract:

Dye-sensitized solar cells (DSSCs) are one of the most promising developments in photovoltaic devices. To obtain high efficiency DSSCs, extensive research has been carried out to consider the effect of dyes, electrolytes, and alternatives to titania.

The performance of photovoltaic devices could be improved by using nano-composites which can provide high electron mobility and efficiently collect photo-generated electrons through a better understanding of their fundamentals. One-dimensional nanomaterials including Single-walled carbon nanotubes (CNT) and metals, such as gold and silver, exhibit very high electron mobility.

However, the incorporation of such materials into DSSCs for efficient photovoltaic performance is still challenging. Also the fundamental of the above system is not well understood.

This study focuses on the synthesis of core-shell nano-composites made of TiO_2, various metals, and CNTs. Several fabrication techniques including chemical vapor deposition, electrochemical deposition, spin-coating, and electro spinning were applied to create various nanostructured composite materials such as one dimensional nanorods, nanowires, and nanotubes. Designing core-shell nanostructures of TiO_2in DSSC help us investigate and compare electron transfer properties of the different core materials.

The Mechanistic study of the DSSCs were carried out by using the following techniques: UV-vis scoptrocopy, time-resolved photoluminescence spectroscopy, impedance analyzer in combination. In addition, characterization methods such as TEM, SEM, XRD, and Raman spectroscopy were used. These studies will help to better understand the principle of DSSCs and better design efficient DSSC devices.

474046 Shaping the Behavior of Colloidal Fluids

Abstract:

The behavior of colloidal suspensions depends upon many factors, including the shape of the colloidal particles. In this talk, we discuss the influence of shape on the thermodynamics and kinetics of hard particle fluids, including how shape is manifested in equations of state and in phase transitions from fluid to solid form. We present new results on the nucleation and growth of colloidal crystals, where subtle changes in particle shape can have profound influence on thermodynamics and kinetic processes.
*with J. Anderson, J. Antonaglia, A. Cadotte, E. Chen, P. Damasceno, J. Dshemuchadse, M. Engel, J. Glaser, E. Harper, E. Irrgang, A. Karas, D. Klotsa, D. Kofke, R. Newman, S. Nola, A. Schultz, and W. Shen.

474912 Modeling Biosynthetic Active Matter

Abstract:

New active materials are being synthesized that use cellular constituents such as biopolymers and motor-proteins, as well as synthetic components such as depletants or drugs that modify polymerization. These self-driven fluids will stretch, roil, or contract, depending upon the polymers and motor-proteins involved, and could be used to drive microscale transport processes. These systems also teach us about complex biological processes such as cell division. I'll discuss how we model these materials as multi-scale complex fluids driven by internal stresses generated by motor-protein activity, how they behave dynamically, and a bit on what they may teach us about biology.

474050 Gas-Solid Clustering: Not Your Ordinary Instability

Abstract:

In high velocity, gas-solid flows, the presence of the clustering instability – collections of highly-concentrated particles that continuously form, rearrange and breakup – are the norm rather than the exception. For decades, such instabilities have been predicted qualitatively by dozens of researchers using a kinetic-theory analogy for the continuum description. The ability to quantitatively predict clusters, however, has received relatively scant attention. Achieving such accuracy is non-trivial given the complexity of the problem – transient, three-dimensional flows with a flow instability that evolves over several length scales – as well as the seemingly inappropriateness of a "small-gradient" starting assumption. In this talk, the mechanisms for clustering will be discussed along with the competing influences that impact stability/instability, as well as recent progress on assessing the quantitative accuracy of a recent theory.

474656 Dynamic Solidification of Dense Suspensions Under Shear

Abstract:

Dense suspensions of small solid particles in a simple liquid exhibit a range of intriguing non-Newtonian flow behaviors [1]. One of the most striking and counter-intuitive of these is the transformation from overall liquid-like to solid-like behavior under rapid impact, shear or extension. This talk discusses some of the mechanisms driving such behavior, emphasizing the differences between steady-state responses to applied forcing, such as shear thickening, and dynamic responses, such as impact-activated solidification. We recently showed that impact-activated solidification involves rapidly moving fronts that emanate from the impact point and convert fluid-like, unjammed suspension into rigidly jammed material in their wake [2]. Measuring the speed of sound, we find that such fronts do not involve any significant increase in local packing fraction, while imaging of the flow field with high-speed ultrasound directly links the evolving front position to regions of high shear intensity. Taken together, this provides evidence for a solidification process we call dynamic shear jamming. We discuss the relationship to shear jamming in frictional granular systems, and we introduce a state diagram that delineates dynamic shear jamming from discontinuous shear thickening (DST) on the one hand and traditional jamming by densification on the other [3].

[1] E. Brown and H. M. Jaeger, Shear thickening in concentrated suspensions: phenomenology, mechanisms, and relations to jamming, Reports On Progress In Physics 77, 046602 (2014).

[2] S. R. Waitukaitis and H. M. Jaeger, Impact-activated solidification of dense suspensions via dynamic jamming fronts, Nature 487, 205-209 (2012).

[3] I. R. Peters, S. Majumdar, and H. M. Jaeger, Direct observation of dynamic shear jamming in dense suspensions, Nature 532, 214â€“217 (2016).

474049 Rheological Modeling of Granular Flows

Abstract:

Granular matter is very common --- sands, soils, raw materials, food stuffs, pills, powders --- but the challenge of predicting the motion of a collection of flowing grains has proven to be a difficult one, from both computational and theoretical perspectives. Grain-by-grain discrete element methods can be used, but these become computationally unrealistic for large systems and long times. A broadly accurate continuum model would be ideal as it could provide a much faster means of calculating flows and stresses in real-world problems, such as those encountered in industry and geotechnical applications. Motivated by this challenge, in this talk we present a new constitutive relation for granular matter, which is shown to produce quantitatively accurate predictions. The model emphasizes the key importance of cooperativity in the granular flow process through the inclusion of a grain-size-dependent length-scale within the rheology. Upon calibration of the model's three material parameters, solutions of the model are shown to be in quantitative agreement with over 160 experiments in several different families of geometries, including the notorious split-bottom cell. The same model is then shown to reconcile a number of other unusual features of granular media, such as the motion-induced "quicksand" effect, wherein flow at one location effectively removes the yield stress everywhere, and the observation that thinner granular layers behave as if they are stronger.

466582 A Unified Framework for Mechanistic Model Development and Validation of High Shear, Fluid Bed, and Twin Screw Wet Granulation Processes

Abstract:

Particle size enlargement by wet granulation can be achieved through three types of processes: fluid bed, high shear, and twin screw granulation. Although these processes have many similarities, they differ at a mechanistic level. For instance, while fluid bed granulation is mainly driven by granule agglomeration and drying, the evolution of granules in a twin screw granulator is primarily governed by breakage and layering. As a result, each type of wet granulation process is defined by a unique set of critical process parameters, and the resulting granules have different properties depending on the means of production. Mechanistic modeling can be used to capture these differences and predict the effects of critical process parameters on granule attributes, facilitating process design and operation. In this work, a unified, mechanistic modeling framework is presented and customized for fluid bed, high shear, and twin screw granulation processes, taking into account the dominant mechanisms in each process and the spatial compartmentalization of the various equipment.

Once a mechanistic model has been developed, model validation poses a key challenge towards applying these tools in practice. These models consider multiple mechanisms that affect particle attributes, with each mechanism containing a number of unknown kinetic parameters. Because of their complex nature, the mechanisms are difficult to decouple experimentally, and the data requirements for model validation are not well-established. In this work, approaches to model validation are developed based on the mechanistic understanding of the relationships between critical process parameters and product attributes. Finally, case studies for model development and validation are presented for fluid bed and twin screw granulation processes.

464962 Impact of Combined Distributive Mixing Elements and Kneading Elements on Granules Produced By Twin Screw Granulation

Abstract:

Impact of Combined Distributive Mixing Elements and Kneading Elements on Granules Produced by Twin Screw Granulation

Nathan B. Davis[1], Carl R. Wassgren[1]

[1]School of Mechanical Engineering, Purdue University, Indiana, USA

The screw elements conventionally used for twin screw granulation (TSG) are conveying elements and kneading elements. The use of kneading elements results in significant tradeoffs between the granule shape and liquid distribution. The use of distributive mixing elements has been show to improve the granule shape while also increasing granule porosity. The flexibility of the TSG means that producing granules with desirable size, shape, and porosity should be possible by combining the use of kneading elements and distributive mixing elements in an appropriate configuration.

In this work, the effects of combining both kneading elements and distributive mixing elements in order to produce rounded granules with a target size and porosity are presented. The kneading elements are tested in both the 60 degree forward and 30 degree reverse configuration to cover different levels of liquid distribution. The distributive mixing elements are placed downstream of the kneading elements and are in the adjacent reverse configuration.

456611 Investigation of the Effect of Raw Material Properties on the Breakage Rate Process in a Twin Screw Granulator

Abstract:

Granulation is often used as a size enlargement unit operation in various solid handling and powder processing industries. Granulation helps to mitigate issues related to poor flowability, dust hazards, and powder segregation. Twin screw granulation (TSG) is becoming increasingly relevant due to its compact size, continuous and robust mode of operation, flexible and customizable design, and flexible production capacity.

Granule breakage has been shown to be an important rate process in TSG. This presentation describes the experiments designed to explore granule breakage in TSG. Cylindrical pellets were prepared from model powders of varying particle sizes and binders of varying viscosity. The powder and binder systems were chosen such that the pellets had a range of dynamic yield strengths (DYS). The breakage probability and daughter size distribution of the pellets was measured as a function of the DYS of the powder/binder system as well as for different screw designs. It was observed that for conveying screw elements, the breakage probability is a strong function of the DYS, whereas for distributive mixing screw elements, breakage is governed mainly by geometric effects.

470992 Discrete Element Modelling of Particle Transformations during Spheronisation

Abstract:

Spheronisation is a process whereby a short rod made of a wet powder mass is transformed into a dense spheroid. The process takes place in a spheroniser which is a fixed cylindrical vessel with vertical axis and a horizontally rotating base called friction plate. Under optimal processing conditions the particulate mass assumes the so-called "roping" regime. During the process the particles undergo a large number of impacts between each other and with the walls of the processing vessel. Each impact produces a small plastic deformation and rounding occurs as the cumulative effect of the impacts suffered by the particle. To date the process has been modelled using three strategies: Finite Element Analysis (FEA), whereby the deformation following each impact is followed explicitly; Multiple Impact Modelling of Particles (MIMP), which involves elasto-plastic contact laws; and Discrete Element Modelling (DEM). In this paper we extend DEM to model the evolution of the shape of each particle by using the algorithm employed by MIMP. This enables using impact conditions in the actual "roping" regime. The paper concludes with a spheronisation process map which identifies optimum material properties and process conditions for spheronisation.

454412 Tailoring Bulk Properties of API By Slurry Agglomeration

Abstract:

Slurry agglomeration is a technique to grow agglomerates from already formed particles using a L-L-S system where one liquid phase preferentially wets the surface of the solid phase.

Active pharmaceutial ingredient (API) is suspended in the continuous phase, an immiscible (or partially miscible) bridging liquid is added that preferentially wets the surface of the API particles.

Slurry agglomeration can be used to produce granular API with improved bulk density, flow properties, filtration rate, and compression properties. This talk will discuss the history of the process, an investigation into the mechanisms at play, a case study which will demonstrate the improvements in bulk properties for an acicular API, and a discussion of the potential for broader applicability to API with varied chemical and physical characteristics.

465806 One-Pot Granulation of Graphene Oxide Aerogels with Hierarchical Porous Three-Dimensional Structures

Abstract:

Recently, three-dimensional (3D) graphene/graphene oxide (GO) networks have attracted much attention, thanks to their high specific surface areas, large pore volumes, strong mechanical strengths and fast mass and electron transport. These unique properties make them widely applied in the fields of absorbents, catalysis, sensors, energy storage and conversion, and biological applications. Many methods were developed to make 3D graphene porous structures in chunks with fixed shapes similar to the template or the container. Further treatments such as crushing and grinding are needed in various applications.

We have developed a one-pot granulation method to assemble graphene oxide (GO) pallets into 3D microballs at liquid-liquid interface, driven by the dissolution of a droplet immersed in a surrounding liquid phase. The assembled GO structures were highly crumpled with exotic morphology. And the detailed structure of GO microballs was closely related to the dynamics of the droplet, which could be adjusted by the composition of the surrounding liquid, the temperature and the reduction degree of GO. The shape of GO assemblies can be tailored by the pinning effect of the contact line between the droplet and the substrate. Rich morphology of the assembled GO structures, including capsule, dim sum, mushroom, peasecod, and olive shaped assemblies were obtained. By adding salts into the GO droplet, the hybrid structures varies with the concentration and salts solubility in the surrounding solvent. Soluble salt (LiCl) have little effect of the GO assembled structure as expected. Insoluble salt (NaCl, KCl) crystalizes on the GO pallets at high concentration, but between pallets at low concentration. During the granulation process, the GO pallets with different sizes were automatically separated and forming the layered shell of the GO microball. Based on these results, it is clear that the structure of 3D GO assemblies is determined by the dissolution dynamics of the droplet. Perfect spherical GO assemblies can be obtained in the absence of the substrate effect.

With the perfect spherical shape, these porous 3D assemblies should have better processibility. And we believe that our method is suitable to granulate all kinds of 2D nanomaterials, besides GO nanopallets. Granulation in large scale is still wanted to meet the demands of wide applications of these functional 3D porous structures.

468605 A Modified Rolling Theory for Granular Solids

Abstract:

A modified Johanson roll compaction model is proposed. The model improves upon the original Johanson model by introducing a solution that satisfies mass conservation at the nip angle, which is achieved through assuming the existence of a transition zone. This supposition of a transition zone enables the description of a progressive change in powder velocity through the nip region, whereas earlier approaches required an abrupt shift. By further assuming a density and velocity profile in the transition zone, solutions for the powder density at the minimum gap width and roll normal stress distribution are presented. Validation of the modified model against available literature roll compaction run data shows that it is capable of capturing the trends observed from the experimental data accurately. Quantitatively, the RMS error between the model predictions and the experimentally measured counterparts is 0.013 and 8.04 MPa, or 1.84% and 8.36% of the average of the experimentally measured ribbon densities and maximum roll normal stresses, respectively.

465209 A Coupled, Constant Heat Flux Boundary Condition for CFD-DEM

Abstract:

The modeling of multiphase flows with a constant heat flux boundary has great practical and fundamental significance. Unlike their single-phase counterpart, multiphase systems experience coupling between phases and thus make the representation of a constant boundary flux non-trivial. Specifically, the partition of the total boundary flux between the phases can vary with space and time. A numerical technique for modeling such a boundary condition is proposed for gas-solids flows, and directly applied within a computational fluid dynamics and discrete element method (CFD-DEM) software. While simulations with discrete, monodisperse particles are considered here, the technique can be extended to polydisperse systems and/or the two fluid model.

465832 Thermal Fluxes in Wall Bounded Sheared Granular Beds Near the Jamming Point

Abstract:

Sheared particle beds are encountered in a broad range of engineering applications, such as geophysics or chemical engineering [1]. In order to simulate these particle systems, Discrete Element Method (DEM) and Computational Fluid Dynamics-DEM tools became state of the art over the past ten years [2]. Only recently, DEM-based tools have been extended to account for intra-particle transport processes, opening the door for full-physics simulations of phenomena occurring in applications such as Chemical Looping Combustion or Reforming. For example, our software tool ParScale [3, https://github.com/CFDEMproject/ParScale-PUBLIC] can be linked to the Euler-Lagrange simulation code CFDEM®, is able to resolve intra-particle properties, and accounts for the transmitted heat to the ambient fluid. The scientific exploitation of these new tools is far from being completed, and here we focus on the delicate topic of thermal transport in sheared gas-particle systems.

Until now little effort was made to study the simultaneous impact of the conductive, convective and transmitted flux (to an ambient fluid) inside a sheared particle bed. Often, fully periodic boxes under vacuum conditions in combination with frictionless particles are used [4], which cannot represent a typical application in chemical engineering. This consequently limits the applicability of the conclusions since these model approaches are mostly developed to study the rheological behavior of sheared beds. Also, the thermal transition during jamming of a sheared particle bed is still unknown.

Thus, it is our goal to close this gap in literature by making use of the coupling capability between the DEM code LIGGGHTS® and our library ParScale. We model heat transfer phenomena over a wide range of dimensionless parameters and resolve for intra-particle temperature gradients. Thereby, we focus on flow situations where jamming of the system is reported in literature, discuss arising questions concerning crystallization, and the effect of inter-particle friction. Finally, we present new regime maps for thermal fluxes in sheared particle beds, and discuss novel scaling laws for predicting these fluxes.

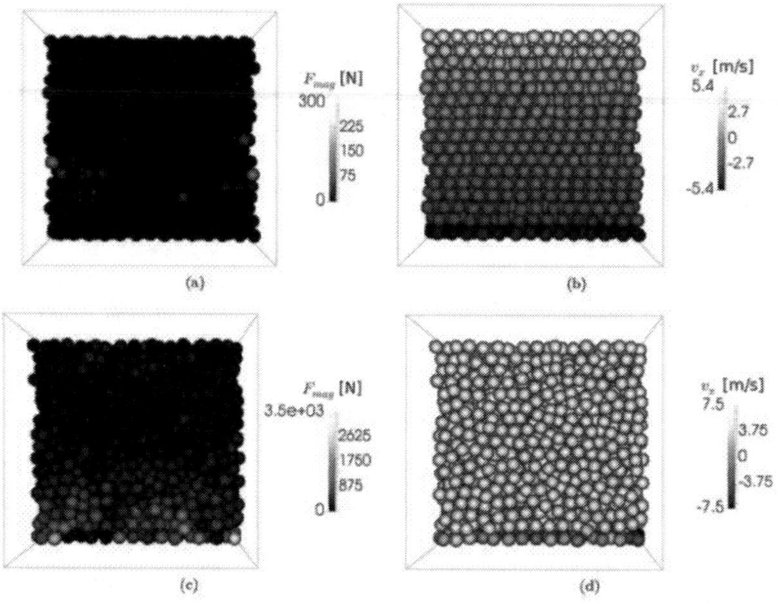

Fig. 1: Typical transition for jammed and unjammed states for $\dot{\gamma} = 10^{-2}$, $Pe = 0.01$, $Bi = 0.1$, $\mu = 0.5$ and (a) $F_{mag}, \phi_P = 0.62$ (b) $v_x, \phi_P = 0.62$ (c) $F_{mag}, \phi_P = 0.64$ (d) $v_x, \phi_P = 0.64$

REFERENCES

[1] J. R. Rice. Heating and weakening of faults during earthquake slip. Journal of Geophysical Research: Solid Earth, 111(5):1–29, 2006

[2] C. Kloss, C. Goniva, A. Hager, S. Amberger, S. Pirker (2012) "Models, algorithms and validation for opensource DEM and CFD-DEM", Progress in Computational Fluid Dynamics, Vol. 12, Nos. 2/3, pp.140–152

[3] S. Radl, T. Forgber, A. Aigner, C. Kloss, ParScale - An Open-Source Library for the Simulation of Intra-Particle Heat and Mass Transport Processes in Coupled Simulations, in: E. Onate, M. Bischoff, D.R.J. Owen, P. Wriggers, T. Zhodi (Eds.), IV Int. Conf. Part. Methods – Fundam. Appl. (PARTICLES 2015), ECCOMAS, Barcelona, Spain, 2015: pp. 1–9

[4] P. Rognon and I. Einav. Thermal Transients and Convective Particle Motion in Dense Granular Materials. Physical Review Letters, 105:1–4, 2010

459643 Biomass Gasification in a Fixed Bed Downdraft Gasifier: Experimental Study & Comprehensive Reactor and Particle Modelling

Abstract:

Gasification of biomass waste is considered as a green and sustainable solution to reduce waste volumes and recover energy by producing syngas. In this study, both experiment and simulation have been conducted to understand the various physical and chemical effects in a fixed bed downdraft gasifier and to provide researchers with a tool to overcome the need to conduct cost intensive experiments for predicting and optimizing energy efficiencies.

A combined transient single particle and fuel-bed model is developed to describe the entire packed bed in the flow direction of primary air and intraparticle phenomena. The reactor is discretized and in each cell one representative particle is chosen to be modelled as shrinking sphere. The reactor was divided into three regions due to the difference of air velocity in each region: natural convection region, mixed convection region and forced convection region. Sensitivity analysis has been conducted to determine the boundary location of each region. The presented model considers drying, pyrolysis, homogeneous gas reactions, and heterogeneous combustion/gasification reactions. In the gas phase eight species (O_2, N_2, CO, CO_2, H_2, H_2O, CH_4 and tar) are considered. The solid phase is biomass waste, in which all the components obtained from approximate analysis (moisture, volatiles, fixed carbon and ash) and ultimate analysis (C, H, O, N) are treated as dependent variables of time and space. This model can predict production rate and quality of both synthetic gas and produced char from gasifier reactor. The optimized biomass properties, equivalence ratio and air inlet location for maximum energy efficiency can also be obtained from this model.

Experiments have been conducted to investigate the effects of different parameters on gasification performance, such as biomass moisture content, particle size and shape, equivalence ratio, etc. The experimental results are also used to validate against simulation results and the deviation between the numerical and experimental results obtained in this study is lower than 10%.

472835 Enhanced Centrifuge-Based Approach to Powder Characterization: Establishment of a Size-Linked Effective Hamaker Constant Distribution

Abstract:

Many types of manufacturing processes involve powders and are affected by powder behavior. It is highly desirable to implement tools that allow the behavior of bulk powder to be predicted based on the behavior of only small (milligram-scale) quantities of powder. Such descriptions can enable engineers to significantly improve the performance of powder processing and formulation steps. An enhancement to the centrifuge technique was proposed for this purpose.

Simulations of the centrifuge technique were performed by calculating the removal of a simulated rough powder containing spheres with sinusoidal surface protrusions and intrusions from both nominally-flat surfaces and surfaces with well-defined indentations. The size distribution of the powder was based on an experimentally-determined distribution from a commercial silica powder. It was discovered that the adhesion of the rough spherical powder could be described in terms of an equivalent powder of smooth spherical particles with a size-dependent â€˜effectiveâ€™ Hamaker constant distribution.

Experimental validation of the simulation technique was performed with a silica powder dispersed onto nominally-flat stainless steel surfaces with no engineered surface features. The observed adhesion and removal of powder from the

plates in the centrifuge was used within the simulation procedure to show its application to real systems. It was determined that the behavior of the rough, spherical powder could be represented with great accuracy by the expected behavior of an equivalent smooth, spherical powder with a size-dependent effective Hamaker constant distribution. In this manner, the roughness effects and size variations of a real powder are captured in this one distributed parameter (effective Hamaker constant distribution) which provides a substantial improvement to the existing technique. This can lead to better optimization of powder processing by allowing the behavior of topographically-complex powders to be described in terms of topographically-simple equivalent ones.

462464 Effect of Friction and Restitution Coefficient on Defluidization and Riser Flows of Cohesive Particles

Abstract:

Fluidized beds and risers are typical systems for dense and dilute gas-solid flows with broad industrial application. Friction and restitution coefficient, as two major sources of energy dissipation, have significant impact on gas-solid flows of non-cohesive particles. The interplay of friction and restitution coefficient with particle cohesion, however, requires further investigation. In this work, we focus on behavior of cohesive particles in terms of defluidization curves in fluidized beds and agglomerate properties in riser flows. Based on the DEM-CFD simulations, a reverse dependence on sliding friction and restitution coefficient was identified: defluidization curves were more influenced by sliding friction while agglomerates in risers were governed by restitution coefficient. By analyzing bed porosity and particle velocities, the reverse sensitivity is understood by the switch of primary particle interactions from lasting contacts to binary collisions between the two systems. A non-monotonic response of increasing sliding friction on defluidization curves was observed, which can be explained by the competing effects of increased slipping and enhanced particle spin on bed porosity. The results suggest the complexity of the continuum modeling for gas-solid flows can be tuned provided the leading factor in the system of interest is properly incorporated. The data provided in this work can also be used in validating continuum model predictions.

470790 Experimental and Simulation of Torque Values of Large and Small Particle Size Powders in Anton Paar Powder Cell Using a Commercial DEM Simulation Software

Abstract:

The flow properties of fine powders at very low consolidation levels are relevant to small scale industrial applications of powder flow, such as in small process hoppers and mixers of pharmaceutical powders, or in everyday applications such as toner flow in cartridges. In addition, many process applications of powder technology involve gas-solids systems, which require the knowledge of flow properties of aerated powders for proper handling.

The equipment used in this research project is the state of the art powder cell purposely developed by Anton Paar Company to fit on an Anton Paar rheometer and carry out torque measurements on both aerated and fluidized powders. In this instrument, the state of consolidation of the powder can be adjusted by changing the impeller depth and the aeration rate and different impeller shape and size provided by the company. Two different size of impeller, 10 and 5 mm, were used to measure the torque values of sand, $d_{bp3.2}$ 225 µm, and Alumina $d_{bp3.2}$ 75 µm, at different powder depth. The same conditions were used for the simulation of torque values with EDEM software.

Simulation are carried out to understand if the main hypotheses made in the model developed by Bruni *et al.* are reasonable in this apparatus under the conditions tested and if limiting conditions occur, for example on the relative dimension between 1) the gap available between the impeller and the container wall and 2) the particle size. Test are carried out and compared with different particle sizes, air flow rates and impeller speeds.

REFERENCES

[1] Bruni, G., Barletta, D., Poletto, M., & Lettieri, P. 2007. A rheological model for the flowability of aerated fine powders. Chemical engineering science, 62(1), 397-407.

462646 The Effect of the Number of Impeller Blades on Granular Flow in a Bladed Mixer

Abstract:

Simulations of granular flow of monodisperse, cohesionless spherical glass beads in a cylindrical bladed mixer agitated by an impeller were performed using the discrete element method (DEM). The number of impeller blades was varied from 1 to 4 blades, and the impact of the number of blades on granular flow behaviors and mixing kinetics was investigated. It was found that particle velocities in an agitated mixer were influenced by the number of the blades used in the mixing system. Higher radial and vertical velocities of particles were observed in the 1- and 2-bladed mixers, which led to more pronounced three-dimensional recirculation patterns. However, the tangential velocity components of particles in the 3- and 4-bladed cases were larger. Additionally, it was found that using two or three impeller blades provided better mixing performance than using one blade or four blades, evaluated by calculation of the relative standard deviation (RSD) and the Lacey index values of the systems. Granular temperature and particle diffusivities obtained for the 2- and 3-bladed cases were also higher than those for the 1- and 4-bladed mixers. Solids fraction analysis showed that dilation of the particle bed occurred to the most extent in the 2-bladed mixer. Contact force network data showed that using different numbers of impeller blades led to significant differences in the force distribution. Finally, the effect of scaling up was evaluated by varying the mixer diameter to particle diameter ratio. Increasing the mixer diameter to particle diameter ratio was found to have little impact on flow and mixing behaviors.

468997 Drag Measurements from Particle-Resolved CFD Simulations of Flow Past Random Assemblies of Stationary Ellipsoidal Particles

Abstract:

Currently-available literature on fluid-particle drag models for non-spherical shapes are focused on reporting the drag coefficients of individual particles, with very few studies on the drag experienced by an assembly of non-spherical particles. For multiphase-flow applications (e.g., fluidized beds), these single-particle drag coefficients are useful at the very-dilute limit but are generally inaccurate at larger solid fractions. New drag models must be developed at larger solid fractionsâ€"as has been done for spherical particles, but not for non-spherical shapes.

In this work, we have performed single-phase computational fluid dynamics (CFD) simulations of fluid flow past random assemblies of stationary ellipsoidal particles. This approach follows the method used in previous studies [Koch and Ladd, J. Fluid Mech. (1997), 349, 31-66; Tenneti et al., Int. J. Multiph. Flow (2011), 77, 1072-1092, to name a few works] to derive drag laws for spherical particles, which is now adapted to ellipsoidal shapes. Our simulations are performed using Gerris, an open-sourced CFD code, which allows automated mesh-refinement of the particle-boundary layer. Using our simulation setup, we probe the influence of varying particle aspect ratio, orientation, solid fraction, and Reynolds number on the fluid-particle drag. We demonstrate that employing a spherical drag model for non-spherical particles misrepresents the actual drag. The particle orientation also plays a key roleâ€"the drag is lowest for ellipsoids aligned perfectly with the flow direction and, as expected, and increases non-linearly as the angle between the particlesâ€™ axes and the flow increases.

Current simulations of non-spherical particles, typically utilizing CFD-DEM models, apply ad-hoc corrections to spherical drag laws to account for non-sphericity. For example, Zhou et al. [Chem. Eng. Sci. (2011), 66, 6128-6145] calculate fluid-particle drag by substituting the drag coefficient C_d in a spherical drag law with a C_d correlation for non-spherical shapes. We show that such corrections are generally not accurate and, therefore, emphasize the need to develop new drag models specific to beds of non-spherical particles.

469532 Enhancing the Physical Picture of Particle Clustering Due to Mean-Flow Effects

Abstract:

Particle laden, gas-solid flows are prone to instabilities that manifest as either bubbles in dense regions, such as a fluidized bed, or as clusters in more dilute regions, such as a riser. One such source causing the clustering instability can be traced to the dissipation of granular temperature, as originally described by Goldhirsch and Zanetti (1993). Thanks in large part to the seminal work of Anderson and Jackson (1968), another source is known to exist â€" independent of the first â€" due to (non-zero) mean relative motion between the gas and solid phases; this source that has been reaffirmed countless times by subsequent linear stability analyses and direct simulations of the continuum governing equations. Despite this large body of work, a clear, physical picture of the underlying mechanism causing the instability remains somewhat lacking. In this work, we take a complete two-fluid model, closed by kinetic theory, and begin by simplifying it drastically. By retaining only the essential components, we are able to show that clustering results from the interplay between a Kelvin-Helmholtz instability and a kinematic instability. For most practical cases, the Kelvin-Helmholtz mechanism is insufficient to produce instability alone, although it is responsible for decreasing the magnitude of the dynamic wave speed, making it easier to be exceeded by the kinematic wave speed â€" the root cause of the kinematic instability. This finding is reminiscent of previous work by Wallis (1969), who first suggested from simple physical arguments that gas-solid particulate instabilities were a type of kinematic instability, a position which was subsequently confirmed by Batchelor (1988) thorough linear stability analysis. Our focus here is adding to the physical interpretation for the causes of the clustering instability. In addition to mathematical interpretation, some numerical simulations will also be provided to aid the discussion.

References:

Anderson TB, Jackson R. (1968) Fluid Mechanical Description of Fluidized Beds - Stability of State of Uniform

Fluidization. *Industrial & Engineering Chemistry Fundamentals* **7**: 12-21.

Batchelor GK. (1988) A New Theory of the Instability of a Uniform Fluidized-Bed. *Journal of Fluid Mechanics* **193**: 75-110

Goldhirsch I, Zanetti G. (1993) Clustering Instability in Dissipative Gases. *Physical Review Letters* **70**: 1619-22.

Wallis GB. (1969) *One-dimensional two-phase flow*. New York, McGraw-Hill.

469989 A Comparative Study of Euler-Euler and Euler-Lagrange Mesoscale Simulations of Moderately Dense Gas-Solid Flows

Abstract:

Accurate modeling of the dynamic behavior of gas-solid flows such as bubbling and circulating fluidized beds is important for the chemical and energy industries. Although Euler-Lagrange (EL) approaches have gained considerable popularity for modeling gas-solid flows in the past decade, the computational cost of fully resolved EL simulations is typically very high. Consequently, Euler-Euler (EE) approaches, such as kinetic-theory-based two-fluid models (TFM), remain the major workhorse in this area. However, the underlying assumption in TFM that the anisotropic part of the particle-phase pressure tensor depends on the deformation rate tensor has been proven invalid by many experiment data and EL simulations, especially when particles are dilute or moderately dense. In our previous work, the EE Anisotropic Gaussian approach (EE-AG) has been shown to produce good agreement for key statistical measures with EL simulations - such as particle settling velocity, volume fraction probability density function, two-point cross-correlations - when particles are dilute (1% by volume). In this work, a novel EE-AG/TFM hybrid solution algorithm, based on different contributions to the particle-phase spatial fluxes, is used to conduct fully resolved mesoscale simulations of cluster-induced turbulence when particles are moderately dense (10% by volume). This new algorithm is implemented in an open-source CFD package, and fully resolved simulation results are compared with EL simulations. The results demonstrate that this new approach of modeling in an EE framework can accurately capture the detailed dynamics of the gas-solid flows and produce results comparable to the EL approach at much lower computational cost.

461929 Coupling of Hydrodynamics and Triboelectric Charging in Fluidized Beds

Abstract:

Coupling of Hydrodynamics and Triboelectric Charging in Fluidized Beds J. Kolehmainen, A. Ozel, C.M. Boyce, S. Sundaresan *Department of Chemical and Biological Engineering, Princeton University, Princeton, New Jersey 08540, USA*

It is well known that materials can develop static charges upon mechanical contact with other materials [1]. This phenomenon, known as triboelectric charging or tribocharging, is at play, for example, when one rubs a balloon against wool cloth and observes the balloon sticking to the cloth. It is also known that granular materials in transport lines and fluidized beds tend to become charged due to contacts with walls and other particles [2,3,4]. In the present study we probe the interplay between tribocharging and fluidized bed hydrodynamics through simulations.

We have simulated gas-solid flow in a small 3D fluidized bed with a conducting wall using the Computional Fluid Dynamics-Discrete Element Method. All the particles are assumed to have the same size and same work function, which is different from that of the wall. The work function based model of Laurentie et al. [5] is employed to evolve the extent of tribocharging. This model has been found be succesful in capturing the extent of particle charging in hopper-chute flows [6,7].

The electrostatic forces on the particles are determined using the hybrid approach of Kolehmainen et al. [8], which finds the long-range contribution to the electric field by solving the Poisson equation, estimates the short-range contribution through a truncated pairwise sum, and adds a correction to avoid double counting.

Simulations reveal that the triboelectric charging behavior can be strongly coupled to particle dynamics. With small levels of charge, the behavior of the bed remains similar to the case without triboelectric charging, and the charge evolution could be predicted based on the work function difference alone. When the electrostatic forces become comparable to the gravity, the system behavior is altered due to particle-particle and particle-wall electrostatic interactions. In this case, the particles gain charges that are larger than the maximum charge anticipated by the work function difference between the particles and the wall alone. This is attributed to bed-induced electric field. Most of the hydrodynamic changes is due to wall-particle interactions that cause highly charged particles to adhere on the walls and also induce a core-annular flow.

[1] Lacks, D., Mohan, S.. Contact electrification of insulating materials. Journal of Physics D: *Applied Physics* 2011; 44(45).

[2] Sowinski, A., Miller, L., Mehrani, P.. Investigation of electrostatic charge distribution in gas-solid fluidized beds. *Chemical Engineering Science* 2010;65(9):2771–2781.

[3] Lee, V., Waitukaitis, S., Miskin, M., Jaeger, H.. Direct observation of particle interactions and clustering in charged granular streams. *Nature Physics* 2015;11:733–737.

[4] Jalalinejad, F., Bi, X., Grace, JR, .. Comparison of theory with experiment for single bubbles in charged fluidized particles. *Powder Technology* 2016;290:27 – 32.

[5] Laurentie, J., Traore, P., Dascalescu, L.. Discrete element modeling of triboelectric charging of insulating materials in vibrated granular beds. *Journal of Electrostatics* 2013;71(6):951–957.

[6] Naik, S., Saurabh, S., Vipul, G., Bruno, H., Abramov, Y., Yu, W., Chaudhuri, B.. A combined experimental and numerical approach to explore tribocharging of pharmaceutical excipients in a hopper chute assembly. *International Journal of Pharmaceutics* 2015;491:58–68.

[7] Naik, S., Sarkar, S., Hancock, B., Rowland, M., Abramov, Y., Yu, W., Chaudhuri, B.. An experimental and numerical modeling study of tribocharging in pharmaceutical granular mixtures. *Powder Technology* 2016. In Press.

[8] Kolehmainen, J., Ozel, A., Boyce, C., Sundaresan, S.. A hybrid approach to computing electrostatic forces in fluidized beds of charged particles. *AIChE Journal* 2016;In Press.

463468 Bed Densities in Cold Flow and Commercial Fluidized Bed Units of FCC Catalyst Particles

Abstract:

Bed density is an important property used in fluidized bed design and operation. Bed density enables estimations of bed heights or solids inventories in fluidized bed units, calculation of standpipe aeration rates, and determination of available pressure driving heads in various components of circulating fluidized bed units. Tests were conducted to measure bed densities in cold flow fluidized bed units of FCC catalyst particles ranging in diameters from 3 in to 5 ft (0.76 to 1.52 m) at gas velocities of up to 3 ft/s (0.9 m/s). The data are compared to literature data obtained from hot commercial FCC fluidized bed units. It was found that fluidized bed densities in cold flow units are significantly higher than in hot units. The roles of temperature and pressure are used to explain this bed density difference.

467147 Development of a Lift Correlation for Solid-Liquid Flows Using Direct Numerical Simulations

Abstract:

Development of a lift correlation for solid-liquid flows using direct numerical simulations (DNS)

The understanding of the transport of solid particles by fluids is important in industrial as well as natural processes like biological flows, sediment transport in rivers, extraction of natural gas using fracking process, and a few more. In the fracking process, the efficiency of natural gas extraction depends largely on the placement of solid particles, also known as proppants, in the fractured rocks. For the proper placement of proppants, it is important to keep these proppants suspended in liquid in the fracture network near horizontal well and sediment them at farther distances. The interaction between solid and liquid plays an important role in the transport of proppants. The main aim of our work is to gain a fundamental understanding of the liquid-solid flow in narrow channels using direct numerical simulations(DNS) and develop a large scale discrete particle model (DPM) for liquid-solid flows which are fundamentally different from gas-solid flows because of lower density ratios (solid to fluid), non-negligible lubrication and lift forces on solid particles.

As an initial step, a fully resolved three-dimensional simulations are carried out for a solid-liquid flow using Finite Volume Method on a staggered grid. An accurate fluid-solid coupling is achieved by incorporating the no-slip boundary condition at particle's surface by means of an efficient second-order ghost-cell immersed boundary method. The immersed boundary method is implemented directly at the level of the discretized fluid equations, unlike some of the other variants in which the no-slip boundary condition is implemented using a momentum source term near fluid-solid interface. A fixed Eulerian grid is used for solving the Navier-Stokes equations and the particle-particle and particle-wall interactions are implemented using Discrete Element Model (DEM). The DEM incorporates the soft sphere collision model and a lubrication force model. In the current implementation, the lubrication force is included as a sub-grid scale model due to its range of influence on a smaller scale than the grid size. The particles considered in this study are non-ideal, where the particle surface has a roughness and the collisions are dissipative. It is important to note that the DEM model for gas-solid flow ignores the lubrication force due to higher Stokes number whereas for solid-liquid flows the lubrication term is dominant and cannot be ignored. Finally, the validated numerical model will be used to gain more insight in the solid-liquid flow in narrow channel and use it for developing the closures. The sample simulation of DNS of fluidization of 24 particles due to lift forces is shown in Figure 1.

In the DPM, the fluid-solid momentum exchange is accounted for in the equations using closure relations for drag force and lift forces. There is a vast amount of literature investigating the drag force and numerous correlations for drag force are developed which are dependent on the Reynolds number and void fraction. However, the momentum exchange due to lift forces is more complicated due to additional parameters that can vary and therefore more work is needed to establish a universal correlation. The lift force on a particle is encountered due to its rotation and the shear rate of the fluid around it. In a narrow channel, the particles are more likely to be in the boundary layer region and experience higher shear rates which in turn gives higher lift force. In DPM, it is not possible to resolve the boundary layer completely and hence, the lift force due to proximity to wall should be included in the correlation. Moreover, the transport of the particles in a narrow channel also depends on the density ratio (solid to liquid) and particle size. In this work, numerous DNS simulations were performed to study the quantitative influence of all the parameters on a lift force for developing a lift correlation.

Keywords : Direct numerical simulations, Immersed boundary method, Solid-liquid flows, Lift forces, Discrete particle model.

459498 Studying the Effect of Vertical Internals on the Radial Pressure Drop in Fluidized Bed Using Probe-Differential Pressure Transducer

Abstract:

In this work, the effect of vertical internals on the radial pressure drop in fluidized beds has been studied by varying the parameters such as superficial gas velocity, solid particles type, solid particles density, internal size and radial position inside the bed (r/R). The radial pressure drop has been measured using probe-differential pressure transducer. The internals size are 0.5 and 1 inch in diameter with cross sectional area of 25% of the column. Two different types of solid particles have been used, glass beads (2500 kg/m^3) and aluminum oxide (3900 kg/m^3), both solid particles are Geldard B type and 350 Î¼m in diameter. The experiments were conducted in a plexiglass column of 0.14 m internal diameter, 1.8 m height and 0.35 m static bed height. The radial pressure drop in glass beads bed was found to decrease with both sizes of internals, thus resulting in lower power consumption. The reduction in radial pressure drop in case of glass beads was increase with increasing the internals size and it was high near the wall and less toward the center. In the case of Aluminum oxide, the radial pressure drop was found to decrease with 1 inch internals and increase with 0.5 inch internals due to the reduction in spaces between the internals and the natural of the aluminum oxide solid particles. The reduction in radial pressure drop for the case of aluminum oxide (1 inch internals) was almost the same for the whole radial positions. In general, the reduction in pressure drop decrease with increasing superficial gas velocity for both solid particles.

457071 Numerical Studies on Effects of Particle Rotation in Riser Flows

Abstract:

Particle-fluid drag correlation is the most essential part in the CFD modeling of Particle-laden flows. Researchers have invested enormous effort to obtain various drag relations. However, none of these drag relations have demonstrating their significant advantages over others in real applications, indicating that the interactions between the fluid phase and particles may be significantly affected by factors other than the drag force. Using an efficient immersed boundary-lattice Boltzmann method (IB-LBM), the Magnus lift force is computed in flows through random arrays of spheres. The results indicate that the lift force is insignificant at the rotational Reynolds number below 1. However, it can be larger than the drag force as the rotational Reynolds number increases up to O(100) especially at low solid volume fractions. The torque exerted on spheres by the fluid phase is also investigated since it is essential to account for addition energy dissipation of the two-phase system due to particle rotation. We then installed the proposed formulas for Magnus lift forces and torques in the open source software Multiphase Flow with Interphase eXchanges (MFiX). Computational fluid dynamics-discrete element model (CFD-DEM) simulations of a riser flow are performed to investigate the effects of particle rotation. Periodic boundary conditions are implemented in the vertical direction while non-slip conditions are used for lateral walls. It is found that the minimum riser width needed to reproduce the core-annular structure is around 100d, where d is the diameter of particles. The size of clusters is relatively smaller compared to results without considering particle rotation. This is believed to be caused by the Magnus lift force produced by fast rotating particles, which makes particles more dispersive in space and accelerates the breakup of clusters. The effects of particle rotation on the granular temperature, skewness and kurtosis of vertical velocity are also analyzed. CFD-DEM simulations of a bubbling fluidized bed are also performed in this study. The effects of particle rotation are found to be very marginal. This is due to that particles in bubbling fluidized bed are most rotating at low and moderating speed, making the Magnus lift force negligible compared to the drag force.

461120 Polyelectrolyte Coating of Nanoparticles for Applications in High Salinity Media

Abstract:

Over the past decade, the interest in exploiting nanotechnology in high salinity media including seawater and brines relevant to environmental science and subsurface oil/gas reservoirs has grown markedly. In such media with high ionic strengths, the electrostatic repulsion between the nanoparticles (NPs) becomes very weak, often leading to immediate aggregation and precipitation of the particles. Herein we present a variety of synthetic strategies to graft water-soluble polyelectrolytes onto NP surfaces in order to overcome the unique challenges associated with NP stabilization in high salinity media. We show that the polyelectrolyte-grafted NPs not only display colloidal stability in high salinity at high temperature but also resist undesirable adsorption on model mineral surfaces. Finally, the nanoparticles be transported through consolidated rock with minimal retention.

469423 A Novel Method for the Analysis of Particle Coating Behaviour Via Contact Spreading in a Tumbling Drum

Abstract:

Spray coating of particles and powders is a commonly used process in many industrial applications, e.g. food, detergents and pharmaceuticals. It involves complex interactions which include three phases; solid, liquid and gas. Consequently, the design and scale up of the process are still based on trial and error. Tumbling drums are typically used for particle coating processes, and the uniformity of the coating layer formed is very important when considering the quality of the final product.

There are two proposed mechanisms included in the spray coating of particles: (I) droplet deposition and (II) contact spreading (liquid exchange between particles). However, research to date has tended to focus on the droplet deposition mechanism rather than the latter. Moreover, in tumbling drums, contact spreading is thought to be very important, and understanding this mechanism is the focus of this study, with the aim of incorporating knowledge of this mechanism into the whole process.

In order to investigate the contact spreading between particles, a new image analysis system, based on colorimetric measurement, has been developed to quantitatively determine the colour uniformity of particles coated with dyed solutions. Here, it is demonstrated that the novel method can analyse a large number of particles in a relatively small time period, and gives data with which to assess inter-particle coating variation and uniformity of particles obtained from tumbling drum coating processes.

In this presentation, we describe results from experiments performed in a tumbling drum using a model system; alumina particles coated with dyed polyethylene glycol solutions of varying viscosities. It is observed that the uniformity of the coating increases with time. Furthermore, the viscosity of the coating solution also has an effect on the contact spreading process; the lower the viscosity, the faster the contact spreading process. This effect is attributed to difference in the formation and stability of liquid bridges between particles which influence the mechanism of liquid transfer via contact spreading.

457437 Development of Population Balance Model and Semi-Mechanistic Layering Kernel for Two Stage Heteroaggregation of Oppositely Charged Micro- and Nano-Particles

Abstract:

Development of Population Balance Model and Semi-mechanistic Layering Kernel for Two Stage Heteroaggregation of Oppositely Charged Micro- and Nano-particles

Anik Chaturbedi, Nina Shapley, Rohit Ramachandran

Rutgers, the State University of New Jersey, Piscataway, NJ, USA 08854

In colloid science, heteroaggregates (aggregates of particles that are different in various aspects such as size, surface charge) are gaining popularity due to their versatile applicability. The properties of these heteroaggregates are affected by the properties of the constituent particles as well as the size of the heteroaggregates, which in turn is influenced by the initial relative concentration of the constituent particles in the system. Different initial relative concentrations lead to different regimes in the system and either facilitates or prevents heteroaggregate formation. Traditionally, core-shell particles, produced by heteroaggregation have been used for xerography, printing ink, drug delivery applications. Customized heteroaggregates for instance can also be used for water purification applications (Yu, et al., 2013), based on the idea that oppositely charged components of the heteroaggregates will adsorb toxic anions and cations such as Cadmium, Mercury, Lead, Chromium, Molybdenum, Vanadium, Arsenic from waste-water.

In this work, the heteroaggregation of alginate microparticles and chitosan nanoparticles has been studied. Since these particles are hydrogels, very different in terms of size (alginate particles are about 130 times bigger in diameter than the chitosan particles), and are oppositely charged, this system is inherently different from the commonly studied systems consisting of charged nanoparticles and negligibly charged microparticles (Zhang, et al., 2008). We assume that the heteroaggregation process, in this case, happens in two stages. Firstly, the positively charged, small chitosan particles are attracted to the negatively charged, big alginate particles dominantly due to the electrostatic force and form a layer around the latter. The particles formed are referred to as monoaggregates. In the second stage, depending on the composition and net surface charge of the monoaggregates, they either repel each other or alternatively attract each other; agglomerate and form big heteroaggregates. There are various forces in play in this system such as van der Waals, electrostatic, hydration. The complex dynamics resulting from the interaction of these forces is modeled using a population balance model with a semi-mechanistic layering kernel developed specifically for the first stage of heteroaggregation to get an accurate particle size distribution at the end of the layering process of chitosan nanoparticles on alginate microparticles, for use as the initial particle size distribution for the second stage of heteroaggregation. Using this model, the relation between the initial relative concentration of alginate and chitosan with the final particle size distribution is studied. Also, the particle size distributions at the end of the process obtained from the model are compared with experimentally obtained particle size distributions from laser diffraction spectroscopy and microscopic images of the particles to validate the model. Results show good agreement between model and experiment. The presence of different heteroaggregation regime is observed in both the experiment results and model calculations.

References

Yu, K., Ho, J., McCandlish, E., Buckley, B., Patel, R., Li, Z., & Shapley, N. C. (2013). Copper ion adsorption by chitosan nanoparticles and alginate microparticles for water purification applications'. *Colloids and Surfaces A: Physicochemical and Engineering Aspects*, 31-41.

Zhang, F., Long, G. G., Jemian, P. R., Ilavsky, J., Milam, V. T., & Lewis, J. A. (2008). Quantitative Measurement of Nanoparticle Halo Formation around Colloidal Microspheres in Binary Mixtures. *Langmuir*, 6504-6508.

461035 Microfluidic Fabrication of Compartmentalized Hydrogel Microparticles By Extractive Gelation

Abstract:

For successful application of compartmentalized microparticles in the therapeutic treatment several requirements have to be met, the size of the microparticles is definitely one of them.
Therefore this work is focusing on the reduction of a characteristic dimension of the composite hydrogel microparticles to that of the red blood cells.

In the present work, compartmentalized microparticles were synthesized from a calcium alginate gel matrix and contain several functional components, namely iron oxide nanoparticles for radiofrequency heating and MRI visualisation, liposomes for the storage and release of active molecules of their precursors. Liposomes are spherical structures formed by a phospholipid bilayer, which are able to release the active matter in response to temperature. The permeability of liposomes can be significantly increased once they are heated above its phase transition temperature. The needed temperature change is achieved by iron oxide nanoparticles upon exposure to external radio-frequency magnetic field.

Monodisperse microparticles were prepared by an extractive gelation process on a microfluidic chip. First, droplets of aqueous solution of alginate in the continuous oil phase are formed in a flow-focusing channel. Second, gelation of the droplets is established in a long wavy channel by diffusion of calcium ions from the continuous phase. At the same time, the volume of the droplets is reduced by partial extraction of water into the continuous phase. Depending on the volumetric flow-rates of the aqueous and the oil phase, different conditions of the production were mapped and the effect of extraction on the achievable size of the microparticles was evaluated. The ideal conditions to produce monodisperse microparticles and to encapsulate iron oxide nanoparticles and liposomes were found. The stability of liposome in the alginate microparticles was investigated and the remotely triggered released experiments were conducted. The possibilities to scale up the microfluidic chip, where the particles are synthesized one by one, to device with higher throughput was also investigated.

464868 Microfluid-Aassisted Hollow Silica Nanoparticle Synthesis through Polyelectrolyte Template

Abstract:

Abstract

Hollow spherical silica nanoparticles (HSNPs) have been widely proposed as a promising material in various fields including catalysis, drug delivery, cell-labeling, semiconductor and optical coatings, due to its unique properties such as biocompatibility, controllable surface areas and pore volumes, as well as chemical and thermal stability. To prepare the HSNPs, template method is a facile and straightforward approach. Hard templates such as polystyrene particles [1] and hydroxyapatite nanoparticles [2], were successfully applied to produce the uniform HSNPs in tunable void space and shell thickness. High temperature (around 500) was required to calcinate out these hard templates thoroughly after the formation of silica shell. Oppositely, soft templates made from emulsion micelles [3] and vesicles [4] can be easily washed away by a selective solvent at room temperature. However, to prepare the soft templates in a uniform shape usually involved at least two surfactants and tedious procedures, especially to obtain and control the uniform HSNPs under 100 nm which have higher surface area is still a challenge. The groups of Du et al. [6] and Wan et al. [7] proposed a simplified and environment-friendly approach by applying a single polyelectrolyte - poly(acrylic acid) (PAA) as the soft-template. However, the PAA globule or necklace monomers start randomly aggregation in microseconds ahead of the silica shell formation and lead to the PAA templates in broad size distribution, which is the main reason of the inconsistent void space of the HSNPs in the batch reaction. Microfluidic system is able to reach the homogeneous environment and elaborate control on the reaction conditions due to its extraordinary mass/heat transfer, expected to improve the management of the colloidal size and uniformity.

Microfluidic system with hydrodynamic focusing micromixer are introduced in this study. The result showed that successfully generated the uniform HSNPs in 50 nm and hollow wormlike nanoparticles were synthesized in a short-time scale. Microfluidic system, scaling down the reactor into micrometer scale, is able to reach the homogeneous environment and elaborate control on the reaction conditions. Its extraordinary mass/heat transfer dramatically improved the management of the colloidal size and uniformity. Hydrodynamic focusing micromixer (HFM) is one of the efficient mixers which can achieve the microsecond-mixing in a low flow rate due to the existence of both molecular and convectional diffusions. HFM possesses a continuous flow in small dead time without any clogging which usually happened in the other chaotic micromixers such as the U-shaped mixers and Tesla mixers. It can provide not only the mixing time in microsecond but also the high volume flow rate ratio (VFRR) between the focused and side streams. This high VFRR is essential in some particular nanotechnology such as our case which only allows the reactants in a small proportion of its solvent. Valencia et al. [10] and Jahn et al. [11] has been applied the HFM to obtain lipid vesicles and micelles. However, controlling the self-assembly of a single polyelectrolyte and conserving its intermediate conformations has not been reported due to their rapid aggregation.

In our microfluidic system with HFM, the fast and homogeneous mixing enable the PAA globules/necklace monomers protected instantly by evenly dispersed silica precursor. Along the further attachment of silica precursor and the rapid growth of silica shell, the aggregation of PAA monomers are efficiently avoided due to the hydroxide bond on the particle surface. A micro-channel following the HFM was applied to accomplish the growth of silica shell. By controlling the residence time of the microfluidic flow, different thickness and density of the silica shell can be obtained. In this study, a numerical model was established by using COMSOL Multiphysics to investigate the mixing profile under different flow condition. Incorporated with this computational model, we describe the experimental operation of the microfluidic chemical reactor to control the continuous synthesis of HSNPs. The effects of the mixer geometry and the operating variables on particle structure and size distribution are discussed.

Reference

1. Ding, Xuefeng, Kaifeng Yu, Yanqiu Jiang, Hari-Bala, Hengbin Zhang, and Zichen Wang. "A Novel Approach to the Synthesis of Hollow Silica Nanoparticles." Materials Letters 58.27-28 (2004): 3618-621.
2. Williamson, Peter A., Philip J. Blower, and Mark A. Green. "Synthesis of Porous Hollow Silica Nanostructures Using Hydroxyapatite Nanoparticle Templates." Chem. Commun. 47.5 (2011): 1568-570.
3. Yildirim, Adem, and Mehmet Bayindir. "A Porosity Difference Based Selective Dissolution Strategy to Prepare Shape-tailored Hollow Mesoporous Silica Nanoparticles." J. Mater. Chem. A 3.7 (2015): 3839-846.
4. Wu, Xue-Jun, and Dongsheng Xu. "Soft Template Synthesis of Yolk/Silica Shell Particles." Adv. Mater. Advanced Materials 22.13 (2010): 1516-520.
5. Li, Yunqi, Bishnu Prasad Bastakoti, Masataka Imura, Jing Tang, Ali Aldalbahi, Nagy L. Torad, and Yusuke Yamauchi. "Dual Soft-Template System Based on Colloidal Chemistry for the Synthesis of Hollow Mesoporous Silica Nanoparticles." Chemistry - A European Journal Chem. Eur. J. 21.17 (2015): 6375-380.
6. Du, Yi, Lunet E. Luna, Wui Siew Tan, Michael F. Rubner, and Robert E. Cohen. "Hollow Silica Nanoparticles in UV–Visible Antireflection Coatings for Poly(methyl Methacrylate) Substrates." ACS Nano 4.7 (2010): 4308-316.
7. Wan, Y., and S.-H. Yu. "Polyelectrolyte Controlled Large-Scale Synthesis of Hollow Silica Spheres with Tunable Sizes and Wall Thicknesses." J. Phys. Chem. C Journal of Physical Chemistry C 112.10 (2008): 3641-647.
8. Dobrynin, A., and M. Rubinstein. "Theory of Polyelectrolytes in Solutions and at Surfaces." Progress in Polymer Science 30.11 (2005): 1049-118.
9. Dobrynin, Andrey V., and Michael Rubinstein. "Counterion Condensation and Phase Separation in Solutions of Hydrophobic Polyelectrolytes." Macromolecules 34.6 (2001): 1964-972.
10. Valencia, Pedro M., Pamela A. Basto, Liangfang Zhang, Minsoung Rhee, Robert Langer, Omid C. Farokhzad, and Rohit Karnik. "Single-Step Assembly of Homogenous Lipid–Polymeric and Lipid –Quantum Dot Nanoparticles Enabled by Microfluidic Rapid Mixing." ACS Nano 4.3 (2010): 1671-679.
11. Jahn, Andreas, Samuel M. Stavis, Jennifer S. Hong, Wyatt N. Vreeland, Don L. Devoe, and Michael Gaitan. "Microfluidic Mixing and the Formation of Nanoscale Lipid Vesicles." ACS Nano 4.4 (2010): 2077-087.

467842 Investigation of the Oxidation Behavior of Copper/Tin and Copper/Carbon Composite Particles Fabricated Via Spray Pyrolysis

Abstract:

Copper particles have potential applications in conductive pastes for printed electronics, metallization of solar cells, and interference packaging. However, copper powders are easily oxidized even at room temperature under ambient air. Therefore, numerous efforts have been made to increase the oxidation resistance of copper powders by addition of a secondary component. Among them, copper/tin and copper/carbon composite particles seem promising, considering the price of the raw materials and the resistivity of the particles. However, the oxidation behaviors of copper/tin and copper/carbon composite powders need to be investigated first.

Up till now, the investigations of copper-based composite particles have been restricted to a small range of secondary components, such as silver. As silver does not react with oxygen, the oxidation processes of these composite particles are simplified. For other composite particles whose secondary components can also be oxidized, research is needed to understand the underlying oxidation mechanism of the composite particles as a whole as well as the oxidations of each individual component.

Here, the oxidation behaviors of copper/tin and copper/carbon composite particles are reported. Spray pyrolysis was utilized to fabricate the dense and solid copper/tin and copper/carbon composite particles. X-ray powder diffraction (XRD) was performed to understand the evolution of the crystal structure in the particles. The Rietveld refinement results showed that with the increase of tin concentration in the composite particles, the relative amount of Cu_2O decreased from 60 wt % for Cu0.95Sn0.05 to 10 wt % for Cu0.9Sn0.1 after the samples had been heated for 10 min at 300 ^{o}C under ambient air. In addition, CuO was only found in the pure copper sample after heating for 10 min at 300 ^{o}C also under ambient air. All the results showed that pure copper samples were severely oxidized compared to the composite particles. To further explore the oxidation processes, *in-situ* XRD was conducted to detect crystal transformations in the composite particles during oxidation. When the samples were continuously heated under ambient air with a ramp of 5 ^{o}C/min, oxide peaks were found in the pure copper powders when the temperature was below 300 ^{o}C. For the composite particles, oxide peaks began to appear in the XRD patterns when the temperatures were higher than 300 ^{o}C. Our preliminary results show that adding a second component such as tin can improve oxidation resistance. Results for carbon as the second component will also be presented.

468473 Physicochemical Properties of Biocompatible Films Containing Concentrated Fenofibrate Particles Prepared Via Liquid Antisolvent Precipitation

Abstract:

Approximately 40% of newly discovered molecules are poorly water-soluble and this class of drugs lacks a formulation strategy capable of producing high loads, fast dissolution rates. Biocompatible films have great potential over other dosage forms for delivery of poorly soluble drugs as they provide distinct advantages such as: easy to swallow, convenient for geriatric and pediatric patients, larger surface area which leads to rapid disintegration and dissolution in the oral cavity and increased bioavailability. In this study, we developed biocompatible strip films by incorporating concentrated suspension of fenofibrate (FNB), a model poorly water-soluble compound, produced by liquid antisolvent (LAS) precipitation. Stabilization of the suspension was established via combination of non-ionic surfactant Pluronic F68 and the polymer hydroxypropyl methyl cellulose (HPMC E3). Centrifugation and subsequent resuspension was used to concentrate suspensions and minimize residual solvent level. The resulting micro suspensions were then formulated with high viscous cellulosic polymer solution of HPMC (E15 LV), anionic polysaccharide of sodium alginate (NaAlg), and a combination of both to make biocompatible film. Glycerin was used as plasticizer and its effect on mechanical strength, dissolution was investigated. The morphology of the drug particles from suspensions and the structure of the films were observed by using scanning electron microscopy (SEM). The crystallinity of drug particles in films was studied by X-ray diffraction (XRD), differential scanning calorimetry (DSC), and Raman spectroscopy. Distribution of drug particles in the films was studied by NIR chemical imaging. The influence of the formulations on the filmâ€™s mechanical properties was studied. The SEM images show that the films exhibit different morphologies depending on film formulation. The cross-sectional images showed that the films with HPMC or combination of HPMC-NaAlg having glycerin had porous structure which could be due to the presence of glycerin that modified the tertiary structure of polymer into porous structure. The NIR chemical imaging indicates that the drug is uniformly distributed without any observable large agglomerates. XRD, DSC and Raman results suggest that the crystalline nature of FNB was largely preserved after film formation and no polymorphic change has occurred. A substantial decrease in Youngâ€™s modulus and tensile strength are observed for film having NaAlg. The drug release profiles for all of the formulated films showed clear distinction between the films and as-received FNB powder. The addition of plasticizer, i.e., glycerin, decreased dissolution time of film containing HPMC as compared to un-plasticized film. The fastest drug release was achieved from film which contained HPMC or combination of HPMC-NaAlg having glycerin. The decreased dissolution time for HPMC film is probably due to increased film hydrophilicity causes by the presence of glycerin. The films with NaAlg and HPMC as the matrix former were extremely weak, which aids in quick disintegration and dissolution of the film. The porous network of both films was beneficial to quick penetration of the dissolution medium resulting in quick drug release. Our study has demonstrated that high drug loaded biocompatible film can be prepared with LAS precipitated drug suspensions and that film properties can be modulated by formulating with NaAlg.

Keywords: Liquid Antisolvent Precipitation, Stabilizer, Biocompatible Film, Sodium Alginate, Fast Dissolution.

456044 Aspects of Good Experimental Practice in Surface Energy Measurements, of Particulate Materials, Using Fd-IGC

Abstract:

Surface energy, arising from the imbalance of the intermolecular forces of the molecules on a surface[1], is gaining momentum as the focus of industry shifts in particles with smaller and smaller dimensions, where surface properties overcome the effects of bulk properties. Traditionally, wettability-based techniques, grounded on Young's pioneer work on the cohesion of fluids[2], were employed in surface energy measurements. Nonetheless, the accuracy of these techniques proved to be limited when encountering particulate materials. The anisotropic nature of individual particles gives rise to a surface energy heterogeneity. This heterogeneous behaviour can not be described by single numerical values, but is more appropriately expressed in the form of surface energy distributions[3]. Inverse Gas Chromatography, an adsorption based technique introduced in mid-70's[4], for surface energy measurements proved to be quite efficient in overcoming the aforementioned limitation. In its finite dilution mode (FD-IGC) it enables

Despite the progress in the field, both from an experimental and a computational perspective, difficulties still exist in the direct comparison of data obtained from different researchers. In this work, the influence of the amount of silanized wool used in the packing of the chromatographic column, the packing structure of the material and the temperature of the measurement were examined using both experimental and computational approaches. The aim was to unveil the mechanism and the conditions under which they influence the measurements and to propose a framework minimizing their effects.

The surface area and surface energy of silanized wool and lactose monohydrate were measured. The dispersive component of the surface energy of wool was determined to be about 35 $mJ*m^{-2}$, whereas the acid-base component was determined to be less than 5$mJ*m^{-2}$.Computational models[6-8] were used to determine the surface energy distributions of the two materials. Then, lactose monohydrate was packed with known quantities of silanized wool at different ratios. Surface energy measurements were performed in these mixtures and the corresponding surface energy distributions were calculated. The surface energy distributions agree well with the predicted surface energy distributions, obtained from direct combinations of the surface energy distributions of silanized wool and lactose monohydrate. The surface area (about 0.25 m^2*g^{-1}) and surface energy of the silanized wool seem to affect the measurements of lactose monohydrate (with $\gamma^d \approx 40$ $mJ*m^{-2}$), even when the surface area of silanized wool is more than the one tenth of the surface area of lactose monohydrate.

Using computational models, a series of in-silico experiments were performed in order to extend the understanding of the influence of silanized wool on IGC measurements. The results suggest that materials can be distinguished in three major categories. The first category includes very high surface energy materials ($\gamma^d > 70$ $mJ*m^{-2}$). In these materials the effects of silanized wool even in large quantities is negligible. The second category includes the materials with similar surface energy as silanized wool. In this category the effects of silanized wool are masked. The third category includes very low surface energy materials ($\gamma^d < 20$ $mJ*m^{-2}$). For this type of materials even a small amount of silanized wool (about 1%) influences the measurements. It is suggested that special care should be taken for this kind of materials. These findings highlight the importance of complimentary techniques which can provide with a rough idea for the surface energy of the material, in order to help the design of more advanced IGC experiments.

Mixtures of carbamazepine and delta-mannitol, at a 1:1 surface area ratio, were used to determine the effects of powder mixtures packing on IGC measurements. Four different packing structures of the mixtures were measured. The surface energy of all of them was found to be identical. Monte Carlo simulations[5,9]performed at different lattice arrangements verify the experimental findings. The experimental and modelling data suggest that packing structure does not influence IGC measurements.

Carbamazepine was, also, used to investigate the effects of temperature on IGC measurements. Using the same column, IGC measurements were performed at four different temperatures. The changes in surface tension and surface area, of the solvent probes, with increasing temperature were taken into account while elucidating the experimental data. The results sow an increase of the measured surface energy. This is in accordance with the theoretical basis provided by the static component of the Lifshitz equation for dispersive interactions. According to this equation there is a possibility of stronger dispersive interactions at increasing temperatures, depending on the dieelectric properties of the material[10].

Overall, this work provided some coherent analysis on parameters influencing IGC measurements, reinforcing the existing rules of good experimental practice. It proved that silanized wool can influence IGC measurements of low surface energy and surface area materials. In addition, it demonstrates that the packing structure of particulate materials does not affect IGC measurements and it, also, highlights the importance of experimental temperature on the measurements. Furthermore, it proved the applicability of computational tools in the field, highlighting their importance when it comes to more advanced types of measurements, including multicomponent mixtures.

Disclosure:

Imperial College London and AbbVie jointly participated in study design, research, data collection, analysis and interpretation of data, writing, reviewing, and approving the publication. Eftychios Hadjittofis is a graduate student at Imperial College London; Jerry Y. Y. Heng is a professor at Imperial College London. They all have no additional conflicts of interest to report. Geoff G. Z. Zhang is an employee of AbbVie and may own AbbVie stock.

References:

1. Israelachvili JN. *Intermolecular and surface forces.* 3rd ed. Oxford: Academic Press; 2011.

2. Young T. An essay on the cohesion of fluids.

3. Heng JYY. *Anisotropic surface properties of crystalline pharmaceutical solids.* [PhD]. Imperial College London; 2006.4. Hudec P. Determination of crystallinity of polyethylene powder by inverse chromatography.

5. Roma F, Ramirez-Pastor AJ, Riccardo JL. Configurational entropy for adsorbed linear species "k-mers".

6. Ho R, Heng JYY. A review of inverse gas chromatography and its Development as a tool to Characterize Anisotropic surface properties of pharmaceutical solids.

7. Jefferson EA, Williams DR, Heng JYY. Computing the surface energy distributions of heterogeneous crystalline powders.

8. Smith RR, Williams DR, Burnett DJ, Heng JYY. A new method to determine dispersive surface energy site distributions by inverse gas chromatography.

9. Smit B. Grand canonical monte carlo simulations of chain molecules: Adsorption isotherms of alkanes in zeolites.

10. Lifshitz EM. The theory of molecular attractive forces between solids.

470241 Pharmaceutical Powder Characterization By Droplet Penetration Technique

Abstract:

Determination of the contact angle between a liquid and a pharmaceutical granular material is crucial for many reasons. On the one hand, wettability of pharmaceutical powders has critical effect on pharmaceutical processes like wet granulation, and product quality attributes like dissolution.Â On the other hand, it allows determining the surface energy of pharmaceutical ingredients by measuring the contact angle between the powder and different liquids. Surface energy is an important particle attribute as it correlates to electrostatic charging, particle cohesion, agglomeration and wall adhesion to name some important properties that determines powder behavior in many pharmaceutical processes.

Determination of the contact angle between a liquid and a granular material is not trivial. In this work, we use the droplet penetration technique which has the advantages over the more traditional â€œWashburnâ€ column technique in that the amount of material necessary to perform the experiment and the time it takes are orders of magnitude smaller.

The droplet penetration technique we use to characterize the contact angle of pharmaceutical powders with a liquid is based on a new development that is described in a different presentation. In essence, it is the comparison between the non-dimensional volume of liquid penetrated at the same non-dimensional times when using two different liquids: the test liquid (water or Diiodomethane) and a reference liquid (PDMS), which completely wets the granular material. This approach is useful when a large part of the penetration process occurs at constant contact area between the external droplet and the powder bed, condition that is amply confirmed for our systems.

Two case-studies are considered: the first one in which the blends have the same formulation and they only differ in the processing and though in the lubrication level, displaying different wettability shown in our results. A second case study, where different active ingredients: Caffeine and Acetaminophen of different particle sizes are measured with respect to a polar and non-polar liquid respectively. In both cases PDMS is used as a wetting reference.

451577 A Practical Framework Towards Prediction of Breaking Force and Disintegration of Tablet Formulations Using Machine Learning Tools and Nondestructive Ultrasound

Abstract:

Enabling the paradigm of quality by design requires the ability to quantitatively correlate material properties and process variables to measureable product performance attributes. Conventional, quality-by-test methods for determining tablet breaking force and disintegration time usually involve destructive tests, which consume significant amount of time, labor and provide limited information. Recent advances in material characterization, statistical analysis and machine learning have provided multiple tools that have the potential to develop nondestructive, fast and accurate approaches in drug product development. In this work, a methodology to predict the tablet breaking force and disintegration time of tablet formulations using nondestructive ultrasonics and machine learning tools was developed. The inputs to the model include intrinsic properties of drug formulation as well as extrinsic process variables influencing the tablet during manufacturing. The model has been applied to predict tablet breaking force and disintegration time of tablets using small quantities of API and prototype formulation designs. The accuracy of the predicted outputs are ± 7 N for tablet breaking force and ± 11 second for disintegration time of the actual value, even with the presence of wide range of measured breaking force and disintegration time. The novel approach presented in this study is a step forward toward rational design of a robust drug product based on insight into the performance of common materials during formulation and process development. It may also help reduce development time, API usage and facilitate the development of a robust drug product.

470147 Systematic Method for Finding Material Surrogates with Matching Flow Properties

Abstract:

Availability of active pharmaceutical ingredient (API) is often very limited, in the early stages of drug development. This limits experimentation, leaving many unidentified challenges that would be encountered later during development and commercial manufacturing. In addition, some materials are highly toxic or tightly controlled, further complicating their use in early development. These and other factors make it highly desirable to find substitute materials with similar properties that could be used to design the formulation and the process, and that could be replaced at a later stage with the actual drug material.

Flow properties are the physical attributes of API and excipients that have the strongest impact on the process in continuous direct compression of solid dosage forms. In order to select an appropriate surrogate material, data obtained from a library of previously characterized surrogate candidates can be analyzed using a statistical toolbox to identify the most "similar" materials. A big part of the problem in doing this effectively is that the practitioner often must select among too many, partially redundant characterization methods of unknown reliability.

Principal Component Analysis (PCA) is a powerful linear projection method used to examine over-specified data sets. It projects multidimensional data into a few orthogonal features, called principal components (PCs). The PCs are constructed as a linear combination of the original variables to maximize the preservation of the data variance. PCA projects a high dimensional dataset into a new coordinate system. The new variable value, based on the observation's component loading and the standardized value of the original variable, is called the component score. Examining the component scores of each sample helps to find similarity between materials and to identify redundant measurements.

The specific example discussed in this talk focuses on the application of PCA to identify placebo materials

exhibiting similar physical properties to that of an API that is a controlled drug substance and hence difficult to use in large quantities. Materials with similar flow characteristics are expected to exhibit similar behavior during manufacturing. Here, PCA is applied to a data set with physical properties spanning diverse physical characteristics from a library of pharmaceutical grade materials, which would enable the identification of materials that are mostly likely to behave similarly to the API of interest. The current dataset includes 21 materials with 33 variables derived from different material property characterization test results. The measurements include particle size distribution, bulk density, FT4 test (bulk, dynamic and shear tests) and electrostatics tests. Unscrambler X software from Camo was used to conduct PCA analysis and the scatter plots of loadings and scores were obtained to examine the relationship between samples by looking at their relative positions in the scatter plot. The relative position of each original variable in the loadings plot can give information about how they relate to one another. For a better understanding of which measurements are driving the trends/groupings seen in the scores plot, the scores and loadings plot need to be examined in conjunction and hence the scores and loadings plots were superimposed in a bi-plot. Materials that are projected close together suggest similarity in terms of all the flow property parameters measured. The scores of each material on the latent variables can be used to calculate statistical distance between the target API and the materials in the library. The statistical distance can be ranked, and the most similar material to the target API is the one that has smallest distance to the target API.

Based on present PCA; the target API was found to close to 3 other API's. This was further verified experimentally by conducting feeder studies on all of the four identified similar materials. Sticking to the feeder was observed for all the API's under study. Moreover, the relative standard deviation and deviation around the set point was found comparable to the target API. Thus, the PCA based systematic approach can be potentially validated by using similarity scoring, and be further expanded by including more common pharmaceutical materials and other material properties.

470537 The Effect of Temperature on Fluidization Behaviour of Ceramic Powder

Abstract:

The relevance of surface impurities on the effect of temperature on powder flow behavior

R. Chirone[1], D. Barletta[2], M. Poletto[2] and P. Lettieri[1]

[1]Department of Chemical Engineering, University College London - Torrington Place - London WC1E 7JE - UK

[2]Dipartimento di Ingegneria Industriale, Università degli Studi di Salerno - Via Giovanni Paolo II, 132 - 84084 Fisciano (SA) - Italy

Cohesive interparticle forces may have a relevant role in several industrial process operations involving particulate materials, such as fluidization, granulation and drying, as well as storage and solids handling units. Several of these operations require process conditions which involve high temperatures which, in turn, may affect the intensity of interparticle forces such as van der Waals, capillary and electrostatic forces. The mean by which the system temperature can affect all these forces, is the change of particle hardness, the generation of liquid phases which determines the formation of liquid bridges or the modification of the particle dielectric properties. A direct measure of interparticle forces is possible but can be affected by large fluctuations that require a great number of repetitions. Interparticle forces, instead, play in averaged ensembles in bulk properties such as powder cohesion. It is of interest, therefore, to have the possibility to measure powder cohesion at the process temperature and to observe possible changes due to temperature variations to infer possible changes at the particle level. Powder shear testing is one of the available methods able to measure powder cohesion and it has the great advantage of measuring well established physical properties and of being able to produce highly repeatable results. It has to be remarked, however that to date no established procedure exists to relate powder cohesion measured at the bulk level to powder fluidization properties.

In this paper a systematic study on the effect of the process conditions on the fluidization quality of ceramic powders is presented. Tests were carried out on powders of industrial interests, characterized by different particle size distributions and by different amounts of surface impurities, ranging from easy-flowing to cohesive flow behaviour. Two different experimental facilities were used: a modified ring shear tester available at the University of Salerno and aX-ray high temperature fluidization facility available at University College of London. The first apparatus was used to characterize powder cohesion at different temperatures between ambient and 500°C. Experimental results have been interpreted in terms of possible changes in interparticle forces as a function of temperature. The powder samples without impurities show an increase of cohesion with temperature as a result of an increase of interparticle van der Waals forces. A larger increase of cohesion was observed in the case of the powder samples with chemical impurities. The behaviour can be explained only by considering a cooperative effect of both van der Waals and capillary forces. It is noteworthy that the amount of surface impurities that is able to determine significant changes of powder flow properties is still so small that no evidence of phase transition could be detected by means of sample thermal analysis. The same powders have been characterized in terms of fluidization quality by using the x-ray fluidization facility available at University College of London under the same temperature range.

The changes by temperature on the flow properties of the bulk solid evaluated with the shear cell and the behaviour of particles under fluidization conditions are analysed and discussed. A direct quantification of the particle-particle interactions in fluidized beds and of their changes under process conditions is very difficult. The paper suggests a method by which powder rheology can be used to indirectly evaluate the effects of the interparticle forces on fluidization.

462296 Using FBRM for Online Monitoring of Particle Size in Radioactive Tank Waste

Abstract:

Focused-beam reflectance measurement (FBRM) is a popular qualitative tool for crystallization control and monitoring. Nevertheless, applying its data quantitatively is non-trivial â€" an FBRM chord length histogram depends on the size distribution of the scanned particles as well as their shape, surface roughness, and opacity. First-principles modeling of these light reflectance properties has proven to be effective, yet cumbersome to apply due to the quantity of parameters that need to be considered[1].

In this talk, we will examine the application of a previously-developed empirical framework[2]to a six-part mixture designed to simulate the particle sizes and densities seen in radioactive waste. FBRM is advantageous for nuclear waste processing because of its ability to carry out online in-situ measurement for a wide range of particle sizes with a short measurement time. Empirical modeling is useful for this application because particle shapes and surface characteristics are not uniform in tank waste. The model considered here takes in FBRM measurements obtained offline of a wide range of particle sizes, then approximates an online chord length histogram via an optimized linear combination of these signatures.

We will demonstrate the modelâ€™s ability to estimate the composition of a mixture containing six differently shaped particles with varied mass fractions. The online monitoring technique developed previously for a single crystal geometry in batch crystallization is extended to mixtures with multiple particle shapes.

[1] Kail, N., Marquardt, W., Briesen, H. 2009. Estimation of particle size distributions from focused beam reflectance measurements based on an optical model. *Chemical Engineering Science* 64, 984-1000.

[2] Li, H., Kawajiri, Y., Grover, M. A., Rousseau, R. W. 2014. Application of an empirical FBRM model to estimate crystal size distributions in batch crystallization. *Crystal Growth & Design* 14, 607-616.

The Influence of Roller Compaction Processing Variables on the Rheological Properties of Granules

Jamie Clayton[1,] Tim Freeman[1,] Hartmut Vom Bey[2], Michael Hanisch[2], Katrina Brockbank[1,]

[1] Freeman Technology, 1 Miller Court, Severn Drive, Tewkesbury, Gloucestershire, GL50 3RP, UK

[2] Gerteis Maschinen + Processengineering AG, Stampfstrasse 74, 8645 Jona SG, Switzerland

Keywords*: Solids Flow, Handling, and Processing, Process Design & Development, Process Control

Pharmaceutical formulations for oral solid dose delivery consist of mixtures of many components, each with a specific role in the optimal delivery of the active pharmaceutical ingredient, (API). In many instances the potency of these API's means that the actual quantity required per tablet/capsule is extremely small and to ensure content uniformity, a granulation process step is undertaken, especially when some or all of the materials in a formulation have poor flow properties. This approach combines the API with one or more of the other components and is frequently carried out as a wet process. The disadvantages of wet granulation are that the resultant wet mass has to be dried and milled to generate a product that can then be tabletted/encapsulated. These downstream steps are time consuming and incur additional costs; equally some active materials will be unsuitable for the wet processing route due to chemical and/or thermal degradation.

The option to use a dry granulation process, based around a roller compactor and integral mill/screen, has significant benefits both in terms of processing cost reduction but also for use with API's that cannot be subjected to moisture/solvents and/or heat. Dry granulation is not suitable for all types of powdered material but there is little in the literature that indicates which properties of a formulation make it suitable/unsuitable for this method of processing, with most equipment suppliers and pharmaceutical manufacturers relying on historical and ad-hoc trial information to identify suitable candidate formulations. Equally there is little to indicate which processing parameters produce optimal granulate quality to achieve interruption free processing and high quality products.

However, it is known that the quality of the dry granulate has a significant impact on downstream processes (including mixing and tabletting) and this has been shown to be governed by the rheological properties of the powder. [1-4] Recent regulatory initiatives on PAT and QbD [5-7]have also emphasised the need for a greater understanding of all pharmaceutical processes and how input variables, such as variation in powder characteristics, influence process performance.

The main barrier to determining these relationships has been the insensitivity of the methods historically used to characterise the feedstock/granulate properties. In many studies, the particle size distribution is the primary method used to quantify granule 'quality' [8, 9], however, it is clear from a number of other studies that powders with the same particle size can have vastly different flow behaviour due to the effects of other properties such as surface texture and shape [10-12]. Other studies have evaluated the quality of the ribbon produced [13] and may simply infer the quality of any granules that would have to be generated or did not extend their evaluation to this point.

Recent developments in automated instrumentation have allowed formulation scientists and engineers to assess powder properties more rapidly and reliably by employing laboratory measurements that simulate the range of process conditions more closely. With these considerations in mind, a multivariate analysis of the behaviour of granules produced by roller compaction will provide a more robust understanding of their downstream behaviour and suitability for tabletting.

Materials and Methods

The test powder that was used in these experiments was a placebo formulation consisting of 70% lactose, 29.5% microcrystalline cellulose and 0.5% magnesium stearate. The test formulation was roll compacted using a Gerteis Mini-Pactor (Gerteis Maschinen+Processengineering AG, Switzerland).

Two aspects of the control of the roll compactor were adjusted to evaluate the effect on the resultant granulate. Phase 1 investigated the effect of compaction force on the bulk powder properties with the roll compaction force increasing from 3 kN/cm to 12 kN/cm in six discrete steps. Phase 2 studied the effect of changing the roll gap from 1.5 mm to 5 mm in six discrete stages. The roll speed and mill mesh were fixed at 2.5 rpm and 1 mm respectively for both phases.

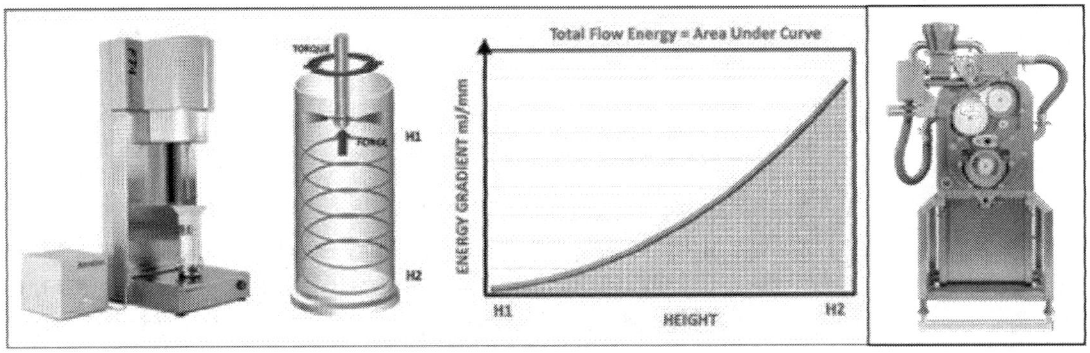

Figure 1: Measurement of flow energy using the FT4 Powder Rheometer and the Gerteis Mini-Pactor

The resultant granulates were evaluated using an FT4 Powder Rheometer (Freeman Technology, UK). The FT4 measurements included dynamic flow, bulk and shear properties. Dynamic properties including the Basic Flowability Energy (BFE) (fig.1) were obtained by means a patented measurement principle that evaluates the resistance to the motion of a specially shaped twisted blade passing through a precise volume of the sample along a prescribed path. The required torque and force are recorded and converted into a flow energy [14]. The repeatability of all measurements was enhanced by the use of a conditioning cycle which removes packing history and operator induced variability.

The particle size distribution was assessed using standard calibrated sieves (Endecotts). The Fine particle fraction was defined as particles below 106µm whilst the Coarse particle fraction was defined as particles above 212µm

Results

Phase 1 – Effect of Roll Compaction Force

Phase 1 investigated the effect of roll compaction force on the bulk powder properties with the roll compaction force increasing from 3 kN/cm to 12 kN/cm generating six different granulates

Clear and repeatable differences were demonstrated between the six batches of granules with good correlation observed between the roll compaction force and several of the granules rheological characteristics (fig. 2) with Permeability (as demonstrated by the Pressure Drop across the powder bed), Compressibility and the Conditioned Bulk Density (CBD) exhibiting the best correlation (highest R^2 values - 0.987, 0.849 and 0.964 respectively).

The Compressibility and Pressure Drop decreased and the CBD of the resultant granulate increased as the roll compaction force increased. This suggests more efficient particle packing for the product granulate formed from higher roll compaction forces. Analysis of the particle size showed that this is due to an increases in the coarse particle fraction.

Good correlation was also observed with a number of the dynamic properties. As the roll compaction force increased, the Aerated Energy (AE) also increased, this is most likely due to the increases in particle size / density as larger particles will require more lift to fully aerate / fluidise. The Flow Rate Index (FRI) also presented a strong dependence on the roller compaction force, samples generated using a higher roll compaction force were less sensitive to changes in flow rate (FRI = 1), suggesting that the granules were more free-flowing.

A reasonable correlation was also obtained with the BFE, with the higher values again suggesting a more efficient particle packing structure, increasing resistance to the passage of the blade. The obvious non-linearity of this relationship may be due to the ability of this parameter to detect the influence of multiple particle properties, for example, particle size distribution, shape, surface texture and density.

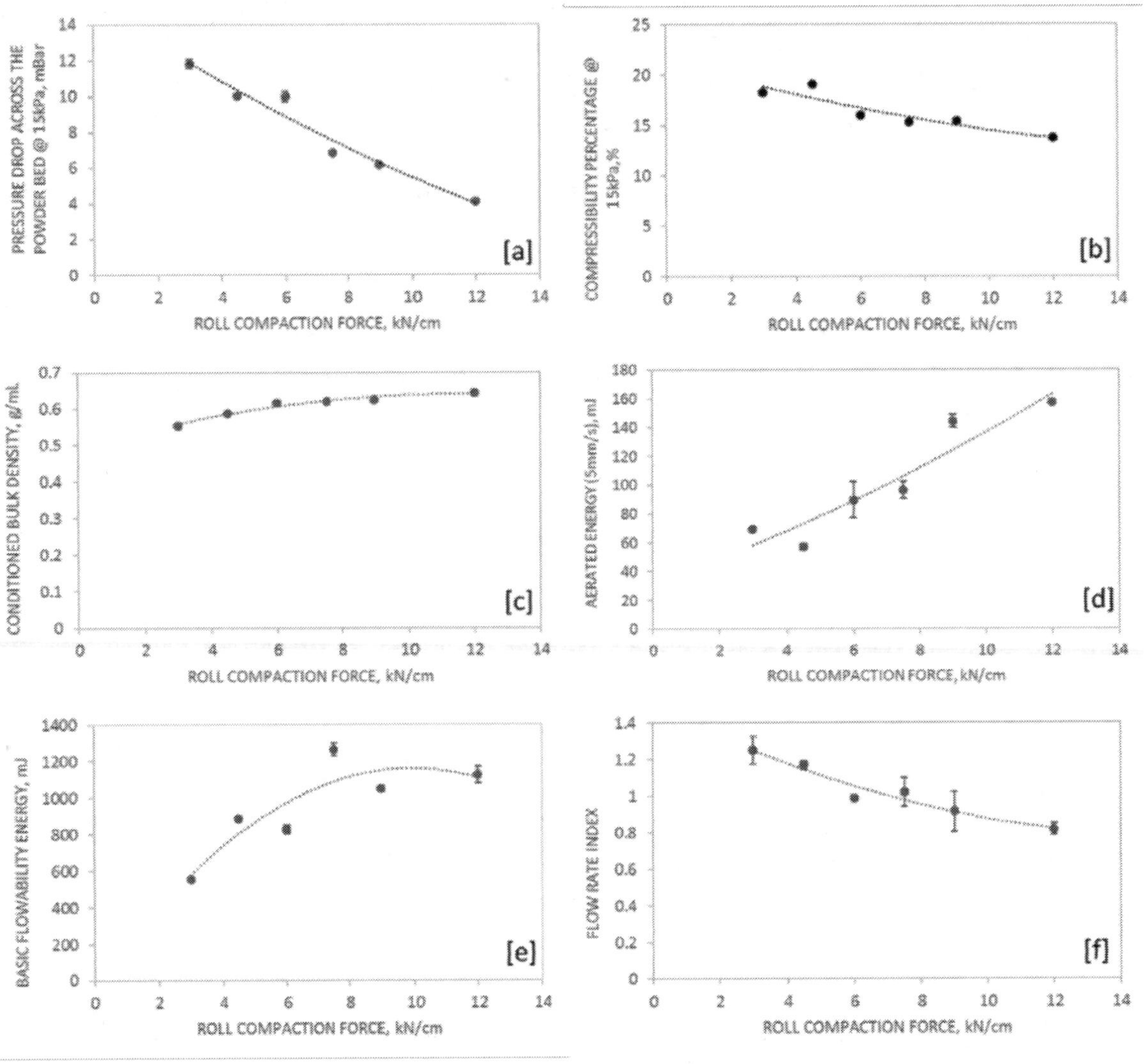

Figure 2: Results for granules produced in phase 1, (a) Pressure Drop, (B) Compressibility, (c) CBD, (d) AE, (e) FRI and (f)BFE

In this instance, the shear cell methodology was demonstrated to be of limited value with the test unable to reliably discriminate between the samples produced using different roll compaction forces.

Phase 2 – effect of roll gap

The results from the rheological testing again indicate a strong correlation with the Pressure drop, Compressibility and the CBD (Fig. 3). Again, the shear cell methodology was demonstrated to be of limited value with the test unable to reliably discriminate between the samples produced using different roll gaps

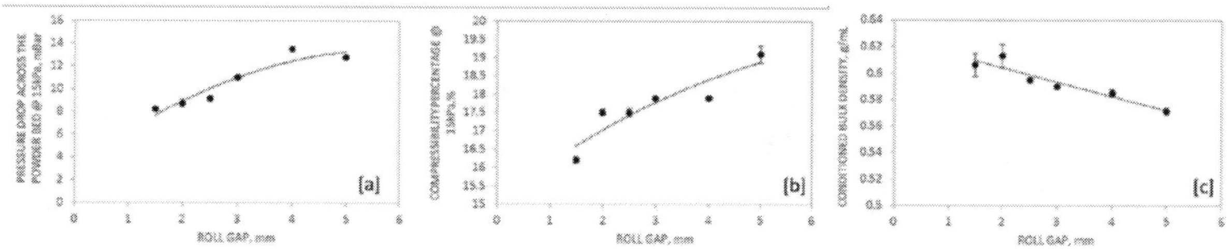

Figure 3: Results for granules produced in phase 1, (a) Pressure Drop, (b) Compressibility and (c) CBD

Overall the results suggest that as roll gap increases, granule uniformity decreases resulting in changes in the particle packing structure and hence the bulk properties. This is likely to be due to the greater variation in the consolidation regime between the rollers as the roll gap increases. This results in a lower consistency in the particle size distribution, shape and surface texture which is manifested by the reduction in the particle packing efficiency as demonstrated by the higher Compressibility and Pressure Drop and lower CBD values.

Phase 2a - comparison of 9 kN/cm and 4.5 kN/cm conditions wrt roll gap

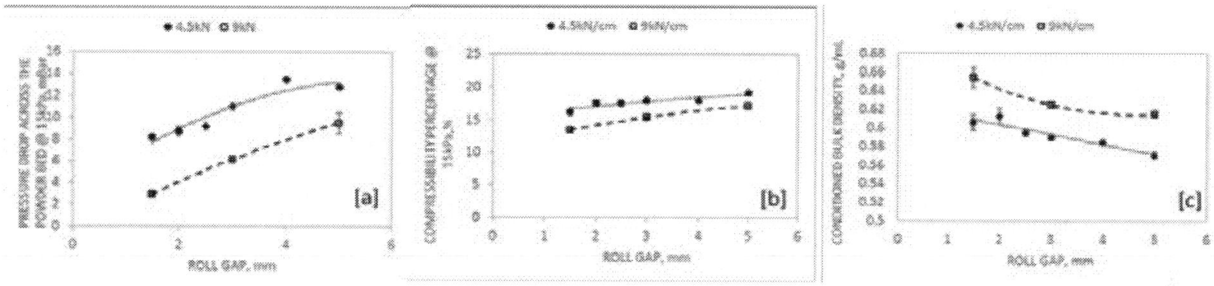

Figure 4: comparison between two different roller compaction forces over a range of gap widths for (a) Pressure Drop, (b) Compressibility and (c) CBD

Comparable curves were obtained for the Compressibility, Pressure Drop and CBD values for the 9 kN/cm and 4.5 kN/cm data, confirming the trends observed in Phase 1 (Fig. 4). As expected, lower Compressibility and Pressure Drop and higher CBD values were generated at the 9 kN/cm roll compaction force.

Multi-Linear Regression

In order to confirm the relationship between the roll compaction force and roll gap and the FT4 parameters, multi-linear regression analysis was performed. As expected Compressibility, Pressure Drop and CBD provided the strongest correlations. In addition to high R^2 values, the reliance factor (p) was found to be significantly below the maximum acceptance value of 0.05 and the predicted values were shown to be comparable with experimental values (Fig. 5). The relationship models are shown below.

$$Comp. \% = 17.816 + -0.573x + 0.820y \qquad (R^2 = 0.900)$$

$$Pressure\ Drop = 9.604 + -0.940x + 1.695y \qquad (R^2 = 0.949)$$

$$CBD = 0.583 + 0.00874x + -0.0107y \qquad (R^2 = 0.873)$$

Where x = Roll Compaction Force and y = Roll Gap

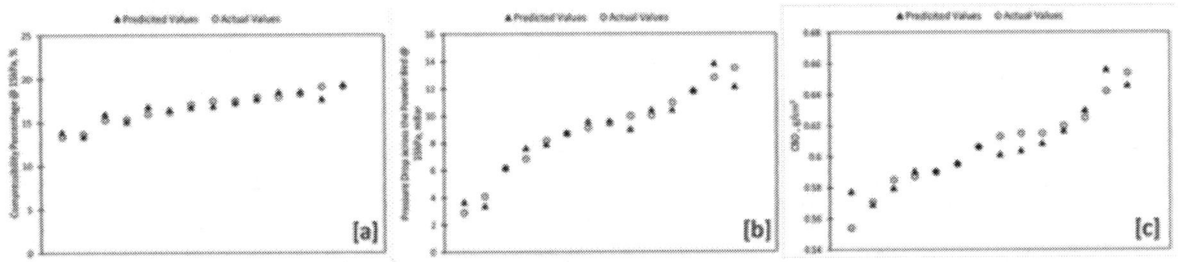

Figure 5: Comparison of experimental values with MLR predicted values for *(a) Permeability (b) Compressibility and (c) CBD.*

Discussion

There is clear correlation between roll compaction force and a number of rheological parameters of the resulting granulate. The strongest correlation is obtained for the bulk properties with good correlations also observed for the FRI and AE values. Whilst there were fewer correlations between roll gap and the rheological parameters, strong relationships were still evident with the bulk properties. This is not unsurprising as a larger roll gap compromises the uniform distribution of force on the powder to form high quality ribbon (and hence granules). However, the results show that the bulk properties of the granulate are still heavily influenced by roll gap.

The results clearly demonstrate how several rheological properties of the product granulate are predictably influenced by the processing parameters – particularly the Pressure Drop; Compressibility and CBD values; which all showed close correlation with several modes of operation of the roller compactor. This was confirmed by MLR analysis, which demonstrated high R^2 values as well as comparable predicted and experimental values for the bulk properties.

Overall the results suggest that a combination of smaller roll gap and higher roll compaction force is more likely to result in more uniform/consistent granules which form a more efficiently packed powder bed typically associated with free-flowing powders. These granulate properties have also been shown to be closely correlated with variance in die filling and tablet strength [2, 3].

Given that many previous studies have relied on shear behaviour to characterise product granulate, another interesting observation was the lack of correspondence between the manufacturing parameters and the shear properties of the granulates. For this particular roller compactor/feedstock combination, the limited differentiation provided by the measured Shear Stress values (rather than mathematically derived parameters) suggests that there is little relationship between shear properties of granulates and the roller compactor settings.

Conclusion

With respect to Quality by Design requirements, these results demonstrate that it is possible to control Critical Process Parameters in order to achieve granule properties - that have been shown to directly influence performance in downstream unit operations and Critical Quality Attributes of the final product [1, 3].

References

1. Freeman, T., A. Birkmire, and B. Armstrong, *A QbD Approach to Continuous Tablet Manufacture.* Procedia Engineering, 2015. **102**(0): p. 443-449.
2. Freeman, T., et al., *The Influence of Powder Flow Properties on the Filling of Dies and Capsules,* in *Particulate Systems Analysis 2011.* 2011: Edinburgh, UK.
3. Osorio, J.G. and F.J. Muzzio, *Effects of powder flow properties on capsule filling weight uniformity.* Drug Development and Industrial Pharmacy, 2013. **39**(9): p. 1464-75.
4. Wu, C.-Y., L. Dihoru, and A.C.F. Cocks, *The flow of powder into simple and stepped dies.* Powder Technology, 2003. **134**(1–2): p. 24-39.
5. ICH, *International Conference on Harmonisation of Technical Requirements for Registration of Pharmaceuticals for Human Use, Quality Risk Management, Q9,.* Federal Register, 2006. **71**(106): p. 32105-6.
6. ICH, *International Conference on Harmonisation of Technical Requirements for Registration of Pharmaceuticals for Human Use, Pharmaceutical Quality System, Q10.* Federal Register, 2009. **74**(66).
7. ICH, *International Conference on Harmonisation of Technical Requirements for Registration of Pharmaceuticals for Human Use, Pharmaceutical Development, Q8(R2).* Federal Register,, 2009. **71**(98).
8. Sun, C. and M.W. Himmelspach, *Reduced tabletability of roller compacted granules as a result of granule size enlargement.* Journal of Pharmaceutical Sciences, 2006. **95**(1): p. 200-206.
9. Herting, M.G. and P. Kleinebudde, *Studies on the reduction of tensile strength of tablets after roll compaction/dry granulation.* European Journal of Pharmaceutics and Biopharmaceutics, 2008. **70**(1): p. 372-379.
10. Fu, X., et al., *Effect of particle shape and size on flow properties of lactose powders.* Particuology, 2012. **10**(2): p. 203-208.
11. Bumiller, M., J. Carson, and J. Prescott, *A Preliminary Investigation Concerning the Effect of Particle Shape on a Powder's Flow Properties*, in *World Congress on Particle Technology 4.* 2002: Sydney, Australia.
12. Freeman, T., et al., *A PAT Approach to the Analysis and Characterisation of Powder Processing Systems*, in *Particulate Systems Analysis 2014.* 2014: Manchester, UK.
13. Nesarikar, V.V., et al., *Instrumented roll technology for the design space development of roller compaction process.* International Journal of Pharmaceutics, 2012. **426**(1–2): p. 116-131.
14. Freeman, R., *Measuring the flow properties of consolidated, conditioned and aerated powders — A comparative study using a powder rheometer and a rotational shear cell.* Powder Technology, 2007. **174**(1–2): p. 25-33.

457288 Semi-Empirical Drag Correlation for a Gas Solid Vortex Reactor Starting from the Radial Momentum Balances

Abstract:

Practical Information

All fields are compulsory

Last name	Friedle
First name	Maximilian
Email address	maximilian.friedle@ugent.be
Keywords (3)	Fluidization, Particle Technology
Abstract title	Semi-empirical drag correlation for a Gas Solid Vortex Reactor starting from the radial momentum balances
Authors	Friedle, M; Niyogi, K; Torregrosa Galindo, M.M.; Heynderickx, G.J., Marin, G.B.
Preferred presentation method	Oral presentation
Division/Forum	Particle Technology Forum
Session	03B06 Fundamentals of Fluidization I

Semi-empirical drag correlation for a Gas Solid Vortex Reactor starting from the radial momentum balances

Friedle, M; Niyogi, K; Torregrosa Galindo, M.M.; Heynderickx*, G.J., Marin, G.B.

Laboratory for Chemical Technology, Department of Chemical Engineering, Ghent University, Technologiepark 914, B-9052 Gent, Belgium

*Corresponding author: Geraldine.Heynderickx@UGent.be

The Gas Solid Vortex Reactor (GSVR) is a novel rotating fluidized bed reactor with a stationary geometry, in which the gravitational force in conventional fluidized bed is replaced by a centrifugal force resulting in process intensification[1]. In a GSVR a stable solids bed can be obtained for high gas flow rates thereby increasing the slip velocity between the phases. Consequently, the overall heat and mass transfer increases as to compared to a gravitational fluidized bed[2, 3]. The GSVR is a disc-like chamber where the process gas is introduced through a series of azimuthally inclined injection slots, uniformly distributed along the circumferential chamber wall. Inside the chamber the gas swirls towards the unidirectional central gas exhaust. When particles are fed inside this swirling flow field the gas transfers part of its momentum to the particles. The particles start to rotate in the chamber and form a stable, dense, rotating bed at the circumferential wall (Fig. 1). The rotational motion generates a radially outward directed centrifugal force on the particles. The injected gas flowing through the solid bed towards the unidirectional central gas exhaust generates a counteracting radially inward directed drag force on the particles. The GSVR compiles a unique set of characteristics typical for reactor technologies that combine high gas-solid slip velocities, good particle mixing and continuous operation under dense bed conditions[4]. As such, the GSVR technology has already been considered for different applications, like drying of biomass[5], biomass pyrolysis[1], SO_2-NO_x adsorption from flue gases[6] or nuclear rocket fuel propulsion[7].

As no universal understanding of the in-depth hydrodynamics of the GSVR technology is readily available, scaling and design of industrially-sized GSVRs are difficult up-to-date. An in-depth understanding of a technology often comes from a first principles approach. A first step in understanding of the multiphase hydrodynamics in the GSVR based on the radial momentum balances is aimed at in the present study.

An extensive set of experimental data has been collected by different researchers[2, 3, 8, 9] over the past years in an experimental semi-batch GSVR setup at the Laboratory for Chemical Technology. The investigated range of operations spans a wide range of gas flow rates, particle sizes, particle densities and bed masses. The experimental data is used to verify the assumptions made and to determine the model parameters of the semi-empirical drag formulation.

Following some simplifying assumptions, like neglecting the weight of the gas, the wall force and particle-particle interaction, the radial solids and gas momentum balances are combined to:

$$\frac{\Delta p}{h_{bed}} V_{bed} = m_{bed} \frac{v_{s,\theta}^2}{R} \tag{1}$$

The resulting one-dimensional, averaged equation shows that the pressure drop over the bed height balances the centrifugal force of the bed. Here h_{bed} describes the radial height of the bed, V_{bed} the overall volume of the bed, m_{bed} the mass of the bed and R the outer radius of the unit. The azimuthal solids velocity $v_{s,\theta}$ is averaged over the height of the bed.

Pressure drop over a conventional fluidized bed is usually estimated using semi-empirical correlations for the drag force[10, 11]. In the present work a semi-empirical correlation for the radial drag force on the GSVR bed is developed, starting from the well-known correlations for single-particle drag[12]. The resulting correlation is:

$$F_{Drag,r} = 282 \frac{(1-\varepsilon)}{d_p} \rho_g v_{g,r,sup}^{1.6} V_{bed} \tag{2}$$

Here ε is the void fraction, d_p the particle diameter, ρ_g is the gas density and $v_{g,r,sup}$ the radial superficial gas velocity in the unit. Equations 1 and 2 are combined and written in dimensionless form :

$$Ar_c = 282 Re_{rad}^{1.6} \tag{3}$$

The correlation parameters are obtained through linear regression of the available experimental data sets. The parity plot for equation 3 is shown in Figure 3. The radial Reynolds number and the centrifugal Archimedes number are calculated from equations 4 and 5, where μ_g is the gas viscosity.

$$Re_{rad} = \frac{v_{g,r,sup} \rho_g d_p}{\mu_g} \tag{4}$$

$$Ar_c = \frac{(\rho_g - \rho_s) \rho_g d_p^3 \frac{v_{s,\theta}^2}{R}}{\mu_g^2} \tag{5}$$

Using this modeling approach a more fundamental understanding of the interaction between gas and particles in a dense rotating bed is obtained. The absence of the void fraction as a classic modeling parameter, states that the gas interacts with every particle in the same way and that particle-particle interaction is minimal. In this respect the GSVR differs from conventional fluidization technologies operating in the dense bed regime, confirming the appropriateness of the GSVR for innovative industrial applications.

Acknowledgments

This work was supported by the European Research Council under the European Union's Seventh Framework Program (FP7/2007-2013) / ERC grant agreement n° 290793.

References

1. Ashcraft RW, Heynderickx GJ, Marin GB, 2012. Modeling fast biomass pyrolysis in a gas-solid vortex reactor. Chemical Engineering Journal 207, 195-208.

2. Ekatpure RP, Suryawanshi VU, Heynderickx GJ, de Broqueville A, Marin GB, 2011. Experimental investigation of a gas–solid rotating bed reactor with static geometry. Chemical Engineering and Processing: Process Intensification 50, 77-84.

3. Kovacevic JZ, Pantzali MN, Heynderickx GJ, Marin GB, 2014. Bed stability and maximum solids capacity in a Gas-Solid Vortex Reactor: Experimental study. Chemical Engineering Science 106, 293-303.

4. De Wilde J, 2014. Gas–solid fluidized beds in vortex chambers. Chemical Engineering and Processing: Process Intensification 85, 256-90.

5. Eliaers P, De Wilde J, 2013. Drying of Biomass Particles: Experimental Study and Comparison of the Performance of a Conventional Fluidized Bed and a Rotating Fluidized Bed in a Static Geometry. Drying Technology 31, 236-45.

6. Ashcraft RW, Kovacevic J, Heynderickx GJ, Marin GB, 2013. Assessment of a Gas-Solid Vortex Reactor for SO2/NOx Adsorption from Flue Gas. Ind Eng Chem Res 52, 861-75.

7. Lewellen W. A study of fluid dynamics of gaseous nuclear rocketsQuarterly Progress Report, July 1-September 30, 1968. Massachusetts Inst. of Tech., Cambridge, 1968.

8. Pantzali MN, Kovacevic JZ, Heynderickx GJ, Marin GB, 2015. Radial Pressure Profiles in a Cold-Flow Gas-Solid Vortex Reactor. AIChE Journal 61, 4114-25.

9. Kovacevic JZ, Pantzali MN, Niyogi K, Deen NG, Heynderickx GJ, Marin GB, 2015. Solids velocity fields in a cold-flow Gas–Solid Vortex Reactor. Chemical Engineering Science 123, 220-30.

10. Gibilaro LG, Di Felice R, Waldram SP, Foscolo PU, 1985. Generalized friction factor and drag coefficient correlations for fluid-particle interactions. Chemical Engineering Science 40, 1817-23.

11. Ergun S, 1952. Fluid flow through packed columns. Chem Eng Prog 48, 89-94.

12. Gidaspow D, 1994. Multiphase Flow and Fluidization. Academic Press, Boston.

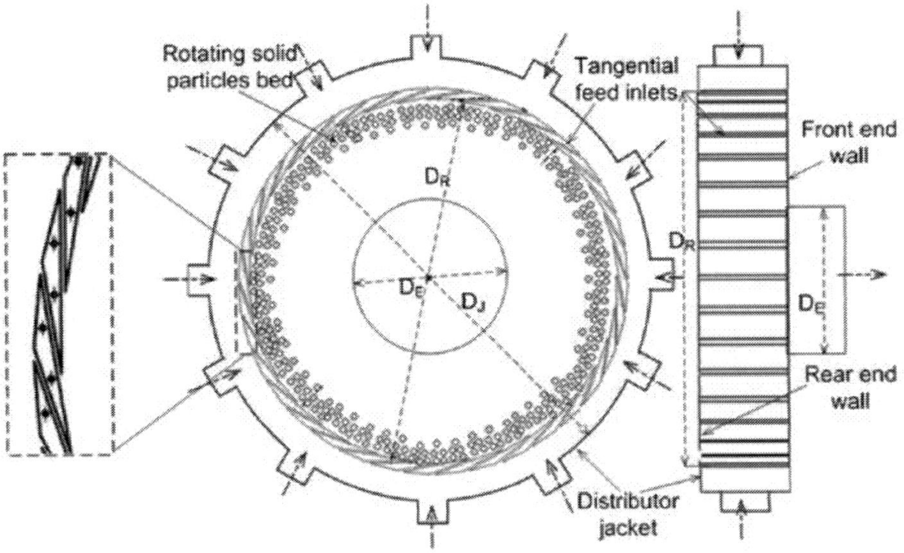

Fig. 1. A schematic representation of the Gas-Solid Vortex Reactor[2]

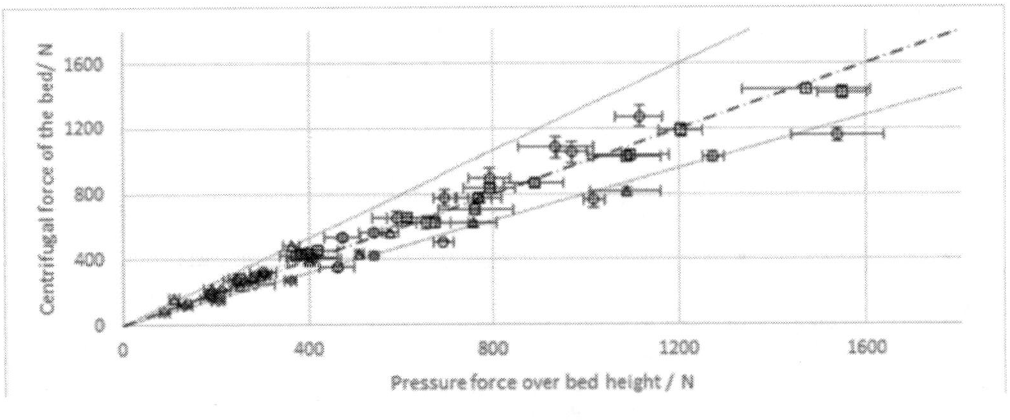

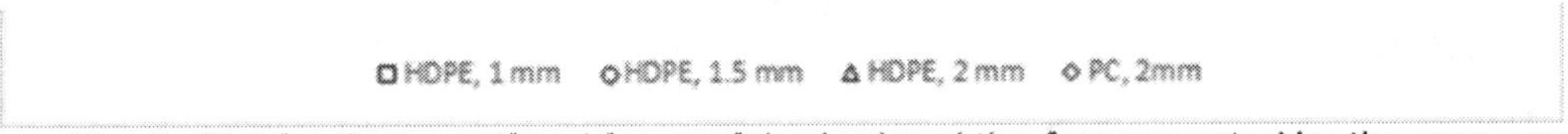

Fig. 2. Plot for the centrifugal force of the bed and the force exerted by the pressure on the bed (Equation 1).

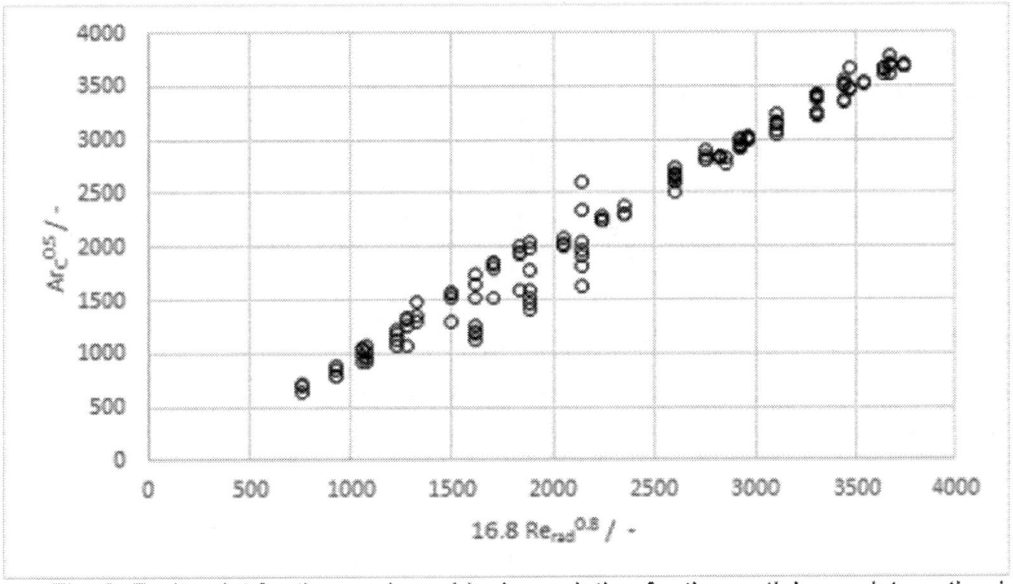

Fig. 3. Parity plot for the semi-empirical correlation for the particle-gas interaction in the GSVR (Equation 3).

460781 Computationally Generated Constitutive Models for Rheology of Frictional Non-Cohesive and Cohesive Particles in Gas-Solid Flows

Abstract:

Two-Fluid Model and Multi-Phase Particle-In-Cell simulations are commonly used to investigate gas-solid flow behaviors in fluidized beds. Both approaches require constitutive model for particle phase stress. The kinetic theory of granular materials [1-3] is widely used to close the particle phase stress in flowing assemblies of monodisperse non-cohesive frictionless particles. On the basis of Discrete Element Method (DEM) simulation results, *ad hoc* modifications to the kinetic theory [4-5] have been proposed to improve the accuracy in dense assemblies and to account for inter-particle friction. To account for inter-particle cohesion, several modifications to kinetic theory have been proposed [6-9].

In the present study, we perform CFD-DEM simulations for gas-fluidization of frictional, non-cohesive and cohesive particles in a periodic domain. By analyzing snapshots gathered from the simulation, quantities of interest in formulating a rheological model are determined. These results then guide refinement to the kinetic theory based stress model.

We first show that this approach, when applied to monodisperse non-cohesive particles, leads to results for pressure and shear viscosity that are consistent with the standard kinetic theory [1] coupled with the radial distribution function at contact from Chialvo & Sundaresan [4]. However, bulk viscosity is found to depend on whether the particle phase is in compression or dilation which is not accounted for in the kinetic theory.

For cohesive particles, it is found that inter-particle cohesion does not noticeably affect the particle phase stress except for at high solid volume fractions and low granular temperatures. This observation is consistent with the previous finding [10] based on simple shear simulations of cohesive particles.

This computational approach can be used to formulate simple rheological models for systems with size distribution, as well as particle-particle interactions due to liquid bridge or electrostatic interactions.

[1] C. K. K. Lun, S. B. Savage, D. J. Jeffrey, and N. Chepurniy, "Kinetic theories for granular flow: Inelastic particles in Couette flow and slightly inelastic particles in a general flowfield," *J. Fluid Mech.* **140**, 223–256 (1984).

[2] D. Gidaspow, *Multiphase Flow and Fluidization: Continuum and Kinetic Theory Descriptions.* San Diego, CA, USA: Academic Press (1994).

[3] V. Garzo and J. W. Dufty, "Dense fluid transport for inelastic hard spheres", *Phys. Rev. E* **59**, 5895–5911 (1999).

[4] S. Chialvo, S. Sundaresan, "A modified kinetic theory for frictional granular flows in dense and dilute regimes", *Phys. Fluids* **25**, 070603 (2013).

[5] D Berzi, D. Vescovi, "Different singularities in the functions of extended kinetic theory at the origin of the yield stress in granular flows", *Phys. Fluids* **27**, 013302 (2015).

[6] D. Gidaspow, L Huilin, "Equation of state and radial distribution functions of FCC particles in a CFB", *AIChE J.* **44,** 279-293 (1998).

[7] H. Kim, H. Arastoopour, "Extension of kinetic theory to cohesive particle flow", *Powder Technol.* **122**, 83-94 (2002).

[8] M. Ye, M.A. Van Der Hoef, J.A.M. Kuipers, "From discrete particle model to a continuous model of Geldart A particles", *Chem. Eng. Res. Des.* **83,** 833–843 (2005).

[9] B. Van Wachem, S. Sasic, "Derivation, simulation and validation of a cohesive particle flow CFD model", *AIChE J.*, **54,** 9–19 (2008).

[10] Y. Gu, S. Chialvo, and S. Sundaresan. "Rheology of cohesive granular materials across multiple dense-flow regimes", *Phys. Rev. E* **90**, 032206 (2014).

463146 Cluster Instabilities in Gas-Solid Flows Using Direct Numerical Simulation and Two-Fluid Model: Quantifying the Mean Sedimentation Velocity and Particle Velocity Fluctuations

Abstract:

The main purpose of this contribution is to present particle-resolved direct numerical simulations (DNS) and two-fluid modeling (TFM) of sedimentation of spherical particles subjected to cluster instabilities. DNS and TFM of freely evolving sedimenting suspensions in periodic domains were performed at two solid concentrations of 15% and 40% and two density ratios ρ_p/ρ_f = 100:1 and 1000:1. Particle Reynolds number based on the terminal velocity was set to 3 and 30. Both elastic collisions and inelastic collisions were studied. Evolution of cluster instabilities can be qualitatively divided into three stages: initial void and clusters randomly distribute, then dense and dilute phases grow and attract each other, leading to large and clearly divided dilute and dense phases the scales of which encompass the entire domain. In a previous study of a "one-dimensional" system ($L_y \approx 34\ d_p$, $L_x = L_z \approx 8\ d_{p\bar{i}}$), when visible clustering occurred it was typically in the form of a horizontally spanning, vertically traveling structure. In "two-dimensional" systems ($L_x = L_y \approx 34\ d_p$, $L_z \approx 8\ d_{p\bar{i}}$) studied here, clustering instability has additional degrees of freedom and additional patterns are observed. The particle clusters (dense regions) constantly swing in the horizontal direction, leading to more turbulent and transient flows than in the "1-D" case. Only at the highest mean flow Stokes numbers are horizontal bands observed which signal a significant reduction in the turbulence statistics.

Sedimentation velocities from DNS are compared with existing drag correlations [1-3]. DNS sedimentation velocities are significantly influenced by the clustering instability, and appear to decrease with increasing density ratio as a result of clustering. The influences of the parameters on the instability and the sedimentation velocity are reported. Velocity fluctuations in horizontal and vertical directions are largely driven by the motion of the clusters and depend significantly on the particle-fluid density ratio and the

mean flow Stokes number. Vertical fluctuations coming from particle clusters characterized by non-dimensional variance are low when Stokes numbers are medium and high. Under most situations, whether the collisions are elastic or inelastic (normal restitution coefficient = 0.9) did not generate significant difference in the mean sedimentation velocity and variance.

Sedimentation velocities and particle velocity fluctuations are also compared to results from continuum simulations using a recently developed kinetic-theory-based TFM [4]. In general, the KT-TFM results agree both qualitatively and quantitatively with the DNS data. Two significant deviations are observed, however. At intermediate mean flow Stokes numbers, the KT-TFM has a tendency to over-predict the degree of fluctuations, which, in turn, leads to over-prediction of the mean sedimenting velocity. It is believed that this behavior stems from an over-prediction in the degree of segregation and studies are currently underway to either verify or refute this hypothesis. At the lowest mean flow Stokes numbers the KT-TFM becomes essentially uniform, i.e., clustering vanishes. Consequently, the KT-TFM fails to capture the departure from homogeneous velocities (mean and fluctuating) observed in the DNS data. This breakdown has been determined to stem from a violation of one of the underlying assumptions of the KT derivation. In other words, it is not necessarily a failure of the model, but a failure to apply it within its range of validity. One of the most promising results of the comparison is the match in the anisotropy of the fluctuating particle velocity. The fluctuations in the streamwise (y-) dimension are always strongest, followed by the wide transverse (x-) dimension and finally the narrow transverse (z-) dimension, which is near the homogenous value. The KT-TFM is able to accurately capture these trends qualitatively and quantitatively where the theory is valid.

[1] J. F. Richardson, and W. N. Zaki. "Sedimentation and Fluidization: Part I," *Trans. Inst., Chem. Eng.*, 32:35-53 (1954).

[2] R.J. Hill, D.L. Koch, J.C. Ladd. "The first effects of fluid inertia on flows in ordered and random arrays of spheres", *J. Fluid Mech.*, 448:213–241 (2001).

[3] Tenneti S, Garg R, Hrenya CM, Fox RO, Subramaniam S. "Direct numerical simulation of gas-solid suspensions at moderate Reynolds number: Quantifying the coupling between hydrodynamic forces and particle velocity fluctuations", *Powder Tech.* 203:57-69 (2010).

[4] Garzó V, Tenneti S, Subramaniam S, Hrenya CM. "Enskog kinetic theory for monodisperse gas–solid flows", *J. Fluid Mech.* 712:129–168 (2012).

466208 Long-Term Transients in Fluidization of Oxide Nanoparticle Agglomerates

Abstract:

Fluidization of nanopowders obtained increasing interest from researchers in the past two decades. Interestingly, it was believed for a long time that nanopowders cannot be fluidized since they are classified as extreme-group-C, very cohesive powders, in the Geldart diagram. For these powders, the acting adhesion forces are too strong to allow fluidization. However, many studies showed that nanoparticles can be fluidized as micron-sized fractal agglomerates with very low densities and very high porosities. The high porosities of the agglomerates are very attractive because most of the particle surface is accessible for mass transfer and reaction.

The formation of the agglomerates in the fluidized bed is a dynamic process which includes events such as collision, unfolding, breaking, and reagglomeration. This makes it likely that the agglomerate size distribution will change over time. On the other hand, long-duration fluidization of nanopowders in the time scale of many minutes to hours is required for industrial applications such as the synthesis of carbon nanotubes, Atomic Layer Deposition (ALD), or photocatalysis. Therefore, a better understanding of, the influence of industrially relevant time scales on the properties of a fluidized bed is mandatory.

Here we present a detailed analysis of the agglomerate size distribution over time during long-time fluidization of oxide nanoparticles. A settling tube set-up was used to investigate the agglomerate size distributions and agglomerate density for multiple hours. We show that agglomerate sizes as well as the bed height decreases while the agglomerate density and the bed density increases. This was verified for different materials. Furthermore, we analysed the influence of the gas velocity and adhesion forces between the nanoparticles on the transient. We explain the long-term transient by two major effects. First: over the time stratification of the nanoparticle agglomerates occurs leading to the fact that larger agglomerates are more likely present in the lower part of the bed. Second: the agglomerate sizes change based on the dynamic events within the fluidized bed which is associated with the decrease in bed height.

Our experimental results clearly indicate that fluidization of nanopowders shows a long-term transient instead of steady state conditions which is highly relevant for industrial scale-up of nanoparticle fluidized-bed reactors.

466277 Energy Dissipation Trends in Tumbling Beds of Geldart A/C and A/B Powder Revealed By Speckle Visibility Spectrometry

Abstract:

The behaviour of powder in rotating drums is proving to be a useful tool for characterizing powder systems and providing insights to fundamental behaviour as well as having direct practical relevance to the process industries. In this investigation we have tracked the movement of avalanches using a drum fitted with a load cell sensor, while simultaneously measuring granular temperature in the powder bed using speckle visibility spectrometry, a laser technique that can measure time-resolved collisional motion of grains in dense systems with fine spatio-temporal resolution. We have taken measurements on two different model lactose powders, one close to the Geldart A/C boundary and the other lying close to the A/B boundary. Our results show high correlation between slumping and intermittent collisional dynamics in the bulk of the bed, particularly for the cohesive particles Geldart A/C particles; this correlation is weaker the A/B powder, apparently a result of fluidization in the bulk of the bed. The use of the SVS technique in combination with synchronous tracking of avalanche movement indicates avalanche magnitude and collisional dissipation as distinct energy dissipation mechanisms in the drum, with collisional dissipation dominant for the cohesive Geldart A/C powder and avalanche dissipation dominant for the Geldart A/B powder.

468242 Large-Scale DEM-CFD Simulation for a Pseudo 2-D Bubbling Fluidized Bed Containing Binary Mixture of Particles with Large Size Difference

Abstract:

In practical applications of fluidized beds, the size of solid paticles is not uniform and large particles coexist with small particles. The large size difference significantly influences the mixing and segregation phenomena in a bed. Despite of its engineering and physical importance, these behaviors are not fully understood. In the present study, large-scale DEM-CFD coupling simulation (Powder Tecnol., **77**(1), pp. 79-87, 1993) is performed for a pseudo-2D bubbling fluidized bed containing binary mixture of particles with large size difference. The momentum exchange between the larger particles and fluid is expressed by the fictitious particle method proposed by Tsuji et al. (AIChE J., **60**(5), pp. 1606-1620, 2014). The small particles are smaller than the computational cell size used in fluid calculation, whereas the large particles are larger than the cell size.

A snapshot of the simulation results is shown in Fig. 1 where twenty-five million small particles with a diameter of 1 mm and density of 2500 kg/m^3 are mixed with five hundred large particles which are 20 mm in diameter and 1600 kg/m^3 in density. In addition to the high concentration of the large particles near the side walls (Fig. 1a), falling motion of the large particles inside the bubbles can be observed. The large particles in the bubbles draw the surrounding small particles and induce characteristic structure formations inside the bubbles (Fig. 1b).

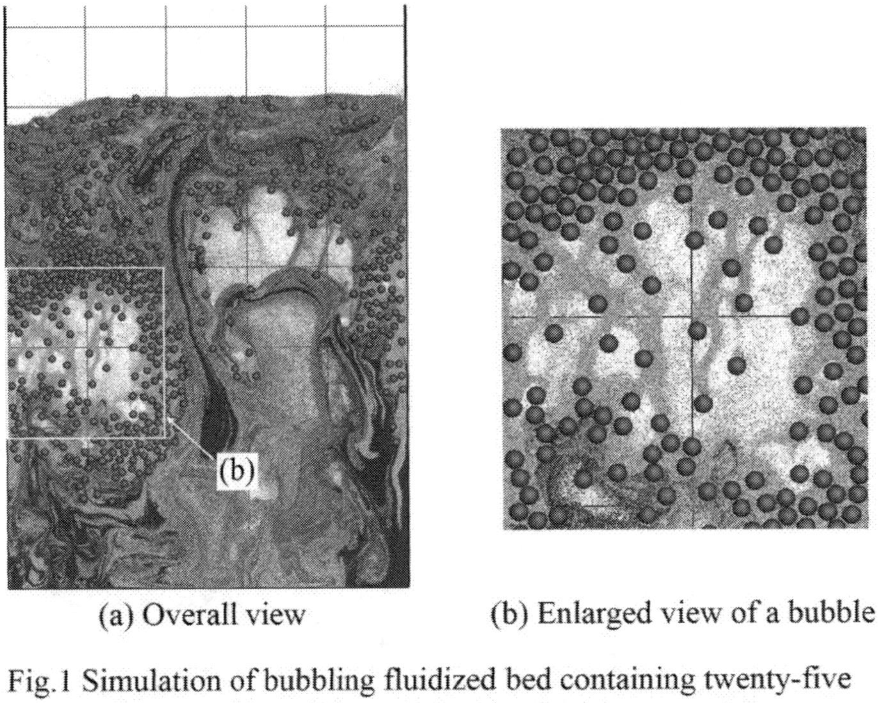

(a) Overall view (b) Enlarged view of a bubble

Fig.1 Simulation of bubbling fluidized bed containing twenty-five million small particles and five hundred large particles

469944 Eulerian-Lagrangian Simulations of Bidisperse Particles in Cluster-Induced Turbulence

Abstract:

Particle-laden turbulent flow is an important feature of many diverse environment and industrial systems such as clouds and fluidized bed reactors. Previous work has focused on an idealized, canonical flow, cluster-induced turbulence (CIT), wherein momentum coupling between a carrier fluid and setting particles leads to turbulent-like fluctuations in various quantities of interest. Two-way coupled Eulerian-Lagrangian simulations of homogeneous CIT with monodisperse particles successfully captured, at a statistically stationary state, large-scale particle clustering, sustained turbulence in the fluid phase state, and significant uncorrelated particle motion. Statistics from these simulations, and simulations of homogeneously sheared CIT (Ireland et al., 2015) and wall bounded CIT (Capecelatro et al., 2016), have been computed.

In this work, we present extended simulations of CIT with bidisperse particle sizes. As was previously done with monodisperse CIT (Capecelatro et al., 2014 and 2015), we track the flow of energy from its generation due to drag until its dissipation due to fluid viscosity and particle collisions. The energy is tracked separately for both particle types and the fluid. As suggested by Fox (2014), we separate the energy from correlated velocity fluctuations, denoted as particle turbulent kinetic energy, and energy from uncorrelated velocity fluctuations, denoted as granular temperature. An energy balance is computed for various energy exchange terms to determine their relative importance and to elucidate the underlying physical mechanisms in bidisperse CIT. Additionally, volume fraction and velocity statistics for both particle types and the fluid are presented. These results are compared to results from Capecelatro et al. (2014 and 2015), and the consequences on closures for Reynolds Averaged stress models of multiphase flows are discussed.

471921 CFD Modeling of Particle Segregation in Gas-Solid Systems

Abstract:

This work highlights ongoing efforts to develop and improve the modeling of particle segregation and mixing in industrial-scale fluidized systems. This is of interest in systems containing particles of different densities or in systems with high levels of particle fines. In both cases, a reasonable prediction of mixing and segregation effects may be necessary for obtaining a suitable CFD model of the fluidized system. Physically, the segregation and mixing of particles in a system is determined by the interplay between differences in drag, gravitational, and buoyancy forces which drive particle segregation and particle collisions and contacting which will restrict it. In CFD approaches that are suitable for industrial-scale modeling, such as Multiphase Particle-in-Cell (MP-PIC), the drag, gravity, and buoyancy forces are readily calculated. The particle collisions and contacting phenomena, however, are not so easily included and CFD modeling of particle segregation and mixing phenomena has historically been difficult as a result. In this work, a new model has been developed for improving the modeling of particle segregation and mixing in simulations with the MP-PIC approach. The development of this model will be presented along with validation against experimental segregation data for gas-solid fluidization.

473233 Fluidized Bed Rheometry Science, Practice and Industrial Applications

Abstract:

The rheological behavior of fluidized beds has been an object of study ever since their first technological use in the 1940´s.

Recently Anton Paar released the Powder Cell for MCR Rheometers , this device is capable of high performance measurements with large ranges of shear rate and shear stress, as well as oscillation measurements for the first time in gas-fluidized beds. Being able to measure extremely small torques (+- 10 nNm) in a reproducible manner enables a proper rheological study of fluidized beds in the style of complex fluids.

For the first time it is possible to directly correlate the airflow through particulate matter to their apparent viscosity (via both Bingham and Stokes assumptions) and therefore reliably measure this value with multiple geometries.

Oscillatory measurements show the gradual decline of elastic behaviour and emergence of a wholly viscous behaviour as soon as the pseudofluid (fluidized bed) forms fully.

While exhibiting some wholly unique effects, the shear rate dependent behaviour of a gas fluidized bed can be for the most part be likened to that of particulate suspension. Effects such as solid fraction dependence, jamming are found and demonstrated in fluidized beds as well.

This provides new avenues for the development/improvement for a multitude of industrial processes.

469503 Electrophoretic Deposition of Nanocrystals Under Flow to Synthesize Semiconducting Thin Films with High Atom Economy

Abstract:

Electrophoretic deposition (EPD) of colloidal nanocrystals (NCs) under flow is explored as a general method for the fabrication of semiconducting thin films that is amenable to high throughput, low cost operation. EPD is readily employed for depositing ceramic or metallic coatings, which typically require hundreds of volts for deposition to occur. However, for photovoltaic applications, a low process voltage is critical to avoid damaging the accreting semiconductor. Here we report a continuous flow reactor design that can operate at reduced voltage compared to a traditional batch reactor. This design preserves the required high electrophoretic velocity of the NCs at low process voltages by utilizing narrow electrode spacing. In a batch reactor, the low volume-to-surface-area ratio dictated by such a narrow spacing of the electrodes would impose a limit on the mass of nanocrystals that are resident in the reactor and therefore the thickness of the films that can be deposited. This limitation is removed by continuously flowing the colloidal dispersion of NCs and thus thick films are deposited. Through modeling and experiment we elucidate the process parameters that control the uniformity of the resulting coating and the utilization of the NCs in the feed solution. The reactor design is compatible with large area substrates and is specifically designed to enable continuous, high-rate fabrication of the active layer of photovoltaic cells with high atom economy.

461291 Influence of Surface Reactions on the Infrared Localized Surface Plasmon Resonance of Indium Tin Oxide Nanocrystals

Abstract:

Heavily doped oxide nanocrystals exhibit a tunable plasmonic response in the infrared, a capability that is promising for future nanoscale photonic technologies. Nanocrystal carrier density, and thus spectral response, is adjustable via chemical reaction; however, the fundamental processes governing this behavior are poorly understood. Here, we study the oxidation and reduction of indium tin oxide (ITO) nanocrystals with O_2 and H_2, respectively, with in situ diffuse reflectance infrared Fourier transform spectroscopy (DRIFTS). We show that the main absorption feature redshifts upon oxidation more than 1000 cm^{-1} and blueshifts to its original position upon reduction. The same magnitude spectral shift is observed over many cycles of oxidation and reduction. A kinetic model, which includes surface reaction and bulk diffusion, allows us to quantitatively rationalize the process, revealing that surface reaction is rate limiting under the conditions studied. Our experiments provide a deeper understanding of the connection between surface reaction and carrier density in oxide nanocrystals, and open the door to *a priori* control of plasmonic response.

462988 Large-Scale Crack-Free Gold Nanoparticle Monolayers with Adjustable Electrical and Optical Properties

Abstract:

Gold nanoparticles are widely studied due to the ease of controlled synthesis, facile surface modification, and interesting physical properties. However, a technique for depositing large-area, crack-free monolayers on solid substrates is lacking. Herein we present a method for accomplishing this. Spherical gold nanoparticles were synthesized as an aqueous dispersion. Assembly into monolayers and ligand exchange occurred simultaneously at an organic/aqueous interface. Then the monolayer film was deposited onto arbitrary solid substrates by slowly pumping out the lower, aqueous phase. This allowed the monolayer film (and liquid-liquid interface) to descend without significant disturbance, eventually reaching substrates contained in the aqueous phase. The resulting macroscopic quality of the films was found to be superior to films transferred by Langmuir techniques. In addition, the interparticle spacing was adjustable from 1.4 nm to 3.1 nm using different length alkylamine ligands. Both the surface plasmon resonance (SPR) and the electrical conductivity of the films show a strong dependence on particle spacing. This approach allows the precise control of interparticle spacing over large-area NP films. The demonstrated self-assembly of the 2D NP superlattice could have an impact on the manufacture of high performance optical and electronic devices.

473585 *In Situ* Characterization of the Effect of Plasma Treatment on Transition Metal Dichalcogenides

Abstract:

Transition metal dichalcogenides (TMDCs) are layered materials with unique material properties that make them attractive choices for electronic and optoelectronic devices. Although plasma processing is commonly used in the fabrication of these devices, the inherent consequences of this treatment have not been fully explored. In this work, *in situ* X-ray photoelectron spectroscopy (XPS) is used to investigate the effect of mild plasma treatment on the chemical composition and material properties of MoS_2 and WeS_2. These results shed light on recent advances in the device performance of metal oxide semiconductor field effect transistors (MOSFETs), demonstrating how defect engineering can be exploited to improve electronic properties for next-generation devices.

461262 Observation of Switchable Photoresponse of a Monolayer WSe_2-MoS_2 Lateral Heterostructure *Via* Photocurrent Spectral Atomic Force Microscopic Imaging

Abstract:

In the pursuit of two-dimensional (2D) materials beyond graphene, enormous advances have been made in exploring the exciting and useful properties of transition metal dichalcogenides (TMDCs)—such as a permanent band gap in the visible range and the transition from indirect to direct band gap due to 2D quantum confinement—and their potential for a wide range of device applications. In particular, recent success in the synthesis of seamless monolayer lateral heterostructures of different TMDCs *via* chemical vapor deposition (CVD) methods has provided an effective solution to producing an in-plane p-n junction, which is a critical component in electronic and optoelectronic device applications. However, spatial variation of the electronic and optoelectonic properties of the synthesized heterojunction crystals—throughout the homogeneous as well as the lateral junction region—and the charge carrier transport behavior at their nanoscale junctions with metals remain unaddressed. In this work, we use photocurrent spectral atomic force microscopy (PCS-AFM) to image the current and photocurrent generated between a biased PtIr tip and a monolayer WSe_2-MoS_2 lateral heterostructure. Current measurements in the dark in both forward and reverse bias reveal an opposite characteristic diode behavior for WSe_2 and MoS_2, owing to the formation of a Schottky barrier of dissimilar properties. Notably, by changing the polarity and magnitude of the tip voltage applied, pixels that show the photoresponse of the heterostructure are observed to be selectively switched on and off, allowing for the realization of a hyper-resolution array of the switchable photodiode pixels. This experimental approach has significant implications toward the development of novel optoelectronic technologies for regioselective photodetection and imaging at nanoscale resolutions. Comparative 2D Fourier analysis of physical height and current images shows high spatial frequency variations in substrate/MoS_2 (or WSe_2) contact that exceed the frequencies imposed by the underlying substrates. These results should provide important insights in the design and understanding of electronic and optoelectronic devices based on quantum confined atomically thin 2D lateral heterostructures.

458340 Effect of Inorganic Nanoparticle Coating on the Optical Transmittance and the Surface Conductivity for Transparent Conductive Carbon Nanotube /Silver Nanowire Hybrid Film

Abstract:

For the application of the touch screen panel and transparent conductive electrode for the flexible display, highly transparent and electrically conductive film is required. To enhance the optical properties and the surface conductivity of carbon nanotube (CNT)/silver nanowire/polyethylene terephthalate (PET) hybrid film, inorganic overcoating has been applied by roll-to-roll coating processing. The maximum total transmittance of the hybrid film without washing and overcoating was less than 90%. The optimum thickness of the overcoating has been determine as 10 micro meter for the highest total transmittance, 92 %, while the sheet resistance of the hybrid film is 180 ohm/sq. For the further enhancement of the sheet resistance of the CNT/silver nanowire hybrid film to get less than 80 ohm/sq the thickness of the overcoating has been controlled as 30 micro meter, while the transmittance is more than 90%. The overcoating of nanoparticle can highly reduce the surface roughness of the CNT/silver nanowire hybrid film and enhance the optical clarity such as the total transmittance and the haze of the hybrid film. This work has been done using large scale roll-to-roll coating machine for the scalable product of the high performance flexible transparent conductive film.

464360 Electrostatic Double Layer Flash Memory Based on Two-Dimensional Crystals

Abstract:

A new type of flash memory will be presented based on the electrostatic doping of two-dimensional crystals using ions. The proposed device consists of two, 2D crystals separated by a 2D electrolyte through which the ions can pass. The top 2D crystal comprises the channel of a field-effect transistor (FET), while bottom 2D crystal is a backgate. When the ions are near the surface of the channel, they induce image charge in the channel and the device is the low resistance, or ON state. When the ions are pulled back to the backgate by an applied field, the device is in the high resistance or OFF state. The 2D electrolyte is cobalt crown ether phthalocyanine (CoCrPc) plus a salt, which can be deposited on the surface of 2D crystals simply by drop casting and annealing. The crown ethers solvate metal ions, and the ions can pass through the cavity of the crowns. Density functional theory (DFT) calculations show that the crown ethers present a small barrier to ion transport required for fast (nanosecond) switching, but the height of the barrier will be increased for long retention both by the image charge induced in the channel and by modulating the gate bias. The current-voltage characteristics of a simplified device architecture (monolayer CoCrPc:$LiClO_4$ on graphene) will be presented. The graphene FET can be reconfigureably programmed by the 2D electrolyte, achieving sheet carrier densities of 4×10^{12} cm^{-2} at a low lithium concentration of 20 crown ethers to 1 Li^+. Based on the geometric packing of the molecules, as determined by scanning tunneling microscopy, the doping density is predicted to increase to 5×10^{13} cm^{-2} at a crown ether to lithium ratio of 1:1. State retention measurements show that the two states can be retained for at least 30 minutes (maximum time measured to date) with a memory window of 10 uA. In future work, the graphene will be replaced with MoS_2 because both the doping density and memory window are predicted to increase.

This work was supported in part by the Center for Low Energy Systems Technology (LEAST), one of six SRC

STARnet Centers, sponsored by MARCO and DARPA, and NSF grant #ECCS-GOALI-1408425.

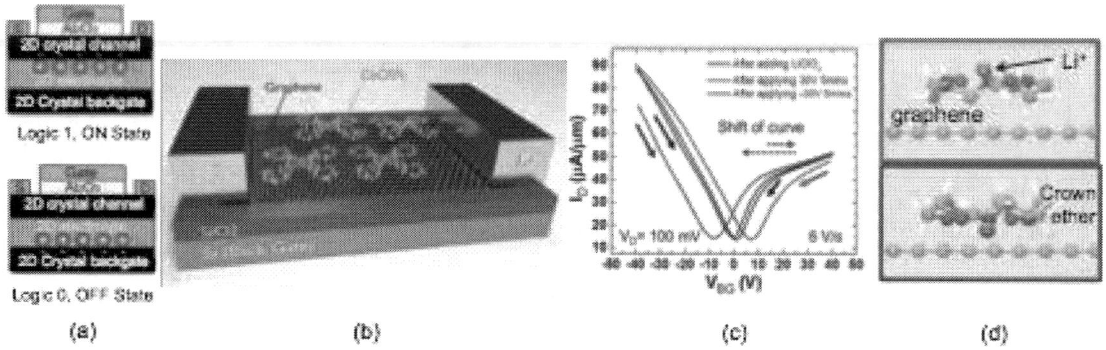

Figure: Electrostatic double layer 2D crystal memory. (a) Cross section of proposed deice; ions induce image charge in the channel in the ON state, and are pulled back to the backgate in the OFF state. (b) Schematic of simplified device; 2D electrolyte and $LiClO_4$ on graphene FET (c) Transfer characteristics of simplified device at varying programming voltages; the shift in the Dirac point indicates reconfigureable doping by the 2D electrolyte corresponding to the two positions shown in (d).

459647 Multiscale Approaches for Modeling the Penetration of Field Effect in Two-Dimensional-Materials-Based Quantum Capacitors

Abstract:

Gate-tunable two-dimensional (2D) materials-based quantum capacitors (QCs) and van der Waals heterostructures involves tuning transport or optoelectronic characteristics by the field effect. Recent studies have attributed the observed gate-tunable characteristics to the change of Fermi energy in the first 2D layer adjacent to the dielectrics, while the penetration of the field effect through the one-molecule-thick material is often ignored or over-simplified. We present a multiscale analysis to model penetration of the field effect through graphene in a metal-oxide-graphene-semiconductor (MOGS) QC, including quantifying the degree of â€œtransparencyâ€ for graphene two-dimensional electron gas (2DEG) to an electric displacement field. We find that the space charge density in the semiconductor layer can be modulated by gating in a nonlinear manner, forming an accumulation or inversion layer at the semiconductor / graphene interface. The degree of transparency is determined by the combined effect of graphene quantum capacitance and the semiconductor capacitance, which allows us to predict the ranking for a variety of monolayer 2D materials according to their transparency to an electric displacement field as follows: graphene > silicene > germanene > WS_2 > WTe_2 > WSe_2 > MoS_2 > $MoSe_2$ > $MoTe_2$, when the majority carriers are electrons. Our findings reveal a general picture of operation modes and design rules for the 2D-materials-based QCs.

464902 Current-Driven Nanowire Formation and Nanopatterning on Crystalline Conducting Substrate Surfaces

Abstract:

The ability to form and precisely manipulate nanoscale features by controlling macroscopic forces is essential to advancing nanotechnology. For example, a macroscopic force in the form of an externally applied electric field can be used to drive atomic motion through the mass transport phenomenon of electromigration providing a paradigm shift in surface engineering and nanofabrication based on physical patterning.

Toward this end, in this presentation, we report a simulation study on formation of nanowires with precisely controlled widths, starting from single-layer conducting islands on face-centered cubic (fcc) crystalline conducting substrates under the controlled action of macroscopic forcing provided by an externally applied electric field that drives island edge electromigration. Our continuum-scale, fully nonlinear driven island evolution model with diffusional mass transport limited to the island edge accounts for edge diffusional anisotropy and island coalescence and breakup; the model is validated by comparison of its predictions with experimental results for Ag islands on Ag substrates reported in the literature. Numerical simulations based on this experimentally validated model and supported by linear stability theory show that large-size islands undergo a current-induced fingering instability, leading to nanowire formation after finger growth. Depending on the substrate surface crystallographic orientation, necking instabilities after fingering lead to the formation of multiple parallel nanowires per island. In all cases, the axis of the formed nanowires is aligned with the direction of the externally applied electric field. The nanowires have constant widths, on the order of 10 nm, which can be tuned by controlling the externally applied electric field strength. Achieving parallel arrays of ~10-nm-wide nanowires, dimensions barely accessible even by e-beam lithography, and demonstrating the ability to control their direction has important implications for developing future nanoelectronic patterning techniques applicable, e.g., to the interconnect structures of future electronic chips.

In addition, we demonstrate based on direct dynamical simulations that under specific conditions, and over a range of edge diffusional anisotropy parameters and nanowire edge orientation, the nanowire straight edge becomes unstable under the action of the externally applied electric field. This current-induced nanowire edge morphological instability leads to breakup of the nanowire into smaller-size daughter islands of practically uniform size and spacing to form regular patterns of nanoscale-size daughter islands. We have conducted a comprehensive parametric study to provide guidelines for generating such regular nanopatterns, which can be classified broadly into two categories depending on the geometrical orientation of the daughter islands in the regular array: aligned along a straight line or aligned along two intersecting straight lines to form a 'V'-shape. Moreover, using a linear stability theory we explain the origin of this nanowire straight edge instability; the theory is successfully validated by comparing its predictions for the number of daughter islands formed and the characteristic spacing between the islands with the results of the dynamical simulations. Our findings have direct implications for the development of novel external-field-driven physical nanopatterning techniques toward device nanofabrication.

467005 Combined Experimental and Modeling Approach to Triboelectric Charging of Polyethylene Powders

Abstract:

The charging of objects caused by their mutual friction is called triboelectric charging or triboelectrification. Triboelectrification accompanies not only our daily activities like brushing hair or walking in shoes, but also several industrial processes. And because the charging is a surface-related phenomenon, it is most apparent in systems with high specific surface like powders (polymers, pigments, flour, drugs). For example, in fluidized-bed polymerization reactor, the polymer particles collide with other particles and with the device wall, thereby generating the excess electrostatic charge on their surfaces. Charged particles, in general, are attracted to the metal walls of devices due to electrostatic induction. Moreover, oppositely charged particles attract each other. Such behavior contributes to the unwanted particle agglomeration and fouling, which can impair the production and manufacturing of polymers. Although triboelectric charging has been documented for centuries, its mechanism hasn't been fully explained yet. Thus we combined the cascade method apparatus (that is a slide followed by the Faraday's pale) with Discrete Element Method (DEM) based model of colliding particles, in order to study the charging of polyethylene (PE) powder particles. Using the apparatus, we measured not only the saturation charge of PE particles, but also the charging dynamics as a function of industrially relevant parameters (temperature, air humidity, material of the slide, presence of sorbent). The charging was less pronounced at low humidities and low temperatures. Moreover, we observed significant charging even for the contact of 'identical' materials (PE particle with PE slide). Our DEM model of particle-wall contact charging was complemented by the capacitor model of charging. Another model was used for the prediction of particle-particle contact charging, in which Surface States theory was implemented. The model predicted fast charging for systems of a wide particle size distribution and/or spatial segregation of 'small' and 'big' particles. Such a result correlates well with the observations of charging in natural and industrial granular systems. Moreover, the model predicted more pronounced charging at high temperatures, which correlates well with our measurements. Combining particle charging and elasticity measurements with modeling, we further address the question of whether

the charging depends only on the contact area of colliding objects (which correlates with the spontaneous release of electrons during collision) or also on the amount of dissipated mechanical energy during the contact (which correlates with the excitation energy necessary for the electron transfer). Our study thus contributes to a better understanding of triboelectric charging fundaments and, at the same time, may provide valuable information for polyolefin industry

461607 The Influence of Surface Chemistry on the Electrostatic Properties of Particles

Abstract:

Background

Electrostatic phenomena are frequently observed whenever powder comes in contact with different surfaces, particularly when the motion of particles is involved. The resulting charge accumulation and its potential prolonged retention become a serious concern when insulating materials, such as pharmaceutical powders, are involved in the process. Due to a large number of factors influencing the process of charge generation, quantification and prediction of electrostatic properties is still one of the greatest challenges in this field.

Aims

The aim of this work is to assess the influence of surface chemistry on the ability of a material to acquire and subsequently dissipate electrostatic charge. The main hypothesis is that surface functional groups can strongly dictate the overall charging behaviour of the material and therefore could provide some indication of materials' electrostatic properties.

Methods

The surface of glass beads, a model material in this study, was modified using a silanisation reaction to produce chemically functionalised samples. Contact angle measurements were used to confirm the extent of surface modification of the samples. Electrostatic properties were studied using a purpose build instrument, which incorporates a capacitive probe[1] within a Dynamic Vapour Sorption instrument. The modified glass beads were tribocharged by contact with a stainless steel surface and their maximum potential as well as decay curves were measured at 25 °C at three different humidity levels, i.e. 20% RH, 50% RH and 75% RH to evaluate the impact of environmental conditions on charge dissipation.

The influence of surface hydrophobicity on the electrostatic behaviour was further investigated for mannitol crystals. Mannitol was recrystallized from water at four supersaturation levels (σ = 3%, 13%, 23% and 33%) to produce needle-shaped crystals with varied aspect ratio and hence different overall hydrophobic character. Strong correlation between crystal shape and crystal surface chemistry for mannitol has been determined previously.[2] The electrostatic properties of the crystals were determined using the capacitive probe method and compared with industrially available instrument called Charge Decay Time Analyser.

Results

Successful surface modification of glass beads produced four samples with distinct surface chemistries, which exhibited differing degrees of surface hydrophobicity. Electrostatic measurements showed that the magnitude of charging followed the trend in hydrophobicity with the most hydrophobic particles acquiring the largest charge. Up to six times larger charge was accumulated by fluoro functionalised glass beads, acquiring charge of -11.2 ± 2.8 V/mg, compared to hydroxyl functionalised sample with overall charge of -1.9 ± 0.3 V/mg. In addition, the unique properties of heteroatoms, such as electronegativity, seem to have further impact on the overall charge adopted as observed by significant charging of the fluoro functionalised sample and charge inversion in case of amino functionalised surface. For all surfaces, increasing the external humidity resulted in faster charge dissipation; however, the actual effect of humidity on charge decay rate seems to be surface dependent. Under insulated charge conditions, increase in humidity from 20% RH to 75% RH resulted in lowering charge half-life time from 21.4 min to 4.9 min for hydrophilic sample and from 45.1 min to 13.3 min for amino functionalised surface.

Crystallisation of mannitol from solutions under varied conditions produced a range of crystals – from thin, long needle crystals at low supersaturation to short wide plate-like crystals at high supersaturation. These showed to have different surface energy properties and hence different electrostatic characteristics.

The data generated was repeatable and reproducible indicating that by controlling particle size, shape and chemistry the variation in electrostatic charge can be limited, but also shows that the capacitive probe setup used for electrostatic measurements is reliable and capable of detecting electrostatic charge of powders.

Conclusion

The results obtained show that surface chemistry and hence surface hydrophobicity play an important role in powder electrostatics. It was shown that materials having different chemical groups on the surface can exhibit different electrostatic behaviour. This includes not only the overall charge acquired under the same conditions, but also charge polarity and charge decay characteristics. Such observations could be beneficial in predicting a material's propensity towards charge accumulation during processing based on its chemical structure and molecular arrangement within a crystal lattice.

Acknowledgments

The authors acknowledge Dr Daryl Williams and Dr Jin Wang Kwek for their comments and useful discussions.

References

(1) Kwek, J. W.; Jeyabalasingam, M.; Ng, W. K.; Heng, J. Y. Y.; Tan, R. B. H., Comparative Study of the Triboelectric Charging Behavior of Powders Using a Nonintrusive Approach. *Ind. Eng. Chem. Res.* **2012,** *51* (50), 16488-16494.

(2) Ho, R.; Wilson, D. A.; Heng, J. Y. Y., Crystal Habits and the Variation in Surface Energy Heterogeneity. *Crystal Growth & Design* **2009,** *9* (11), 4907-4911.

470762 Catastrophic Failure of Agitated Filter-Driers Due to Triboelectric Charging of Certain APIs

Abstract:

Agitated filter driers, typically consisting of a dished vessel with a perforated plate, and jacketed walls, bottom dish, and impeller, are commonly used in the process of solid API isolation. This work will present the investigation of the physical processes underlying a catastrophic failure during the drying of a certain API at the pilot plant scale. Prior to the failure, the powder resistance to flow would increase dynamically during the course of the agitation resulting in an increase in the agitator torque (as measured by the hydraulic pressure required to maintain the rotational speed set point). Eventually, this increase in flow resistance would lead to a solid body-like rotation of the powder bed, translating the agitator's mechanical energy, through the non-slip contact at the dryer bottom, into shear on the filter cloth sufficient to tear it out of its retaining o-ring.

Additional experimentation at scale showed that this dynamic increase in flow resistance was repeatable across multiple batches, and was also evident in other APIs and intermediates. Further investigation identified tribo-electric charging as the underlying cause for this behavior. Characterization of the dependence of the phenomenon on the process parameters and cake properties allowed for the development of a procedure for the early detection of potential "problem" compounds at the lab-scale.

COMPARISON OF MAGNETICALLY ASSISTED IMPACTION COATING (MAIC) WITH TRADITIONAL MIXING TECHNIQUES FOR THE ADDITION OF A SILICA FLOW AID

Tim Freeman[1], William A. Hendrickson[2], Charles R. Bowman[2], Christopher J. Rueb[2], Robert G. Bowman[2], Katrina Brockbank[1], Jamie Clayton[1]

[1]Freeman Technology, Tewkesbury, Gloucestershire, UK
[2]AVEKA Inc., Woodbury, Minnesota, USA

Keywords*: Solids Flow, Handling, and Processing, Mixing, Process Design & Development

Introduction

Many powders have poor basic flow properties, they block in hoppers and die feed frames, exhibit inconsistent or pulsatile discharge rates, adhere to equipment surfaces or do not mix readily with other materials. Additionally, many powders tend to cake on storage or during transport due to changes in environmental conditions and/or consolidation. As such, flow aids are frequently added to a blend to enhance flow properties and/or inhibit caking.

In order to optimise the influence of a flow aid, it must be dispersed in a fine, uniform layer on the surface of the substrate particles. However, most flow aids, for example fumed silica or magnesium stearate, consist of very fine particles that have a tendency to form agglomerates. The mixing/coating process employed can therefore have a major influence on how efficiently the flow aid is distributed. The process must be intense enough to disperse the agglomerated flow aid without damaging the particles of the substrate or the flow aid itself. This is particularly true for many food and pharmaceutical ingredients which are often relatively friable and easily deformed by extreme mechanical forces [1].

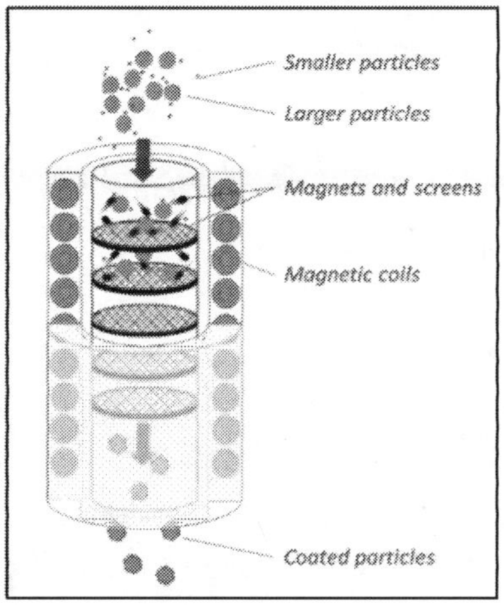

Figure 1 - Magnetically Assisted Impaction Coating (MAIC) process

Magnetically Assisted Impaction Coating (MAIC) has been demonstrated as a technique for coating host particles without causing major changes to particle shape and size [1,2]. This system relies on a pounding process in which an oscillating magnetic field is used to accelerate and spin large magnetic particles (600-800 µm, though size may vary with application), causing the magnetic particles to fluidise, through which the host and guest particles (flow aid) are passed in a continuous fashion (fig 1). The agitated magnetic particles impart energy to the host and guest particles, and coating is then achieved by means of impaction or peening of the guest particles onto the host particles as collisions occur between particles and between particles and the vessel wall [1].

Materials and Methods

Jet milled citric acid (Hawkins Chemicals, jet milled by *AVEKA*) was pre-blended with 0.05%, 0.5%, and 2.0% (w/w) fumed silica (Aerosil 200 Pharma, Evonik) using a V-blender with intensifier bar for 5 minutes. The blends were then fed through the MAIC process (US Patent 5962082 [2]). The samples were run through the MAIC (*AVEKA*, USA) in increasing mass percent of silica, so as to minimize the impact of potential contamination of subsequent samples. Prior to this, citric acid was passed through the MAIC without any treatment with fumed silica, and without being placed in the V-blender. In addition to the MAIC-treated samples, two v-blended samples not subjected to MAIC were created, with 0.5% and 2.0% (w/w) fumed silica. The untreated citric acid made for the seventh and last of the evaluated samples.

After being randomized, the samples were labeled without any indication of processing method (V-blend only or MAIC-treated) or silica loading, which facilitated a blind study of the flow characteristics. All packaged samples were then shipped from Minnesota, USA by air to Tewkesbury, UK. Each sample was then evaluated for flow on an FT4 Powder Rheometer (Freeman Technology, UK) before the identity of each sample was revealed and matched to the relevant data.

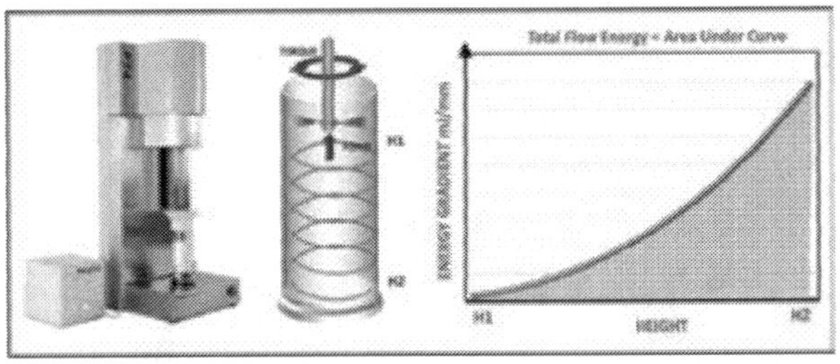

Figure 2 – Measurement of the flow energy using the FT4 Powder Rheometer

The FT4 measurements included dynamic flow, bulk and shear properties. Dynamic properties, including Basic Flowability Energy (BFE) (fig. 2), were obtained by means of a patented measurement principle that evaluates the resistance to the motion of a specially shaped twisted blade passing through a precise volume of the sample along a prescribed path. The required torque and force are recorded and converted into a flow energy [3]. The repeatability of all measurements was enhanced by the use of a conditioning cycle which removes packing history and operator-induced variability.

Results and Discussion

This study compared MAIC with V-blending through comparison of the flow properties of the substrate after the addition of silica. In addition, this study also investigated the effect of silica concentration on samples blended via MAIC.

MAIC vs. V-Blending

The MAIC blends at both silica concentrations (0.5% and 2.0% (w/w)) presented lower BFE, Aerated Energy (AE), Cohesion, Consolidation Index (CI) and Pressure Drop values indicating improved flow properties compared with the V-blended samples (fig. 3).

BFE measures the resistance to flow in a dynamic, constrained forced flow environment, for example feeding via a screw conveyer. Whilst all the silica blends exhibited improved flow compared to the raw material, the MAIC blends exhibited the largest decrease in flow energy, suggesting the greatest improvement in flow. This is due, in part, to the fact that MAIC creates a more uniform coating with the flow aid than can be achieved by V-blending alone [4].

The silica blends also exhibited a decrease in Specific Energy (SE) compared to the raw material suggesting reduced mechanical interlocking and friction. However, there were limited differences between the MAIC and V-blended samples at 0.50% (w/w) silica; although at the higher load of 2.0% (w/w) silica, the MAIC blend presented a greater improvement in flow compared with the V-blended sample.

The lower AE values presented by the MAIC samples not only indicate that they were more responsive to the introduction of air, but also demonstrated a clear reduction in the absolute cohesion compared with the V-blended samples. This is also indicated by the lower cohesion values obtained during the shear tests suggesting that the MAIC samples were also less resistant to flow at higher (applied) loads. This behaviour has been previously observed on a batch MAIC process, and used to improve the ability of a powder to fluidize [5].

Powders can also undergo compaction when exposed to external vibrational forces, for example during transport or due to proximity to (other) process equipment, resulting in reduced flowability. The V-blended samples both exhibited much larger increases in flow energy with notably higher Consolidation Index (CI) values following tapping, demonstrating that these samples were much more sensitive to vibrational forces and indicating reduced flow when exposed to external forces. Overall these results suggest that the MAIC blends exhibited improved flow properties at low, moderate and high stress regimes, and may be more stable during storage.

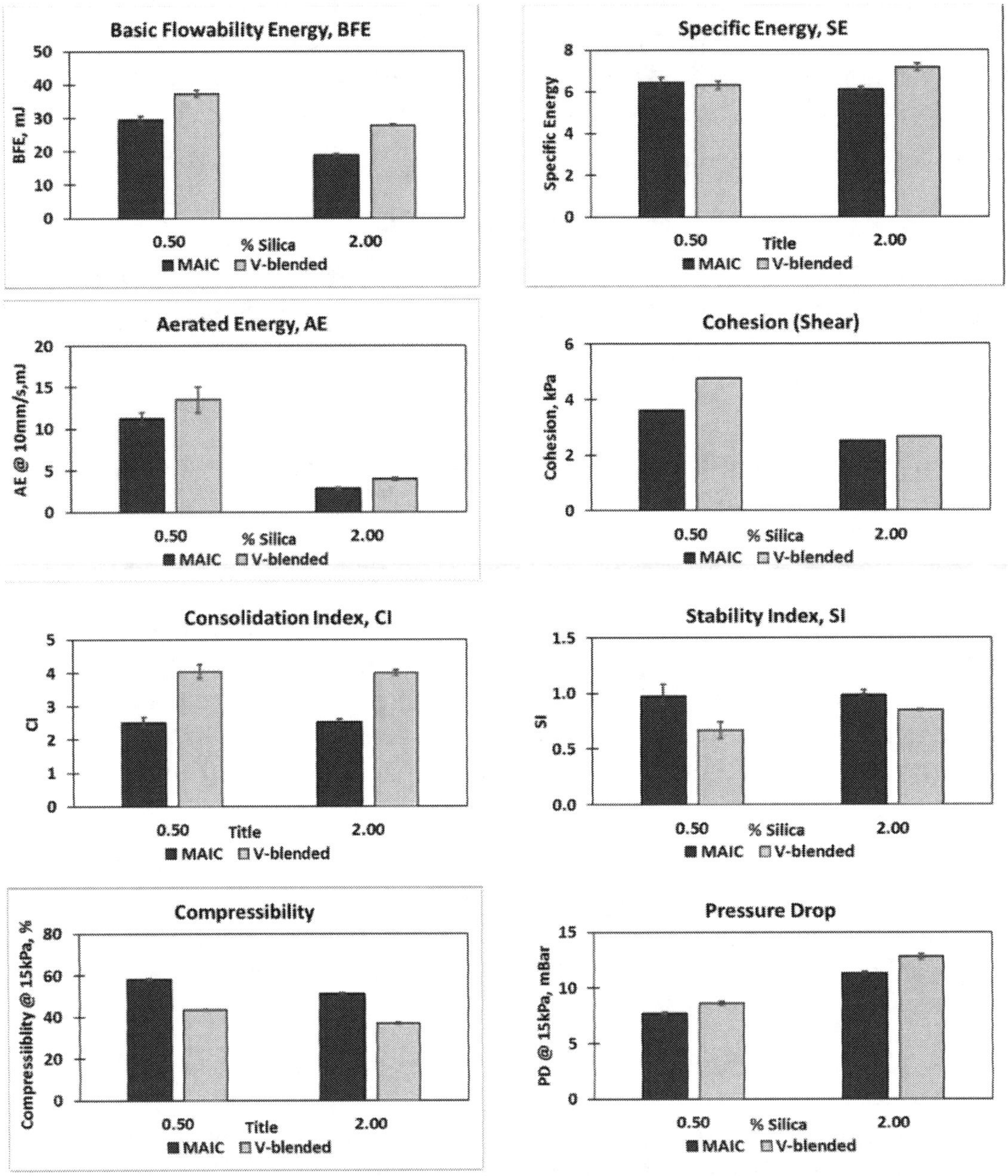

Figure 3 – FT4 Powder Rheometer Results for MAIC and V-Blended samples with 0.50% and 2.0% (w/w) fumed silica

The FT4 dynamic flow tests also provide an indication of the physical stability of the powder by measuring the change in flow energy over repeat test cycles. Whilst the MAIC blends were demonstrated to be physically stable over 7 test cycles, the V-blended samples presented a marked decrease in flow energy with Stability Index (SI) values less than 1, indicating that resistance to flow decreased as the sample was tested. This suggests that the fumed silica was not fully dispersed at the start of the test, with the mixing action of the blade as it traversed the powder bed resulting in a more uniform and wider dispersion of the fumed silica over the duration of the test. This in turn resulted in the decrease in the measured flow energy. It should be noted that these measurements were made after the samples were shipped via air from the USA to the UK. This may have generated stress conditions in the powder samples prior to testing which the MAIC process appears less affected by.

While the MAIC samples clearly showed reduced cohesion during flow, they were also demonstrated to contain more entrained air compared to the V-blended samples with lower Bulk Density (in both the Dynamic Flow and Compressibility tests) and Compressibility values. Flow aids improve particle flow behaviour by acting as a surface lubricant, reducing mechanical interlocking and friction, and/or increasing interparticular spacing, thereby removing agglomerates. Increased particle spacing may partially account for the increases in entrained air, however, the bulk of the entrained air is more likely to be due to improved separation of both the host and guest particles. Whilst this will allow for a more uniform distribution of the fumed silica on the surface of the citric acid particles as more of the host surface will be exposed, it will also cause the powder to mix with air which will subsequently be trapped in the powder bed.

% Loading

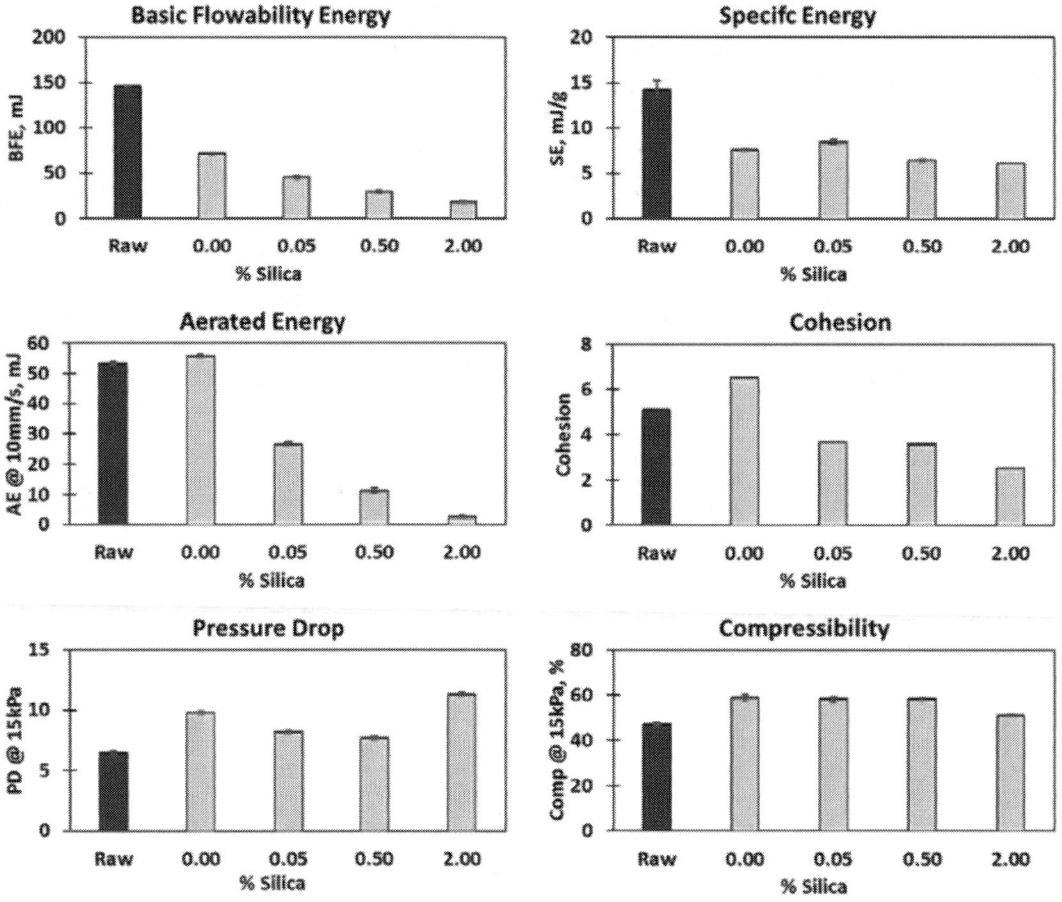

Figure 4 – FT4 Powder Rheometer Results for MAIC blended sample with increasing % silica (w/w)

As demonstrated by the results in Fig. 4, the MAIC blends exhibited a clear decrease in the BFE with increasing silica loading, with the 2.0% (w/w) presenting the lowest BFE. However, a marked decrease in flow energy was also demonstrated between the raw sample and the 0% (w/w) blend with a similar decrease demonstrated by the SE. This indicates that the MAIC process can improve reduce the resistance to flow in a low-moderate dynamic regime without the addition of silica. Indeed, there were only minor reductions in SE with addition of silica at higher loads (at 0.05% (w/w) the SE was actually slightly higher), suggesting the mechanical interlocking and friction had minimal influence following MAIC.

Improvements in the flow properties of the raw material following MAIC may be due to a number of factors. As previously mentioned, both the Bulk Density and Compressibility values indicate that the treated samples contain more entrained air, suggesting improved separation of the particles and wider interstitial spacing. The higher Pressure Drop values also suggest less efficient particle packing. Improved separation may result in the breakage / reduction of liquid and solid bridging, whilst wider interstitial spacing will decrease short range forces, such as van der Waals forces (effective range of approximately 100 nm) [6], and mechanical interlocking and friction. Collisions between the agitated magnetic particles and substrate may also result in the loss of surface asperities, smoothing the particle surfaces and again reducing mechanical interlocking and frictional effects, although SEM has shown limited changes in the particle physical properties. It may also be that in the absence of flow aid, the MAIC process attaches the fine particle fraction (of the citric acid) to the larger citric acid particles in the bulk, not only removing the smaller and therefore typically more cohesive particles from the bulk but also increasing interparticular spacing between the larger particles.

With respect to the effect of silica loading on the particle packing properties (as demonstrated by Compressibility and Permeability), the silica concentrations appear to have minimal effect below 2.0% (w/w) silica and even at 2.0% (w/w) there is only a slight improvement. This may be due to the reduced cohesion allowing for improved particle packing, or excess silica filling interstitial spaces. Either way, the number/size of air pockets and channels which allow for the passage of air would be reduced, explaining why the Pressure Drop increases at 2.0%(w/w) silica, whereas it presented minor decreases at lower concentrations.

Interestingly, when the effects of particle packing are negated, either by fully aerating the powder bed or through compaction (shear), the 0% (w/w) blend is marginally more cohesive compared to the raw sample with a slightly higher Aerated Energy (AE) and Cohesion values. With the addition of silica both tests present a clear improvement in flow properties with the lowest AE / Cohesion values presented at 2.0% (w/w) silica. The Aeration test presents a similar trend to the BFE with a clear reduction in the AE with increasing silica loading whereas Cohesion presents limited variation between the 0.05% (w/w) and 0.5% (w/w) blends despite a 10-fold increase in silica concentration (this is also shown by other shear parameters including Angle of Internal Friction and Flow Function). Indeed, a 10-fold increase in silica did not result in similar magnitude of change in flow properties for any of the measured parameters, as such while higher concentrations may exhibit the largest improvements, as long as the powder is fit for purpose at lower silica concentrations, further benefits may not outweigh the added costs and complexities of increasing the silica concentration. Furthermore, previous studies have shown that continuing to increase the flow aid concentration can be detrimental to flow as once the point of surface saturation is reached, the flow aid remains loose in the powder bulk adding to the total cohesion [7].

Conclusion

Although both the V-blended and MAIC samples exhibited a clear improvement in flow properties compared to the raw materials, the results were more pronounced for the MAIC samples, which were shown to be less cohesive across a range of stress regimes and modes of flow. The MAIC also resulted in a more conditioned uniform blend that is more resistant to changes in its flow properties, even after extensive shipping and handling. In terms of sample loading, whilst the 2% (w/w) silica blend was shown to be the least cohesive in a number of tests, the marked increases in silica concentrations (from 0.05% (w/w) to 2.0% (w/w)) did not yield similar magnitudes of improvement in the flow properties. As such, once the added costs and complexities of substantially increasing the silica concentration are considered, a lower silica loading may be more beneficial as these still return a marked improvement in flow behaviour. Interestingly, even without the addition of silica, the MAIC process appear to have some positive effects on the flow properties, however the reason for this requires further investigation. Overall, it can be concluded that MAIC shows an advantage over V-blending when adding flow aid.

[1] M. Ramlakhan, C.Y. Wu, S. Watano, R.N. Dave, R. Pfeffer, Dry particle coating using magnetically assisted impaction coating: modification of surface properties and optimization of system and operating parameters, Powder Technology, 112 (2000) 137-148.

[2] W.A. Hendrickson, J. Abbott, Process for applying liquid coatings to solid particulate substrates, US Patent 5962082 A (1997)

[3] R. Freeman, Measuring the flow properties of consolidated, conditioned and aerated powders — A comparative study using a powder rheometer and a rotational shear cell, Powder Technology, 174 (2007) 25-33.

[4] J. Yang, A. Silva, A. Banerjee, R.N. Dave, Dry particle coating for improving the flowability of cohesive powder, Powder Technology, 158 (2005) 21-33

[5] R. Pfeffer, C.H. Nam, R.N. Dave, G. Liu, J.A. Quevedo, Q. Yu, C. Zhu, System and method for nanoparticle and nanoagglomerate fluidisation, US Patent 7658340 B2, (2010)

[6] J.Q. Feng, D.A. Hays, Relative importance of electrostatic forces on powder particles, Powder Technology, 135–136 (2003) 65-75.

[7] T. Freeman, J. Yin, M. Delancy, J. Clayton, Characterising and optimising the use of flow additives in powders and powder formulations, AIChE Annual Meeting (2014)

471620 Surface Chemical Modification and Its Effect on Dynamic Flow Properties

Abstract:

Surface Chemical Modification and its Effect on Dynamic Flow Properties

C.G. Jange; R. P. K. Ambrose

Department of Agricultural and Biological Engineering, Purdue University, West Lafayette, IN

Powder flow properties are crucial parameters in handling and processing operations, including discharge from hoppers and silos, transportation, mixing, compression and packaging. Particle size, shape, density, and particle size distribution are some of the factors that influence the flow properties. The interparticle forces are highly dependent on surface properties of particles and may impact flow behaviour. van der Waals, electrostatic and capillary forces, and mechanical interlocking are the possible interparticle interaction that could exist between particles. This research quantified the effect of surface chemical composition on the powder flowability. Qualitative and quantitative understanding on the effect of surface chemical modifications on interparticle interactions and flow behavior remain poor in published literature. To study the general effect of surface composition on bulk flow properties of powders, samples of uncoated and coated glass beads with size range of 125 µm and 180 µm were used. Flow property measurements were performed using the FT4 Powder Rheometer with pre-consolidation stress from 1kPa to 9kPa. The samples were coated with gluten, starch, and fat before flow property analysis. The average values of unconfined yield strength (UYS), major consolidation stress (, cohesion index (C), flow function coefficient (ff_c), and angle of internal friction, for 125 µm size uncoated glass beads were 0.25kPa ± 0.03, 2.50kPa ± 0.31, 0.09 ± 0.01, 20.27 ± 3.96, 18.66 ± 0.32, respectively. For 180 µm samples, the values were 0.23kPa ± 0.06, 2.35kPa ± 0.02, 0.08 ± 0.02, 30. 46 ± 10.58, 19.66 ± 0.52, respectively. Results indicated 'easy flowing' characteristics for both size ranges followed by a slight increase in ff_c for particles in the range of 180 µm. Particle size of coated and uncoated glass beads were measured using G3-ID-Raman Spectrometer (Malvern Instruments, UK). In completion of this study, we expect to quantify the influence of surface chemical composition on the dynamic flow properties of particulate material.

Key words: Surface Composition; Flowability; Interparticle Forces

472544 Effect of Surface Roughness and Particle Size on Wall Friction

Abstract:

Friction between particulate materials and the process, storage and transfer equipment wall surface is of critical importance for design and operation of solids processing systems. A typical example would be hoppers. The wall friction angle between the particulate material and the hopperâ€™s material of construction is measured and used along with the particle-particle friction to calculate the required outlet diameter of a hopper for unobstructed flow (i.e. without forming solid bridge). The type of the material, the surface finish (i.e. the degree of roughness), and the coating used inside the equipment may vary greatly and affect the wall friction. Despite the vast amount of experimental data collected for various surfaces and particulate materials over the last couple of decades, the effect of surface roughness on wall friction is unpredictable and depends on the particulate material. In this study, we investigate the effect of wall roughness systematically for model materials such as glass spheres and polymer particles. The effect of particle size will be also presented.

466258 Innovations, Interactions and Integrations in Fluid Particle Systems; Memories and Future Tasks

Abstract:

Innovations, Interactions and Integrations in Fluid Particle Systems

Memories and Future Tasks

Masayuki Horio

Tokyo University of Agriculture and Technology

masa06@cc.tuat.ac.jp

1. Introduction

The early history of fluidization was inseparable from the huge innovation of the catalytic cracking technology. Such a tradition has been active until recently and we have been committing quite a few challenging developments in environment, energy, materials and pharmaceutical fields. As discussed elsewhere [1, 2], researchers of the fluid-particle engineering field were entering the second 30 years' period of scientific breakthrough at the time of early 1970s. During this second 30 years period our peer engineers have developed several commercialized and some partially commercialized processes and a few candidate processes that could be deployed in the coming future. However, shapes of industries are expected to change drastically for year 2050 and further from both necessities for environmental sustainability by mitigating global warming and for economic sustainability by developing a compromise among global, national and local economies. Also important is the sustainability of fluid particle education to keep our knowledge and technology active, interesting and renewable. Or in other words, fundamental scientific and engineering knowledge on fluid particle systems need to become more popular so that our classical achievements in the technical/engineering field can be inherited safely for long. Therefore, it would be worth discussing the shape of industries in the future and the shape of academia to prepare for a big transformation of the present social energy metabolism and the transition to the low-carbon and humanistic society. In this respect it should be stressed that innovations in the near future are becoming not just technical but also social. Both engineers and academic researchers would need more knowledge on social issues in the near future.

The era started early in the 1970s, which was the time I enjoyed most with my students and with many friend researchers in the world, was indeed a critical period of fluidization and fluid particle engineering in terms of innovations, interactions and integrations, but in several different senses as indicated above.

Concerning interactions, we have been concerned about several different categories of interactions: fluid-particle interactions, particle-particle interactions, hydrodynamics-plant interactions, chemical reaction-hydrodynamic interactions, and even the interactions among different branches of academia and between industries and academia.

Concerning integrations, we have achieved the integration of previously separated modes of fluid-particle systems including fixed/moving bed, bubbling fluidized bed, turbulent and fast fluidization, entrained bed and pneumatic transport, the integration of microscopic phenomena into mesoscopic and macroscopic ones and integrations of different branches of fluid-particle reaction engineering.

In the present paper let me take some memorable experiences of mine and discuss the three 'I's by leaving my thoughts.

2. Interactions — Incident of L-valves

When the license of multi-solid FBC boiler was imported and scaled-up by Mitsui Engineering & Shipbuilding (MES) from Battelle Columbus Laboratories, one of the most difficult issues was L-valve control. The original picture of the L-valve was beautiful in recycling solids from the external heat exchanger either by cold recycle or hot recycle to control the combustor temperature as desired. However, L-valves for the hot recycle were quite like a group of unruly children flowing free even in the no-aeration conditions. At that time the plant was waiting for the final transfer to the customer at the site. While discussing with an engineer from MES a suspicion came to my mind that some gas generation took place by chemical reactions. We started making a rough evaluation of CO formation and found out it was sufficient to fluidize the stand pipe.

There I learned that interactions between hydrodynamics and chemical reactions are two-way interactions, not just one way from hydrodynamics to chemistry, and chemical reactions to hydrodynamics could be critical particularly in high temperature processes. But from this incidence I was also convinced that any business risks have chances to be tightly related to some very fundamental phenomena and that relevant scientific knowledge has a potential to save them. Moreover, I re-acknowledged the importance of industry academia collaboration for the progress of technologies as well as engineering sciences.

3. Integrations — Two topics

1) DEM simulation

It is a talent required for researchers to look at phenomena always as integrated things. So, it is their common habit to decompose things into elements or essential laws and to find some structures composed of them behind the superficial phenomena. The decomposition is not always easy. And even when it is done well, we should not use the style of reductionists, but search for solutions or suggestions on the macroscopic level of phenomena re-constructing things by integration.

In the mid-90s we developed DEM (Discrete Element Model) to simulate macroscopic bed behaviors from laws governing microscopic phenomena. However, the computer capacity of that time was much smaller than now and it was difficult to conduct realistic scale computation. However, we conducted a variety of demonstration for beds of sticky particles to see agglomeration behavior, for beds of immersed tubes to see contact packet renewal over the tube surface, for beds of polymer particles in reaction to see distribution of temperature of individual particles, and for beds char combustion to see surface temperature of burning char which is important for NOx formation and reduction. With these small scale computation we could have tested innovative ideas in the macroscopic level, which is produced from freewheels thinking. Actually the real scale computation may not induce innovative ideas nor quick design as long as it takes much computational effort. Macroscopic models and correlations would be still valuable for quick actions as well as deep understanding of the phenomena of concern. Furthermore, with scaling laws we can conduct scaled down experiments as well as scaled-up computations for the integration.

2) Cluster and agglomerate size prediction

Prediction of mean cluster or agglomerate sizes is another challenge of macroscopic modeling. Here, I use the term a 'cluster' as a group of particles brought together by external (not particle-particle) effects and an 'agglomerate' as a group of particles held together by direct interparticle forces as proposed by myself and Professor R. Clift [3].

In the dilute region of CFBs and in the freeboard of bubbling beds we have clusters that are group of drifted particles. In bubbling beds of sticky particles we have agglomerates. In both cases the force fields changing dynamically are the key factors, one in a dilute suspension with turbulent gas motion and another in a dense bed with frequent compaction and sharing due to bubble passage. We have proposed models [4] and [5] for prediction of cluster and agglomerate sizes, respectively, but both models should be examined in more detail for validation.

4. Innovations

The key technological approaches for sustainability are low carbon transportation, zero emission buildings with highly insulated walls, biomass material utilization, solar heating, lighting and PV power generation, renewable energy utilization and process intensification, integration for low exergy losses. To reduce GHG emissions from transportation we need to avoid unnecessary carbon foot prints and induce local production and consumption. In terms of energy the scheme of decentralized local generation and consumption should be encouraged more to reduce emissions from centralized mineral energy conversion systems. Moreover, in many countries, their megalopolises consume much more energy than they can locally produce. In this respect encouraging people and industries move to areas rich in renewable energies is promising [6]. It is a counter action against the over-urbanization of the last 50 – 60 years. All of these indicate that decentralization and community empowerment should be the necessary trend in the coming decades for countries having centralized systems.

Destination of decentralization is also confirmed by thoughts on the global economy. So far the size of financial economy is said to be much larger than real economy. More money is invested for the sake of profit and there is a definite world trend of widening disparity and increase of poverty. Circulation of local investment in the local economy is one of the current issues to stabilize modern society. So, from economic viewpoint de-centralization of some extent and community economy empowerment are necessary.

Keeping these destinations in mind, the future practices of chemical and fluid-particle engineering should be newly defined. In the last part of my paper I would discuss their features-to-be.

References

1. M. Horio, Fluidization Science, its Development and Future, Particuology, 8, No.6, 514-524, 2010

2. M. Horio, 1. Overview of Fluidization Science and Fluidized Bed Technologies, 'Fluidized-bed Technologies for Near-zero Emission Combustion and Gasification,' ed. by F. Scala, Woodhead Publishing, 3-41, 2013

3. M. Horio, R. Clift, A Note on Terminology: 'Clusters and agglomerates', Editorial, Powder Technology, 70, pp.196, 1992

4. M. Horio, M. Ito, Prediction of Cluster Size in Circulating Fluidized Beds, J. Chem. Eng. Jp., 30, pp.691-697, 1997

5. Y. Iwadate, M. Horio, Prediction of Agglomerate Sizes in Bubbling Fluidized Beds of Group C Powders, Powder Technology, 100, pp.223-236, 1998

6. M. Horio et al., The Potential for Massive GHG Reduction by Mass Rural Remigration (The Renewable Energy Exodus): A Case Study for Japan, Applied Energy, 160, 623-632, 2015

474968 The Difference Between L-Valves and Loop Seals

Abstract:

People often do not understand the difference between operation of non-mechanical L-valves and loop seals, and this can lead to solids transport design failures. In their most common operating modes, an L-valve is used to control the flow rate of solids in a system, while the loop seal does not control the solids at all, but automatically moves solids automatically from a cyclone back into a fluidized bed.

Loop seals can operate with Geldart Group A, B or D solids if they are designed correctly. Only a little aeration is required on the upleg of the loop seal to maintain this part of the loop seal in a fluidized mode. The L-valve, however, is generally operated to control the solids flow rate around a system. Adding a little aeration to the L-valve at the correct location causes solids to flow through the Lâ€"valve. Adding more aeration causes a higher solids flow rate through the L-valve. While operating in this solids-control mode, the L-valve is practically limited to operating with solids greater than about 100 microns. Thus, only Geldart Group B and D solids can be used with an L-valve operating in the control mode. Group A solids do not work well with L-valves operating in the solids-control mode.

Another difference between the two devices is that the loop seal operates with an overflow, fluidized bed standpipe above it. The L-valve operating in a control mode cannot operate with this type of standpipe, but must operate only with an underflow, non-fluidized (packed bed) standpipe above it.

Why the L-valve operates differently than the loop seal will be explained by describing the differences in operation between the types of standpipes feeding the two non-mechanical devices.

468950 CFB Flow Regimes with Application to Fluidized Bed Combustion

Abstract:

Based on more than 25 years of fluidized bed (FB) research on mixing and fluid-dynamics with application to FBs in fuel conversion, we give an overview of the characteristics of gas and solids flow in fluidized beds for combustion and gasification. In particular, we discuss the flow and mixing characteristics of the four main regions identified in FB units for fuel conversion, with focus on CFB furnaces; the dense bed, the splash zone, the transport zone and the exit zone. We discuss the main characteristics of these zones and present recent results on how they relate to the net solids flux (G_s), i.e. the externally recirculated solids flux, which is a key parameter for CFB boilers and other dual bed systems, such as applied in chemical looping combustion for CO_2 capture and indirect biomass gasification. We draw our conclusions based on measurements and modeling of cold models as well as large commercial FB boilers and gasifiers.

CFB furnace characteristics

The furnace of fluidized beds for fuel conversion - combustion and gasification â€"typically have the following characteristics:

- Aspect ratio of the furnace (H_0/D_{eq}) of the order of or less than 10;
- Aspect ratio of the settled bed, i.e., the bed formed if the solids are not fluidized, $H_{b,settled}/D_{eq}$ of less than 1;
- Solids belonging to Group B in the Geldart classification; and
- Net solids flux that typically ranges from 0.5 to 20 kg/m^{2}s (CFB boilers).

Especially when it comes to circulating fluidized beds (CFB), experimental data on fluid dynamics and mixing reported in the literature were often carried out for other chemical engineering applications than fuel conversion, such as cracking, coating and polymerization. Such data may therefore not be used to draw conclusions of relevance for fuel conversion systems. In contrast to the characteristics listed above, the latter systems typically employ Geldart Group A particles and operate at net solids fluxes often far exceeding 20 kg/m^{2}s and for riser inventories which, if operating at low velocities, correspond to a tall bed operating in slugging regime - a regime which is not found in commercial-scale FBC boilers and gasifiers.

CFB furnace mixing and fluid-dynamics - an overview

In short, the fluid dynamics and mixing of the main fluid dynamic regions in a CFB furnace can be characterized as follows.

The dense bed: The dense bed is present provided there is a sufficiently high solids inventory and it is characterized by strong fluctuations in the gas flow due to the bubble passages. The dense bed can be maintained also when the superficial velocity exceeds the terminal velocity of a major part of (or all of) the bed solids (under CFB conditions). This is mainly due to that the primary gas passes through the dense bed in a heterogeneous way, partially in the form of large bubbles, which, at high velocities, create a shortcut of gas between the gas distributor and the region above the dense bed. Thus, this flow results in bubbles with oxidizing conditions and an emulsion phase under reducing conditions. Although the amplitude of the gas flow as induced by the bubble flow is drastically reduced above the dense bed, the gas flow fluctuations have important implications for the characteristics of the gas-solids flow in the freeboard.

The splash zone: Above the dense bed, there is a so-called splash zone that consists of solids, which is characterized by strong back mixing mainly due to ballistic movement of clustered solids thrown up by the bubble eruptions.

The transport zone: Under CFB conditions, the gas velocity is sufficiently high to cause a significant portion of the solids to be entrained up through the furnace. Thus, above the splash zone, a transport zone is formed which is characterized by that most of the back-mixing occurs at the furnace walls. The transport zone flow forms a core-annulus structure, with the up-flow mainly in the core region and with a continuous back-mixing at the furnace walls, thereby forming downward flowing wall-layers of solids. Such continuous back-mixing represents a solids size segregation up through the transport zone, i.e. the coarser the solids the higher the probability that they are separated at the furnace walls. Therefore, there is a continuous decrease in average solids size with height in the freeboard of a CFB furnace.

The exit zone: Near the inlet to the cyclone, an exit zone is formed which is governed by gas and solids flow properties and the design of the fluidized bed furnace. A certain portion of the solids which reach the level of the inlet to the cyclone are internally recirculated due to back-flow effects. Thus, whereas the above described flow characteristics of the bed, the splash and the transport zones seem to only depend on furnace (riser) pressure drop and fluidization velocity given a certain set of gas and solid properties, the externally recirculated solids â€" the net solids flux (G_s) â€" depend on exit geometry and the configuration and operation of the cyclone side of the CFB loop (cyclone dipleg and particle seal).

470434 Biomass Gasification in Fluidized Beds: Current Challenges and Advances

Abstract:

Fluidized bed technology applied to gasification of woody biomass can produce heat, power, liquid fuels, hydrogen and value-added chemicals with positive environmental impacts. Recent advances in this area include a culmination of fundamental and applied research leading to industrial development at pilot and demonstration scales. One of the hurdles affecting biomass utilization is the cost related to pre-treatment, biomass gasification, gas clean-up and downstream process. Novel technologies such as improved gasification conversion and destruction of tarts are required to improve the economics of the processes. Sorbent enhanced gasification advantageously combines CO2 capture with gasification. This talk highlights the state of the technology with experiences in tar cracking catalyst development, lime-enhanced gasification and community engagement through operation of two gasifiers at the University of British Columbia, Vancouver campus, while tracing the extraordinary contributions Prof. Horio has had in many aspects of the biomass utilization, fluidization and broader energy systems innovation.

454094 Flow Patterns in High Density Fluidized Beds Used As Solar Energy Carrier Systems

Abstract:

Dense gas-solid suspensions have the potential to be applied as heat transfer fluids (HTF) for energy collection and storage in concentrated solar power plants [1]. In the frame of the European CSP2 project [2], a solar receiver using a Dense Particle Suspension (DPS) upward flow in a bundle of tubes was studied. The interest of using as Heat Transfer Fluid (HTF) solid particles transported by an air flow is that the created two-phase medium, exhibits fluid-like flow properties while being able to withstand high temperatures up to the solid sintering temperature. Moreover, the DPS can be used as direct heat storage medium. In the design investigated here, the particles form a dense upward-flowing suspension. Both the density of the SiC particles suspension and their movement have a strong influence on the heat transfer, as demonstrated by Positron Emission Particle Tracking (PEPT) experiments [3].
The NEPTUNE_CFD computational code was used to simulate the DPS circulation both at ambient and high temperature. The numerical results are compared with those obtained on the cold mock-up and with PEPT experiments. The solid reflux evidenced by PEPT experiments was corroborated by simulations and during on-sun experiments [4].
Finally, the DPS flow is analysed highlighting the impact of the air velocity and the density variation along the tube height and the strong influence of the various operating parameters on the temperature distribution. The results of this study will be useful when planning for solar energy capture systems scale-up.

References:

[1] Gilles Flamant, Daniel Gauthier, Hadrien Benoit, Jean-Louis Sans, Roger Garcia, Benjamin Boissière, Renaud Ansart, Mehrdji Hemati, Dense suspension of solid particles as a new heat transfer fluid for concentrated solar thermal plants: on-sun proof of concept; Chemical Engineering Science 102, 567-576

[2] FP7 EC project CSP2. http://www.csp2-project.eu.
[3] Pablo Garcia-Trinanes, Jonathan Seville, Benjamin Boissiere, Renaud Ansart, Thomas Leadbeater, David Parker. Hydrodynamics and particle motion in upward flowing dense particle suspensions: Application in solar receivers. Chemical Engineering Science, (2016) 146:346-356.
[4] H. Benoit, I. PÃ©rez LÃ³pez, D. Gauthier, J.-L. Sans, G. Flamant. On-sun demonstration of a 750 C heat transfer fluid for concentrating solar systems: Dense particle suspension in tube. Solar Energy, (2015) 118:622â€“633

*Updated contact details: P.GarciaTrinanes@greenwich.ac.uk

469655 Time-Resolved X-Ray Tomography of Fluidized Beds

Abstract:

The dynamics of fluidized beds are challenging to study: the voidage distribution is changing quickly over time (i.e., a high temporal resolution is needed) and dense gas-solids flows are typically opaque to visible light. X-ray tomography can cope with these challenges. It is a non-intrusive technique that can be used for the reconstruction of the cross-sectional distribution of the different phases in fluidized beds. In Delft, we have an X-ray system consisting of three X-ray sources [1,2]. Originally, this system was equipped with 192 detectors, but we recently upgraded it to three vertical 30 x 30 cm detector plates with each 1524x1548 pixels, resulting in a resolution of about 1 mm.

We have applied our tomographic systems to a nanopowder fluidized bed. This is challenging, because of the much lower bulk density than for regular fluidized bed, as a result of the very high voidage [3]. We will show that the bulk density is a function of the height. Moreover, we will demonstrate that the history of the powder plays a strong role in nanopowders fluidization: stepwise increase of the gas velocity gives much poorer fluidization behaviour than starting with a high gas velocity, and stepwise decreasing it.

We also applied X-ray tomography to study the bubble hold up and its distribution in bubbling fluidized beds with and without vertical internals such as heat exchanger tubes. The goal is to understand the influence of vertical internals on the hold up. We show that with employed internals, the bubbles are more homogeneously distributed in the bed [4]. More, we have studied the influence of column diameter on bubble size and velocity in such a system [5].

[1] Van Ommen, J.R., Mudde, R.F., 2008, "Measuring the Gas-Solids Distribution in Fluidized Beds - A Review", International Journal of Chemical Reactor Engineering 6 (2008) R3.

[2] Mudde, R.F., Bubbles in a Fluidized Bed: A Fast X-Ray Scanner, AIChE Journal 57 (2011) 2684-2690.

[3] Quevedo, J.A., Omosebi, A., Pfeffer, R., "Fluidization enhancement of agglomerates of metal oxide nanopowders by microjets", AIChE Journal 56 (2010) 1456-1468.

[4] Maurer, S., Wagner, E. C., Schildhauer, T. J., van Ommen, J. R., Biollaz, S. M., & Mudde, R. F. (2015). X-ray measurements of bubble hold-up in fluidized beds with and without vertical internals. International Journal of Multiphase Flow 74 (2015) 118-124.

[5] Maurer, S., Gschwend, D., Wagner, E.C., Schildhauer, T.J., van Ommen, J.R., Biollaz, S.M., & Mudde, R.F., "Correlating bubble size and velocity distribution in bubbling fluidized bed based on X-ray tomography", Chemical Engineering Journal 298 (2016) 17-25.

461024 Integrated Exergy Recuperative Biomass Gasification and SOFC System for Hydrogen and Power Coproduction

Abstract:

Integrated Gasification Fuel Cell (IGFC) power generation system is being developed as high-efficiency power generation technology. However, the highest theoretical gross thermal efficiency of conventional IGFC is still below 52% because of large exergy destruction mainly due to partial oxidation gasification, air cooling of SOFC stacks, and the combustion in a gas turbine combustor. In order to obtain higher power generation efficiency, a novel hydrogen and power coproduction system based on the exergy recuperative biomass gasification integrated with solid oxide fuel cell was proposed. In the proposed system the exothermic heat from micro-tubular SOFC stacks immersed in the fluidized bed gasifier at a temperature range close to the SOFC operating temperature is recuperated and utilized for the endothermic heat of biomass gasification, preventing the exergy destruction due to the partial oxidation in gasifier and heat transfer for cooling SOFC. In addition, providing the recirculation system of unreacted hydrogen separated from SOFC anode exhaust, gas turbine can be eliminated. Since the exergy loss of SOFC is much smaller than that of gas turbine, the net power generation efficiency of the combination of gasification and SOFC power generation system can be considerably improved. Corresponding to this system, a novel triple-bed combined internal circulating fluidized bed gasifier, composed of a moving bed high-temperature pyrolyzer with the function of tar cracking, a bubbling bed char gasifier equipped with micro-tubular SOFC stacks, and a fluidized bed combustor of remained char, was proposed for 40kW class pilot plant. In the present study, the conceptual design of a integrated exergy recuperative biomass gasification and SOFC system for hydrogen and power coproduction by using a triple-bed combined internal circulated fluidized bed as a gsifier is described.

471392 Metal Oxide Reaction Engineering and Its Connection to Fluidization Systems for Energy Conversion Applications

Abstract:

Metal oxide reaction engineering has been widely applied in energy industries. A typical example of considerable current interest is chemical looping technology. Fossil fuel chemical looping applications were used with the steam-iron process for coal from the 1900s to the1940s and were demonstrated at a pilot scale with the carbon dioxide acceptor process in the 1960s and 1970s. There are presently no chemical looping processes using carbonaceous fuels in commercial operation. The enabling technology for chemical looping applications has been and will be the fluidization systems. A key factor that hampered the continued use of these earlier processes was the inadequacy of the reactivity and recyclability of the looping particles in the circulating fluidized bed operation. This factor led to unsustainable operation of the process. With CO2 emission control now of great concern, interest in chemical looping technology has resurfaced due to their unique ability to generate a sequestration-ready CO2 stream.

Chemical looping technology is a manifestation of the interplay among all the key elements of particle science and technology including particle synthesis, reactivity and mechanical properties, circulating fluidized bed stability and contact mechanics, gas-solid reaction engineering and particulates system engineering. This presentation will describe the fundamental and applied aspects of modern chemical looping technology that uses circulating fluidized bed processing fossil and other carbonaceous feedstock. Specifically, it will discuss the reaction chemistry, ionic diffusion mechanisms, metal oxide synthesis and thermodynamics, fluidized bed reactor design, and system engineering along with energy conversion efficiency and economics of the Coal-Direct Chemical Looping Process and Syngas Chemical Looping Process being developed at Ohio State University at a pilot level. Further, CO2 emission control using the chemical looping technology will be illustrated and compared with other CO2 capture methods. Potential for selective oxidation in the production of fuels and chemicals, as well as solar based chemical looping technology, will also be discussed.

450876 A Size-Controllable Preparation Method for Indium Tin Oxide Particles Using a Membrane Dispersion Micromixer

Abstract:

A chemical co-precipitation process was developed for the preparation of indium tin oxide (ITO) particles using a membrane dispersion micromixer with indium chloride, tin chloride and ammonia as the reagents. The traditional co-precipitation process make it hard to control the particle size in a beaker, and therefore a novel membrane dispersion micromixer, which could control particle size, was used. At the same time, the particle size was significantly reduced with improved mixing performance in the membrane dispersion micromixer.

The effect of the pH value and concentrations of reagents solutions on the size of ITO particles were investigated. The particle size increased from 18nm to 83nm and the specific surface area decreased from 55.44 m^2/g to 16.32 m^2/g, respectively, where the pH value changed from 9.56 to 6.70. But the particle size decreased when the reagents concentrations were increased. Furthermore, the effect of the aging process on the crystal structure of ITO particles was investigated. It was proved that the aging process is beneficial for the preparation of cubic ITO particles, excluding hexagonal ITO particles. This study provides a size-controllable ITO preparation method to control particle size and enhance the cubic phase purity.

463077 Optimization of Batch and Continuous Cooling Crystallization of High Aspect Ratio Crystals Under Nucleation, Growth, Dissolution and Breakage for Shape Control

Abstract:

Batch and continuous cooling crystallization of potassium dihydrogen phosphate in an impure media is evaluated through the implementation of various optimization frameworks to assess the achievable crystal size and shape. The continuous system studied is a two-stage mixed-suspension mixed-product removal (MSMPR) crystallization process. The systems are modelled using a multi-dimensional population balance model under nucleation, growth, and dissolution kinetics. The impact of agitation rate is studied by considering a binary crystal breakage model. A significant decrease in the achievable aspect ratio is observed by controlling the dissolution of crystals (batch scenario) and limiting the residence time and growth (continuous scenario). Further improvement of the final shape is achieved by considering the agitation rate in the optimization framework. However, the increase in the variance of the crystal mean size and aspect ratio indicates a trade-off between the aspect ratio and spread of the crystal size distribution. Nonetheless, this work illustrates promising results for further development of process optimization frameworks and control approaches for batch or continuous cooling crystallization operations by considering multiple crystallization kinetics and binary crystal breakage kernels, which can be used as a generic model-based design framework for batch and continuous crystallization processes of needle shaped particles to achieve desired mean aspect ratio.

453501 Evidence of Crystal Nuclei Breeding in Laboratory Scale Seeded Batch Isothermal Crystallisation Experiments

Abstract:

Experimental investigations of the batch seeded crystallisation of paracetamol in 2-propanol were carried at 200, 300 &375 RPM agitation rates, using a large seed size (355-500 μm) and a low level of initial supersaturation (S_0=1.2) in a laboratory scale reactor. Such experiments are normally conducted for the indirect measurement of crystal growth, contingent on the assumption of negligible nucleation, agglomeration and breakage. In the present work a copious increase in crystals nuclei was noted shortly following seed addition. The formation of substantial numbers of new nuclei was substantiated through FBRM, laser diffraction and SEM. Secondary nucleation was proposed as the origin of the new crystals and a Secondary Nucleation Threshold (SNT) of between 1.09-1.11 determined, beyond which crystal growth only was apparent. A study was undertaken to investigate the origin of secondary nucleation. Crystal nuclei breeding, as a mechanism of secondary nucleation, has being theorised for many years, however it is only very recently that definitive molecular dynamics simulations have provided mechanistic insight as to its action. The mechanically driven attrition and breakage mechanism of secondary nucleation remains prominent in literature. Stirred vessel experiments were conducted using paracetamol seed crystals suspended in a non-solvent indicated. Despite three hours of continuous agitation, no significant change in particle number or size was detected. Only after a threshold of four hours were significant crystal fatigue and fragmentation evident. Shadowgraphy investigations of crystal jet wall impingement revealed the squeeze film as a key protective element in preventing crystal attrition and breakage. A low temperature (283.15 K) crystallisation was conducted which indicated a significant temperature dependency, entirely inconsistent with the attrition and breakage mechanism of secondary nucleation. It was shown that through the use of smaller seed crystals (125-250 μm), a high agitation rate and elevated solution temperature that the rate of secondary nucleation could be enhanced thereby creating the potential for confounding rapid secondary nucleation with growth. The current work elucidates the potential impact of cluster breeding in laboratory scale crystallisations and furthermore, provides additional experimental support for the crystal breeding mechanism of secondary nucleation.

467222 CFD Modelling of Spray Drying of Biopharmaceuticals

Abstract:

Spray Drying (SD) has attracted lots of interest in pharmaceutical industry as an efficient one-step drying and formulation process of biopharmaceuticals. The complexity of the influential factors on final product properties make its optimization process a costly, time consuming procedure. Also, use of temperatures higher than 100° C is another concern in processing of thermally unstable drugs with spray drying technique.

Modelling techniques have shown a huge promise in defining predictive models of SD process. Although modelling of SD is still in its infancy, it has now become an inevitable part of continuous manufacturing process.[1,2] Computational Fluid Mechanics (CFD) simulations is a powerful method of modelling air flow of SD. Also, it is capable of predicting particle size, outlet temperature and interparticle interactions. However this is all pertinent that the model is defined and interpreted correctly.

In this research CFD modelling was employed to investigate the air flow in a co-current laboratory scale Buchi B-295 spray dryer. A mixed Eulerian/Lagrangian approach, in which the single phase Reynolds Averaged Navier Stokes (RANS) equations are solved were used to determine the flow field. Also, the effect of different type of nozzle and the swirl angle (ultrasonic *vs.* two fluid nozzle) on air flow pattern was examined.

Particle tracking of hundreds of particles was used to check the temperature variance between the particles at different stages of drying process in drying chamber. Using the Lagrangian approach the fate of individual representative droplets was determined when they reach the walls. By having the knowledge on moisture content and temperature the expected contact regime, i.e. sticking or bouncing could be defined.

Finally, the design of experiment approach on a ternary system of API(Acetaminophen)/excipient(HPMCP) was used

to validate the modelling approach results. A good agreement between the inlet-outlet temperatures was found between modelling and experimental results. Also, particle size distribution validated the outcome of the CFD modelling approach.

References

(1) Oakley, D. E. Spray Dryer Modeling in Theory and Practice. Drying Technology 2004, 22 (6), 1371â€"1402. DOI: 10.1081/DRT-120038734.

(2) Spray Drying Technology (2); M.W.Woo, A.S. Mujumdar, W.R.W. Daud, Ed. 1; Singapore, 2010.

471248 High-G Drying in Vortex Chambers

Abstract:

Vortex chambers allow high-G operation and intensifying interfacial transfer of mass, heat and momentum. In the presented experimental work, multi-zone vortex chambers are studied in which gas at different temperature can be fed to each zone. A first part of the presentation deals with the flow pattern in the absence of a particle phase and mixing between different zones. In a second part, the interaction with a drying particle phase is studied, as well in terms of mass, heat and momentum transfer. Different design aspects and the impact of some important operating conditions are discussed. The potential process intensification that can be achieved by operating high-G is evaluated.

454606 Novel Strong Starch and Salt Composites Structures, What Are They and What Characterises Them

Abstract:

Novel bio-based composite materials attract a lot of attention at the moment. In this work we have looked in combinations of starch, salt and water which under the right heating conditions form a very strong composite material which could be used for building purposes. By doing a design of experiments of the different variables of the composite material the strength of the material was assessed. Also the plastic and elastic deformation was studied. The composites material was optimized to the best strength possible and compared to other materials used for building. This work showed that a remarkable new composite material was developed consisting of simple in nature available material could be easily made and showed good strength characteristics. The deformation under stress showed remarkable new insights. Examples of possible use of the material will be shown and discussed.

464335 A Continuous Crystallization Process to Separate Mixtures of Enantiomers Exploiting Two Coupled Fluidized Bed Crystallizers

Abstract:

Crystallization is one of the most effective ways to separate mixtures of enantiomers. So far, primarily batch processes are used. In particular, Preferential Crystallization (PC) has proven to be applicable in batch processes to crystallize selectively pure enantiomers from conglomerate forming racemic mixtures [1]. This contribution presents results of applying a fluidized bed crystallizer configuration for continuous PC of pure enantiomers from a racemic mother liquor [2]. To crystallize both enantiomers simultaneously two fluidized bed crystallizers are coupled via their liquid phases. Saturated racemic mother liquor is supplied from a feed tank with a specific flow rate into the bottom sections of the two double jacketed fluidized bed crystallizers. To initiate the crystallization process in both crystallizers seeds of the opposite enantiomers are added. The preferential growth of the seed crystals reduces the supersaturation in both vessels. In order to allow for continuous production of crystals with a defined size, the lower sections of both crystallizers are conical to offer the opportunity of classified product removal at a certain height. The depleted solutions leaving both crystallizers are returned into the feed tank. In this way always racemic solution is supplied to both crystallizers. Nuclei of the counter enantiomer and other small particles, which have a negative effect on the purity of the product, are taken out with the fluid phases at the tops of the two crystallizers and subsequently dissolved in the heated recycle lines. To assure that the process can be continued continuously, larger particles and formed agglomerates are withdrawn at the bottom, crushed and reinjected to provide permanently new seeds. As a proof of feasibility we will describe the set up developed [3] and present experimental results for the separation of the enantiomers of D/L-asparagine monohydrate [4]. The status of theoretically describing the process using detailed [5] and reduced models [6] will be discussed as well as challenges and opportunities for the separation of other chiral substances. The financial support of German Research Foundation (DFG) within the priority program "Dynamic simulation of interconnected solids processes DYNSIM-FP (SPP 1679)" is gratefully acknowledged.

1. H. Lorenz, A. Seidel-Morgenstern, Angewandte Chemie- International Edition, 53, 2014, 1218.

2. H.-H. Tung, E. L. Paul, M. Midler, J. A. McCauley, Crystallization of Organic Compounds: An Industrial Perspective, Wiley, Hoboken, New Jersey, 2009.

3. D.. Binev, H. Lorenz, A. Seidel-Morgenstern, Chemical Engineering Science 133, 2015, 116.

4. D. Binev, H. Lorenz, A. Seidel-Morgenstern, Crystal Growth & Design, 16, 2016, 1409.

5. K. Kerst, L. Medeiros de Souza, A. Bartz, A. Seidel-Morgenstern, G. Janiga, Computer Aided Chemical Engineering , 2015, 263.

6. M. Mangold, L. Feng, D. Khlopov, S, Palis, P. Benner, D. Binev., A. Seidel-Morgenstern, Journal of Computational and Applied Mathematics 289, 2015, 253.

466054 Tuning the Properties of Crystalline Particles â€"Temperature Cycling in a Continuous Tubular Crystallizer

Abstract:

Tuning the properties of crystalline particles –Temperature Cycling in a Continuous Tubular Crystallizer

P. **Neugebauer**[1], M. **Besenhard**[2], J.G. **Khinast**[1,2]

1 Graz University of Technology, Institute for Process and Particle Engineering, 8010 Graz, Austria

2 Research Center Pharmaceutical Engineering, 8010 Graz, Austria

Correspondence:

e-mail: peter.neugebauer@tugraz.at; tel: +43 (316) 873 - 30424

e-mail: khinast@tugraz.at; tel: +43 (316) 873 – 30400; fax: +43 (316) 873 - 1030400

The utility of tubular reactors for the continuous production of crystalline particles by now has been demonstrated for a wide range of different substrates. Mostly, these systems profit by the ease of supersaturation control by the installation of sections at different temperatures.

Experimental designs employing water baths or jackets have shown to be adequate to benefit from the high surface:volume ratio offered by tubular crystallizers, guaranteeing efficient heat flow from or to the crystallization slurry[1–3].

In the present study we established an easy-to-use platform which allows the modification of crystals from ingredients of pharmaceutical interest under controlled conditions. Our installation facilitates easy changing of process settings such as flow rate, crystal loading and temperature trajectory. Simultaneously, it provides immediate results enabling us to experimentally test a broad range of settings in a short period of time.

Within our setup (Figure 1) a tubular crystallizer with hot and cold zones was shown to be adequate to oscillate dissolution and growth efficiently via temperature cycling. By this, the feasibility of crystal engineering could be shown for pharmaceutical ingredients. By establishing a segmented flow mode interfering effects such as aggregation and breakage could be minimized and by thorough characterization of the temperature cycler unwanted nucleation could be overcome.

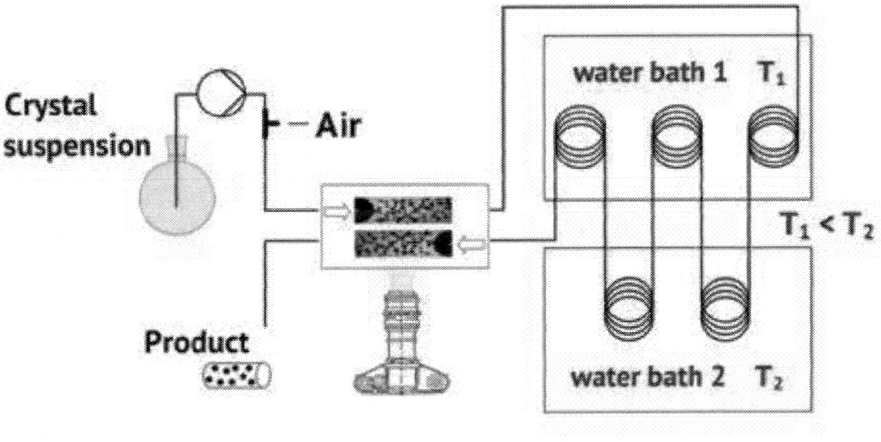

Figure 1: Schematic drawing of the experimental setting

(1) Neugebauer, P.; Khinast, J. G. Continuous Crystallization of Proteins in a Tubular Plug-Flow Crystallizer. *Cryst. Growth Des.* **2015**, *15*, 1089–1095.

(2) Besenhard, M. O.; Neugebauer, P.; Da-Ho, C.; Khinast, J. G. Crystal Size Control in a Continuous Tubular Crystallizer. *Cryst. Growth Des.* **2015**, *15*, 1683–1691.

(3) Jiang, M.; Zhu, Z.; Jimenez, E.; Papageorgiou, C. D.; Waetzig, J.; Hardy, A.; Langston, M.; Braatz, R. D. Continuous-Flow Tubular Crystallization in Slugs Spontaneously Induced by Hydrodynamics. *Cryst. Growth Des.* **2014**, *14*, 851–860.

453235 Continuous Crystallization Using a Sonicated Tubular System for Controlling Particle Size in an API Manufacturing Process

Abstract:

A pH swing crystallization that was very sensitive towards scale-up and required specific equipment due to the non-Newtonian characteristics of the fluid, has been investigated using continuous crystallization (CC) as a means to improve particle size control, scalability and potentially reduce manufacturing cost. A sonicated tubular crystallizer was designed and shown to allow the pH swing crystallization to be operated robustly without blockage, to give a wide range of desired particle size. The results were compared with crystallization in a premixing semi-batch system and a simple addition semi-batch system under similar conditions. Interestingly, it was found that the sonicated tubular system was the only one of the three systems that gave the desired control for small particle size (1~7 Î¼m). Even though the productivity of the sonicated tubular crystallizer still needs to be increased to allow CC and numbering-up to become cost competitive with batch crystallization, we have demonstrated the potential of continuous tubular crystallization as a useful alternative method to conventional batch operation.

463983 Continuous Reactive Crystallisation for Control of Salt Formation Processes

Abstract:

Salt crystallisation is of great importance to the pharmaceutical industry (majority of pharmaceutical products are marketed as salts) but it is often not a well-controlled process. This is because salt formation via a pH shift can be highly sensitive to mixing, and under batch or semi-batch conditions this can lead to severe issues with reproducibility and consistency of quantitative particle attributes, including solid form, crystal size distribution and crystal habit. The salt crystallisation process typically takes the form of a batch titration where the basic drug is dissolved in a solvent before the acid is slowly added. The issue with this procedure is that when the two solutions first come into contact the local supersaturation will build up very quickly and uncontrolled nucleation may occur leading to an undesired salt stoichiometry or crystal size distribution. For this reason, greater control over mixing to achieve greater control over the supersaturation build up is required to improve the consistency of the process and the quality of the particulate product. The aim of this research is to develop a quantitative reactive crystallisation model including a solution speciation model, a mixing model and a population balance model to guide the operation of continuous reactive crystallisation processes. We show the development of this quantitative reactive crystallisation model and how it is used to select operating conditions for the continuous crystallisation of the example 2:1 salt of 3,5-dinitrobenzoic acid and ethylenediamine. By doing this we can control which of the two polymorphic forms (monoclinic & triclinic) is crystallised, the crystal habit and the crystal size distribution.

456813 Process Intensification for Continuous Slug-Flow Crystallization

Abstract:

In the pharmaceutical and chemical industries, the continuous generation of crystals of target size distribution has the potential to improve efficiency for post-crystallization operations [1]. The control of crystallization processes can be challenging when undesirable phenomena such as particle attrition and breakage occur. Slug-flow crystallization has been demonstrated to allow enhanced control of crystal properties such as size and shape [2][3]. The slurry flow is combined with an air flow and fed to a tube to induce a multiphase hydrodynamic instability that spontaneously generates well-mixed slugs where the crystals continue to grow. These slugs are well-mixed without having the mixing blades as in traditional crystallizer designs that induce undesirable particle attrition.

This presentation will elaborate: (1) the process design and strategies for continuous slug-flow crystallization that controls the crystallization phenomena; (2) nucleation systems based on multiphase hydrodynamics from joining two streams in dual impinging jet, coaxial, or radial micromixers, which has been demonstrated to produce seed crystals of uniform crystal size distribution (CSD) [4][5], with a nucleation rate that depends on the liquid flow rate [4][5][6]; and (3) nucleation facilitated by advanced ultrasonication (e.g., focused and indirect) which provides extra degrees of freedom for controlling the CSD by decoupling the nucleation rate from liquid flow rate [3]. Process intensification strategy to improve the crystal growth control is also discussed which makes the whole process more robust. Experimental validation confirms that the proposed crystallizer designs reduce production time and equipment cost by orders of magnitude while suppressing secondary nucleation, attrition, and aggregation—dominant but undesired phenomena that worsen the ability to control the crystal properties.

References

[1] Z. Sun, N. Ya, R. C. Adams, and F. S. Fang, "Particle size specifications for solid oral dosage forms: A regulatory perspective," *Am. Pharm. Rev.*, vol. 13, no. 4, pp. 68–73, 2010.

[2] R. J. P. Eder, S. Schrank, M. O. Besenhard, E. Roblegg, H. Gruber-Woelfler, and J. G. Khinast, "Continuous sonocrystallization of acetylsalicylic acid (ASA): Control of crystal size," *Cryst. Growth Des.*, vol. 12, no. 10, pp. 4733–4738, 2012.

[3] M. Jiang, C. D. Papageorgiou, J. Waetzig, A. Hardy, M. Langston, and R. D. Braatz, "Indirect ultrasonication in continuous slug-flow crystallization," *Cryst. Growth Des.*, vol. 15, no. 5, pp. 2486–2492, 2015.

[4] M. Jiang, Z. Zhu, E. Jimenez, C. D. Papageorgiou, J. Waetzig, A. Hardy, M. Langston, and R. D. Braatz, "Continuous-flow tubular crystallization in slugs spontaneously induced by hydrodynamics," *Cryst. Growth Des.*, vol. 14, no. 2, pp. 851–860, 2014.

[5] M. Jiang, M. H. Wong, Z. Zhu, J. Zhang, L. Zhou, K. Wang, A. N. Ford Versypt, T. Si, L. M. Hasenberg, Y. E. Li, and R. D. Braatz, "Towards achieving a flattop crystal size distribution by continuous seeding and controlled growth," *Chem. Eng. Sci.*, vol. 77, pp. 2–9, 2012.

[6] A. J. Mahajan and D. J. Kirwan, "Micromixing effects in a two-impinging-jets precipitator," *AIChE J.*, vol. 42, no. 7, pp. 1801–1814, 1996.

455203 Continuous Cooling and Membrane-Assisted Crystallization with Airlift Crystallizer

Abstract:

Crystallization plays a substantial role in separation and purification of chemical and pharmaceutical compounds. The common problems in batch crystallisation processes often include large variations in the product quality per batch, inflexibility of the process and complex process controllability. The reason could be difficult handling of wet or semidry crystalline materials and high necessity of operating labour force in batch processes [1]. Furthermore, a common problem in conventional stirred crystallizers is the entanglement of multiple phenomena such as primary nucleation, growth and secondary nucleation.

Initially, batch experiments with airlift crystallizer confirmed that it can suppress secondary nucleation to a higher level than conventional stirred crystallizer [2].

Subsequently, experiments with membrane unit showed its feasibility in assisting crystallization processes by providing sufficient rate and level of supersaturation.

Based on results from membrane unit experiments effective parameters are known as feed temperature and air flow rate. The rate and level of supersaturation can be directly controlled and manipulated by changing these parameters for different process requirements.

In this study the performance of a continuous operated airlift crystallizer is discussed and secondly it is shown that membrane distillation can be applied as an alternative technology for the generation of supersaturation in combination with an airlift crystallizer.

Next, membrane unit and airlift crystallizer were combined. The setup is capable of operating in batch and continuous mode, cooling and membrane-assisted operation. Membrane-assisted Crystallization (MaC), firstly meet the requirements for decoupling the crystallization phenomena toward a more flexible process optimization and secondly it has the potential to operate in continuous mode and to significantly reduce energy consumption and operating labor force.

In this contribution feasibility of performing cooling and membrane-assisted crystallization using airlift crystallizer is presented. It is shown that a continuous membrane-assisted airlift crystallization is believed to be more controllable, flexible and energy efficient.

A model is developed and validated which is capable of predicting the membrane flux and level of supersaturation generation and the crystallization kinetics. The model is used to design and optimize the process.

1. Myerson, A., *Handbook of industrial crystallization*. 2002: Butterworth-Heinemann.

2. Lakerveld, R., J.J. Van Krochten, and H.J. Kramer, *An air-lift crystallizer can suppress secondary nucleation at a higher supersaturation compared to a stirred crystallizer*. Crystal growth & design, 2014. **14**(7): p. 3264-3275.

472727 Continuous Heterogeneous Crystallization of Active Pharmaceutical Ingredients on Excipients Surfaces

Abstract:

Novel continuous heterogeneous crystallization processes were developed, which the active pharmaceutical ingredient is crystallized directly on the surface of an excipient within the crystallizer. The product is then dried, and formed into tablets without the need for complex downstream processing steps such as milling, sieving, granulation, and blending. The APIs and excipients systems selected by study of heteroepitaxial mechanisms, and surface functionality of substrate. The effect of various process parameters, such as temperature, residence time, mode of operation, and excipient loading were studied. Three different process types in two categories of direct crystallization on surface of crystalline excipient and on surface of a carefully selected polymer (with surface modification) were studied. The excipient selection, and process parameters design have significant impact on drug loading, avoiding bulk nucleation and crystallization, control of API crystals shape and size, and control of process. In general, a decrease in temperature as well as the increase in residence time, both increased the drug loading of excipient with API. Within the range of our analysis, proportional drug loading can be easily tailored for a wide range depending on required drug dosage per tablet. The fully continuous processes, and eliminating the downstream steps result in producing crystalline compounds and final form (tablets) in significantly faster, more efficient, with lower footprint, and more economical.

468841 Mixing Models for the Prediction of Fundamental Properties of Powder Blends

Abstract:

Predictive models of the flow behavior and bulk density of powder blends have been developed to enable a priori determination of these attributes from measured values of individual blend components. The models collectively serve as an effective means of evaluating the interplay between physical properties of novel APIs and accompanying bulk excipients, which serve as the basis of any prospective formulation. In practice, they provide the basis of an advanced formulation design tool which eliminates the need for time and material consuming studies to determine fundamental system attributes. Characterization of API properties is instead the only requirement. Binary systems were used to build the models; however, the mixing rules which serve as their basis enable the prediction of the behavior three- and four-component systems as well. Here we demonstrate both the development of the models themselves as well as their application to early stage development.

471114 CFD-DEM Model for Fluidization Segregation during Transfer of Pharmaceutical Blends

Abstract:

This work presents a case study where detailed coupled computational-fluid-dynamics and discrete-element-method CFD-DEM models are used to identify and address powder segregation during a commercial-scale pharmaceutical tablet manufacturing operation. In this process, a well-blended powder mixture of active pharmaceutical ingredient (API) and several excipients is discharged from a bin blender through a long (~3 m) transfer chute before feeding into a tablet press. The tablets produced exhibited a systematic variation in content uniformity and evidence of segregation during the discharge process.

A coupled CFD-DEM modeling approach was selected to virtually simulate the process and probe the effects of material properties and process conditions in order to recognize the mechanisms responsible for segregation. The material properties were first calibrated using experimental measurements from the FT4 Powder Rheometer and the Schulze Ring Shear Tester. Next, the model (and input material parameters) were validated in a number of ways, including a comparison with a small-scale fluidization-segregation tester. Subsequently, simulations of the industrial-scale transfer chute were performed and the results successfully reproduced the segregation trends observed in the actual physical systemâ€”air-driven segregation of dissimilar-sized API and excipient particles was identified to be the root cause of content uniformity variations.

This calibrated and validated CFD-DEM model offers an opportunity to virtually test various strategies to mitigate segregation. Specifically, alternative API particle size distributions were simulated with results indicating that an optimum API particle size distribution can help to reduce the risk of segregation. Thus, these CFD-DEM computational models have helped to improve the understanding of the source of segregation issues and guided the strategy to reduce the segregation risk during the manufacture of a commercial pharmaceutical product.

458178 Experimental and Numerical Simulation Study of Elutriation Segregation Phenomenon during Gravity-Driven Flow of Powder

Abstract:

Elutriation segregation is a key concern in dry powder handling processes. This phenomenon becomes critical during the handling and transportation of a powder blend that contains appreciable amount of fines in presence of air. Elutriation results in a vertical segregation of particles caused by counter-current air flow in a gravity driven flow of powder. It adversely affects the final product quality, and might also result in huge financial losses due to scrapping of whole batch of the product due to non-uniformity. Recently, Procter & Gamble and ANSYS collaboratively started looking into understanding and quantifying the elutriation segregation phenomenon through experiments and simulations. The goal is to quantify the effect of powder and process parameters that affect particle segregation. A pilot scale elutriation rig is constructed to experimentally simulate this phenomenon and then further guide the modeling efforts. Eulerian-Eulerian framework is used to model this segregation phenomenon. Particle-particle interactions are modeled using Kinetic Theory of Granular Flow (KTGF), while interaction between the phases is modeled through drag and exchange terms. Experimental and simulation results of elutriation segregation will be presented.

469514 Discrete Particle Simulations of Particle Segregation in Rotating Fluidized Beds in Vortex Chambers

Abstract:

In vortex chambers, the tangential injection of the gas via multiple slots in the outer cylindrical wall and the central evacuation via a chimney generates a fast rotating flow in which particles can be suspended. Due to their inertia, the particles concentrate near the outer cylindrical wall, forming a rotating fluidized bed. High-G operation allows gas-solid slip velocities exceeding the terminal velocity of the particles in the earth gravitational field while maintaining a uniform and dense particle bed [1]. Intensified gas-solid contact can be combined with intensified gas-solid separation and particle segregation. The latter is of interest for certain applications, such as fine particle coating and granulation [2] or chemical looping or oxy-fuel combustion [3], but had not been studied in detail so far [4]. With a rotating fluidized bed of mono disperse particles, intense particle bed mixing in the radial and axial directions was observed [5], but a clear indication of particle segregation followed from the separation of uncoated and coated particles in fine particle coating experiments [2]. In this study MFIX-DEM simulations are used to quantify segregation with respect to particle size and density in a vortex chamber generated rotating fluidized bed. A 15 cm diameter vortex chamber with 8 x 4 mm gas inlet slots was used. Batch operation with binary particle systems is simulated for different particle size and density ratios and starting from a well-mixed particle bed. Density ratios between 1.125 and 10 and size ratios between 1.06 and 4 have been considered. The total bed height was 2 to 4 cm, depending on the operating conditions. Both the rate of segregation and the steady state quality of segregation were analyzed. Within a relatively wide size and density range, fast (0.2-3 sec depending on the conditions) and efficient segregation takes place, with an inner and slower rotating bed of the lighter particles forming within the outer and faster rotating bed of the heavier particles, each bed only few cm high only. Contamination of the outer bed with lighter particles occurs more easily than contamination of the inner bed with heavier particles and increases with decreasing size or density ratio. With a too high size or density ratio, on the other hand, the inner bed is seen to expand, indicating a decrease of the bed rotation speed, and some bubbling in the inner bed is observed in certain cases. Too

much bed expansion can lead to solids losses via the chimney, only observed in extreme cases in this study. The velocity of fine particles is seen to fluctuate due to the formation of a wavy interface between the segregated layers. Porosity plots show that vortex chambers with more than 8 slots have to be used to guarantee particle bed uniformity and improve the vortex chamber performance. Finally, the flexibility with respect to the gas flow rate is discussed, a unique feature of vortex chamber generated rotating fluidized beds.

References

[1] Eliaers P, De Wilde J. Drying of biomass particles: Experimental study and comparison of the performance of a conventional fluidized bed and a rotating fluidized bed in a static geometry. Drying Technol. 2013;31(2):236 245.

[2] Eliaers P, de Broqueville A, Poortinga A, van Hengstum T, De Wilde J. High-G, low-temperature coating of cohesive particles in a vortex chamber. Powder Technol. 2014;258:242-251.

[3] Volchkov EP, Dvornikov NA, Lukashov VV, Abdrakhmanov RKh. Investigation of the flow in the vortex chamber with centrifugal fluidized bed with and without combustion. Thermophys Aeromech. 2013;20(6):663 668.

[4] De Wilde J, Richards G, Benyahia S. Discrete Particle Simulations of Segregation in a Bi Disperse Rotating Fluidized Bed in a Vortex Chamber (accepted for publication in Advanced Powder Technology, in press).

[5] De Wilde J. Gas-solid fluidized beds in vortex chambers. Chem Eng Process. 2014;85:256-290.

458406 *Efficient DEM-Simulation of Granular Mixing Behavior for Large-Scale Applications*

Abstract:

Due to increasing computational power and rapid development of graphical cards in the last decades, Discrete Element Methods (DEM) is becoming popular computational technique even for simulations of large-scale granular systems. Industrial needs are fast and reliable simulation results, which should be obtained in reasonable computational time. Simulations of several weeks or months can be only carried out within some research projects. Process and device optimizations, where huge number of simulation variants is needed, is currently almost done using fast calculating 0D and/or 1D codes. Highly resolved 3D simulations are almost performed for several operating points, primary due to long computational time. Nevertheless, using smart simulation strategy together with high efficient codes, simulation time can be dramatically reduced and the complex simulation technique, such as DEM, can be successfully applied in product development process. This presentation shows an efficient simulation strategy to reduce computational time of a DEM simulation of granular mixing process. Using a scaling law to scale geometry and particle size by keeping the Froude-number constant, the time step of the simulation can be increase by factor between $O10^{-2}$ and $O10^{-3}$ and the duration of the simulation from several weeks to couple of days. The scaling law ensures that the particle behavior is still remaining unchanged even when the simulation time-step is increased. The simulation results are compared to standard simulation setup and to measurement data. A good agreement between simulation results and measurement data confirms applicability of the simulation strategy in the product development process where the huge number of simulation variants is needed in order to improve product quality.

461959 Experimental and Numerical Study of Granular Size Segregation in 3D Bounded Wedge Flows

Abstract:

Segregation due to particle size in flowing granular materials is an important phenomenon observed in a wide variety of industrial flowsâ€"in particular, hopper filling and discharge. As a bidisperse mixture is continuously fed onto a heap, small particles percolate downward between large particles within a thin flowing layer (order $O(10)$ particle diameters) and are deposited upstream onto the static heap, whereas large particles segregate to the top of the free surface and are advected downstream. The final segregation pattern is determined by the mass flow rate and width of the silo, as well as the particle size ratio. A recent continuum size segregation model quantitatively agrees with experimental spatial segregation patterns for bidisperse granular mixtures in simple quasi-two-dimensional (quasi-2D) heaps, where span-wise gradients in the velocity field are neglected. The generalization of this model to fully 3D geometries like cylindrical silos remains unclear. In this study, we investigate the kinematics and scaling behavior of flow during heap formation in an intermediate three-dimensional (3D) bounded wedge geometry, which introduces non-trivial divergence in the advection term involving the flow velocity. Results from 3D heap flow experiments of mono- and bidisperse granular mixtures in this geometry are compared and validated with numerical data from discrete element method simulations (DEM) for a range of wedge angles, feed rates, and particle mixtures. Frictional wedge sidewalls produce non-uniform azimuthal variation in concentration profiles consistent with results from DEM simulations using frictional sidewalls. Azimuthally periodic boundary conditions were also implemented in the DEM simulations and seem to accurately reflect a fully 3D bounded heap. Key considerations concerning the use of the 3D bounded wedge as an experimental tool in the study of segregation in a fully axisymmetric heap geometry are highlighted. Funded by The Dow Chemical Company.

Towards a Markov chain modelling for a planetary mixer: a preliminary study in a model mixer with blades

L. Legoix[1] , C. Gatumel[1], M. Milhé[1], H. Berthiaux[1], V. Mizonov[2]

[1] Université de Toulouse, centre RAPSODEE, Ecole des Mines d'Albi-Carmaux, campus Jarlard, 81013 Albi Cedex 09, France

[2] Department of Applied Mathematics, ISPEU, Rabfakovakaya 34, 153003 Ivanovo, Russia

cendrine.gatumel@mines-albi.fr / leonard.legoix@mines-albi.fr/ mathieu.milhe@mines-albi.fr / henri.berthiaux@mines-albi.fr / mizonov46@mail.ru

1- Introduction

The annual amount of powder involved in mixing operations in the industry is high and the quality level required for the blends is increasing as formulations become more complex. Planetary blenders are of wide interest because of their versatility and ability to process free-flowing and cohesive powders efficiently but the relation between powder characteristics, process parameters, mixing quality and energy consumption in such blenders still need to be clearly attested. As far as there is no constitutive relation predicting the result of agitation from materials properties and operating conditions, mixing process parameters are often derived from experiments. This process conception task may become tedious and dispendious for companies. In this context there is a need to develop models to predict agitation and mixing of powders, taking into account their physical properties. For this purpose, Markov models offer the advantage of short calculation times and have proved their efficiency in many powder mixing studies. We can cite for example simulations of a hoop mixer [1], static mixer [2] and continuous mixer [3]. Modelling mixing operations in planetary blenders by Markov Chains model is one of the issues of interest of the research center RAPSODEE.

Mixing in the Triaxe® blender is carried out by combining movements of rotation around a 15° inclined axis and gyration around a vertical axis. A study of in-mixer rheology in this blender (Figure 1) showed drastically different results depending on the flowing properties of the powders [4].

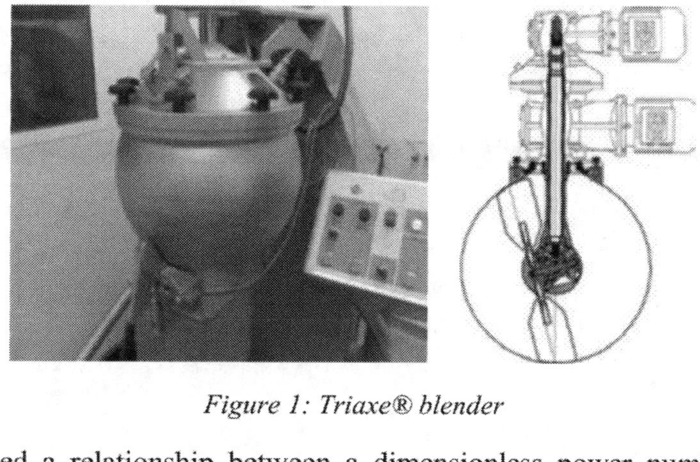

Figure 1: Triaxe® blender

We had established a relationship between a dimensionless power number and a Froude number involving two coefficients. A study above 23 powders with different mechanical characteristics showed that these coefficients depend on the cohesion of powders and operating parameters. In order to model the agitation and mixing in such a blender the rheology observed has to be linked to the flow of powders and the mechanisms of agitation - homogenization.

In this objective and because of the complex stirring pattern and the opacity of the tank, the study is decomposed into several steps. The first issue concerns the study of a simpler stirring device in a transparent tank. A transparent multi-position cylindrical mixer agitated by means of a four bladed propeller has been built for the needs of this study. Its purpose is to figure rotation when put in horizontal position.

In a first part we map the different flow regimes observed with different powders in this device, in relation to those observed previously in the planetary blender [4]. Stirring experiments are carried out with semolina as a free flowing powder and with lactose as a cohesive one in order to identify flow regimes, according to observations and in-mixer rheology. Following, tracer's experiments lead to identify the agitation mechanisms that are involved in mixing operations, in relation to the different flow regimes. In a last part we build a Markov chain model of the agitation by rotation in the mixer. The mixer volume is divided into states and the objective is to determine the rate at which a particle can move from a state to another in order to build a matrix of transition probabilities. For a given flow regime and a powder cohesion, the transition coefficients are determined thanks to the powder tracer's experiments.

2- Materials and methods

2-1 Powders

Two powders have been studied, a free flowing one and a cohesive one. The free flowing powder is fine semolina and the cohesive powder chosen is lactose available in two sizes (lactose Granulac140 and fine lactose Granulac 230).

Table 1 gathers the main properties of these two powders, d_{50} being the median diameter measured by LASER diffraction with a Mastersizer3000 (Malvern) under an air pressure of 3.5 bar. The particle true density ρ_p has been measured thanks to an Accumulator® Pyc 1330 (Micromeritics) with the 10 cm^3 cell. The bulk and tapped densities ρ_b and ρ_t have been measured with a volumenometer (Erweka®) for a powder mass of 110 g. The tapped density has been measured after 1000 taps, which is enough to be sure that the powder cannot be more packed by the apparatus since the volume stabilizes around 300 taps. The Carr Index for semolina is smaller than 15 %, which makes it an easy flowing powder. For lactose, the Carr Index is greater than 15 % which is characteristic for powders that do not flow easily [5]. This attests that semolina can be considered as a free-flowing powder, whereas lactose is a cohesive system.

Table 1: Powder properties.

Powder	d_{50} (µm)	ρ_p (kg.m^{-3})	ρ_b (kg.m^{-3})	ρ_t (kg.m^{-3})	CI (%)
Semolina	312	1463	679	720	5.8
Lactose	61	1533	661	795	16.9
Fine lactose	26	1539	495	638	22,5

2-2 Model bladed blender

A prototype convective mixer has been designed and built as a multi-functional tool in order to get a better understanding of powder flow regimes and governing mechanisms under very different geometrical configurations (Figure 2). The cylindrical vessel can be positioned with different inclinations from horizontal to vertical position. It is maintained horizontal for this study. The volume of the tank is delimited by two removable lateral walls as to vary the distance from the stirring system that, as a consequence, is not necessarily centered.

The tank is made of PMMA because of its mechanical resistance and so as to observe the progression of the mixing process. The vessel diameter is 24 cm, blade to wall distance is 1 mm and blade width is 6 cm. It has been designed to be in geometric similitude with the Triaxe® in order to compare results from dimensionless analysis. In particular, the ratios between the blade widths of the two mixers are the same as the ratios of the blade lengths and ratios of the vessel diameter.

Between 1 and 4 blades can be mounted on the shaft, in any of the four locations available. The blade angle α with respect to the shaft can also be adjusted blade by blade, between 0° and 180°. In the present study, the four blades will be mounted on the shaft, all at the same angle. Two angles will be studied: 0° and 33°. For α= 0°, the blades will be qualified as "straight blades".

The rotational speed can be varied between 2 and 20 rad.s^{-1} thanks to a gear-motor. A torque-meter allows recording the motor torque between 0 and 15 N.m. Rotational speed and torque are acquired thanks to a Labview® program that records the torque at 1 Hz. The powder can be loaded thanks to a rectangular aperture in the cylinder.

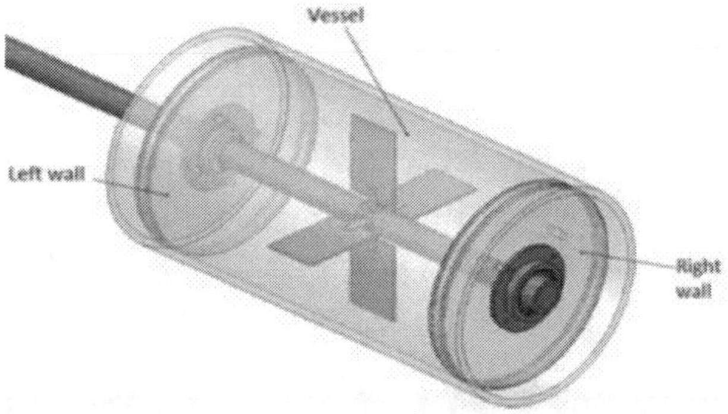

Figure 2: Model blender with straight blades

2-3 Experimental procedure for observation of flow regimes and rheology recording

As stated above, the torque-meter allows performing in-process rheology by measuring the torque exerted on the shaft during stirring. Each experiment begins by running the blades 5 min at a speed of 8 rad.s^{-1} in the empty vessel. Torques are then measured at a speed increasing from 3 to 18 rad.s^{-1} with a 3 rad.s^{-1} step. For each speed, the agitation is maintained during 5 to 10 min to be sure that the torque is stabilized, after which its mean value is calculated over 1 min and will be considered as the recorded empty torque. Then, the vessel is filled with the powder studied and after a conditioning step that lasts 5 min at 8 rad.s^{-1}, the same torque measurement procedure is repeated. For all studies, 3 filling ratios f are investigated: 0.42, 0.63 and 0.83. These are calculated using the bulk density of the powder ρ_b [kg.m^{-3}], the mass of powder m_p [kg] and the volume of the blender V_{tank} [m^3] which is about 8.85 l (eq. 1).

$$f = \frac{m_p}{\rho_b V_{tank}} \qquad \text{eq. 1}$$

The power consumption P[W] can be calculated knowing the blade rotational speed ω[rad.s^{-1}], the empty torque T_0 [N.m] and the torque with filled vessel T_f [N.m] (eq. 2).

$$P = \omega\,(T_f - T_0) \qquad \text{eq. 2}$$

Dimensional analysis of this process leads to the definition of different numbers, such as the Froude number F_r and the power number N_p [6,7]. These two numbers can be calculated with g [m.s^{-2}], ω [rad.s^{-1}], the blade radius R [m], ρ_b [kg.m^{-3}] and P [W] (eq. 3 and 4). This approach is useful for scale-up calculations, or to predict power consumption at industrial scale, thanks to lab-scale or pilot-scale experiments that require less powder and consume less energy. With the rotational speeds considered in this work, the Froude numbers involved are 0.1, 0.4, 1.0, 1.8, 2.8 and 4.0.

$$F_r = \frac{R.\omega^2}{g} \qquad \text{eq. 3}$$

$$N_p = \frac{P}{\rho_b \omega^3 R^5} \qquad \text{eq. 4}$$

The different process parameters of each experiment are gathered in table 2. For these 12 combinations of powder cohesion, blade inclination and filling ratios, the 6 rotational speeds listed above, ranging from 3 to 18 rad.s^{-1}, will be studied.

Table 2: Powder and process parameters investigated.

Powder	Semolina						Lactose					
Blade angle (°)	0			33			0			33		
f	0.42	0.63	0.83	0.42	0.63	0.83	0.42	0.63	0.83	0.42	0.63	0.83
ω (rad.s^{-1})	3-18	3-18	3-18	3-18	3-18	3-18	3-18	3-18	3-18	3-18	3-18	3-18
Fr	0.1-4	0.1-4	0.1-4	0.1-4	0.1-4	0.1-4	0.1-4	0.1-4	0.1-4	0.1-4	0.1-4	0.1-4

Flow observations have been performed using a Canon 650D camera with a 18-55 mm lens. The same rotational speeds and filling ratios are studied as for rheology measurements. An exposure time of the digital sensor of 5 ms is used, which is the greatest duration available with the photoflash for this device. This is fast enough to strap up the powder flow, even for greatest rotational speeds, which is useful to observe flow regimes. All the pictures are taken in the front view of the blender, meaning that the vector of the angular speed goes from the left to the right. The direction of the rotation is reminded on each figure with a flat white arrow, while black arrows highlight powder flows.

2-4- Procedure for tracer's experiments

According to [8], "a tracer is an ingredient that is meant to trace out movements within a powder mixing system … One uses tracers to find out directions and pathways within a powder mixer". In the present work, we aim to understand powder flow inside a prototypal convective blender that can be representative of the whole spectra of convective blenders. The situation of this study is therefore corresponding to a single powder A that is representative of a flow category, i.e. cohesive or free-flowing [9, 10, 11]. In this case, a particulate product A' is chosen for his flow properties that are assumed -or demonstrated to be- identical to those of A, and is employed as a tracer of A. Particles of A' can be placed in specific locations in the mixer's volume, their concentrations in any locations (or cells) of the whole volume being detected with time. This also allows to calculate the RSD of A' with respect to A, and further determine "homogenization" kinetics. The term "homogenization" is employed despite the fact that it does not really fit with the situation, which refers more to "stirring".

Two tracers of different nature have been chosen separately for semolina and lactose, the main criteria being the size of the particles to limit segregation. Because of the different choices of tracer, the experimental protocol to determine the mass proportion in tracer inside the different zones of the mixer will be different.

The tracer that has been considered for semolina is made of semolina itself, colored by an iodine solution that has been previously prepared by dilution of solid iodine in a 96% ethanol-water solution. After impregnation of semolina, the black-colored semolina is stirred with a spatula and dried with hot air, in order to avoid agglomeration of grains. An image analysis technique has been set up and calibrated to quantify the proportion of colored semolina.

Talc has been employed as the tracer for fine lactose. Median diameter of talc is 16 μm whereas that of fine lactose is 26 μm. It will be assumed that talc is convenient for the study despite of its difference of shape –platelets- as far as lactose –polygonal- is a cohesive material that doesn't promote segregation. The analysis of the tracer content of a fine lactose-talc mixture is carried out by thermal degradation of lactose that occurs around 650°C.

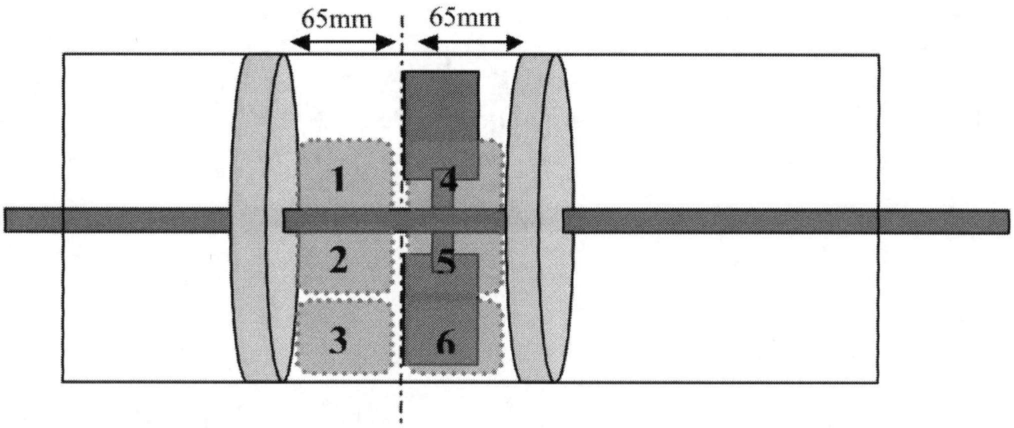

Figure 3: Front view of the prototype and approximate position of the six zones. The white arrow represents the real direction of rotation of the system.

The configuration of the blender chosen here consists in four straight blades for a vessel positioned horizontally. The distance between lateral walls is 13 cm, one of the walls being close to the blades with a 0.5 cm gap. Agitation speeds can reach 20 rad.s^{-1}. The blender volume is artificially divided into six zones or cells. These zones are of identical volumes and are represented schematically in Figure 3. Five of these six zones will be filled with the powder whose flow is studied (either semolina or lactose), the last zone being initially occupied by a tracer. For each experiment, the mass of powder poured in the blender is about 3232 g for semolina and 2357 g for fine lactose (including the tracer), corresponding to a volume bulk filling ratio f of 0.83. The powder inside each cell can be sampled layer by layer with a system of aspiration. If some powder is present over the top of the zones 1 or 4, it will be sampled and considered as belonging to these.

A typical experiment of powder stirring consists in filling the blender with a powder and its corresponding tracer, agitating the powder bed at a given rotational speed ω for a duration t, and then sampling the whole powder bed according to the six cells defined previously. For each of the two powders studied, two agitation speeds and several stirring times are chosen (Table 3).

For each experiment, three different initial positions of the tracer are studied: close to the free-surface (zone1), far from blades and surface (zone 3), in the blades direct action zone (zone 4).

Table 3: Product and process parameters investigated for experiments.

Powder	Semolina					Fine lactose				
Blade angle (°)	0					0				
f	0.83					0.83				
ω (rad.s^{-1})	6				15	6				15
Stirring time (s)	4	10	60	300	4	4	10	60	300	4

The mean powder weights contained into each cell, for each experiment of stirring made in this work, are represented in table 4. The fluctuating values of the mass are mainly due to the different state of compaction of the powder, depending on its position in the vessel. The angle of repose of the powder near the blades and near the walls has an impact on the mass sampled too. For semolina, the mass do not vary a lot from zone to zone, nevertheless it can be seen that the mean values are greater for zones 1 and 4 that are closer from the free-surface. Concerning lactose, the cells 1 and 4 contain also more powder, as well as zone 3 for which powders may be compacted at the bottom of the blender.

Table 4: Powder weights measured in each zones of the blender (g).

Zone	1	2	3	4	5	6
Semolina (g)	565 ± 55	509 ± 36	490 ± 57	602 ± 29	552 ± 58	484 ± 57
Fine lactose (g)	425 ± 73	344 ± 29	408 ± 63	414 ± 39	361 ± 74	379 ± 64

Knowing the mass of powder in a given zone and the mass proportion of tracer in this zone as given by the analytical techniques described above, it is possible to calculate the corresponding mass of tracer. The tracer mass fraction in zone i, will be denoted X_i, the sum of these fractions being equal to 1 at any time.

The Relative Standard Deviation (RSD) is the index chosen here to quantify the homogeneity of the tracer, for a given time and rotational speed. The RSD can be calculated from the standard deviation σ of the tracer mass proportion in each zone and the tracer mean composition in the mixer, which is equal to 0.167 since only one zone is filled initially with it (eq. 5). The more the tracer concentrations are close to each other, the more the RSD approaches zero.

$$RSD = \frac{\sigma}{0.167} \qquad \text{eq. 5}$$

3- Identification of flow regimes from observations and torque measurements

3-1 Flow observation and analogy with tumbling blenders

3-1-1 Free-flowing material

Qualitative observations of different flow regimes in the agitated zone of the blender, were made thanks to the pictures taken with the camera. Two main flow regimes have been distinguished for the free-flowing particulate system: the rolling regime and the cataracting regime. These names are given in analogy with those employed for tumbling blenders operating with free flowing material. For these flow regime determination, observations are characteristic of the flow in front of the blades, at the free-surface level. The classification given by Mellmann [12] can be adapted for the convective blender, since the powder flow has some similarities.

Concerning the rolling regime, the powder follows the walls or the blades until it reaches a vertical position, and then rolls down by avalanches (see figure 4) meaning that the centrifugal force is less influent than gravity. The Froude number, that compares the centrifugal acceleration with gravity, helps characterizing the flow regime in the blender. For a filling ratio of 0.42, 0.63 and 0.83, a Froude number of 1 is always the border of a flow regime change (Table 5).

At higher speeds, when the Froude number is greater than 1, a cataracting regime can be observed. Once the particles are not in contact with the walls or the blades, they follow the same trajectory and do not roll by avalanching. This is characterized by projections of particles in the air (see figure 5).

Table 5: Map of flow regimes for semolina agitation with the convective blender.

$f \backslash F_r$	0.1	0.4	1.0	1.8	2.8	4.0
0.42	Rolling	Rolling	Cataracting	Cataracting	Cataracting	Cataracting
0.63	Rolling	Rolling	Cataracting	Cataracting	Cataracting	Cataracting
0.83	Rolling	Rolling	Cataracting	Cataracting	Cataracting	Cataracting

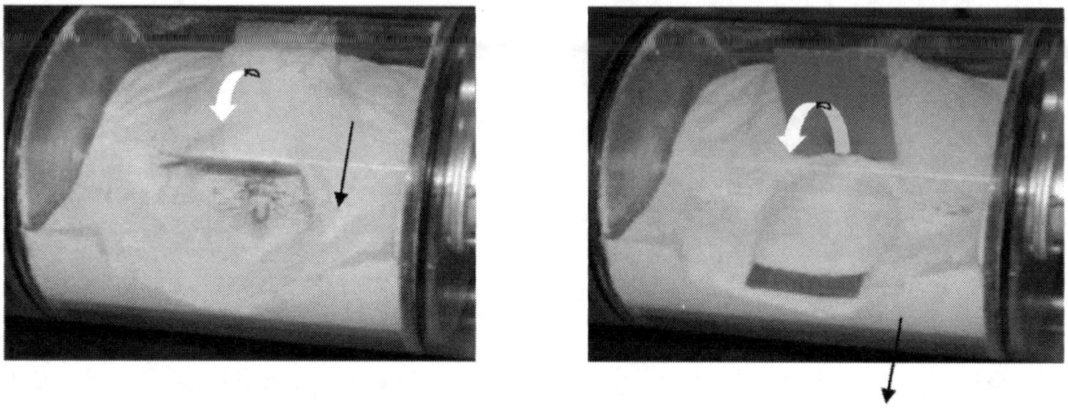

Figure 4: Observation of the rolling flow regime for semolina inside the convective blender equipped with straight blades, at Fr=0.1 and f=0.63. The white and black arrows show respectively blade rotation and powder motion.

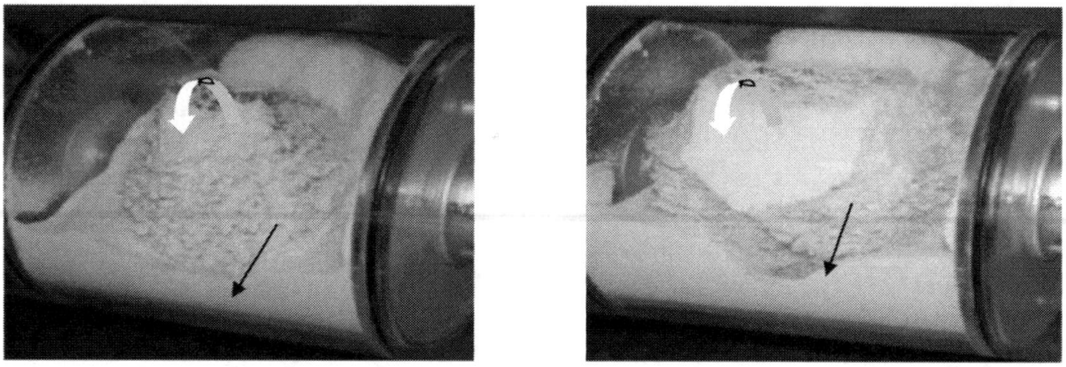

Figure 5: Cataracting flow regime of semolina inside the convective blender equipped with straight blades, at Fr=1.0 and f=0.63 (left), and at Fr=1.8 and f=0.83 (right).

3-1-2 Flow of cohesive powder

Experiments carried out with lactose show the same behavior to what concerns Froude numbers. A rolling regime and a cataracting regime can be identified when the Froude number is smaller or greater than 1 respectively. As for semolina, the filling ratio has no influence on these regimes.

However a difference between semolina and lactose flow can be detected in the rolling regime. For semolina, all the powder carried by blades rolls down, but for lactose, the powder avalanching occurs by blocks that fall one after the other. This is due to cohesion forces between particles that do not allow them to fall separately (Figure 6).

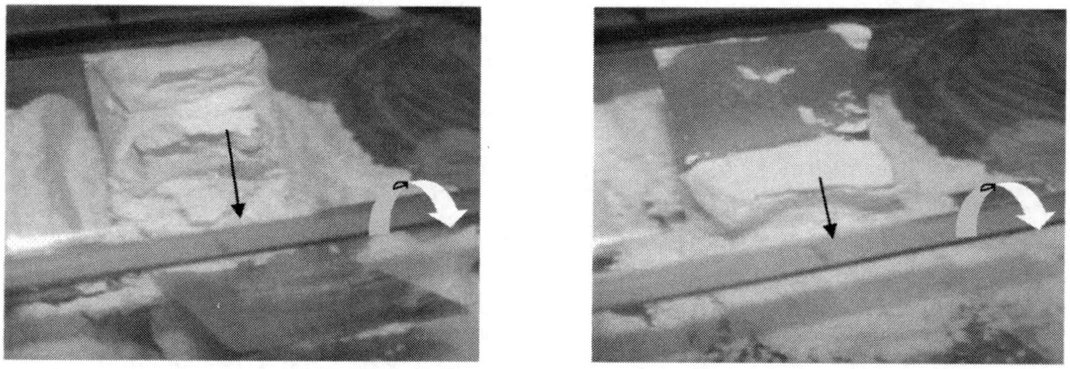

Figure 6: Consecutive falling of lactose powder by blocs, at Fr=0.1 and f=0.42

3-2 In-mixer rheology

The results of power measurements can be represented thanks to correlations between dimensionless numbers. The power is expressed as a power number N_p and the agitation speed as a Froude number F_r. As reported for powder stirring experiments in other studies of convective powder blenders [6,7], the typical correlation that links those numbers is of the following form (eq. 6):

$$N_p = a.F_r^b \qquad \text{eq.6}$$

The dimensionless correlation for semolina, with straight blades and a filling of 0.42, is given as an example in figure 7. This representation shows that, according to the two different flow regimes observed, for Froude numbers smaller than 1 and for Froude numbers greater than 1, separated correlations should be considered. The slope represents the *b* coefficient, the intercept being *ln(a)*. Concerning Froude numbers smaller than 1, meaning that the flow regime is rolling, the coefficient *b* is close to -1 and the coefficient *a* has a value about 0.65. For the cataracting regime, *b* is close to -0.7 and *a* value is about 0.63. The coefficient *b* is therefore significantly different between the two regimes, while the coefficient *a* can be said to be the same.

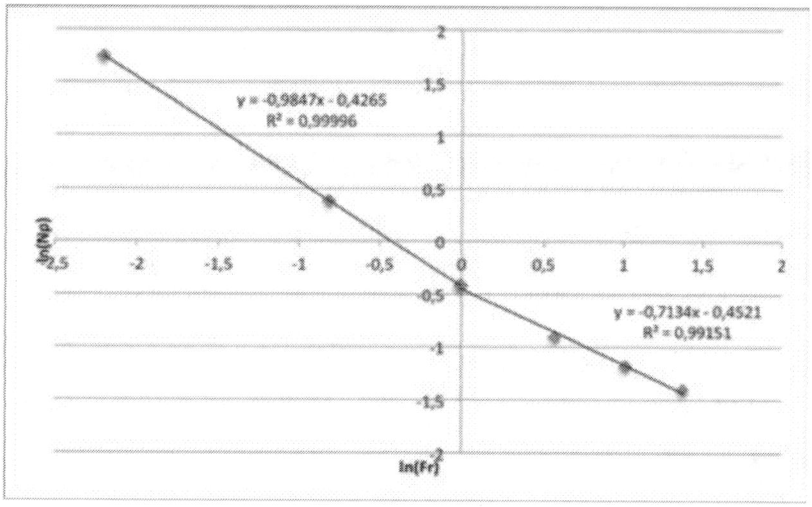

Figure 7: Dimensionless correlation between Fr and Np in the horizontal convective blender equipped with straight blades, f=0.42, semolina case.

Similar correlations have been investigated for the other filling ratios and blade angle, for the free-flowing and cohesive material (Table 6). The coefficient b is globally always equal to -1 in the rolling regime, except for the greatest filling ratio and when the blender is equipped with reclined blades (equals to -1.4). In the cataracting regime, while the correlation is of a regular quality, b is still close to -0.7 for straight blades and -0.8 for reclined blades. Once again, a filling ratio of 0.83 for reclined blades apparently drives to a different situation. As for semolina, the coefficient *a* is remarkably independent on the flow regime and the blade inclination, its value being the fact of the filling ratio, and the type of material: approximately three times less important for the cohesive powder than for the free-flowing powder.

Globally, it can be said that the coefficient *a* depends on the powder (kind, mass), whereas the coefficient *b* is more sensitive to a change in flow regime.

Table 6: Correlation $N_p=a.F_r^b$ coefficients for powder agitation with the convective blender.

parameters			**b**		**a**		**R^2**	
powder	**blades**	**f**	**rolling**	**cataracting**	**rolling**	**cataracting**	**rolling**	**cataracting**
semolina	**straight**	**0.42**	**-0.98**	**-0.71**	**0.65**	**0.63**	**0.999**	**0.991**
		0.63	**-0.95**	**-0.67**	**1.15**	**1.18**	**0.999**	**0.999**
		0.83	**-0.89**	**-0.75**	**1.84**	**1.79**	**0.999**	**0.978**
	reclined	**0.42**	**-0.92**	**-0.84**	**0.64**	**0.67**	**0.991**	**0.992**
		0.63	**-0.93**	**-0.79**	**1.30**	**1.32**	**0.998**	**0.998**
		0.83	**-0.86**	**-0.81**	**2.12**	**1.99**	**0.997**	**0.942**
lactose	**straight**	**0.42**	**-0.99**	**-0.51**	**0.22**	**0.22**	**0.999**	**0.986**
		0.63	**-0.96**	**-0.72**	**0.50**	**0.51**	**0.998**	**0.991**
		0.83	**-1.18**	**-0.58**	**0.41**	**0.44**	**0.985**	**0.952**
	reclined	**0.42**	**-1.08**	**-0.69**	**0.24**	**0.24**	**0.999**	**0.995**
		0.63	**-1.07**	**-0.84**	**0.47**	**0.49**	**0.998**	**0.999**
		0.83	**-1.40**	**-0.49**	**0.33**	**0.33**	**0.999**	**0.935**

Results are compared to those of the study performed in the Triaxe® [4] operated in the rotation mode only (see figure 8). The suppression of the gyration leads to a movement of the four blades close to the rotation provided by the horizontal convective blender. It must be said that for this reason the agitation speeds are not in the same range of Froude numbers, the Triaxe® covering smaller values while the polyvalent blender covers greater ones. However, there is a partial overlap particularly around Fr=1. Whatever the particulate system considered and the filling ratio, the log-log representation considered in the previous section leads to similar correlations for both technologies. This attests the ability of the polyvalent convective blender designed for this work to be used for scale up or power prediction in blenders that cannot be equipped with torque-meters.

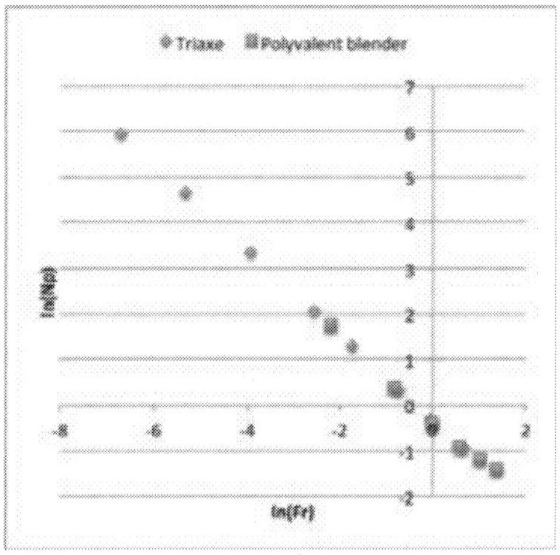

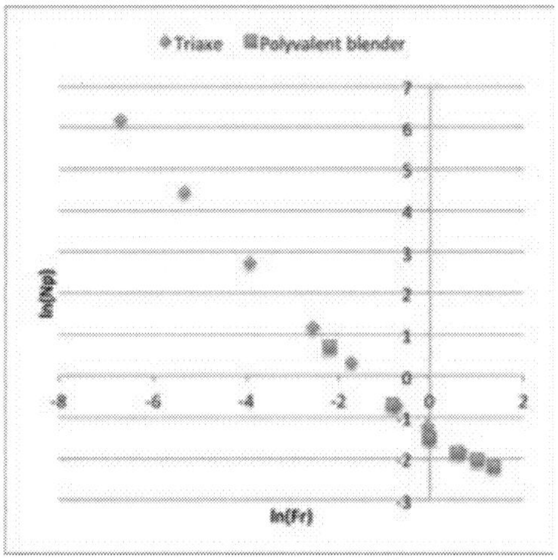

Figure 8: Comparison of the power law correlations obtained in the Triaxe® and in the actual horizontal convective mixer: semolina (left) and lactose (right), f=0.42

4- Identification of agitation mechanisms from tracer experiments

Three main transport mechanisms are identified during powder mixing operations: convection, diffusion and shearing [13]. Convection can be described as the movement of adjacent groups of particles. Diffusion is a movement of the powder at a scale closer to that of the particle size. Shearing is characterized by the slipping of planes within the mixture. Shear zones can be observed between the clumps formed by mixing through convection, so that shear cannot occur alone [14]. These transport mechanisms are responsible for mixing of particles. It is therefore not a surprise to see that many studies have focused on flow understanding through the use of tracers and the development of tracking techniques in mixing operations over the past 20 years or so.

In the following all experiments have been performed with straight blades.

Convection

When the initial position of the tracer is in zone 4, a fast dissipation of the tracer of semolina to the other zones of the blender can be observed (Figure 9). After four seconds of agitation, the mass proportion of tracer is almost divided by three, while the contents in tracer for zones 5 and 6 are increasing. After 60 s of stirring, the tracer contents in zones 4, 5 and 6 do not change anymore and are close to each other. This is typical of a convection mechanism where the blades, which are directly in contact with zone 4, move important amounts of powder. Convection has also been evidenced with a video camera recording (Figure 10). It is clear that after each blade pass, an important proportion of the powder is exchanged between zones 4, 5 and 6.

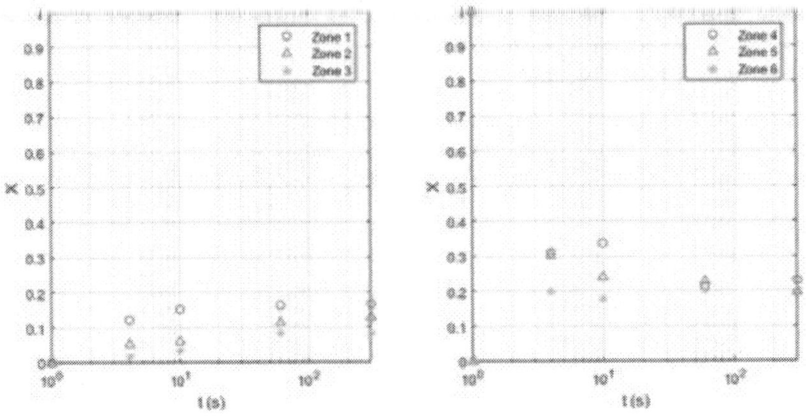

Figure 9: Evolution of composition of each zone in the mixer for semolina, with a blade speed of 6 rad.s^{-1} and tracer initially placed in zone 4.

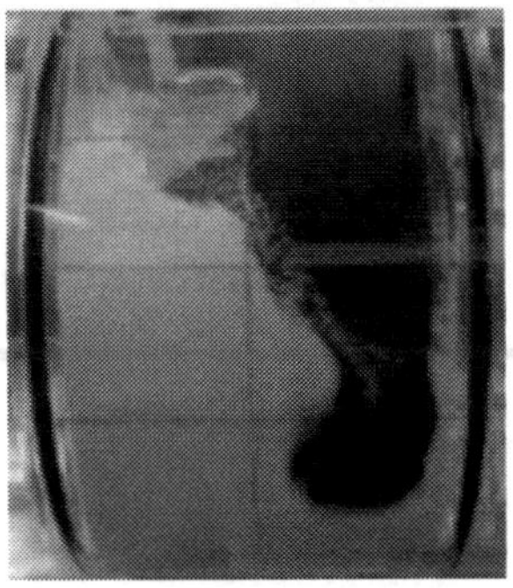

Figure 10: Front view of the mixer after one blade pass, tracer has been placed initially in zone 4, the blade rotational speed is 6 rad.s^{-1}.

Avalanching

When the tracer is initially placed in zone 1, its concentration stabilizes in each zone after 60 s of stirring (Figure 11). The tracer concentrations in zones 4, 5 and 6 are evolving the same way: if the tracer is present in one of these zones, it is quickly homogenized with the two others, due to the convection mechanism identified above. The tracer concentration in zone 1 decreases sharply, but not as fast as for the previous situation. In this case, this is not convection that moves powder from the zone 1 to others, but avalanches. This can be seen on figure 12 and explains the fast transfer of particles from zone 1 to other zones, despite the blades are not passing through this zone. Part of the powder bed is conveyed higher than zone 4 in the back of the blender and partly falls in zone 1, while part of the powder in zone 1 goes back on the front of the zone 4. This mechanism is slower than convection and allows the tracer present initially in zone 1 to interact with the agitated zones.

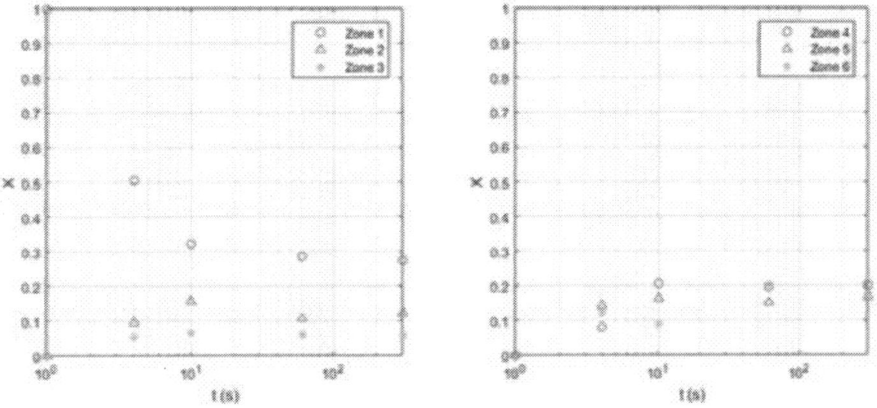

Figure 11: Evolution of composition of each zone in the mixer for semolina, with a blade speed of 6 rad.s^{-1} and colored semolina initially placed in zone 1.

Figure 12: Front view of the mixer after 4 s of mixing semolina at 6 rad.s^{-1}.

Diffusion

The dissipation of the tracer when initially placed in zone 3 is presented in figure 13, and may help to gain more insight into the slower mass transport mechanisms acting in this zone. A fast decrease of concentration in zone 3 is measured during the first ten seconds, after which the mixing of the tracer become much slower. It reaches a value close to 0.5 after 5 min of stirring. A more important quantity of powder remains in zone 3, as compared to the situations reported in the two previous sections. The concentrations in zones 1, 4, 5 and 6 increase and are all smaller than 0.1. A certain quantity of tracer seems to transit from zone 3 to zone 2, since after 60 s of stirring the decrease of concentration in zone 3 is followed by an increase of concentration in zone 2. This suggests the presence of a mass transport mechanism which is slower than convection and avalanching, between zones that far from the free-surface and in which blades do not pass through.

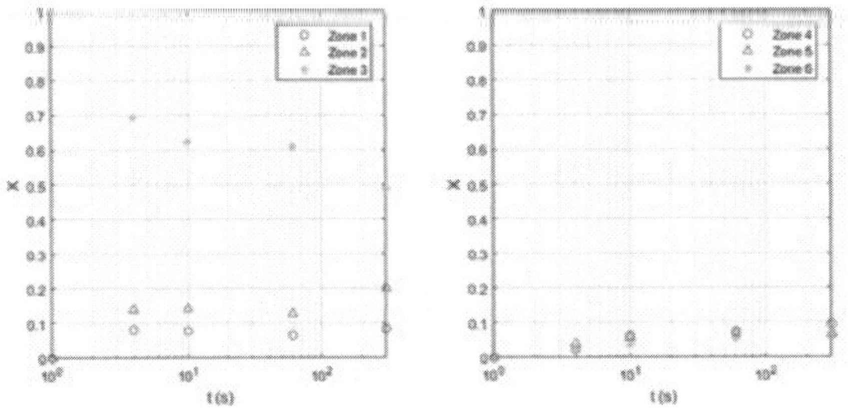

Figure 13: Evolution of composition of each zone in the mixer for semolina, with a blade speed of 6 rad.s^{-1} and colored semolina initially placed in zone 3.

The decrease of tracer proportion in zone 3, at the beginning of agitation, is mainly due to the fact that a part of zone 3 is dragged by shearing to the zones where blades are passing. Since this shearing mechanism is relatively slow and may be viewed at a scale close to that of the particles, it will be considered as a diffusion mechanism in this work. Indeed, the powder repartition inside zone 3 is far from being homogenous, a part of it being occupied by the tracer only (Figure 14).

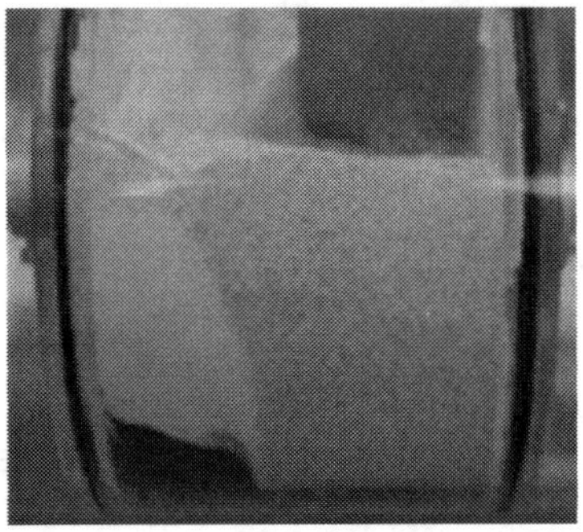

Figure 14: Front view of the blender after 5 min of stirring of semolina at 6 rad.s^{-1}, tracer initially placed in zone 3.

Kinetics of homogenization

RSD has been calculated according to equation 5 for semolina-tracer mixtures and lactose-talc (figure 15).

When tracer is filled in cell 4 (fig. 15 –left) the evolution of the RSD is similar between the two particulate materials, with a sharp decrease of the RSD during the first seconds, characteristic of the convection mechanism. However, for the cohesive powder, the RSD is greater whatever the stirring time is. This may be attributed to the unmixed powder that remains in zones 2 and 3, for which the zero content in tracer gives rise to high RSD's.

When tracer is filled in cell 1 (fig. 15 –middle) the initial RSD decrease is almost as stiff as in the precedent case. Avalanches of semolina that take place at this moment are efficient to distribute the tracer to the blades zones. This mechanism is less achieving for lactose. Considering the RSD as a parameter to quantify mixing, the cohesive powders are clearly more difficult to blend than free-flowing powders when the tracer is placed initially in zone 1. It is also worth noting that stirring times longer than 60 s do not improve homogeneity that is finally worse than that obtained if the tracer is initially placed in zone 4.

When tracer is initially filled in cell 3 (fig. 15 –right) the powder is never perfectly homogenized since the steady-RSD is far from zero.

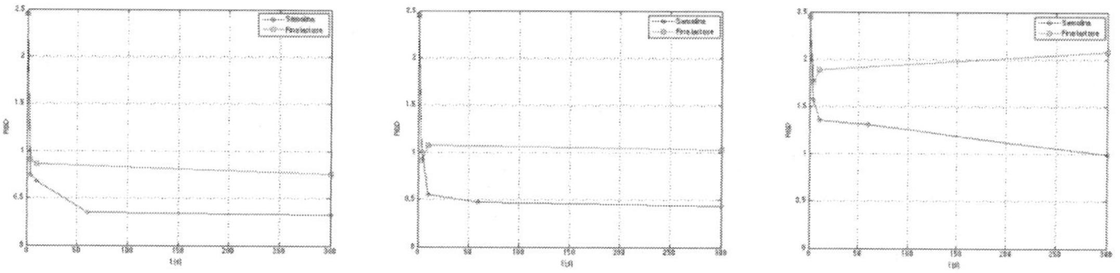

Figure 15: Homogenization kinetics for semolina and lactose, rotational speed of 6 rad.s^{-1} tracer initially placed in zones 4 (left), 1 (middle) and 3 (right).

The minimum is reached with a RSD value of 0.33 for the free-flowing material, when the tracer is initially placed in the blade zone (cell 4). It is clear with the RSD calculations that the cohesive powder is stirred with more difficulties, the worst case being that of a tracer placed at the bottom of the vessel, out of the blade zone (cell 3). This zone can be assimilated as a stagnant zone for which an important proportion of the powder will never transit to the other zones of the vessel.

Influence of flow regime

For rotational speeds greater than 9 rad.s^{-1}, the flow regime is cataracting. Cataracting is characterized by a movement of projection of the powder carried by the blades at the free surface, since centrifugal force is greater than gravity. In the figures 16 and 17, we report tracer distributions obtained after four seconds of mixing at a rotational speed of 15 rad.s^{-1} and compare it with the results presented for the rolling regime (6 rad.s^{-1}) for 4 seconds of agitation.

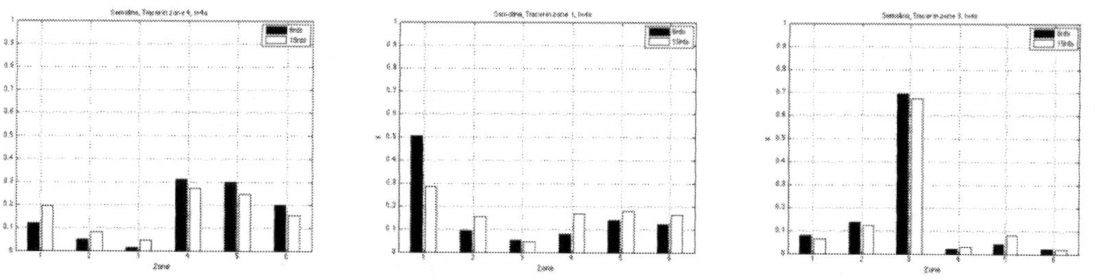

Figure 16: Impact of the rotational speed on the distribution of the free-flowing powder over the whole mixer, 4 s stirring time, tracer initially placed in zone 4 (left), zone 1 (middle), and zone 3 (right).

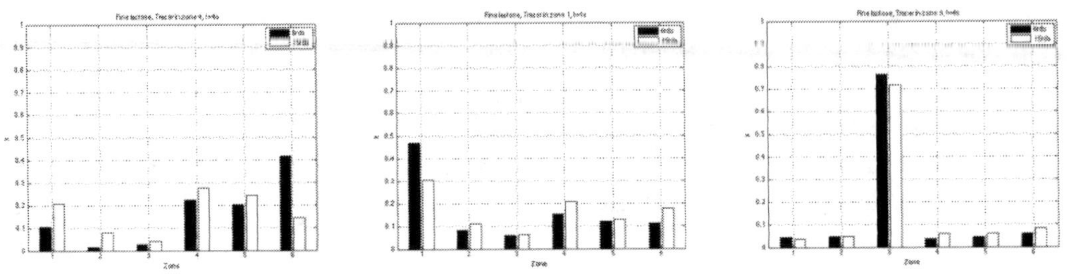

Figure 17: Impact of the rotational speed on the distribution of cohesive powder over the whole mixer, 4 s stirring time, tracer initially placed in zone 4 (left), zone 1 (middle), and zone 3 (right).

The flow regime that is taking place shows a major influence on the mixing kinetics. In particular, the cataracting regime looks more efficient to homogenize the bulk, whatever the particulate material studied.

A closer perspective of this work lays in the building of a model of powder agitation, able to catch all the tendencies revealed here, qualitatively and quantitatively. This may be based on a Markovian model by the inclusion of the mechanisms identified in terms of transition probabilities.

5- Markov model of powder flow

Applied to powder blending operations, Markov chains are based on a separation of the mixer into several parts called cells. The Markov chain model proposed is made of 6 cells of equal volume, which are similar to the zones defined previously (Figure 3). This structure originates from the observations of avalanching mechanisms taking place at the surface of the powder bed, this phenomenon being particularly active in the upper half of the blender: zones 1 and 4 are thus supposed to behave differently from the other ones where this mechanism is not observed. As well, a vertical separation seems appropriate to describe the convection mechanism present in zones occupied by blades: zones 4, 5 and 6 may exchange higher amounts of powder than zones 1, 2 and 3 for which blades are not passing through.

The amount of material in each cell is supposed to remain constant, i.e. the model describes mass transfer in the system for a steady-state operation. Thus, it can simulate either the evolution of quantities or concentrations of a given material in the system. The cells can exchange powder and the proportion of powder in cells j that goes to cells i is denoted p_{ij}. This transport of powder between cells takes place after every transition, that is representing one blade pass (a quarter of shaft rotation). A matrix notation of the different proportions of tracer in different cells S_n after n transitions, and of the transition probabilities P, with p_{ij} coefficients placed in line i and row j, allows the calculations of the proportion of tracer present in cells after the $n+1$ transition (eq. 7). Since P is supposed to be linear and homogenous in this work, which means that the transition probabilities do not depend on the cells compositions and on the time, S_{n+1} can be calculated directly with the initial vector of cells S_0.

$$S_{n+1} = PS_n = P^n S_0 \qquad \text{eq. 7}$$

The matrix of transition probabilities P is based on the three main flow mechanisms that have been identified in this mixer, namely convection, avalanches and diffusion.

The mechanism of convection is directly induced by the blades. It is predominant in cells 4, 5 and 6 (Figure 18 -left) that are in direct contact with the blades. This mechanism stands for the transport of powder by blades. It can be represented by a transition coefficient c linking these cells. In the present model, we will consider that this mechanism doesn't take place between other cells.

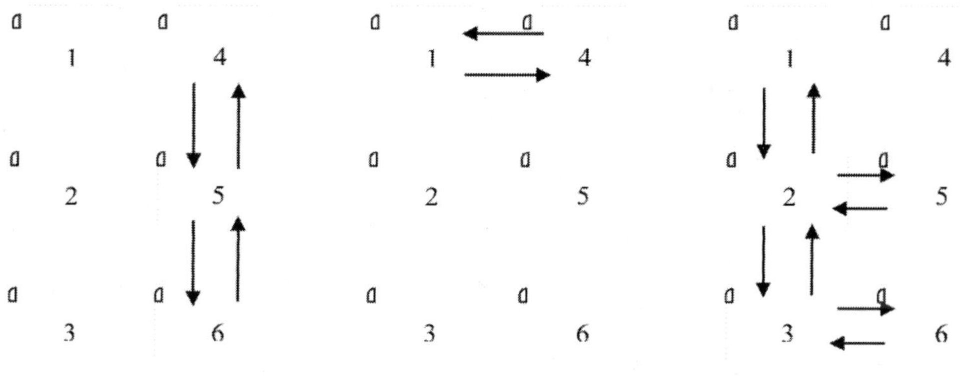

Figure 18: Convection (left), avalanching (middle) and diffusion (right) mechanisms transitions between cells

The mechanism of avalanching can be observed at the surface of the powder bed. It is actually composed of two distinct powder flows: one from cell 4 to cell 1 where the blades are uplifting the powder bed, and another from cell 1 to cell 4 where the blades and the powder are going downwards. It can be modeled as an exchange of powder between the cells 1 and 4 (Figure 18 -middle). In addition, we postulate that both mechanisms have the same magnitude and are thus represented by the same coefficient a that represents the proportion of powder transiting from cell 1 to cell 4 due to this avalanching mechanism at each transition.

Diffusion gathers all powder movement that cannot be observed easily during agitation. It is characterized by small amounts of powder that move from a cell to its neighboring cell (Figure 18 -right). This mechanism has been evaluated experimentally by the slow evolution of tracer content in cells 2 and 3, that are not directly in contact with the blades, and for which no avalanching mechanism can be observed. As vertical and horizontal mass transports are probably independent, it is chosen to represent the diffusion with two distinct coefficients: d_v for vertical diffusion and d_h for horizontal diffusion.

The matrix of transition probabilities P constructed with these different coefficients is given in eq. 8. Diagonal coefficients (the proportion of powder remaining in its current cell) are calculated to ensure that the sum of each row is equal to 1. Coefficients c, a, d_v *and* d_h have to be positive values ensuring that diagonal coefficients of P are not negative. The parameters should not be greater than 0.5. This is obvious for coefficient c because it would imply a negative coefficient in the matrix P. Concerning a, d_h and d_v, it is assumed -based on experimental observations- that these mechanisms involve less material than the convection mechanism.

$$P = \begin{pmatrix} 1-d_v-a & d_v & 0 & a & 0 & 0 \\ d_v & 1-2d_v-d_h & d_v & 0 & d_h & 0 \\ 0 & d_v & 1-d_v-d_h & 0 & 0 & d_h \\ a & 0 & 0 & 1-a-c & c & 0 \\ 0 & d_h & 0 & c & 1-d_h-2c & c \\ 0 & 0 & d_h & 0 & c & 1-d_h-c \end{pmatrix} \quad \text{eq. 8}$$

Coefficients c, a, d_v and d_h are identified thanks to the tracer's experiments. The corresponding operation conditions are presented in table 7. For both powders, homogenization has been determined with a tracer initially placed in zone 1,3 and 4. These homogenizations have been represented with the evolution of the tracer quantities in the 6 cells of the blender as a function of agitation time. The result of each experiment is a cells vector X_n, describing the distribution of tracer concentration in the 6 cells at the transition n. Transition number n is deduced from the speed ω [rad.s-1] and the stirring time t [s] (eq. 9). In this case, the agitation speed is about 6 $rad.s^{-1}$, which is associated to Froude numbers lower to 1 and to a rolling flow regime.

$$n = \frac{2.\omega.t}{\pi} \qquad \text{eq. 9}$$

Table 7: Operating conditions applied for mixing kinetics determination.

Powder	Semolina				Fine lactose		
ω ($rad.s^{-1}$)	6				6		
t (s)	4	10	60	300	4	10	300
n	15	38	229	1146	15	38	1146

An algorithm has been implemented to determine the best set of parameters suitable to fit experimental data for each initial position of the tracer.

The four parameters of the model for semolina and lactose have been evaluated (Table 8). The value of c is very close to 0.5: this means that cells 4 and 6 are supposed to exchange an amount of powder representing half of their volume with cell 5 at each transition by convection alone. Cell 5 sees almost of its powder content exchanged at each transition. Coefficient a shows that about 3.5 % of the material present at the surface of the powder bed is exchanged between cells 1 and 4 at each transition step, which is realistic. The diffusion is here only represented by d_v, model optimization suggesting that horizontal diffusion does not exist. The value of d_v represents an exchange of 0.7 % of the powder between zones 1, 2 and 3 that are not in direct contact with blades.

The model parameters have been also calculated for lactose. The coefficient of convection calculated is also close to 0.5, but is smaller than the c value obtained with semolina. This means that the convection mechanism due to blade's movement is less efficient for cohesive powders. There is a similar trend for the coefficient a: 2.8 % of the material contained in cells 1 and 4 is exchanged at each transition. Contrarily to semolina, the only diffusion supposed to be acting is the horizontal diffusion. The value of this coefficient is about 70 times less important than for the diffusion coefficient of semolina and close to the optimization uncertainty. The smaller mobility of the tracer in the lactose bed is certainly due to the presence of cohesion forces gathering particles that cannot move individually. The compressibility of lactose is more important than that of semolina. This can be quantified with Carr Index values, that is about 5.8% for semolina and 22.5% for fine lactose. This can explain the smaller mobility of powder in "non-agitated cells" (1, 2 and 3) as compared to the agitated ones (4, 5 and 6). In non-agitated cells, the powder may be slightly packed, whereas in agitated cells the powder may be aerated by the blades.

Table 8: Model parameters at 6 rad.s^{-1}.

	c	a	d_v	d_h
Semolina	0.4999	0.0347	0.0071	0
Lactose	0.4511	0.0279	0	0.0001

Comparisons between simulations and experimental kinetics of homogenization are presented on figures 21 to 23. Experimental points correspond to experiments used to identify the l parameters of the model. The charts are plotted in semi-logarithmic scale. For a better view of the kinetics, the initial state n=0 has been added to the charts.

Tracer initially fed in cell 4:

At the firsts transitions the convection mechanism moves large amounts of tracer from cell to cell and lead to homogenize rapidly cells 4, 5 and 6 whatever the powder is free-flowing or cohesive.

For semolina (fig. 19 -left), avalanching and vertical diffusion take place slower to finally distribute homogeneously the tracer in every zone. For lactose (fig. 19 -right), the concentration in cell 1 increases slowly due to avalanches but cells 2 and 3 are almost stagnant. This is correctly described by the model.

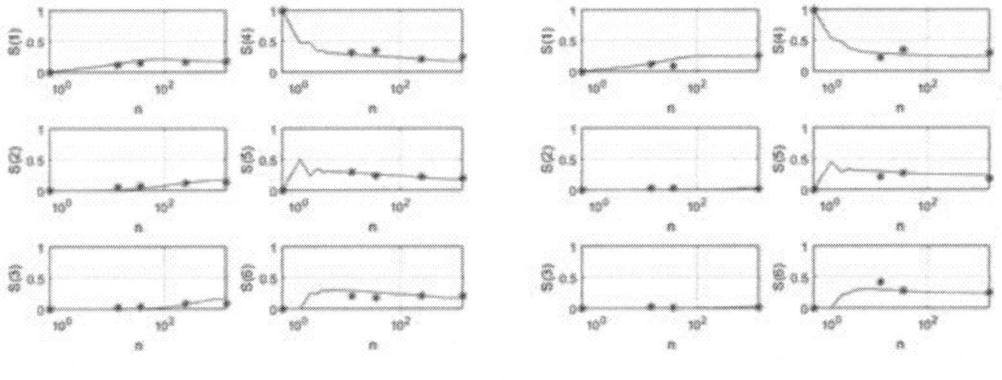

Figure 19: Tracer proportion evolution in each cell for semolina (left) and lactose (right) at 6 rad.s^{-1} with initial placement of tracer in cell 4, experimental values (points) and model (lines).

Tracer initially fed in cell 1:

For semolina (fig. 20 -left), when the tracer's initial position is cell 1, its experimental content is divided by two during the first 15 transitions since the avalanching is distributing the tracer towards cell 4. Convection takes place simultaneously to feed 5 and 6. According to the model, vertical diffusion takes place between 1, 2 and 3. The maximum time of simulation -5 minutes- doesn't allow to reach perfect mixing in this case.

For fine lactose (fig. 20 -right), experiments suggest that an asymptotic value of 0.5 is reached in cell 1 after 15 transitions whereas the model predict a diminution in tracer concentration (like for semolina). Avalanches of cohesive powders are discontinuous fallings of consolidated block, they don't have the same homogenization efficiency than avalanches of free-flowing powders. Moreover, in this case the model doesn't feature any vertical transport to feed cells 2 and 3.

The consolidation of cohesive powders may change during stirring operations, and in this case, non-linear model, where P is time dependent, should be more appropriate.

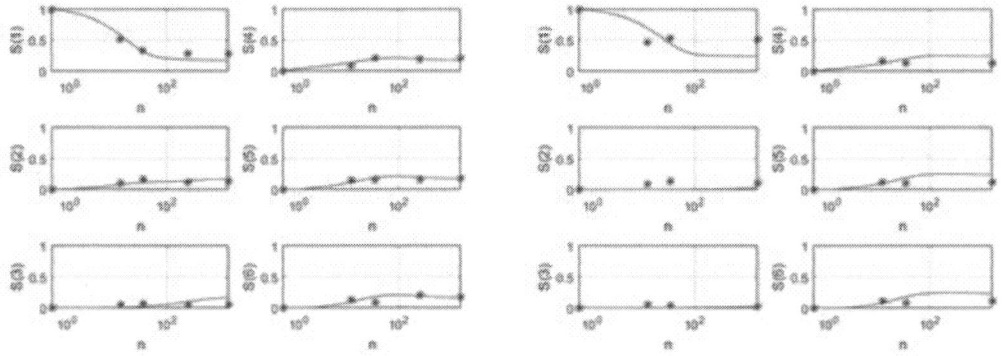

Figure 20: Tracer proportion evolution in each cell for semolina (left) and lactose (right) at 6 rad.s^{-1} with initial placement of tracer in cell 1, experimental values (points) and model (lines).

Tracer initially fed in cell 3:

Initial placement of tracer in cell 3 leads to a slower repartition of the tracers for semolina and lactose. Nevertheless, even if the values of the diffusion coefficients are small, the model does not fit very well the decrease of tracer's concentration in cell 3 neither for semolina nor for lactose.

In the case of semolina (fig. 21 -left) vertical diffusion is the only mechanism connecting cell 3 with the others. Experiments suggest that the decrease of tracer's concentration is faster during the 100 first transitions and then slows drastically, which is the opposite trend of the model. One wonders whether horizontal mixing by micro-avalanching mechanisms from cell 3 to cell 6 behind the passage of the blades would take place, but it is not identified by the model. However, the quantification of the coefficients is not precise because of the small values of tracer's concentration in most of the cells and the slow evolution of the concentrations.

In the case of lactose (fig. 21 -right), most of the tracer remains in cell 3, it is quite stagnant. The horizontal diffusion mechanism suggested by the model underestimates the transport of tracer from cell 3.

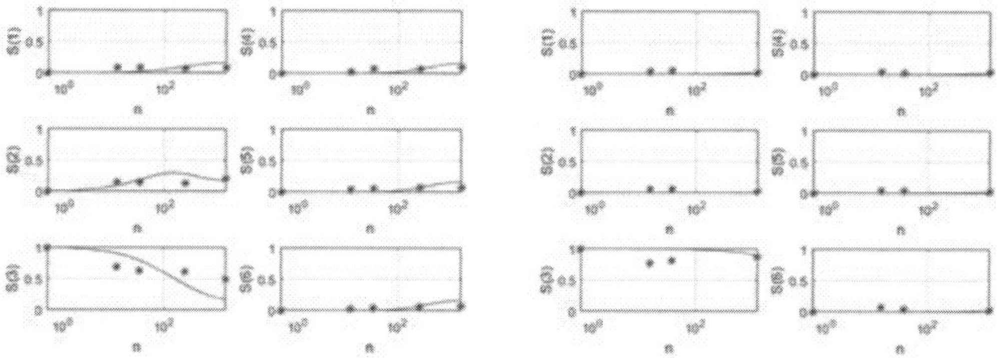

Figure 21: Tracer proportion evolution in each cell for semolina (left) and lactose (right) at 6 rad.s^{-1} with initial placement of tracer in cell 3, experimental values (points) and model (lines).

6- Concluding remarks and perspectives

The polyvalent convective blender, built for this work, allows to investigate in-situ powder rheology induced by blades, and may be helpful in scale-up procedures and understanding of powder flow in different systems. This device has been studied for two powders of different kinds: fine semolina that is free-flowing, and lactose, which is cohesive. Flow regimes have been identified for these two powders, three filling ratios (0.42-0.63-0.83). For Froude numbers smaller than 1, a rolling regime is evidenced and characterized by avalanches of powders at the free-surface. For Froude numbers greater than 1, a cataracting regime is observed: the powder is projected when it reaches the surface of the powder bed. However, the great differences in power consumption for the two powders highlight that the flow mechanisms in the powder beds might be completely different.

Correlations between a power number and a Froude number through a simple power law have been established for free-flowing and cohesive powders. These relations involve two coefficients depending on the powder type (coefficient a) and on process parameters as the blades inclination or the flow regime considered (coefficient b) and are in good agreement with a previous work in the planetary Triaxe ® blender agitated in the rotation mode only.

Tracer's experiments helped identifying the different mechanisms of homogenization that are appearing in a horizontal convective blender, according to the type of powder and the rotational speed of the stirring system that is the main process parameter. It has been demonstrated that the convection mechanism is predominant in zones in which blades are passing through while the avalanching mechanism takes place at the free-surface. Diffusion is a key mechanism to transport powder over zones that are not in contact with the free-surface or the blades. The flow regime that is taking place shows a major influence on the kinetics of homogenization. In particular, the cataracting regime looks more efficient to homogenize the bulk, whatever the particulate material studied is.

A model of powder mixing in the convective prototypal blender has been established thanks to a Markov chain modelling scheme. The three main mechanisms of agitation that have been established experimentally, namely convection, avalanches and diffusion, are implemented in the model as coefficients of transition. It concerns stirring times smaller than five minutes and a rotational speed of the impeller of 6 rad.s^{-1}. These parameters have been fitted thanks to experimental data. For both products, the convection coefficient value is the larger, followed by the avalanching coefficient. For free-flowing mixtures, the diffusion is only vertical, whereas for cohesive mixtures, the diffusion is only horizontal. Convection, avalanche and diffusion coefficients are always smaller for cohesive material as compared to coefficients for the free-flowing powder. This is due to cohesive forces between particles and powder bed compaction that counteract the movement of the powder, especially in cells for which blades are not passing through.

A close perspective to this work is to identify the model parameters in the case of the cataracting regime and go further into the diffusion mechanism which is not correctly represented by the current parameters d_v and d_h.

Furthermore, a non-homogenous and non-linear (where the matrix *P* depends on time and *S* vectors) Markov chain modelling should be suggested to provide a model that fits better the experimental kinetic data, especially for cohesive powders where the actual model shows difficulties. Also more experimental data are required to describe more precisely the evolution of tracer concentrations.

As far as the model mixer was found to be able to figure the rheology of the planetary Triaxe® blender, limited to the rotation movement, it has to be extended to the whole movement. This should be done considering that gyration distributes the zone of influence of the rotation blades all around the tank of the mixer.

Moreover, future works should focus on the influence of different configurations of blades, vessel and wall position, or vessel inclination, as well as combinations of these. This will give a better insight into the flow mechanisms at play, especially for cohesive systems and allow to model complex convective blenders.

Acknowledgements

We would like to thank Asma Toualoua and Luana Grillet, internship students in our laboratory, for their help during rheology measurements that are presented in this work. Many thanks to Pierre Bertorelle for his technical ideas and dedication, both necessary for the elaboration of the blending prototype.

References

[1] M. Aoun-Habbache, H. Berthiaux, V. Mizonov, An experimental method and a Markov chain model to describe axial and radial mixing in a hoop mixer, Powder Technology, 128/2-3 (2002) 159-167

[2] D. Ponomarev, V. Mizonov, H. Berthiaux, C. Gatumel, J. Gyenis, E. Barantseva, A 2D Markov chain for modelling powder mixing in alternately revolving static mixers of Sysmix® type, Chemical Engineering and Processing: Process Intensification, Volume 48, Issue 11-12, November 2009, 1495-1505.

[3] C. Ammarcha, C. Gatumel, J.L. Dirion, M. Cabassud, V. Mizonov, H. Berthiaux, Predicting bulk powder flow dynamics in a continuous mixer operation in transitory regimes. Advanced Powder Technology, Volume 23, Issue 6, November 2012, 787-800

[4] L. Legoix, C. Gatumel, M. Milhé, H. Berthiaux, Rheology of cohesive powders in a pilot scale planetary blender, Powder Technol. 10.1016/j.powtec.2016.09.021 (2016).

[5] R. Carr, Evaluating flow properties of solids, Chem. Eng. J., 72 (1965) 163-168.

[6] C. André, J.F. Demeyre, C. Gatumel, H. Berthiaux, G. Delaplace, Dimensional analysis of a planetary mixer for homogenizing of free flowing powders: Mixing time and power consumption, Chem. Eng. J., s 198–199 (2012) 371–378.

[7] P.C. Knight, J.P.K. Seville, A.B. Wellm, T. Instone, Prediction of impeller torque in high shear powder mixers, Chem. Eng. Sci., 56 (2001) 4457–4471.

[8] B.H. Kaye, Powder mixing in Technology and Engineering, Chapman and Hall ed., London, 1997

[9] M. Aoun-Habbache, H. Berthiaux, V. Mizonov, An experimental method and a Markov chain model to describe axial and radial mixing in a hoop mixer. Powder Technol., 3rd French Colloquium on Powder Science & Technology 128 (2002) 159–167.

[10] K. Marikh, H. Berthiaux, V. Mizonov, E. Barantseva, D. Ponomarev, Flow analysis and markov chain modelling to quantify the agitation effect in a continuous powder mixer, Chemical Engineering Research and Design,Trans IChemE Part A, 84-A11 (2006) 1059-1074

[11] D. Barling, D.A.V. Morton, K. Hapgood, Pharmaceutical dry powder blending and scale-up : maintaining equivalent mixing conditions using a coloured tracer powder, Powder Technology, V 270-B (2015) 461-469

[12] J. Mellmann, The transverse motion of solids in rotating cylinders—forms of motion and transition behavior, Powder Technology, 118 (2001) 251–270.

[13] P.M. Lacey, J Appl Chem 4 (1954) 257.

[14] J. Bridgwater, Fundamental Powder Mixing Mechanisms, Powder Technology 15 (1976) 215–236.

471776 Simulating the Wheat Flour Particle Separation Process during Sieving

Abstract:

Size based separation of flour is an important processing operation during wheat milling. The flour particles are highly cohesive that leads to discontinuous flow and agglomeration during the size separation process. Differences in particle and bulk behavior exist between the wheat flour from different classes and varieties. In general, wheat is classified as either hard wheat or soft wheat with inherent differences in chemical composition. Noticeable loss in throughput has been observed during size based separation of soft wheat flour compared to that of hard wheat flour due to differences in inter-particle cohesion. Similar to other particulate materials, the cohesion of wheat flour is highly dependent on the moisture content, particle size and on compaction. In this research, our objectives are to develop a prediction model for wheat flour flow behavior and to simulate the flour particle separation process using discrete element method (DEM). A granular bond number (GBN) model, based on Rumpf equation, was developed to predict the particle flow during the size based separation process. The GBN model, predicted the flow pattern of wheat flour with 0.05-0.2 standard error of prediction (SEP). The cohesion values predicted using GBN model was used as particle input parameters during DEM simulations. The DEM model was verified by the SEP values between the numerical and experimental results in terms of the percentage particles passing through the sieve screens. The effect of variables such as particle size, moisture content, and particle cohesion were studied with respect to overall separation efficiency and rate of separation for different sized particles. Based on the SEP values (0.02-0.07), it is demonstrated that the size based separation of wheat flour could be simulated using DEM method. The DEM model can satisfactorily estimate the sieving performance in terms of separation efficiency and the sieve blinding phenomenon specifically when there is significant interaction between particle-particle and particle-screen material.

461101 API Particle Formation: Technological Approaches to Forming Unique API Characteristics

Abstract:

The formation of highly differentiated APIs to enable specific downstream filtration, drying and drug product performance, in terms of flowability, blending, content uniformity and in vitro disintegration and dissolution is becoming an ever important feature in pharmaceutical process design.

Within this presentation the application of technological platforms, such as continuous processing, hybrid batch/continuous wet milling enabled system, will be discussed in detail. A series of case studies will show the work flow and design details associated to engineering and population balance models, reactor design, supersaturation monitoring / control to enable the delivery of highly differentiated API particle characteristics. The interface between the drug substance and drug product will be discussed at length to demonstrate the ability of engineering strategies to enable predictable downstream process performance.

466385 Seeding Policy and Optimization of Batch Crystallization Processes

Abstract:

Two important components of a recipe for the operation of a seeded batch crystallization process are the seed properties and the temperature (or supersaturation) profile. When seeds are used in batch crystallization, in most cases the objective is to relieve as much of the generated supersaturation as possible by the growth of the seeds, which corresponds to minimizing the nucleated mass.

Doki et al. (2002) showed that at least for some systems, nucleation could be effectively suppressed by using a sufficient seed mass, called the critical seed loading, and that the required seed mass depends on seed size. Later, Tseng and Ward (2014) showed that the critical seed loading could be determined knowing only crystal nucleation kinetics, which permitted the determination of the critical seed loading ratio for a large number of chemical systems for which nucleation kinetics have been measured. In this contribution we show that similar analysis can be used to determine a feasible range of seed sizes and corresponding seed loadings that will generate product crystals of a desired size (or size distribution) with negligible nucleation.

Furthermore, extending previous work by Hofmann et al. (2010), we show that nearly-analytical solutions for the optimal supersaturation trajectory (and Praeto fronts for multi-objective cases) can be determined for a variety of different objective functions. The nearly-analytical nature of the results allows greater understanding of the effect of the objective function on the shape of the optimal trajectory. This method also requires only knowledge of the nucleation kinetics, and therefore can be applied to determine the optimal supersaturation trajectory for a large number of systems.

Combining the two methods together, we develop a systematic procedure for rapidly determining operating recipes for seeded batch crystallization with a small number of experiments.

References

Doki, N., N. Kubota, M. Yokota and A. Chianese (2002). Determination of critical seed loading ratio for the production of crystals of uni-modal size distribution in batch cooling crystallization of potassium alum. Journal of Chemical Engineering of Japan 35(7): 670-676.

Hofmann, S., and J Raisch (2010). Application of optimal control theory to a batch crystallizer using orbital flatness, 16th Nordic Process Control Workshop, Lund, Sweden, 25â€"27 August 2010; Nordic Working Group on Process Control.

Tseng, Y. T. and J. D. Ward (2014). Critical seed loading from nucleation kinetics. AIChE Journal 60(5): 1645-1653.

466256 Defining a Design Methodology for Conversion of a Batch API Crystallization to a Continuous Crystallization

Abstract:

Crystallization is an essential unit operation for small molecule active pharmaceutical ingredient (API) manufacturing. In recent years there has been an increasing focus on continuous crystallization of such APIs. Advantages of continuous processing include increased control of product attributes, reduced equipment foot print, potential elimination of the requirement for scale-up activities, greater flexibility of equipment infrastructure and reduced expenditure.

This study focuses on creating a defined methodology for converting an existing batch crystallization process to a continuous system. Many previous studies have focused on characterizing continuous crystallization equipment and addressing processing challenges arising from continuous operation such as fouling and attaining steady state operation. The equipment used for this study was a paired CSTR system operated as an MSMPR. The use of this type of system for continuous crystallization has been previously demonstrated in literature [1-4].

The main goal of the design methodology was to transition an existing batch cooling crystallization process to a continuous format, while (at least) maintaining the yield and desired PSD of the existing batch crystallization. The API chosen as the case study for application of the design methodology was produced by a cooling crystallization following a salt formation in solution.

The initial stages of the design methodology consisted of a deeper investigation of the existing batch process, to include solubility screening, solvent selection optimisation, metastable zone width investigation and solid form characterisation. Concurrently, investigations of the existing crystallization process were performed with online monitoring of the process using FBRM and FTIR at 1L scale. From this work an improved understanding of induction times, nucleation temperature and the effect of the cooling rates and hold times on PSD as well as the crystal form was generated.

Temperature selection for continuous cooling crystallization is crucial, and a number of temperature set points were evaluated in batch format to improve confidence for continuous operation. A temperature strategy was selected for first implementation in continuous format, to be sequentially modified as understanding grew.

Following this work a two-step continuous cooling crystallization process was implemented using 2 glass jacketed 1 L reactors. Hot solution was fed to the first crystallizer using a peristaltic pump. Pressure transfer zones were used for intermittent transfer of slurry from the first crystallizer to the second crystallizer and on to a holding vessel for isolation. The effect of start-up strategy on particle size was investigated and was optimized to give the fastest time to steady state for the desired particle size. Various start up strategies were evaluated, including start up with a saturated solution, start up with a seed bed (with various seed loadings), as well as start up with crystal slurry from a previous batch respectively in the first tank. Through systematic trials the initial temperature strategy was sequentially modified until the desired operation and particle size production was achieved. Particle size was measured offline using a Malvern Mastersizer. Crystal form of the product was measured using XRD. The effect of various process parameters such a residence time, agitation and slurry transfer regime on the final PSD were investigated. The product yield was found to compare well with the existing batch process.

This project was made possible by funding from Enterprise Ireland Innovation Partnership Project IP-2014-0356

1. Hou, G., et al., *Development and Characterization of a Single Stage Mixed-Suspension, Mixed-Product-Removal Crystallization Process with a Novel Transfer Unit.* Crystal Growth & Design, 2014. **14**(4): p. 1782-1793.

2. Morris, G., et al., *Estimation of Nucleation and Growth Kinetics of Benzoic Acid by Population Balance Modeling of a Continuous Cooling Mixed Suspension, Mixed Product Removal Crystallizer.* Organic Process Research & Development, 2015. **19**(12): p. 1891-1902.

3. Power, G., et al., *Design and optimization of a multistage continuous cooling mixed suspension, mixed product removal crystallizer.* Chemical Engineering Science, 2015. **133**: p. 125-139.

4. Polster, C.S., et al., *Pilot-Scale Continuous Production of LY2886721: Amide Formation and Reactive Crystallization.* Organic Process Research & Development, 2014. **18**(11): p. 1295-1309.

467662 Tailoring Crystal Size Distribution with Sonocrystallization

Abstract:

Control over the product crystal size distribution (CSD) is very important in industrial crystallization for efficient downstream processing and improved formulation properties[1] Sonocrystallization is a prominent technique to reduce the crystal size and to generate uniform crystal size distributions[2, 3]. Current state of the art includes the sonocrystallization assisting PAT(Process Analytical Technology) to better understand the sonocrystallization process to better control the crystal size distribution to meet the specific requirements of an active pharmaceutical ingredient (API). Various approaches to use ultrasound for size distribution control in a batch cooling crystallization have been investigated. Solute concentration is monitored by In-situ React IC-IR and FBRM used to identify the metastable zone width and crystal chord length distributions. Crystallization of Piracetam in Isopropyl alcohol (IPA) was studied in a batch crystallizer with a working volume of 500ml. A calibration based approach was used for the solute concentration measurement. Crystal size measurement using light microscopy and Image-processing is a promising technique to determine the actual CSD rather than measuring with Laser diffraction technique. Width of the crystal size distribution characterized as Span = (D(90)-D(10))/D(50) determines the final product size distribution. It was observed that, conventional cooling crystallization with operating conditions of agitation rate and cooling rate resulted in bigger and broader crystal size distribution. However, reduced Meta stable zone width (MSZW), lower crystal size and lower Span were observed in case of continuously sonicated crystallization. Application of ultrasound in MSZW at low power of ultrasound and lower cooling rate leads to creation of lower nucleation rate and bigger crystals with narrow distribution. The application of ultrasound prior to nucleation leads to higher nucleation rate (which was observed from FBRM count rate) and smaller crystals with a more narrow distribution. Careful supersaturation control after the nucleation step can lead to the formation of uniform product crystals. The only drawback of application of ultrasound in crystallization process is that the heat liberation due to high power ultrasound, which will be difficult to control by cooling systems. The temperature rise because of liberation of heat of crystallization will also lead to partial dissolution of nuclei which also helps in serve as a feedback control system of the nucleation. Total supersaturation was consumed due to ultrasound induced nucleation and concentration followed the solubility curve, where applied intermittent sonication serve as heating cycles rather than creating more nucleation. For optimization, the influence of time and power of ultrasound, and cooling profile were investigated. Crystal size distribution and span were compared between different control strategies including uninsonated, continuous sonicated, sonicated prior to nucleation and sonicated in MSZW followed by intermittent sonication. Additionally continuous crystallization with ultrasound in a two reactor system of Sono-Nucleator (50ml) and subsequent growth of crystals in a 3liters of reactor was studied.

The results of this research provides better understanding of the sonocrystallization process because of the inclusion of in-situ monitoring systems along with ultrasonic probe in a batch crystallizer. An insight is obtained about how to choose process trajectories to produce smaller crystals with narrow CSD and bigger crystals with narrow CSD, and about implementation in a larger scale batch and continuous crystallization.

References:

1. S. Kim, B. Lotz, M. Lindrud, K. Girard, T. Moore, K. Nagarajan, M. Alvarez, T. Lee, F. Nikfar, M. Davidovich, S. Srivastava and S. Kiang, *Organic Process Research Development*9 (2005), 894–901.

2. Luque de Castro, M. D.; Priego-Capote, F. Ultrasonics sonochemistry 14 (2007), 717–24

3. Jieqiong Li,Ying Bao, Jingkang Wang, Chemical engineering technology 36 (2013), 1341–1346

This work has been supported by Science Foundation Ireland, Grant number: 12/RC/2275.

469553 Process Parameters in the Purification of Curcumin By Cooling Crystallization

Abstract:

Structurally related impurities (often by-products) are of great importance in the pharmaceutical industry. Such impurities are hard to remove completely by crystallization, because of their structural similarity to the solute molecule. Impurities may become incorporated into structures and influence nucleation[1,2] and growth[2], and thus polymorphic outcome, crystal shape and size.[2-4] Often multi-step successive crystallizations are needed to reach a desired purity, significantly lowering the product yield.

Curcumin, a major active component of the Indian spice Turmeric (Curcuma Longa), has a wide range of bioactivities (antimicrobial, anti-inflammatory and anticancerogenic properties). Extract obtained from Turmeric comprises a mixture of curcumin, demethoxycurcumin (DMC) and bisdemethoxycurcumin (BDMC), known as curcuminoids. Commercially available crude curcumin contains significant amounts of the other two curcuminoids, typically about 17% DMC and 3% BDMC. Separation and purification of curcumin is quite challenging, since curcuminoids are structurally very similar, with differences only in the number of methoxy groups.

In this work, purification of crude curcumin by successive cooling crystallizations has been investigated.[5] Statistical experimental design techniques were used to identify the effect of cooling rate, seeding, seed polymorph and agitation on the yield, crystal purity, particle size, shape and structure as well as determining the minimum number of crystallization steps needed to arrive at a particular product specification. Isopropanol was selected as the solvent for the curcumin crystallization based on determined solubilities of the individual curcuminoids. The composition of solid curcumin samples and concentration of individual curcuminoids dissolved in the mother liquor were quantified by HPLC. The polymorphic composition of the crystals was determined by Raman spectroscopy and by X-ray powder diffraction.

Particle size and shape distribution was determined by light microscopy and image-processing analysis, using Feret's diameter and aspect ratio as particle characteristics. Molecular dynamics simulations were performed to calculate curcumin lattice energy as a function of curcuminoid impurity concentration (GAFF 600 molecules, Gromacs).

The results showed that cooling crystallization of curcumin is sensitive to process parameters including seed polymorph, stirring method, cooling rate and the number of crystallization steps, the latter being closely related to impurity concentrations. Manipulation of these process parameters allows for polymorph selection, particle habit and size control, purity and yield control. It was found that removal efficiency of DMC, the major impurity present in crude curcumin, exponentially decreases as a function of the number of successive crystallizations. Crystallization operated in seeding mode and slow cooling rate increases the purity of the obtained curcumin crystals. In particular, using seeds of the metastable Form II improves the purification of curcumin significantly. The computational results showed that incorporated impurity decreases the lattice energy, and that this decrease is more pronounced for the metastable Form II, indicating a weaker affinity of the impurity molecules for the metastable Form II curcumin. At substantially optimal crystallization conditions high purity curcumin (>99.1%) can be obtained by a reduced number of successive crystallization steps. Such a decrease in the total number of successive crystallizations increase the overall curcumin yield from 28% to 50%.

1. Pino-Garcia, O.; Rasmuson, A. Cryst. Growth Des. **2004**, 4, 1025–1037.
2. Shtukenberg, A.; Lee, S.; Kahr, B.; Ward, M. Annu. Rev.Chem. Biomol. Eng. **2014**, 5, 77–96.
3. Cashell, C.; Corcoran, D.; Hodnett, B. K. Cryst. Growth Des. **2005**, 5, 593–597.
4. Liu, J.; Svärd, M.; Hippen, P.; Rasmuson, A. C. J. Pharm. Sci. **2015**, 104, 2183–2189.
5. Ukrainczyk, M.; Hodnett, K. B.; Rasmuson, A. C. submitted to Org. Process Res. Dev. (2016).

This work has been supported by Science Foundation Ireland, Grant number: 12/RC/2275.

460409 Influence of the Composition of Water/Methanol Mixtures and Temperature on the Crystal Form and Solubility of Sodium Dehydroacetate

Abstract:

The slurry transformation experiments for the monohydrate of sodium dehydroacetate with different compositions of water and methanol at varying temperatures were investigated in this study. It was found that the temperature and water content affected not only the solubility but also the crystal form of the precipitated material. The sodium dehydroacetate anhydrate was obtained with a low water content; however, the monohydrate of sodium dehydroacetate was the more stable form with a high water content. At the same time, the solubility of anhydrate and monohydrate of sodium dehydroacetate were measured with different compositions and temperatures. From powder X-ray diffractionï¼ˆPXRDï¼‰patterns, thermal gravimetric analysis (TGA) data and scanning electron microscopy (SEM), a new needle hydrate form of sodium dehydroacetate was shown on the surface of anhydrous phase when the water content was between 2 % and the critical transformation values. In situ Raman spectra and microscope images demonstrated that this needle hydrate was not the metastable intermediate nor was it the result of concomitant crystallization during the transformation. Pure needle hydrate of sodium dehydroacetate was obtained using evaporation crystallization. Moreover, the PXRD patterns and TGA results revealed that this needle hydrate was another new form of sodium dehydroacetate monohydrate. A possible explanation for these observations was found in the relationship between the change of water content, the interaction between solute and solvent and the influence of supersaturation.

463058 Reaching Optimal Productivity in Batch Viedma Ripening Processes Using Minimal Data

Abstract:

In 2004 Viedma reported that nearly racemic mixtures of L-sodium chlorate and D-sodium chlorate undergo a deracemization process in the presence of an attrition source, i.e., the initially racemic mixture reaches chiral purity within several days of simply stirring the suspension[1]. This discovery, subsequently termed Viedma ripening (VR), was subsequently shown to also be applicable to organic molecules that exhibit a conglomerate phase diagram. It was successfully applied to a host of different molecules, among them amino acids[2,3] and pharmaceuticals[4]. VR thus represents a new and promising way to arrive at enantiomerically pure products.

One of the key observations made in early studies on VR was that the rate towards enantiomeric purity proceeds in an autocatalytic fashion, i.e., starting from a low enantiomeric excess there is considerable lag time before the conversion rate speeds up. The rate is finally slowing down again towards the end of the process.

The overall Viedma ripening process was later identified to consist of a complex interplay between racemization in solution, enantioselective agglomeration, attrition/breakage, and size-dependent solubility/Ostwald ripening.[5,6] Being a rather slow process, attempts at optimizing deracemization processes focused on shortening the time needed to reach enantiomeric purity and hence to increase productivity.[7,8] However, no study so far has attempted a thorough optimization of process conditions with the goal of maximizing productivity.

The incentive to optimize a process towards optimal productivity and shortened batch times is evident, but the flexibility to exchange equipment on existing plants is limited. Consequently, the focus of process optimization should be shifted towards the remaining process conditions, such as amounts of seeds (and enantiomeric purity of seeds), seed distribution, and suspension density. Hence, the key objective of this contribution is to introduce a methodology that achieves optimal productivity (maximizes mass of pure enantiomer produced per time) for batch Viedma ripening processes with the use of as little experimental data as possible.

The methodology presented here seeks to optimize a batch-after-batch production of enantiopure crystals. Within such a processing strategy, the current batch can be conveniently run by re-using the mother liquor from the previous batch (minus appropriate purging to avoid impurity build up), adding a defined amount of a racemic mixture of solids and by seeding the process with some enantiomerically pure seeds (part of the product from the previous batch). Due to VR, such a batch will reach enantiopurity after some time. The goal of this methodology is to keep this time as short as possible. It can be shown that such a processing strategy is mathematically equivalent to an ideal continuous plug flow reactor with a recycle stream consisting of part of its effluent stream. Since Viedma ripening exhibits autocatalytic features, this analogy can be used to find an optimal seeding policy for batch Viedma ripening processes. It will be shown that the only data needed for this strategy is a curve of enantiomeric excess versus time, obtained from a single experiment by standard techniques (e.g. HPLC). It will further be shown that an optimal suspension density can be established without additional data.

The strategy is first shown to work based on extensive model results employing population balance equation models and various kinetic expressions (to ensure a wide range of applicability) and then exemplified using an experimental study on Naproxenethylester crystallized from ethanol, which is racemized in solution using a base (sodium ethoxide in this case).

References:

[1] Viedma, "Chiral symmetry breaking during crystallization: complete chiral purity induced by nonlinear autocatalysis and recycling", Phys. Rev. Lett., 2005, 94, 065504

[2] Viedma et al., "Evolution of Solid Phase Homochirality for a Proteinogenic Amino Acid", J. Am. Chem. Soc., 2008, 130, 15274–15275.

[3] Noorduin et al., "Emergence of a Single Solid Chiral State from a Nearly Racemic Amino Acid Derivative", J. Am. Chem. Soc., 2008, 130, 1158–1159.

[4] Noorduin et al., "Fast attrition enhanced deracemization of naproxen by a gradual in situ feed", Angew. Chem. Int. Ed., 2009, 48, 4581–4583.

[5] Iggland and Mazzotti, "A population balance model for chiral resolution via Viedma ripening", Cryst. Growth Des., 2011, 11, 4611–4622.

[6] Gherase et al., "Experimental and Theoretical Study of the Emergence of Single Chirality in Attrition-Enhanced Deracemization", Cryst. Growth Des., 2014, 14, 928–937.

[7] Iggland et al., "Complete solid state deracemization by High Pressure Homogenization", Chem. Eng. Sci., 2014, 111, 106-111.

[8] Noorduin et al., "Scaling Up Attrition-Enhanced Deracemization by Use of an Industrial Bead Mill in a Route to Clopidogrel (Plavix)", Org. Process Res. Devel., 2010, 14, 908-911.

458903 Determining Quality of Binary Mixtures from a 2D Hopper

Abstract:

Granular mixing has many applications in the pharmaceutical and food industries. Understanding the quality of the mixture is critical for end-use applications. There are many methods used to define mixing quality, with the Lacey Index being the most prevalent. However these mixing definitions or indices do not convey if clusters of one particulate has formed at a scale above the scale of scrutiny, or if some regions of the mixture can be considered well-mixed while other regions may have clustering. In this study, alternative methods will be investigated to identify a better tool (or tools) than the Lacey Index to characterize mixing quality. A transparent 2D hopper and silo are used to mix and hold a binary particulate mixture with varying material density, diameter, and color. Checkerboard, vertically segregated, or horizontally segregated mixtures are initially held in the hopper. Mixing is gravity driven as the particles fall from the hopper into the silo. MATLAB® image analysis tools are utilized to extract spatial data from the resulting mixture. Extracted spatial data are analyzed using statistical methods such as point pattern analysis that include quadrat analysis, clustering, and nearest neighbors. The different analysis methods are then compared with established mixing indices to confirm viability of method. Preliminary results show that using point pattern analysis provides a better characterization tool than the Lacey Index. While more complex than an index that provides a single mixing number between 0 and 1, this new method can convey more information about the mixture that was lost in previous mixing indices.

461382 Mixing in Double Screw Mixers

Abstract:

The mixing of granular materials has a significant influence on the yield and/or quality of the desired products in numerous industrial processes including energy generation, concrete manufacturing, food processing, and pharmaceutical production. The operating conditions of the mixing equipment require careful selection because of their impact on the homogeneity of the mixture, especially when handling materials which vary in size, shape, and/or density. Mixer scale can also have a significant impact on the desired level of mixing. In this study, the mixing effectiveness of three double screw mixers is compared for several operating conditions. The mixers have screw diameters of D = 2.54 cm, 3.81 cm,Â and 5.08 cm, characteristic mixing lengths of L/D = 10, and are identified as 1x, 1.5x,Â and 2x scale mixers. Two different size ranges of glass beads (300-500 Î¼m and 800-1000 Î¼m) are mixed using screws with one of three dimensionless screw pitches (p/D = 0.75, 1.25, and 1.75) and one of three rotation speeds (Ï‰ = 20, 40, and 60 rpm). Three different screw orientations are also investigated and include co-rotating (CoR), counter-rotating up pumping (CrtR UP), and counter-rotating down pumping (CrtR DP). The mixers were designed such that four equal area exit ports separate the exit stream; these exit streams were captured to determine the mass fraction of each material exiting the port. The coefficient of variation of the mixture composition across the exit ports was then determined for various operating conditions. Preliminary results indicate that counter-rotating down pumping has better mixing than the counter-rotating up pumping or the co-rotating configurations, as the coefficients of variation are low and consistent. The counter-rotating up pumping configuration has the highest coefficients of variation, particularly at the highest mixing speeds used in this study.

460674 Flow Modulation Driven Stratification of Size Bidisperse Granular Mixtures in Quasi-2D Heaps

Abstract:

Modelling transient size segregation in granular flow has important applications in various industrial processes such as hopper and silo filling. In this research, we demonstrate, using experiments, discrete element method simulations (DEM), and a continuum model, how temporal modulation of the fill rate can generate ordered stratification in a quasi-2D bounded heap and investigate how the modulation parameters determine the wavelength and streamwise extent of the stratification layers. By temporally modulating the fill rate between fast and slow phases for a size-bidisperse particle mixture, stratified layers of large- and small-particle-rich bands form parallel to the free surface of the heap. The relative influence of the various mechanisms driving the stratification are not yet well understood. To gain physical insight into modulation driven stratification, we develop a simplified transient continuum model for granular size bidisperse modulated flow segregation. The model depends on three kinematic dimensionless obtained from experiments or DEM simulations: the Peclet number, which captures the interplay of advection and diffusion, a second dimensionless parameter that describes the interplay between segregation and advection, and a time scale that represents the ratio of the flow modulation time to the characteristic residence time of the particle in the flowing layer. The model captures the leading order characteristics of the stratification patterns and agrees well with DEM simulations and experiments under similar operating conditions. Funded by The Dow Chemical Company and NSF Grant CBET-1511450.

472448 Transport Analogies for Segregation in Granular Shear Flows

Abstract:

Shear induced segregation is a problem of considerable import both industrially and academically. As a granular material is sheared, the bed dilates and heavier (or smaller) particles move perpendicular to the mean flow. In the present work, we perform a computational study -- based on the Discrete Element Method -- of the flow in a simple boundary-driven planar shear cell. Inside the cell is a mono-disperse light particle packing that includes several heavier intruder particles which are used to measure density segregation for a variety of flow conditions: differing density ratios, shear rates, confining pressures (for constant pressure boundary conditions) or solids' fractions (for constant volume boundary conditions). We find that the underlying granular rheology has a profound effect on the rate of the segregation process. In particular, we note that by choosing the proper scaling we can identify a transition in segregation rate between a "quasi-static" and "intermediate" regime. These results suggest that one can construct an analogy between segregation and granular flow that is akin to transport analogies (like the Colburn or Reynolds analogy) in continuum fluids processing.

460657 Modeling Segregation in Granular Flows with Overlapping Particle Size Distributions

Abstract:

Unwanted demixing is a widespread problem in bulk solids processing. In the past, a combination of trial and error, small scale experiments, and continuum models have been used to estimate and negate the effects of bulk material segregation. Recent advances in continuum modeling of segregation in size multidisperse and polydisperse systems have only been applied to systems whose species particle size distributions (PSDs) are non-overlapping, whereas mixtures in industrial bulk solids processing are composed of constituent species of broad and overlapping size distributions. We study changes in segregation behavior as a result of overlap between species PSDs by using a series of discrete element method (DEM) simulations of a bounded heap with two species with overlapping log-normal size distributions. Surprisingly, a continuum based model that treats the mixtures as consisting of just two different particle sizes corresponding to the means of the log-normal distributions is able to quantitatively reproduce the degree of mixing from the DEM simulations for overlapping PSDs. We test this simple bidisperse model in cases of broad, overlapping, and varied size distributions to determine when it fails, and we resolve any discrepancies between it and the overlapping continuous distributions using a minimum set of independent model parameters. Funded by Procter & Gamble.

470400 Quantifying the Root Causes of Segregation in Multi-Mechanism Segregation Systems

Abstract:

Segregation is a multi-component and multi-mechanism phenomenon. Most data in current literature is for simple bimodal systems. These systems are far too simple to predict real segregation of the complex mixtures created in industry. One of the key problems is a lack of theory to handle real mixtures of more than two materials where more than one cause of segregation is present. It would be very useful to understand and quantify the amount of each type of segregation present within a given material on a component basis. This understanding will allow formulators to perform simple tests and determine the root cause(s) of segregation and then determine what to change in the formulation to mitigate the segregation. This paper suggests a framework to accomplish that task. Modified versions of the convection dispersion equations can be used to understand segregation of a particular mechanism on a component-by-component basis. In these equations segregation is often handled by adding a segregation flux term to offset the symmetric dispersion term. This segregation flux velocity is always directional and there are different segregation flux terms for each type of segregation occurring in the system.

This paper examines the relationship between segregation velocities when there are multiple causes of segregation. The main question to answer is: what percent of the overall segregation dispersion coefficients are due to the individual mechanisms of segregation? In dispersion, the local convective velocities distribute in a symmetric pattern around the overall average convective velocity. When segregation flux terms are added to the dispersion equations they cause a skewed concentration distribution off of those caused by average convective velocities for each component in the system. We assume that the segregation flux velocities are additive. Thus, to develop a theory that incorporates multi-mechanisms, the relationship between segregation velocity terms for each type of segregation must be understood. The overall segregation profile is a sum of a series of skewed concentration distributions from the average dispersion velocities. These distributions can be combined using the concept of multi-modal distribution functions to yield

equivalent overall segregation variance numbers for each component in the mixture. The overall segregation variance numbers can be measured from segregation potential tests done with spectral reflectance techniques (SPECTester). The overall segregation variance data can then be used with convective dispersion models to quantify the segregation of each component by mechanism, providing a detailed look at the cause of segregation in multi-component and multi-mechanism situations. The resulting technique allows a component-by-component characterization showing how much segregation is caused by a prescribed set of individual mechanisms in a mixture consisting of more than two components.

Improving the Mixing Process of Dry Powder Inhaler Formulations with the Iron Oxide Colour Method

Kahlil Desai[1], *Professor Peter Stewart*[2], *Associate Professor David Morton*[2], *Professor Karen Hapgood*[1]

[1] *Monash Advanced Particle Engineering Lab, Monash University, Clayton, VIC, Australia*

[2] *Monash Institute of Pharmaceutical Science, Monash University, Royal Parade, Parkville, VIC, Australia.*

Abstract

Blending of one or more powders is critical for many pharmaceutical powders, including formulations used for Dry Powder Inhalers. Dry powder inhalers (DPIs) are mixed dry powder medications (asthma) administered via the lungs (pulmonary route). DPIs generally comprise 1% of API and 99% lactose and are mixed in a tumbler mixer or a high shear mixer. Due to empirical nature of commercial production there are production issues such as failed batches, content uniformity issues, and product performance failures, especially low inhalation efficiency values. This leads to production and economic losses but also a lack of understanding in the blending process.

The iron oxide colour method (IOCM) is a recent technique that involved mixing 1% of red iron oxide and 99% of lactose in different mixers and observing the colour [1]. The colour of the blend indicates how mixed the blend is; as the mixing time continues the blend colour changes from pink to red to orange; this method helps to understand the mixing process and the quality of mixing. This work follows on with the method by mixing and testing DPI formulations. Actual DPI blends, 99% DPI grade lactose and 1% API (salbutamol sulphate) are mixed in a high shear mixer (1L KG5) and samples are taken for time points of the blend and are checked with their corresponding colour points.

This study investigates whether this colour method (IOCM) can predict or determine how much mixing is required for a Dry Powder Inhaler (DPI) blend (lactose/drug) to attain content uniformity (Real standard deviation (RSD) < 5). A High Shear Mixer (HSM) is used to mix DPI blends and samples were taken at different time points for different speeds and the content uniformity was tested. These time points were then compared with their corresponding colour points (lactose/iron oxide) to see if there is a correlation between colour points and content uniformity.

Introduction

Iron Oxide Colour Method

The iron oxide method employs the CIEL*C*h system to measure and quantify colour. The CIELCh system categorises colour into three parts: L* for lightness, C* for Chroma, and h for hue [2]. Lightness (L*) describes how light/dark the colour is; Chroma (C*) is the intensity of colour (saturation); and hue (h) is pure colour or frequency [3]. The iron oxide method uses CIELCh to measure 1%wt red hematite iron oxide in lactose blends [1]. Figure 1 is a visualisation of the CIELCh system.

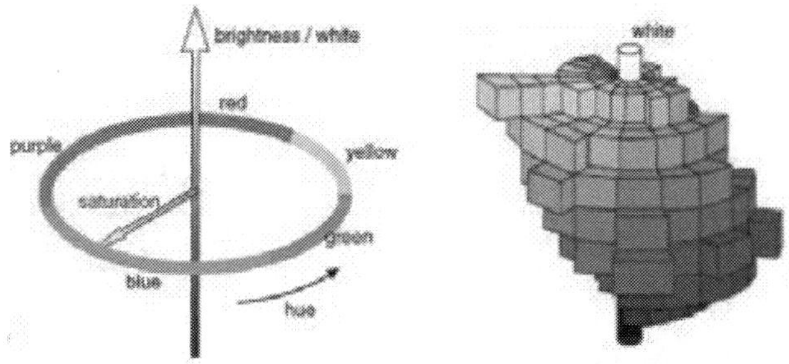

*Figure 1 - CIEL*C*h System [3]*

It was observed when mixing lactose and 1% iron oxide there are three stages of colour: pink, red, and orange [1] (seen in Figure 2). The 1% iron oxide is used to understand the extent of mixing behaviour of the bulk powder. The resulting colour of the lactose blend shows the quality of mixing.

Figure 2 - Three colour stages of Lactose mixed with 1% Iron Oxide [4]

The pink-red blend colour transition is considered *dispersion dominant* because the iron oxide agglomerates move throughout the bulk powder attaching to the larger lactose particle surface; there is large changes in both Chroma (C*) and lightness (L*) and a small change in hue (h) in this colour transition [1]. The red to orange transition is considered *deagglomeration dominant* as further mixing energy further breaks down agglomerates into smaller agglomerates and primary particles; there is consistent hue (h) change with much lower C* and L* changes; a visual graphic of the colour transitions is shown in Figure 3 and Figure 4.

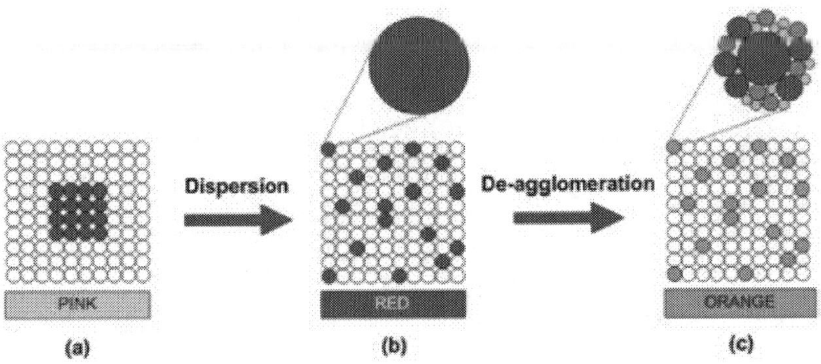

Figure 3- Schematic of Iron Oxide (Hematite) with Dispersion (a-b) and Deagglomeration (b-c) in bulk lactose [1]

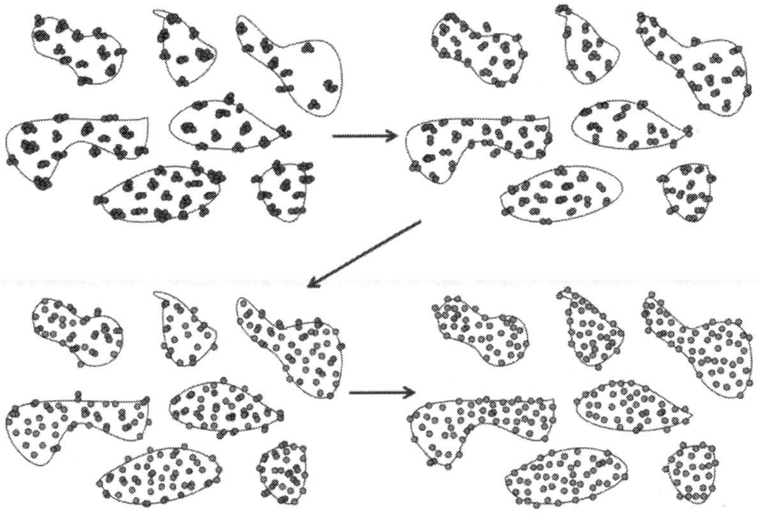

Figure 4 - Proposed Colour Change Mechanism

Materials and Experimental

This study extends to drug mixing of lactose and salbutamol sulphate (API) by testing for its uniformity content over mixing time and how it compares with its mixing colour point. The first section of these experiments was to blend 1 wt% (4g) of the tracer powder and 396g of lactose in a 1L Key high shear mixer at a specified speed. Colour samples were taken from the powder bed over a sixty minute time period (see Table 1 for sample times). This was done over a range of speeds: 185, 250, 325, 400, and 475rpm.

Table 1 *- Time Intervals for Colour Sampling*

Time	Sample Rate	Number of Samples
0 - 1 minutes	Every 30 seconds	2
1-5 minutes	Every 1 minute	4
5-15 minutes	Every 2.5 minutes	4
15-30 minutes	Every 5 minutes	3
30-60 minutes	Every 15 minutes	2

The second section of the investigation was to perform the lactose and drug blending. So, 99% w/w lactose (396g) and 1% w/w salbutamol sulphate (4g) were placed in the same 1L high shear mixer and mixed for a maximum of 60 minutes. For each time point 10 samples of the formulation were taken from the powder bed; five on the surface and five below the surface (~3cm below). One run was performed for each speed 185, 250, and 325rpm and 4-5 mixing time points were sampled for each run. The 10 samples were then analysed in a salbutamol sulphate HPLC assay to identify the concentration of the API in the formulation for each time point. The average, standard deviation, and real standard deviation were calculated to determine and assess the uniformity content (RSD < 5) for each mixing time point. Each blend was 'sandwich filled' when loaded into the high shear mixer.

Powders – Carrier Particle & API

Lactose is a cohesive carrier particle in Dry Powder Inhaler (DPI) formulations [5] and was used as the excipient in these experiments. The lactose selected was Lactohale 200 and is a gently milled monohydrate lactose powder.

Figure 5 – SEM Images of Lactohale

A PSD (particle size range) of Lactohale shows it has a high proportion of particles in the range of 50-100µm but also contains fines as seen in Figure 6.

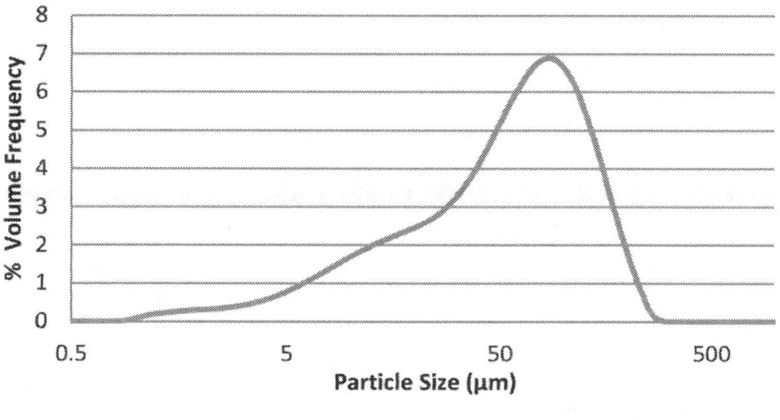

Figure 6 - Particle Size Distribution of DPI Lactose (Lactohale)

The drug used for blending was salbutamol sulphate and is a common API in DPI formulation blends [6].

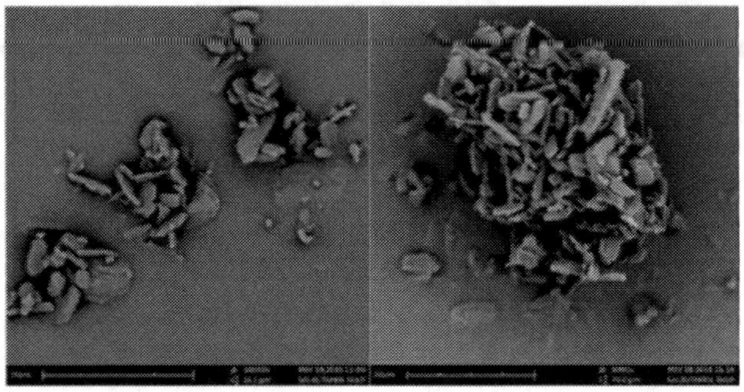

Figure 7 – SEM Images of Salbutamol Sulphate Agglomerates

From the SEM images Figure 7 it is shown that Salbutamol sulphate is much smaller than the lactose particles, which is a requirement for DPI blends (<5µm) [6].

Results

Iron Oxide Blends

Figure 8 is the colour range graph for Lactohale and iron oxide for different speeds. The blends fit onto one curve. It shows that the blends mixed at higher speeds have greater hue change (more orange) and higher Chroma (surface coverage) over the sixty minute blending period.

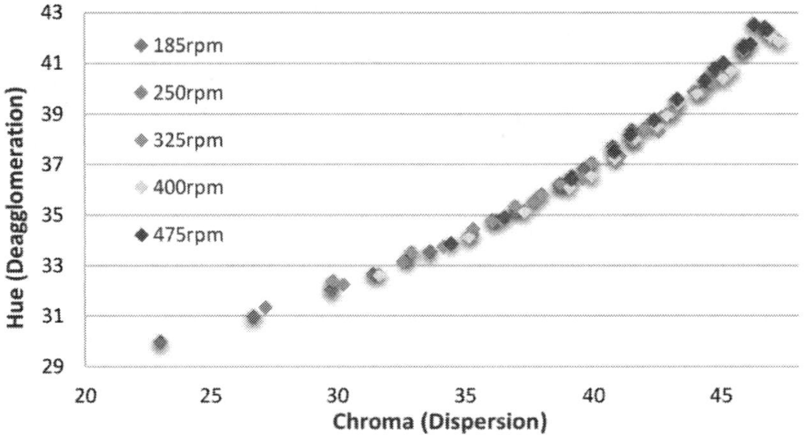

Figure 8 - Colour range of different speeds of lactose & red iron oxide (1%) blends for 60 minutes

Therefore the more energy you put into a blend the more orange and higher intensity of colour the powder blend will have. This colour is then used to compare lactose (99%) and salbutamol sulphate (1%) blends at particular speeds.

Drug Blends

A table of each run with the time of sampling and its corresponding Chroma and hue is shown. The 60 minute blend at 185rpm had the lowest RSD value for this speed but its average is lower than 1%, which also fails the mean requirement. The RSD range for this blend was 14-68.

*Table 2 - Mixing of Lactose (99% w/w) and Salbutamol Sulphate (1% w/w) at **185rpm***

Mixing Time (minutes)	5	15	20	60
Average (%)	0.85	1.16	1.09	0.88
Standard Deviation	0.47	0.79	0.27	0.13
RSD	55.07	67.94	24.79	14.78
Chroma	33.62	37.67	38.68	41.49
Hue	33.55	35.52	36.14	38.36

The RSD values on average for the 250rpm blend decreased for the same time and ranged between 17 and 25.

*Table 3 - Mixing of Lactose (99% w/w) and Salbutamol Sulphate (1% w/w) at **250rpm***

Mixing Time (minutes)	5	10	15	20	60
Average (%)	1.10	0.85	0.89	0.88	0.85
Standard Deviation	0.20	0.21	0.16	0.19	0.15
RSD	17.72	24.72	18.40	21.34	17.90
Chroma	36.45	39.59	41.02	42.55	44.43
Hue	34.78	36.43	37.30	38.31	40.44

For the 325rpm blends RSD values decreased, which was expected as there is an increased amount of energy placed into the blend. The hue and Chroma for the lowest RSD value (9.17) is the highest. The RSD range was 9-22.

*Table 4 - Mixing of Lactose (99% w/w) and Salbutamol Sulphate (1% w/w) at **325rpm***

Mixing Time (minutes)	20	25	30	45	60
Average (%)	0.98	1.00	1.05	0.97	0.99
Standard Deviation	0.15	0.21	0.21	0.12	0.09
RSD	15.24	21.11	19.90	12.06	9.17
Chroma	44.23	44.58	44.99	45.83	45.85
Hue	39.86	40.26	40.68	41.47	41.73

Figure 9 plots all the RSD values and their corresponding hue values. The graph shows the general trend that higher hue values have lower variability and for hue values generally found below 37 the RSD values were above 20.

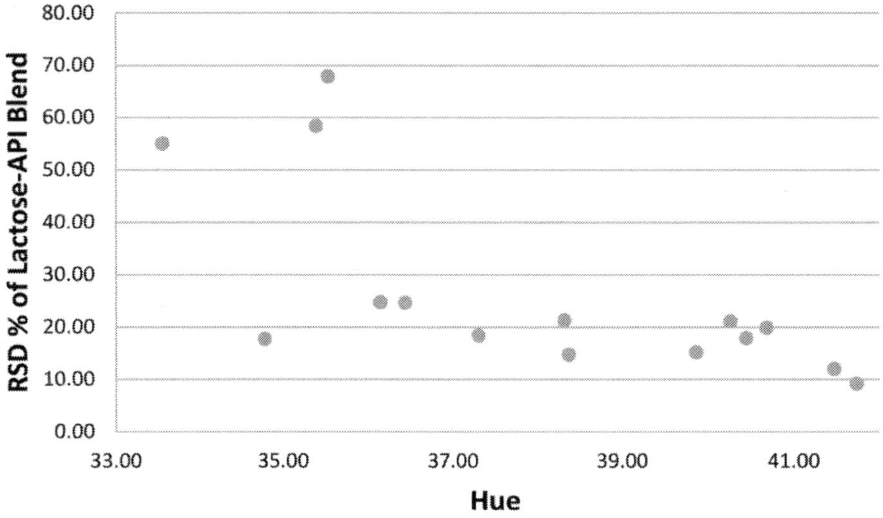

Figure 9 - Comparison of Hue points and RSD%

There is some consistency with RSD values for particular hue points, which will need to be investigated with repeated experiments.

Future Work

These preliminary findings suggest that there is similar content uniformities based on their colour points. Further investigation would require mixing at higher speeds to obtain a satisfactory uniformity content and again comparing with their corresponding colour point. Repeat tests would need to be performed to confirm reproducibility.

Conclusion

For this study the colour blends were obtained for a range of different speeds to see what colour points these represented in the CIELCh colour space. Drug blends were then performed at three of these speeds and tested to see if equivalent colour points could predict the RSD values of drug content uniformity. The results did show that higher hue values followed with lower RSD values and this is expected as higher speeds and more energy should achieve better mixed formulations. They also indicated RSD consistency with hue points. However, these findings are preliminary and further investigation and more testing is required to obtain satisfactory uniform content blends and high precision results.

Acknowledgements

The authors would like to acknowledge the support from GlaxoSmithKline, Monash University, and the Australian Research Council.

References

1. Barling, D., D.A.V. Morton, and K. Hapgood, *Pharmaceutical dry powder blending and scale-up: Maintaining equivalent mixing conditions using a coloured tracer powder.* Powder Technology, 2014.
2. Gupte, V.C., *3 - Expressing colours numerically A2 - Gulrajani, M.L*, in *Colour Measurement*. 2010, Woodhead Publishing. p. 70-87.
3. Tilley, R.J.D., *Colour due to Scattering*, in *Colour and the Optical Properties of Materials*. 2011, John Wiley & Sons, Ltd. p. 175-196.
4. Barling, D., et al., *Analysis of the Effect of Impeller Type and Speed on the Rate and Quality of Mixing in a High Shear Mixer*, in *6th International Granulation Workshop*. 2013, Sheffield University: Sheffield.
5. Noordman, W., *Product Specification: Lactohale 200*, D. pharma, Editor. 2013. p. 1.
6. Dalby, R.N., S.L. Tiano, and A.J. Hickey, *Medical devices for the delivery of therapeutic aerosols to the lungs.* Inhalation Aerosols: Physical and Biological Basis for Therapy, 2007: p. 417-444.

467923 Differences in the Dynamics of Cylindrical and Spherical Particles in a Rotating Drum Using Multiple Radioactive Particle Tracking

Abstract:

Differences in the Dynamics of Cylindrical and Spherical Particles in a Rotating Drum Using Multiple Radioactive Particle Tracking

Majid Rasouli, François Bertrand*, Jamal Chaouki*

Department of Chemical Engineering, École Polytechnique de Montréal, P. O. Box 6079, Succ. Centre-ville, Montréal, QC, Canada, H3C 3A7, jamal.chaouki@polymtl.ca, francois.bertrand@polymtl.ca

Rotating drums are widely used to process granular materials in the chemical, pharmaceutical, food-processing, polymer, and waste treatment industries, among others, for applications such as mixing, size reduction, sintering, coating, heating, cooling, drying, and chemical reactions. This broad array of applications is made possible by the ability of rotating drums to handle heterogeneous feedstock and to ensure satisfactory mixing and heat transfer of the solid phase[1]. Since the flow dynamics of particles determines to a great extent the mass and heat transfer rates, it plays a critical role in controlling and/or limiting the processes mentioned above[2]. The optimal design and operation of rotating drums require in-depth fundamental knowledge of the phenomena that occur inside them, including the transverse flow that controls primary effects such radial particle movement and particle mixing and/or segregation, as well as secondary effects such as bed temperature, reaction rates, and the rate of axial particle movement.

The flow behavior of spherical particles has been often studied. However, investigations of cylindrical particles, which are not numerous in the literature, are required since these particles have many uses (biomass pellets, capsules, rice, candy, etc.) and, more importantly, since their shape is known to affect their flow behavior[3] in terms of shear resistance, dilation under shear, compaction efficiency, transfer of momentum between translational and rotational movements, and their ability to form arches and block the flow[4].

The goal of the present study was to compare the flow behavior of spherical and cylindrical particles in a rotating drum. We used the RPT technique to determine the positions of the spherical particles over time. We used the MRPT technique[5] to track the positions and orientations of the cylindrical particles and calculate the components of their transverse flow dynamics in the rotating drum. To illustrate, Fig. 1 shows velocity vectors of the transverse flow of spherical and cylindrical particles at the same rotational speed (10 RPM) in the rotating drum.

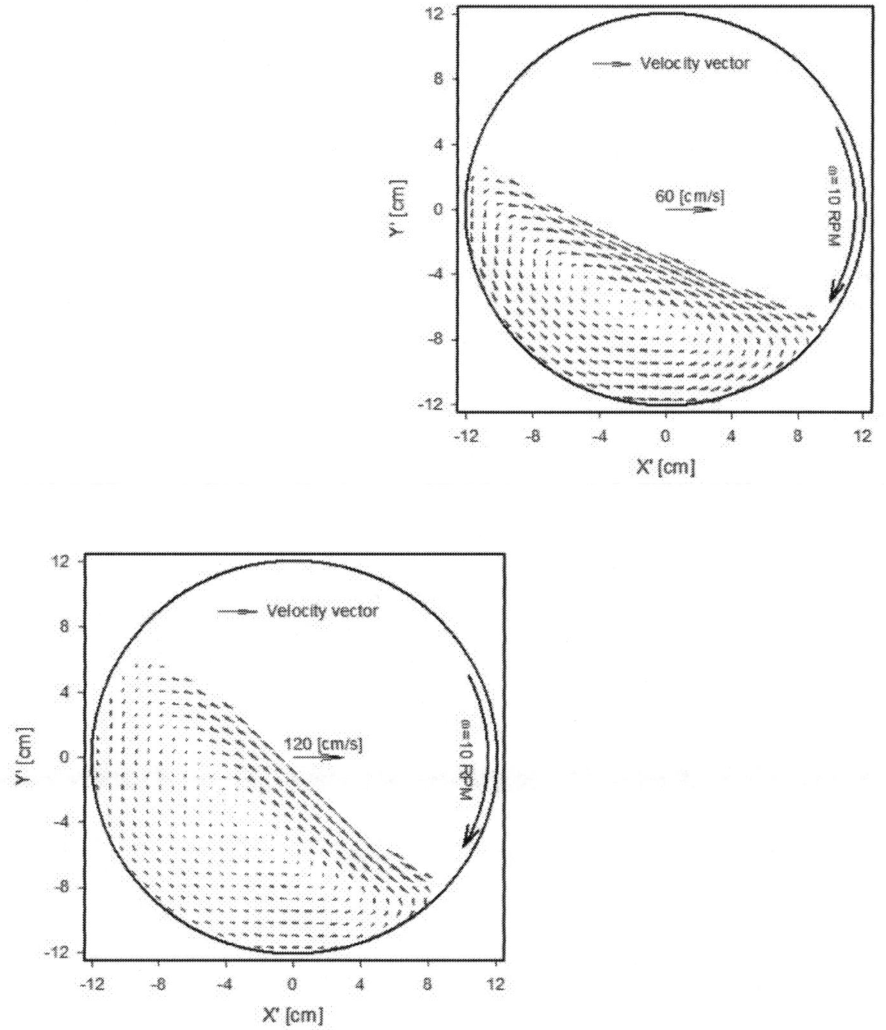

(b)

(a)

Figure 1: Velocity vectors for (a) spherical and (b) cylindrical particles at a rotational speed of 10 RPM[6]

Literature cited

1. Dube O, Alizadeh E, Chaouki J, Bertrand F. Dynamics of non-spherical particles in a rotating drum. *Chemical Engineering Science.* Sep 20 2013;101:486-502.

2. Mellmann J, Specht E, Liu X. Prediction of rolling bed motion in rotating cylinders. *AIChE Journal.* 2004;50(11):2783-2793.

3. Ridgway K, Rupp R. Mixing of powder layers on a chute. The effect of particle size and shape. *Powder Technology.* 1971;4(4):195-202.

4. Cleary PW. DEM prediction of industrial and geophysical particle flows. *Particuology.* 2010;8(2):106-118.

5. Rasouli M, Bertrand F, Chaouki J. A multiple radioactive particle tracking technique to investigate particulate flows. *AIChE Journal.* 2015;61(2):384-394.

6. Rasouli M, Dubé O, Bertrand F, Chaouki J. Investigating the dynamics of cylindrical particles in a rotating drum using multiple radioactive particle tracking. *AIChE Journal.* 2016.

Acknowledgments

This work was made possible by financial support from Praxair and the Natural Sciences and Engineering Research Council of Canada (NSERC). The authors are grateful to Cornelia Chilian and Cristina Cimpan (Institute of Nuclear Engineering) for activating the tracers, Sepehr Hamzelouia for helping to measure the static repose angle, and Amin Esmaeili for helping to determine the positions of the RPT detectors.

450281 Invited - Solar Light NO Removal By Flame-Made Sub-Nano Pd Clusters on Nano-TiO_2

Abstract:

Noble metal clusters deposited on metal oxide supports are broadly used. Size reduction of such noble metal particles is a classic way to enhance their activity with minimizing the use of expensive noble metals. Recently, loading of extremely small Pd clusters (~ 0.5 nm) on TiO_2 exhibits superior photocatalytic activity for NO removal[1] and its further size reduction from the clusters to the isolated atoms have a great potential to maximize the NO_x removal. Here, TiO_2 with different Pd loadings (0-3%) were prepared by scalable flame aerosol technology to attain atomically dispersed Pd atoms on TiO_2. The photocatalytic NO (1 ppm) removal efficiency of those Pd/TiO_2 under solar light (100 mW/cm^2) was evaluated based on standard ISO procedures.

Under solar light irradiation, these materials remove NO_x 3 or 7 times faster than commercial TiO_2 (P25, Evonik) with or without photodeposited Pd on it. X-ray photoelectron spectroscopy (XPS) reveals that such photodeposited Pd consists of metallic Pd along with several Pd oxidation states. In contrast, flame-made Pd subnano-clusters on TiO_2 dominantly consist of an intermediate Pd oxidation state between metallic Pd and PdO. In that intermediate state, the Pd subnano-clusters are stable up to, at least, 600 °C for 2 hours in air. However, a fraction of them is reduced into relatively large (> 1 nm) metallic Pd nanoparticles by annealing in N_2 at 400 °C for 2 hours, as elucidated by XPS and scanning transmission electron microscopy. The Pd subnano-clusters interact with oxygen defects on the TiO_2 surface as shown by Raman spectroscopy. This interaction suppresses CO adsorption on Pd as observed by diffuse reflectance infrared Fourier transform spectroscopy (DRIFTS), analogous to strong metal support interactions (SMSI) of nano-sized noble metals on TiO_2.

1. Fujiwara, K.; Müller, U.; Pratsinis, S. E., Pd Subnano-Clusters on TiO_2 for Solar-Light Removal of NO, *ACS Catalysis,* 6, 1887-1893 (2016).

470756 *Visible Light Photocatalytic Degradation of Organic Pollutant Using N Doped TiO_2 nanoparticles Synthesized By One Step Liquid Flame Aerosol Method (LFSP)*

Abstract:

Currently, there is a growing interest in the synthesis of nanomaterials. Nitrogen doped titanium dioxide is attracting a continuously increasing attention because of its potential as material for environmental photocatalysis. Conventional wet synthesis methods need multiple processing steps, high energy input and long processing times, In contrast, flame synthesis, especially liquid flame spray pyrolysis (LFSP), is a versatile one-step process for synthesizing a variety of anion doped nanoparticles. In the present work, we have developed a novel, inexpensive, green and highly reproducible method of the making the liquid precursor for the N incorporation in TiO_2 by using the rapid LFSP method. A combination of various physico-chemical characterizations were performed. The photocatalytic activity and stability of the N-doped TiO_2 samples were evaluated through using the decomposition of acetaldehyde as model reaction under visible light irradiation. Characterization results show that the nitrogen dopants have a significant effect on the crystallite size and optical absorption of TiO_2. It is found that the N-doped TiO_2 catalysts have enhanced absorption in the visible light region, and exhibit higher activity for photocatalytic degradation of VOCs. We have produced The UV-vis spectroscopy results of N incorporated TiO_2 showed enhancement of light absorption in the visible range by selected composite materials in the UV range (250-400 nm). We have found the Ti/N ratio of 13% for the materials synthesized from our XPS data. These interesting results will be discussed in the presentation.

452386 Invited - DNA Encapsulated within Silica Nanoparticles for Environmental Monitoring and Tracing

Abstract:

From an analytical chemistry perspective short DNA sequences are unique as they can not only be detected at extremely low concentrations (1 molecule by PCR), but that they can carry identifying information (barcodes). In order to make DNA more resilient to application in environmental monitoring and tracing, we have developed a method to encapsulate DNA into amorphous silica nanoparticles by a sol-gel method.(1,2) Within the particles the DNA is protected from hydrolysis and can withstand increased temperatures and prolonged exposure in oxidizing conditions.

Within this presentation we will discuss how such encapsulated DNA barcodes can be utilized to a) monitor the faith of silica nanoparticles in the environment, and b) to monitor the flow/distribution of liquids, foodstuff and commodity products within the environment and throughout a product life cycle. Examples include the use of the particles in understanding nanoparticle clearance in waste water treatment plants (3), the tracing of pesticides in the environment (4) and the tracking of milk and gasoline(5,6).

We will also show how the concept of DNA tracers can be utilized for the remote and distributed sensing of optical irradiation and temperature (7,8) and discuss future possibilities of combining bio analytics and inorganic chemistry for applications in the field of chemical engineering.

(1) D Paunescu, R Fuhrer, RN Grass, Angew. Chem. Int. Ed 52, 4269 (2013)

(2) D Paunescu, M Puddu, JOB Soellner, PR Stoessel, RN Grass, Nat. Protoc. 8 , 2440 (2013)

(3) RN Grass, J Schälchli, D Paunescu, JOB Soellner, R Kaegi, WJ Stark, ES&T Lett. 1, 484 (2014)

(4) CA Mora, HJ Scharer, T Oberhänsli, M Ludwig, R Stettler, PR Stoessel, RN Grass, WJ Stark, ES&T Lett. 3, 19 (2016)

(5) MS Bloch, D Paunescu, PR Stoessel, CA Mora, WJ Stark, RN Grass, J. Agr. Food Chem. 62 , 10615 (2014)

(6) M Puddu, D Paunescu, WJ Stark, RN Grass, ACS nano 8 , 2677 (2014)

(7) G Mikutis, CA Mora, M Puddu, D Paunescu, RN Grass, WJ Stark, Adv. Mater. 28, 2765 (2016)

(8) M Puddu, G Mikutis, WJ Stark, RN Grass, Small, 12, 452 (2016)

459882 Textile Dye Removal from Water Using Catalytic Nanoparticle-Carbon Composites

Abstract:

In recent years, increasing consumer spending and new emerging markets in the world trade have caused a boom in production in the textile industry. With this growth comes the contamination of the environment through substances called azo dyes. In the presence of digestive tracts, these azo dyes can cleave creating aromatic amines, which, in some cases have been proven to be carcinogenic. Also, the azo dyes often cannot be completely filtered out of water naturally due to their highly hydrophilic nature and very small size. Studies have shown that metal nanoparticles reduce the concentration of these azo dyes in aqueous solution to a certain extent, but many studies have focused on precious metals such as gold, copper, and palladium. Due to the cost of these metals, there is interest in developing reactive nanoparticles that are based entirely on non-precious metal combinations. For this study, we have chosen to focus on iron-nickel bimetallic nanoparticles.

In this study, our research focuses on demonstrating the effectiveness of bimetallic iron-nickel nanoparticles in the treatment of Orange G azo dye for different molar ratios of nickel to iron and for the case where these nanoparticles are deposited onto a high surface area carbon support material. If reactive nanoparticles are to be used in realistic treatment scenarios, the nanoparticles must be immobilized within a treatment system, and high surface area carbon materials are a potentially ideal choice for immobilization. Carbons, such as activated carbon, are often used in water treatment to adsorb and remove water contaminants and thus may serve not only as an immobilization strategy but as an additional available surface for adsorptive contaminant removal.

In our experimental work, nanoparticles were created by using an aqueous nanoparticle synthesis technique and using an alloyed nanoparticle morphology. The process uses an iron sulfate heptahydrate salt along with a stabilizer (ATMP) and a nickel chloride hexahydrate salt with its respective stabilizer (PVP). These metals are mixed under argon gas, then reduced using sodium borohydride. This procedure results in the formation of nanoparticles and hydrogen gas. The nanoparticles are then added to an aqueous solution of Orange G in a batch reactor setup and mixed while samples are taken at specific time intervals to evaluate Orange G removal. Our results thus far indicate a strong correlation between the decrease in Orange G concentration and the molar ratio of nickel to iron. Experiments performed with monometallic nanoparticles (10 g/L iron and 10 g/L nickel) showed little to no decrease in Orange G concentration while the bimetallic counterparts showed promising results in decreasing the concentration of Orange G. Results showed that at a lower molar ratio of nickel to iron increased the effectiveness of the bimetallic nanoparticles in decreasing the concentration of Orange G. In comparison to nanoparticles alone, nanoparticles were deposited onto several high surface area carbon support materials and tested in the same batch reactor setup for removal of Orange G. Preliminary results suggest that nanoparticle immobilization onto carbon results in an increase in the rate of removal of Orange G and an increase in the overall amount of Orange G removed.

In this talk, we will present results for immobilized iron-nickel nanoparticles on several key high surface area carbon support materials. In particular, two types of commercially-available activated carbons will be compared to several novel surface-functionalized carbons. The novel carbon materials are functionalized with specific heteropolyacids thought to have potential oxidizing capabilities in aqueous solution. Results for immobilized nanoparticles will be compared to results for non-immobilized nanoparticles in a batch reactor setup. The influence of nickel to iron molar ratio on Orange G removal for both immobilized and non-immobilized nanoparticles will be discussed. Characterization results for nanoparticle and nanoparticle-carbon composite morphology, composition and phase will be presented. Characterization methods include electron microscopy, power x-ray diffraction, and x-ray photoelectron spectroscopy. Finally, a comparison of the different tested carbon support materials will demonstrate that there are differences in Orange G dye removal depending on what carbon support is chosen for nanoparticle immobilization.

468651 Grafted Nanopyroxene for the Removal of Polar Organic Compounds from Produced Wastewater

Abstract:

A large amount of water is needed for bitumen extraction processes and, subsequently, a significant amount of Oil Sand Process-affected Water (OSPW) is produced and discharged into tailing ponds. The OSPW contains organic compounds such as Naphthenic Acid (NA), which is considered the main organic pollutant occurring naturally in tailing ponds. The removal of NA from OSPW has been of paramount importance in order to meet government environmental regulations and has been challenging due to its high solubility in water, which in turn makes it resistant to the conventional oil water separation processes. The current technologies used for the removal of NA present several shortcomings, such as high capital requirements and operating costs, and they are largely ineffective in meeting government regulations. Despite the fact that the adsorption process is promising for the removal of dissolved contaminants in the OSPW, classical adsorbents lack several properties that can be best established using nanoparticles.

Nanoparticles poses unique properties that make them desirable, such as large surface area, high efficiency, high reactivity, dispersibility, low cost and selectivity. Nanoparticle technology is becoming the new frontier in wastewater treatment due to its high efficiency, low cost and easy to implement processes.

In this study, silica-based nanoparticles were prepared in the lab for their proven advantages over classical adsorbents. These nanoparticles needed to be modified in order to attract polar organic compounds. After several studies, grafting was chosen, this entails the modifying and functionalizing of the surface of the nanoparticle with a naturally occurring and environmentally friendly polymer in order to achieve high affinity towards the polar organic compounds present in the OSPW. The preparation of the Grafted Nano Silica (GNIS) was accomplished and optimized using different polymer molecular weights as well as different preparation procedures.

A synthetic wastewater solution was prepared by dissolving NA in water. Microscopic batch adsorption took place to carefully study several factors influencing the adsorption process; including: contact time, initial concentration, GNIS amounts and the pH of the solution. The Sips, Langmuir and Freundlich adsorption isotherm models were employed using the experimental data to describe and understand the adsorption behavior.

This study confirmed that GNIS could be employed as efficient adsorbents for naphthenic acid removal from wastewater where classical adsorbents could be ineffective or costly.

456741 Invited-2D Materials for Packaging and As Nanoparticle-Surfactants

Abstract:

Our group has carried out systematic study on 2D materials, including liquid crystal phase formations, nanocomposite fabrication, and understanding of Janus platelets as nanoparticle surfactants for Pickering emulsification. I will discuss the discovery of discotic smectic phase, which can be used in more effective industrial sealants, improved food packaging, enhanced electronic displays, and improved fuel cell technology. We also developed method to mass produce Janus nanoplates, first using the model system of Zirconium Phosphate nanoplates, then using clay and graphene Quantum Dots with reduced cost. I will discuss the applications of these nanoplate surfactants to make stable Pickering emulsions and foams, to stable graphene suspensions, for enhanced oil recovery, and for nano-encapsulations.

1. "Hydrothermal synthesis of layered a-zirconium phosphate disks: control of aspect ratio and polydispersity for nano-architecture", M. Shuai, Andres F. Mejia, Y. Chang and Z. Cheng **Cryst. Eng. Comm. 15,** 2013, 1970-1977.

2. "Aspect Ratio and Polydispersity Dependence of Isotropic-nematic Transition in Discotic Suspensions" A.F. Mejia, Y. Chang, R. Ng(undergraduate), M. Shuai, M.S. Mannan, and Z. Cheng, **Physical Review E** 85, 2012, 061708. Figure selected as part of PRE kaleidoscope images, June 2012.

3. "Stable smectic phase in suspensions of polydisperse colloidal platelets with identical thickness" D. Sun, H-J. Sue, Z. Cheng, Y. Martinez-Ratonm and E. Velasco, **Physical Review E**, 80, 2009, 041704

4. "Gelation via Ion Exchange in Discotic Suspensions" Y. Chang, A.F. Mejia, Z. Cheng, D. Di, and G.B. McKenna, **Phys. Rev. Lett. 108,** 2012, 247802.

5. "Pickering Emulsions Stabilized by Amphiphilic Nano-sheets" A.F. Mejia, A. Diaz, S. Pullela, Y. Chang, M.S. Mannan, A. Clearfield, and Z. Cheng, **Soft Matter 8,** 2012, 10245-10253 (**Journal cover**).

6. "Stabilization of Pickering Foams by High-aspect-ratio Nano-sheets" J. S. Guevara (undergraduate), A.F. Mejia, Y. Chang, M. Shuai, M.S. Sam Mannan, Z. Cheng, **Soft Matter** 9, 2013, 1327-1336.

7. "Nano-encapsulated PCM via Pickering Emulsification", X. Wang, L. Zhang, Y. Yu, L. Jia, M. S. Mannan, Y. Chen & Z. Cheng, **Scientific Reports** 5, 2015, 13357.

465761 Ionophore-Decorated Magnetic Graphene Oxide As a Composite Adsorbent for Precious Metal Recovery

Abstract:

Recovery of lithium from aqueous streams has gained significant attention as it can contribute in resolving the anticipated challenges on shortage of lithium supply in the future. Given the variety and complexity in the composition of aqueous Li^+ sources, more advanced materials must be developed for selective Li^+ capture. It is also desirable to synthesize multi-functional materials which can be readily regenerated and reused for long term application.

Herein, a multifunctional adsorbent (CE-rGO-Fe_3O_4) was successfully synthesized and used as a selective lithium ion (Li^+) adsorbent. The adsorbent was decorated with crown ethers (CE) as Li^+-specific ionophores, magnetite (Fe_3O_4) for its easy separation and recyclability, and graphene oxide (GO) as a two-dimensional, high-aspect ratio support of the CE and Fe_3O_4.

The GO was prepared by oxidizing graphite powder through a modified Hummer's method. The Fe_3O_4 nanoparticles were then immobilized on the surface of the GO through a solvothermal technique which also partially reduced the GO nanosheets (rGO-Fe_3O_4). Alkyne moieties were grafted on the rGO-Fe_3O_4 surface using 4-ethynylaniline The alkyne terminals in 4-ethynylaniline served as "clickable" CE-specific attachment sites. Meanwhile, Li^+-selective CE (i.e. 2-hydroxymethyl-12-Crown-4 Ether) was prepared via a two-step mesylation-azidation reaction to obtain 2-(azidomethyl)-12-Crown-4 Ether. The azidated CE was then "clicked" on the alkyne- rGO-Fe_3O_4 at 60 °C via the 1,3 cycloaddition click chemistry reaction.

The synthesized composite adsorbent (12CE4-rGO-Fe_3O_4) was characterized using Transmission Electron Microscopy (TEM), Boehm Titration, X-ray Diffraction Spectrometry (XRD), Fourier-Transform Infrared

Spectroscopy (FT-IR), Thermogravimetric Analysis (TGA), Raman Spectroscopy, and X-ray Photoelectron Spectroscopy (XPS). TEM was used to observe the morphology of the adsorbent, which revealed the presence of Fe_3O_4 NPs on the surface of the GO nanosheets. The XRD spectrum and FTIR conservatively confirmed the presence of different functional groups on the GO support, which was also consistent with the TGA results. Raman Spectroscopy also reflected the degree of functionalization of the GO as indicated by the changes in the D- and G-bands and their corresponding I_D/I_G ratios. The changes in the chemical states of the elements present in the adsorbent upon subsequent modification and functionalization was also confirmed via XPS. XPS results revealed the presence of the Fe_3O_4 in the adsorbent (12CE4-rGO-Fe_3O_4) as indicated by the Fe2p signals whereas the successful attachment of the CEs via click reaction was confirmed with the presence of the peak attributed to the triazole functional group found in the N1s spectra.

The adsorption results show the Langmuir-type Li^+ uptake of the 12CE4-rGO-Fe_3O_4, with the maximum Li^+ uptake at 7.33 mg g^{-1}. The results also show that the material has high selectivity towards Li^+ as compared to the other cations. The integration of magnetite allowed the easy separation and recyclability of the adsorbent. Overall results demonstrate the suitability of the 12CE4-rGO-Fe_3O_4 for long-term Li^+ adsorption application.

This research was supported by the Basic Science Research Program through the National Research Foundation of Korea (NRF) funded by the Ministry of Science, ICT and future Planning (2015R1A2A1A15055407) and by the Ministry of Education (No. 2009-0093816).

475122 Development of Filtered Particulate Eulerian Modeling Approach for the Prediction of Bi-Disperse Gas-Solid Fluidized Bed

Abstract:

Eulerian gas-solid fluidized bed simulations of industrial processes are usually performed with mesh sizes much larger than the smallest meso-scale structure size because of computational resource limitation. Thus, these effective simulations do not fully account for the particle segregation effect (cluster or bubbles formation) and the lack in prediction of the meso-scale structures generally causes poor prediction of bed hydrodynamic [1]. According to previous studies, this effect seems to be very effective in bi-solid mixture with large inertia difference between the solid species.

Following Igci et al. [2], filtered approach may be developed where the unknown terms accounting for the influence of unresolved structures, called sub-grid contributions, have to be modelled in terms of the computed (filtered) variables. In the work presented here, the development of such modelling approach is based on a priori analysis of 3D periodic circulating bi-disperse gas-solid fluidized bed simulations using computational grids cell size of a few particle diameters to obtain mesh-independent results. Using the 3D N-Euler multiphase code NEPTUNE_CFD, separate transport equations are computed for the number density, velocity and random kinetic energy of the two solid species, coupled by collisions terms developed in the frame of kinetic theory of granular media supplemented by interstitial gas effect. These mesh-independent results are filtered using volume average to support the development of closure models for the subgrid fluid-particle and particle-particle interactions terms in bi-disperse gas-solid fluidized bed. These closure models are implemented in the multiphase code NEPTUNE_CFD and tested by performing coarse grid simulation of the periodic circulating bi-disperse gas-solid fluidized bed.

[1] Agrawal, K., Loezos, P., Syamlal, M., Sundaresan, S., 2001. The role of meso-scale structures in rapid gasâ€“solid flows. J. Fluid Mech. 445, 151â€“185.

[2] Igci Y., Adrews AT., Sundaresan S., Pannala S., Oâ€™Brien T., Filtered two-fluid models for fluidized gas-particle suspensions, AIChE J., 2008;54:1431-1448.

460982 On the Relevance of Closures Laws for Momentum, Heat and Mass Transfer in Gas-Particle Suspensions

Abstract:

Nowadays, computer simulations allow us to develop closures for momentum, heat, and mass transfer rates on different levels: first, one can use Direct Numerical Simulations (DNS, [1,2]) of fluid flow in the interstices of dense particle beds to determine, e.g., the forces acting on individual particles. Second, one can use highly-resolved Eulerian-Eulerian, or Eulerian-Lagrangian (EL) approaches to quantify the effect of structure formation in these suspensions [3,4]. Naturally, the question arises which of these closures dominates the predictions of the overall performance of a full scale device like a fluidized bed. Unfortunately, this question has not been answered with sufficient rigor yet.

In the present contribution we give a tentative answer to this question with a focus on polydisperse systems, as well as wet fluidized beds. In part I we present results of DNS to quantify particle-based heat transfer rates in dense bi-disperse suspensions (see Figure). Specifically, we highlight that existing correlations developed for monodisperse systems cannot give a precise estimate of the heat transfer rate. Also, we determine critical Peclet numbers (as a function of the particle concentration) for which heat transfer rates are extremely fast – in these systems the exact form of the closure for the heat transfer rate will not influence the overall dynamics of the system.

Part II is devoted to full-physics simulations of heat and mass transport in wet fluidized beds with continuous droplet injection. We analyze (i) the deposition rate of droplets on the particles, (ii) the evaporation rate of droplets suspended in air, as well as (iii) evaporation from droplets deposited on the particle surface. By systematically de-activating individual closures for these phenomena, we (i) explore their relative importance, and (ii) give guidance for the future improvements of these closures.

Last, we summarize recent developments in the field of computational tools for the simulation and analysis of dense suspension flow which were heavily influenced by Prof. Sundaresan's ideas. Specifically, we present the latest version of the CFDEM® EL simulator, which has been significantly upgraded by the inclusion of cutting-edge models for momentum, heat and mass transfer. Also, we detail on some applications of the filtering tool "CPPPO": this tool should help researchers and engineers to follow the footprints of Prof. Sundaresan, and apply his seminal ideas to a variety of challenges in the field.

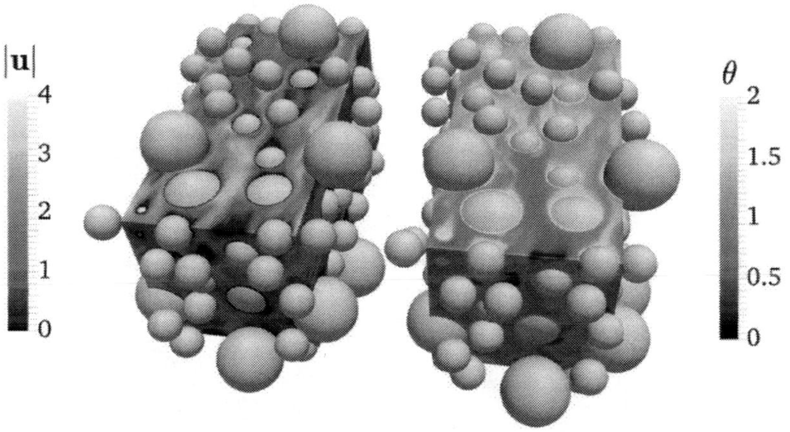

Figure: Direct numerical simulation of heat transfer in a bi-disperse gas-particle suspension in a fully periodic cuboid. Dimensionless velocity (left panel) and temperature fields (right panel; the overall particle volume fraction is 0.35, the Reynolds number is 250, and the Prandtl number is 1).

References

[1] W. Holloway, J. Sun, S. Sundaresan, Effect of microstructural anisotropy on the fluid–particle drag force and the stability of the uniformly fluidized state, J. Fluid Mech. 713 (2012) 27–49.

[2] B. Sun, S. Tenneti, S. Subramaniam, Modeling average gas–solid heat transfer using particle-resolved direct numerical simulation, Int. J. Heat and Mass Trans. 86 (2015) 898–913.

[3] K. Agrawal, W. Holloway, C. C. Milioli, F. E. Milioli, S. Sundaresan, Filtered models for scalar transport in gas–particle flows, Chem. Eng. Sci. 95 (2013) 291–300.

[4] S. Radl, S. Sundaresan, A drag model for filtered Euler-Lagrange simulations of clustered gas-particle suspensions, Chem. Eng. Sci. 117 (2014) 416-425.

475105 Clustering in Gas-Solid Flows: How Are Clusters Modified By Shear?

Abstract:

We present simulation and model results for sheared, cluster-induced turbulence (CIT). In CIT, heavy particles settling under gravity induce velocity fluctuations in their carrier fluid as a result of momentum coupling, and in turn form large-scale clusters. To understand the effect of shear on the system dynamics, we perform Euler-Lagrange simulations of homogeneous shear flow, using a discrete element method for the particle phase and direct numerical simulation for the fluid phase. We notice that for low values of the shear Stokes number (i.e., the ratio between the particle response time and the characteristic shear time scale), particle clustering persists and the system dynamics are similar to those of fully developed CIT in the absence of shear. However, at a sufficiently large shear Stokes number, particle clustering disappears entirely. We develop a Reynolds-stress model for sheared CIT to understand these effects, and show that this model also predicts vanishing clustering above a critical shear Stokes number. We compare model results with simulation data, and use both to better understand the effect of shear on the dynamics of fully developed CIT. These results (in an idealized, homogeneous system) are then used to provide insight into the fundamental physical processes affecting particle motion in more realistic, wall-bounded configurations such as those in fluidized beds and risers.

462187 Cohesive Particle Flows: On the Relevance of Cohesive Energy Vs. Cohesive Force Descriptions

Abstract:

Particle cohesion arising van der Waals force, liquid bridges, etc. is known to vary as a function of separation distance. Although the incorporation of such a dependency is straightforward to accommodate in DEM simulations, the same is not true for continuum models since individual particles are not tracked. To overcome this challenge, previous models have used either (i) a force balance at particle contact or (ii) cohesive energy â€“ the integration of the force profile with distance â€“ as a basis for incorporating the effects of cohesion. This force vs. energy conundrum is also evidenced by the various dimensionless numbers used to describe various systems: cohesive force/gravitational force (Bond number), cohesive energy/kinetic energy, etc. In this talk, we consider a number of systems to illustrate the appropriateness of force-based vs. energy-based descriptions in a variety of circumstances.

466748 A Dilation-Driven Secondary Flow in Sheared Granular Materials, and Its Rheological Signature

Abstract:

The cylindrical Couette device is commonly used for viscometric measurements of fluids, but its use for granular materials is problematic: previous studies showed anomalous wall stress profiles[1,2], wherein the vertical shear stress changes sign and all components of the stress increase nearly exponentially with depth. In a recently study[3], we have shown the presence of a remarkable dilation-driven secondary flow when a dense granular material is sheared slowly in a cylindrical Couette cell. The secondary flow appears in the form of a single toroidal vortex that spans the entire granular column, and is fundamentally different in origin and manifestation from the Taylor-Couette vortices seen fluids. Most importantly, the vortex explains the anomalous, nearly exponential, rise of the stress with depth.

In this presentation, we report the robustness of the vortex to variations in the parameters that determine inter-grain interactions, and to confining pressures exerted at the top of the granular column. We present the results of DEM simulations and experiments for granular columns confined at the top by a freely movable weight. We show that the strength of the vortex decreases with increasing confining pressure, but its essential features, such as the sense of rotation and symmetry, remain unchanged. We show how the character of the vortex changes from anti-centrifugal to centrifugal as the shear rate is increased, establishing that the vortex is indeed driven by dilation in the dense, slow flow regime. Finally, we propose a continuum plasticity model that accounts for dilation in viscometric flows, and provides a plausible explanation for how the vortex may arise as an instability.

[1]Mehandia, V. ,Gutam, K. J. & Nott, P. R., Anomalous stress profile in a sheared granular column. *Phys. Rev. Lett.* **109**, 128002 (2012).

[2]Gutam, K. J., Mehandia, V. & Nott, P. R., Rheometry of granular materials in cylindrical Couette cells: anomalous stress caused by gravity and shear. *Phys. Fluids* **25**, 070602 (2013).

[3]Krishnaraj, K. P. & Nott, P. R., A dilation-driven vortex flow in sheared granular materials explains a rheometric anomaly, **Nat. Commun. 7**, 10630 (2016)

475104 The Big Impact Small Particles Have on Fluidized Beds

Abstract:

Fluidized bed design and scale-up is relying more on computational fluid dynamic modeling. Such models can now capture the strong dependence on particle characteristics such as size and shape. However, recent research has shown that particle clustering has a significant effect on fluidized bed hydrodynamics, which impacts how these units should be designed and scaled-up. Indeed, successful scale up depends on the accurate estimates of the entrainment rate to determine the location and size of the cyclones. Yet, the amount of fines, particle shape and surface morphology play a role on the level of particle clustering in a fluidized bed which directly affects the entrainment rates. The fine particles are an excellent conduit for moving charge as electrons or ions which appear to be the dominant mechanism of electrostatics for Geldart Group A material in a bubbling fluidized bed. This electrostatic force trades off with particle momentum relaxation and rotational to translation momentum transfer with regard to forming a particle cluster. Preliminary work on particle shear in a packed and fluidized beds, suggest that particle clustering can be measured and may provide a quantifiable metric for the level of particle clustering.

475118 Scale up of Heat Transfer of Particles in a Rotating Cylinder

Abstract:

Calcination and drying are widely used in the production of concrete, catalysts and other materials. Calcination and drying are known to be extremely energy intensive processes and calcination alone is estimated to consume around 3% of the worldâ€™s energy. Drying and calcination of particles often takes place in a rotating cylinder heated at the wall. An ongoing challenge is the scale-up of calcination and/or drying in rotating cylinders from the laboratory and pilot plant scales to the manufacturing scale. Developing such fundamental understanding of rotary calcination and drying can improve product quality and cut energy and material costs. Our research seeks to provide a methodology for scale-up through understanding of the effects of material properties, operating conditions and calciner/dryer size on temperature distributions. Two important time scales for continuous calcination/drying in rotating drums are: (1) in the axial direction, the residence time of the particles inside the calciner/dryer, and (2) in the radial direction, the time required for heating of the particles to a target temperature. To optimize calciner/dryer performance, the particle residence time must exceed the time required for heating to the target temperature. Discrete element method (DEM) simulations are used to explore the influence of these competing timescales on scale-up. We have carried out DEM simulations for the case where particles are heated through a hot wall and conduction dominates over radiation. We are able to collapse all the results into three heating regimes: 1) a regime where the bed heats slowly at a nearly uniform temperature, 2) a regime where the system heats as a cool core with warm outer layers and 3) a regime where the system heats as a solid body with temperature decreasing with distance from the wall. Based on the different heating regimes, we are able to derive equations that predict the particlesâ€™ average temperature and temperature distribution. The results of this work have implications for improving the design and operation of calciners and dryers.

456361 Towards Large Scale Production of Core-Shell Microparticles: Multi-Scale Modelling and Experimental Studies

Abstract:

Core-shell microspheres exhibiting great advantages in drug delivery can be produced by coaxial electrohydrodynamic atomization (CEHDA) process. Through this process, uniform size distribution and high drug encapsulation efficiency can be achieved. However, this process shows a monotonic dependence of droplet size on flow rate, therefore the mass flow rate is extremely low if small particles are desired. As a result, the productivity is limited by the low flow rate. Multi-nozzle spraying would be indispensable to overcome the drawback of the low throughput. On the other hand, CEHDA is a complex process including multiphysics and multiscale sub-process. In this study, to simulate the multiscale CEHDA process including the cone-jet formation, droplet formation, solvent evaporation, spraying, and deposition, a multi-scale model (hybrid CFD-PBM-DPM model) was developed. Multi-nozzle coaxial electrohydrodynamic atomization (MCEHDA) was investigated by the developed multi-scale model and experiment. The interaction and deposition pattern of multi-nozzle electrosprays were studied. In addition, the structures of the core-shell microparticle in different regions were analysed. The effects of the distances among the nozzles for different nozzle array configurations were also investigated. The developed multi-scale model can conveniently guide the design and operation of multi-nozzle electrosprays for a large scale production of core-shell microparticles.

471370 Reduced Bubbling and Effect of Cohesive Forces through High-G Operation in Rotating Fluidized Beds in Vortex Chambers

Abstract:

The performance and application range of classical fluidized beds is limited by the role of earth gravity. The formation of meso-scale structures is detrimental for the gas-solid contact. Large (Geldart-D type) and fine (Geldart-C type) particles are poorly fluidized, the latter because of the dominance of the intra-particle van der Waals forces. In rotating fluidized beds in vortex chambers, high-G operation allows increased gas-solid slip velocities and intensification of interfacial mass, heat and momentum transfer. It is shown that the latter allows reducing the formation of meso-scale structures, further intensifying interfacial mass and heat transfer and eventually reactions. Intensified interfacial momentum transfer also reduces the impact of cohesive forces, allowing fluidizing and processing fine cohesive particles and facilitating fluidization of wetted particles. The gas driving the particle bed rotational motion in the static vortex chamber also results in a unique flexibility with respect to the gas flow rate. The fluid dynamics and design of vortex chambers are discussed and the relation with the particle characteristics is explained. Examples of experimental studies demonstrating the advantages and disadvantages of vortex chamber technology in particle drying, coating and granulation are given.

462787 Chemical Looping Combustion for Carbon Capture Efficiency

Abstract:

Among the well-known state-of-the-art technologies for CO_2 capture, Chemical Looping Combustion (CLC) stands out for its potential to capture with high efficiency the CO_2 from a fuel power plant for electricity generation. CLC involves combustion of carbonaceous fuel such as coal-derived syngas or natural gas via a red-ox chemical reaction with a solid oxygen carrier circulating between two fluidised beds, air and fuel reactor, working at different hydrodynamic regimes. Avoided NO_x emissions, high CO_2 capture efficiency, low CO_2 capture energy penalties and high plant thermal efficiency are the key concepts making worthy the investigation of the CLC technology. One of the main issues about the CLC technology might concern the cost of the solid metal oxides and therefore the impact of the total solid inventory, solid make-up and lifetime of the solid particles on the cost of the electricity generated. In this work, macro scale models of fluidised beds (i.e. derived applying macroscopic equations) are developed and implemented in Aspen Plus software. Kinetic and hydrodynamic phenomena, as well as different operating conditions, are taken into account to evaluate their effect on the total solid inventory required to get full fuel conversion. Additionally, a 2D â€œmicroâ€ scale model of the fuel reactor (i.e. derived applying partial differential equations), making use of a CFD code, is also developed. The results, in terms of the effect of the different kinetic and hydrodynamic conditions on the outlet gas conversion, drive the improvement of the macro scale model to capture the main physics influencing the performance of the fuel reactor. The latter macro scale model is embedded into a natural gas fired power plant configuration where mass and energy balances are solved simultaneously. Two CFD platforms (namely MFIX and OpenFOAM) are implemented to devise the influence that the reactorâ€™s internals have on the hydrodynamics and then on the conversion. Finally, a thermal efficiency evaluation of the CLC power plant is carried out together with a detailed economic analysis. The effect of the fuel price and lifetime of the solid particles on the Levelised Cost Of Electricity (*LCOE*) is investigated and the resulting outcomes are critically discussed in comparison with post-combustion technology employing amines.

469869 Cold Flow Model Studies of a Counter-Current Moving Bed Syngas Chemical Looping Pilot Unit for High Purity Hydrogen and Electricity Co-Generation with Carbon Capture

Abstract:

The syngas chemical looping (SCL) process developed at the Ohio State University (OSU), which used Fe-based oxygen carrier and a counter-current moving bed reactor design, has the capability of achieving nearly 100% carbon capture and the flexibility of co-generating high purity hydrogen and electricity. A 250 kW_{th} â€“ 3 MW_{th} pilot unit operated under 10 atm pressure was designed and operated. The pilot unit consists of three major reactors, the reducer and the oxidizer operated as counter-current moving beds, and the combustor operated as a fluidized bed, to perform reductionâ€“oxidation reaction cycles for converting gaseous fuels to highly concentrated discrete streams of hydrogen (H_2) and carbon dioxide (CO_2). An L-valve is used to transport the solids between the moving bed reactors and the fluidized bed reactor with a controllable solids circulation rate. To investigate the hydrodynamic characteristics and optimize the system design, a one-to-one ratio cold flow model is built and tested. The cold flow model is operated under ambient temperature and pressure with both simulated Geldart Group D particles (glass beads) and actual oxygen carrier particles. The operational conditions of the cold flow model tested are to simulate the expected relative differential pressures and gas-solid flow regimes of the three reactors at operating temperature and pressure. Multiple solid circulation rates and gas processing capacities were tested to understand range of operating conditions achievable in the pilot unit while maintaining proper gas sealing between each reactor and solid circulation rate. The solids circulation rate with respect to the aeration gas flow rate was tested to determine a calibration curve. For solid flow measurement, an iso-kinetic device was designed and tested. To date, over 200 hours of operational results were completed with SCL cold flow model with over 19 test cases completed.

467632 Development of a Carbon Stripper Particle Separation System for Chemical Looping Applications

Abstract:

Chemical looping combustion (CLC) is a promising new technology for reducing carbon emissions from coal-based energy generation. The most significant limiting factor for implementation of this technology is the ability to remove unburnt carbon (char) from the oxygen carrier material as it is cycled between the fuel and air reactors within the CLC unit, thus preventing formation of CO_2 formation in the air reactor. In an effort to address this critical issue, a novel carbon stripper unit has been developed for the separation of unburnt char from oxygen carrier compounds for use with chemical looping combustion (CLC) systems. The carbon stripper unit consists of a conical spouted bed with a thin annular ring located between the spout cone and the bed wall. A sweep gas is injected into this annular region such that the sweep gas velocity within the annulus falls between the terminal velocities of the oxygen carrier and char particles. This facilitates separation of particles by weight, as the heavier oxygen carrier particles are allowed to fall down through the annular region, while the lighter char particles are prevented from entering the annulus. The char particles are ultimately entrained out of the dense bed region and into the freeboard region by the spouting gas, where they are carried out of the carbon stripper unit via elutriation. Preliminary results from experiments carried out in a cold flow prototype unit have demonstrated favorable separation characteristics.

Validation Attempts of Proposed Particle-level Growth Mechanism of Ash Agglomerates in Fluidized Bed Combustors

Aditi B. Khadilkar[1], *Peter L. Rozelle*[2], *Sarma V. Pisupati*[1]

[1] *John and Willie Leone Family Department of Energy and Mineral Engineering, and The EMS Energy Institute, The Pennsylvania State University, University Park PA 16802, USA*

[2] *Office of Fossil Energy, U.S. Department of Energy, FE-221/Germantown Building, 1000 Independence Avenue, S.W., Washington, DC 20585-1209 USA*

emails: abk165@psu.edu; peter.rozelle@hq.doe.gov; sxp17@psu.edu

Increase in particle size occurs during several processes that use fluidized beds. While it is desirable in processes such as granulation and pellet-making its occurrence can cause undesirable operational issues in combustion and gasification processes. Models that incorporate interdependencies and variations in both bed-hydrodynamics and binder chemistry are required for accurate predictions of agglomerate growth kinetics. Penn State has developed a unique two-particle collision-based model that combines these two aspects. The model is tested with poly-disperse fluidized bed combustion and gasification systems.

In fluidized bed gasification and combustion systems, agglomeration occurs by sticking of fuel ash particles that are wetted by slag-liquid. The non-uniform temperature, gaseous atmosphere and heterogeneity in ash chemical composition need to be accounted for in the prediction of the slag-liquid binder formation and chemistry. Additionally, the poly-dispersity of bed ash, particle velocities and collision frequencies are the important physics-based factors that affect particle growth rate. In the Penn State model, thermodynamic equilibrium calculations are used to estimate the amount of slag-liquid in the system, and the changes in particle collision frequencies are accounted for by continuously tracking the number density of the various particle sizes. The particle number density is also affected by the ash chemistry as solid particles melt to form slag. Computational fluid dynamics modeling is used in conjunction to obtain the initial granular physics inputs to the model. Thus, the heterogeneities in ash chemistry and granular physics are integrated in this model. The model development was presented previously and the present study focuses on validation of the modeling results and the methodology developed. For this purpose, ash agglomerates were generated in a laboratory-scale fluidized bed combustor at Penn State. Ash agglomerates were produced by operating the reactor at a superficial gas velocity close to the minimum fluidization velocity of the particles, using rejects from a bituminous coal with about 82 % ash content, under oxidizing conditions. Agglomerate samples were also obtained from a fluidized bed combustor at Canada. Polished cross-sections of these agglomerates were studied using scanning electron microscopy with energy disperse x-ray scattering (SEM-EDX) to relate the particle-level slag formation and sticking to the chemical composition. FactSage simulations of the slag-forming components were used to estimate the agglomeration temperatures.

The results of this study indicated that agglomerate growth in fluidized bed combustors (FBC) is initiated at the particle-level by low-melting components rich in iron- and calcium-based minerals. Although the bulk ash chemical composition does not indicate potential for agglomeration, study of particle-level heterogeneities revealed that agglomeration can begin at lower temperatures than the FBC operating temperatures of 850 °C. After initiation at the particle-level, more slag is observed to form from alumino-silicate components at slightly higher temperatures caused from changes in the system, and agglomerate growth propagates in the bed. A post-mortem study of ash agglomerates using SEM-EDX helped to identify stages of agglomerate growth.

Keywords- fluidization, agglomerate growth, SEM-EDX, slag, hydrodynamics

1. Introduction

Prediction of agglomeration is challenging since the heterogeneity in ash chemical composition, gaseous atmosphere and distributive granular physics properties affect the process simultaneously. Several instances of solids agglomeration have been reported, even at the relatively low operating temperatures in bubbling fluidized bed (BFB) and circulating fluidized bed (CFB) systems[1, 2]. Mineral matter in fuel ash is distributive and the bulk chemical composition is not representative. Based on mineral processing operations used, such as grinding, the mineral content is distributed amongst particle classes. Thus, the bulk fuel contains classes of particles that may be rich in specific minerals such as a class that contains heavy, iron-based minerals or calcium-based minerals. The particles of such a class can be assumed to be reasonably homogeneous in composition, while each particle class would have a composition distinct from another. The term "particle-level" refers to the differences that exist at the level of these ash particle classes. The term "bulk" composition refers to the overall composition of the whole coal that contains these specific mineral-rich particle classes. The particle-level differences in size and composition have been ignored in using only the composite fuel for analysis and prediction of agglomerate growth. A consideration of particle-level non-uniformities in the prediction of ash behavior and agglomeration is warranted. Khadilkar et al.[3, 4] showed that the bulk ash chemical composition did not indicate slag formation at low temperatures, while the study of particle-level heterogeneities revealed that it can begin at lower temperatures than the FBC operating temperatures of 850 °C.

2. Research objective

To develop a mechanistic understanding of agglomerate growth in fluidized bed combustors, based on the modeling results and to discuss attempts to validate the proposition.

3. Materials and Experimental Methods

The fluidized bed agglomeration modeling methodology that combines particle-level heterogeneities in ash chemistry and granular physics is based on two-particle collisions, tested using the Stokes' criterion. If the particles' kinetic energy is dissipated by the viscous slag, they remain stuck on collision. The kinetics of agglomerate growth depends on the collision frequency of the particles, which is calculated based on the kinetic theory of granular flow. The slag chemistry and surface properties determine the surface coverage and also affect the rate of agglomeration. The model development has been described in detail in previous work[5].

A study of the slag formation tendencies at the particle-level using density separated coal gravity fractions, was conducted in previous work, under oxidizing[4] as well as reducing[3] conditions. A lower slag formation onset temperature occurred for particles rich in iron- and calcium-based phases. The amount of slag formed at the onset temperatures was less than 10 %. The extent of agglomerate growth was estimated using the agglomeration model at different slag amounts to gain a better understanding of the conditions that promote agglomeration.

Additionally, agglomerate samples obtained from a laboratory-scale combustion facility in Canada were analyzed using SEM-EDX. Agglomerate samples were also generated in a laboratory-scale fluidized bed combustor at Penn State as described below using Pittsburgh No. 8 coal rejects for analysis. The chemical composition of these rejects is shown in Table 1.

Table 1: ICP-AES and sulfur analysis of Pittsburgh No.8 coal rejects

Species	SiO_2	Al_2O_3	Fe_2O_3	CaO	TiO_2	K_2O	MgO	Na_2O	BaO + SrO	MnO	S
Wt. %	62.57	24.82	5.74	2.17	1.26	2.5	0.83	0	0.09	0.02	1.61

3.1. Laboratory scale fluidized bed reactor

A laboratory-scale fluidized bed[6] was used to produce agglomerates from a fraction of Pittsburgh No. 8 seam coal. The rejects of this coal were used for this purpose as the rejects have a high ash content of 81.6 % and would reduce effects of the presence of carbon, as that is not accounted for in the model. The reactor was operated as a combustor by using 3% oxygen, 15% carbon dioxide, 10% water and 72% nitrogen as the fluidizing gas. The particle size distribution used consisted of about 48 % in the size range of 250-355 μm and 52% between 149-250 μm. The size and velocity distribution were chosen such that the superficial gas velocity would be greater than the minimum fluidization velocity and less than the terminal velocity of majority of the particles. A 200 g sample was used and the experiment was run for 2 h 15 min, using a gas flow rate of about 10 L/min.

3.2. Scanning electron microscopy coupled with energy dispersive x-ray spectroscopy (SEM-EDX)

The samples were tested in a FEI Quant 200 SEM coupled with a 10 mm^2 EDX detector (Oxford Instruments, Concord, MA). This technique was used to conduct post mortem analysis of agglomerate samples. They were given a conducting iridium coating to avoid excessive charging of the sample surface with electrons during the experiments. This helped to identify the elements in the regions that formed slag bridges as compared to the solid particles. The elemental analysis was performed using the software Aztec, version 2.4.

4. Results and Discussion

Slag formation onset temperature was as low as 850 °C for density fractions of Pittsburgh No. 8 and Illinois No. 6 coals and Skidmore anthracite. The amount of slag formed at initiation was less than 10 %. In order to determine the extent of growth at low slag amounts, a sensitivity analysis to the amount of slag was performed using the agglomeration modeling methodology. Particle agglomerate growth was limited by slag-liquid amounts up to 20 wt. % slag after which it plateaued as shown in Figure 1. When the slag is less than 15 wt. %, the diameter does not increase significantly since the slag viscosity is too high and amount too low. As the slag-liquid amount increases further, although the extent of particle wetness increases, the number of solid particles decreases as they melt to form the slag-liquid. So, it is believed that the collision frequency of particles decreases although the wetness of particles increases. However, the amount of slag that can undergo solidification into a larger agglomerate is higher, as explained in regime 3 described below.

Three regimes of agglomerate growth are proposed as follows-
Regime 1: Slag-liquid amount less than 10%- In this regime, the agglomerate growth is limited by the amount of slag-liquid available for sticking. Thus, the probability that the particles undergo a wet collision is low. The rate of growth is slow and the time required for the particles to reach the defluidization size would be very long.

Regime 2: Slag-liquid amount between 10-30 %- In this regime, the agglomerate growth rate is maximum. An adequate number of particles are available in the system and there is enough slag-liquid for particle sticking on collision. The particles in this regime are likely to be completely coated with slag-liquid layers and the probability of wet collisions will be high. The particle growth rate is high and significant defluidization would occur.

Regime 3: Slag-liquid amount greater than 50%- In this regime although a large amount of slag is available for wetting ash particles, since this slag is formed by the melting of solid particles, the number of particles in the system is low. The collision frequency of particles is likely to be low and although there is adequate amount of slag-liquid, the particle-particle interaction is limited. In

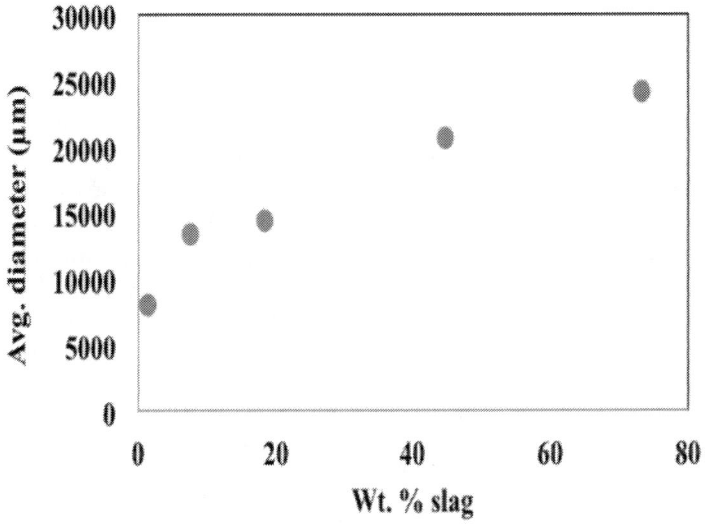

order to demonstrate the effect of the large slag amounts present in regime 3, the average agglomerate diameter was calculated assuming the extreme case that all the slag coalesces and deposits into a large solid agglomerate. Thus, Figure 1 shows an increase in average diameter with increasing slag amount. In the case of bituminous coals and anthracite, at the low operating temperatures in fluidized bed combustors, regime 1 and 2 are more likely to occur in the bed. Additionally, regime 1 and 2 are of interest in understanding the initiation of agglomerate growth and are the focus of the modeling study.

Figure 1: Effect of slag-liquid amount on agglomerate growth with solidification of slag formed

*4.1. **Proposed mechanism of the process of agglomeration in fluidized beds***

The heterogeneous behavior of bed ash helps to explain the occurrence of agglomeration at temperatures lower than operating temperatures in fluidized bed combustors. At low temperatures, certain low-melting components such as those in particle classes with high iron- and calcium-based components studied, can initiate the slag formation in regime 1. The agglomerate growth is likely to propagate in Regime 2.

Experimental studies by Ataköl et al.[7] showed that agglomeration occurs at 125-200 °C lower than characteristic IDT and about 300-400 °C lower than the flow temperature (FT) in a laboratory-scale fluidized bed combustor. They identified the temperature at which agglomeration was initiated based on the detection of bed temperature fluctuations and pressure drop measurements, as well as pictures of the active bed surface. Since the iron- and calcium-based mineral rich SG4 fraction has been identified as the initiator of agglomeration, in the present study, a comparison of IDT of bulk coal with the slag-liquid formation onset temperature obtained for the SG4 density fraction of Pittsburgh No. 8 coal and Illinois No. 6 coal is shown in Table 2. This comparison shows that the initiation of slag formation occurs at temperatures around 400-600 °C lower than the measured IDT temperature. This indicates that agglomerate growth is likely to be initiated around particles of certain compositions such as of the SG4 particle class, at temperatures significantly lower than AFT predictions that are based on bulk coal properties.

Table 2: IDTs of Bituminous coals from AFT measurements under reducing conditions[8]

	Pittsburgh No. 8	Illinois No.6
Initial deformation temperature (IDT) using whole coal (°C)	1,237	1,201
Particle-level Slag Formation Onset Temperature (°C)	840	650

The proposition that agglomeration is initiated around a few particles and then propagates throughout the fluidized bed reactor is supported by the study of agglomerates presented in a report by the U.S. Department of Energy (DOE)[9]. In this study undertaken by the U.S. DOE, agglomerates formed in thirteen fluidized bed combustion units were observed using SEM, which showed the presence of dark cores with rims. The agglomerates were also found to have a high calcium sulfate and iron oxide content. These observations further support the onset of conditions favorable for agglomeration by sticking of particles with composition similar to the SG4 fraction, which could have a dark iron core at locations of initiation of agglomerate formation with rims of alumino-silicate rich slag. Below 850 °C, the SG4 fraction in all the three coals studied began to form slag under reducing conditions[10]. The amount of slag is less than 10 weight % near the onset temperature. Agglomerate growth is likely to be initiated around these particles but as shown by the modeling results in Figure 2, it is not likely to lead to defluidization at these low slag-liquid amounts. At higher temperatures that can occur in bed, the SG4 fraction can form up to 90 % slag. Once initiated, agglomerate growth is likely to propagate in the bed due to one or more of the following reasons:

1) Temperature excursions due to changes in system hydrodynamics- Once particle size increases, and defluidization begins, degree of mixing will be reduced and temperature fluctuations increase. Local hot spots can form in a poorly mixed bed[11]. An empirical equation to obtain the critical velocity needed to avoid dead zone formation for particles of a certain size in fluidized beds was developed by Wen et al.[12]. As per their work, an increase in dead zone formation occurs with an increase in particle size. Similarly, a decrease in mixing at the onset of defluidization, resulted in temperature fluctuations and differences in temperatures measured by thermocouples at different heights in a fluidized bed combustor, in the study by Scala et al.[13]. A discussion of the effect of hydrodynamics on heat transfer in fluidized beds by Bock et al.[14] showed that heat transfer decreases with an increase in particle size and subsequent changes in bubble motion. Hence, with an increase in particle size with agglomerate growth and also dead zone formation at the onset of defluidization, heat transfer is likely to decrease causing an increase in particle temperatures. Bottom ash particles that are agglomerating are primarily of Geldart type A or B and as they grow in size, they will eventually move into type D. For BFBs, an increase in size of type A and B particles causes a decrease in the heat transfer co-efficient. As the particle size increases from 100 to 1,000 μm, the heat transfer co-efficient of heat transfer to heat exchanger surfaces decreases by about 4.5 times, mainly due to decrease in particle convection that causes decreased renewal of particles from heat transfer surfaces[15]. Additionally, after agglomeration is initiated, as particle size increases, the ratio of the particle velocity to the minimum fluidization velocity of the larger particles can decrease. This will increase the possibility of particle segregation. Hence it is possible that once agglomeration is initiated by particles melting at low temperatures, such as those rich in iron- and calcium-based minerals, the change in bed hydrodynamics would lead to further agglomeration due to decreasing particle velocities and decreasing bed voidage. As agglomeration proceeds, the volume of the expanded bed decreases due to the lesser fluidization of the heavier particles formed. Thus, with decreasing voidage and subsequent increasing collision frequency, agglomeration rate may increase initially. As it progresses, decrease in granular temperature and particle number density and the resulting collision frequency decrease would also decrease agglomeration rate. Thus particles grow quickly once the process of agglomeration is initiated until a stable size is reached as the agglomeration rate begins to decrease.

2) Temperature rise due to deposition on heat-transfer surfaces- One more plausible mechanism of the propagation of agglomeration in the system is due to decrease in heat transfer efficiency after initiation of agglomeration. Once particles of larger sizes form by sticking of mineral matter with low slag-formation temperatures, the particle velocity decreases and they can deposit on to heat transfer surfaces, thereby decreasing the efficiency of heat transfer. The decrease in heat transfer can cause a rise in bed temperature leading to melting of alumino-silicate rich components, which have higher slag-formation temperatures as shown from the analysis of

density fractions. The increase in slag amount would then cause an increased agglomeration rate. This mechanism is prone to occur when there are in bed heat exchanges, which are generally restricted to bubbling fluidized bed systems.

3) Presence of local pockets of reducing conditions in the bed- Besides a local increase in particle temperatures, agglomeration is also likely to occur due to changes in the local gaseous atmospheric conditions. The temperature required for the formation of a certain amount of slag-liquid is lower in a reducing gaseous environment than an oxidizing gaseous environment[4]. Local reducing conditions can exist in a fluidized bed reactor around regions in which char particles are concentrated. The literature shows that mineral matter within a carbonaceous particle experiences a high temperature and a reducing atmosphere[16]. Once a few particles such as those from the SG4 density fraction, rich in iron- and calcium-based minerals become sticky at low temperatures, char particles can stick to their surface. The concentration of char particles within mineral matter and slag can create these local reducing conditions. The included minerals in these char particles are also then subjected to the reducing environment and begin to form slag at lower temperatures.

In order to support the mechanism of agglomerate growth in fluidized bed reactors, microscopic analyses of agglomerate samples was performed as follows-

4.2. SEM-EDX analysis of agglomerates obtained from a laboratory scale FB combustion facility at Canada

Agglomerates that formed within two hours during combustion in a laboratory-scale combustion facility in Canada were used as samples to conduct a post-mortem study to understand the agglomerate growth mechanism. In order to avoid a biased analysis, microscopic imaging was performed in randomly selected regions across the entire sample surface. In this way, the entire sample surface was scanned to identify particles with different morphologies. Also, EDX of the whole sample cross section was done at a low magnification (50x to 200x) to identify regions with higher concentration of a particular type of element. Regions distinct in elemental composition were then magnified for detailed study with microscopic images and elemental maps, so as to account for all different types of particle behaviors. Primarily two distinct morphologies were observed across the sample surface as shown in Figure 2- small particles adhering to form a larger sphere and slag-coated particles with bridges. The latter showed that greater slag amounts were formed and coated the particles. Morphologies showing discrete adhered particles indicated lower slag amounts. The formation of higher amount of slag was believed to have occurred at higher temperatures. Hence, the slag-coated particles were likely to be a later stage (causing propagation at higher temperatures) in the agglomeration process, after initiation by the smaller adhered particles. In order to test this implication of the two distinct morphologies observed, the elemental composition of each of them was studied using EDX. EDX was performed on 8 to 10 spots of the two morphologies identified in Figure 2.

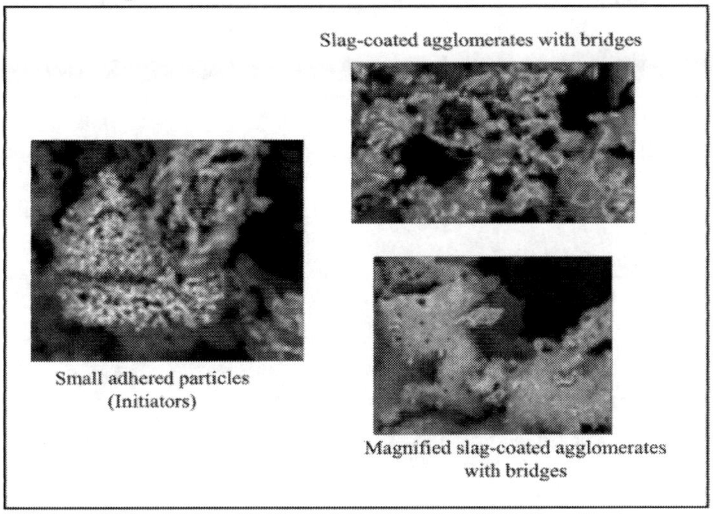

Figure 2: Morphologies of initiators and slag-coated particles observed by SEM of agglomerates

The EDX results are shown in Figure 3. The orange sphere in the figure is mapped to high concentrations of iron- and calcium-containing phases. Previous work using FactSage, TMA and XRD also indicated that density fractions rich in calcium- and iron-based minerals

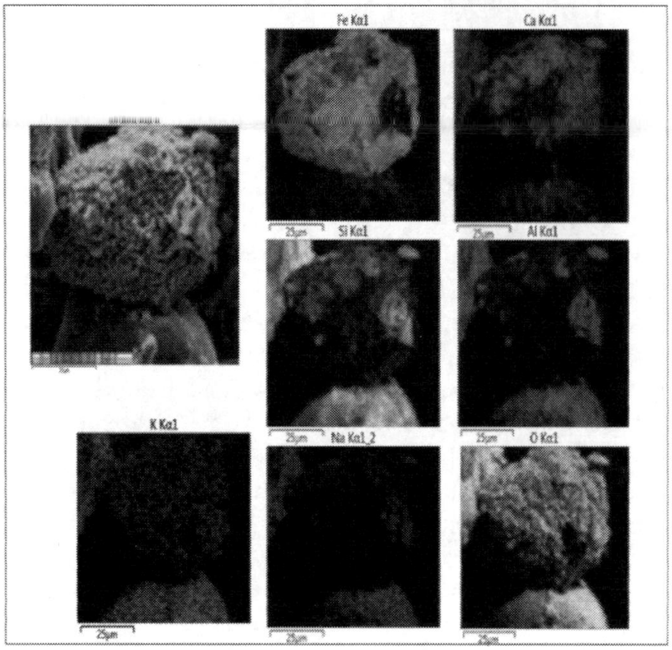

Figure 3: SEM-EDX of initiator and alumino-silicate slag coating

showed slag formation onset temperature less than 850 °C. The amount of slag obtained at initiation for these density fractions was less than 10 %. When the slag amount is less than 15%, the modeling results in Figure 1 showed that the agglomerate growth was restricted and particles did not reach their defluidization size in 10 hours. These classes rich in calcium- and iron-based minerals were thus believed to initiate agglomerate growth at the particle-level, which would then be followed by propagation. Further, molten slag (blue) seen to adhere to these initiators, is rich in silicon and aluminium with some sodium and potassium content as seen in Figure 3. In order to obtain a semi-quantitative estimate of the composition of the initiators and slag-coated particles, EDX was used and the average composition of the studied spots is reported in Table 3.

In order to verify this, FactSage thermodynamic simulations were performed to predict the initiation temperature of slag formation in these regions. It was found that the initiators, which are rich in iron phases, began slag formation at 820 °C as shown in Figure 4, while the alumino-silicate rich slag was formed at a higher temperature of 920 °C. Slag-coated agglomerates began to form bridges with this higher slag amount. The amount of slag formed from the components in the initiators was also less than 10% up to 950 °C. Since the particles do not melt completely, the SEM images show small discrete particles of these initiators adhering to form larger spheres.

Table 3: EDX compositional analysis of stage 1 and stage 4

	Small adhered particles (Initiators/Stage 1)		Slag melting at higher temperature (Slag on slag-coated agglomerates/stage 4)	
Element	Wt.%		Wt.%	
	Mean	Standard deviation	Mean	Standard deviation
O	36.9	8.4	57.5	4.4
Na	0.2	0.2	1.8	0.3
Mg	1.1	0.6	1.5	0.5
Al	1.9	0.5	10.7	1.5
Si	4.4	2.1	21.5	2.7
K	0	0.1	1.8	0.5
Ca	3.1	1.9	1.8	1.4
Mn	2.2	1.1	0	-
Fe	50.3	8.2	3.1	0.8

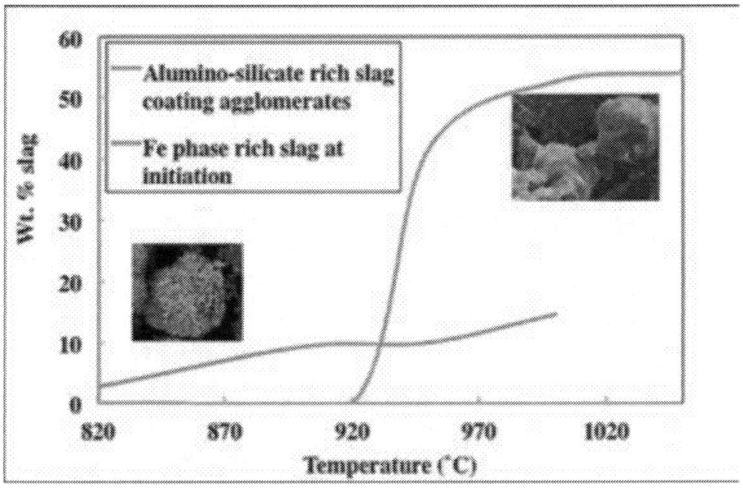

Figure 4: Slag formation tendency of components in stage 1 and stage 4 using FactSage simulations

Through this SEM-EDX study of agglomerates, four different stages of agglomeration were identified as shown in Figure 5. Stage 1 shows small particles rich in iron–based minerals adhered together to form a larger sphere. These particles begin to form slag at a lower temperature of 820 ˚C, as shown in Figure 6. The amount of slag formed from this composition is less than 20 % at temperatures up to 1,000 ˚C. Hence, this is adequate to initate agglomeration and lead to the formation of bigger spheres that are rich in iron, but not sufficient to make the particles grow up to their defluidization size. Stage 2 shows that slag formed at the higher temperatures, begins to coat these iron spheres. Stage 3 shows a completely coated agglomerate from stage 1. Finally, with the formation of up to 40 % slag at higher temperatures of above 920 °C, the coated agglomerates begin to stick to one another as shown in stage 4, wherein bridging is clearly seen between two slag-coated agglomerates.

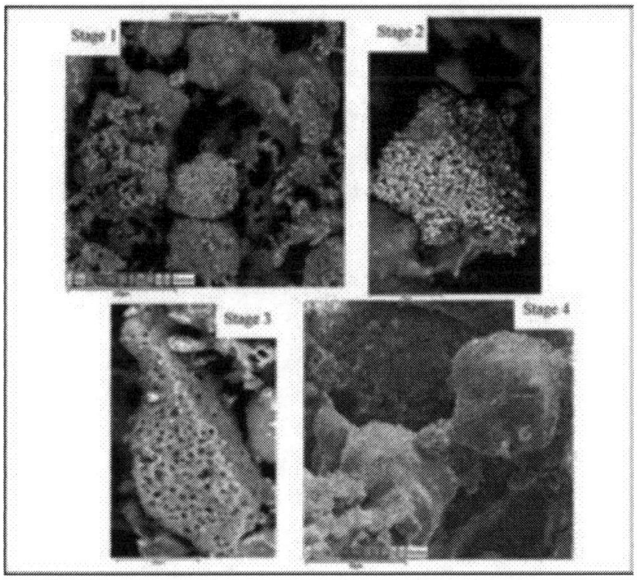

Figure 5: Stages of agglomerate growth observed through SEM-EDX of agglomerates

4.3. SEM-EDX analysis of agglomerates produced in laboratory fluidized bed reactor at PSU

Agglomerates of Pittsburgh No. 8 seam coal rejects (Figure 6) were produced in the laboratory-scale reactor at Penn State. The composition of this material is shown in Table 1.

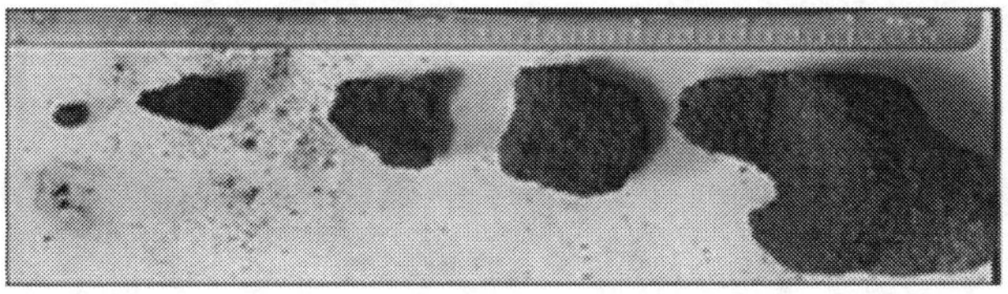

Figure 6: Agglomerates of Pittsburgh No. 8 coal rejects produced in PSU reactor

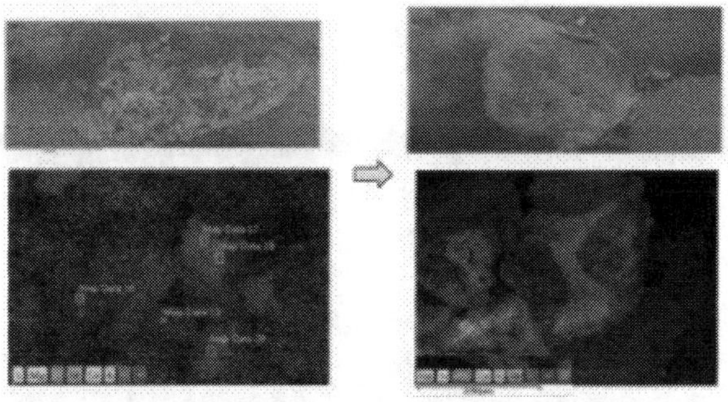

Figure 7: SEM-EDX images of agglomerate of Pittsburgh No. 8 coal rejects

The SEM-EDX analysis of the agglomerates shows two particles rich in siliceous minerals (violet) adhered by a slag bridge (orange) rich in iron and calcium containing minerals. SEM-EDX analysis of these agglomerates also showed regions with smaller, discrete particles adhering together as shown in Figure 7. Across the entire cross section of the sample, 25 regions including those with bridging and with smaller particles adhering were observed. 9 spots in regions with initiators (small, discrete particles adhering) were studied using EDX and found to be rich in calcium-containing compounds with the composition as shown in Table 4. Similarly, 15 spots in the regions that formed slag bridges as shown in Figure 7 were also analyzed for their elemental composition. The composition of the highly molten slag is shown in Table 4.

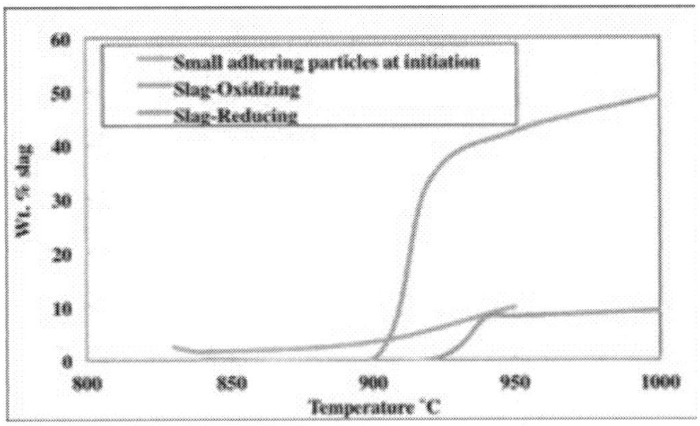

Figure 8: Slag formation tendency of components in small adhering particles and alumino-silicate slag-forming components using FactSage

FactSage simulations under oxidizing conditions were performed with the ash compositions in Table 3 to identify the temperature at which slag formation begins for the initiators rich in calcium-based minerals. The results shown in Figure 8 indicate that the agglomeration was initiated at 830 °C by the calcium-rich particles. The temperature of particles in bed is then likely to rise by one of the plausible mechanisms discussed, causing propagation of agglomeration with the formation of liquid bridges above 930 °C. This onset temperature of slag-formation from the alumino-silicate components that lead to bridging was calculated using FactSage simulations under oxidizing conditions.

Table 4: EDX of initiators and alumino-silicate, high temperature melting slag

Element	Initiator		Alumino-silicate Slag	
Wt.%	Mean	Standard Deviation	Mean	Standard deviation
C	2.1	0.94	1.3	0.39
O	40	0.64	40	9.5
Na	0.3	0.12	0.4	0.21
Mg	0.4	0.02	0.4	0.16
Al	4.4	0.88	9	1.8
Si	6.7	1.11	12.3	2.2
S	0.9	0.19	0.1	0.07
K	0.5	0.41	1.6	0.62
Ca	5.1	0.17	0.7	0.43
Ti	0.3	0.86	0.5	0.08
Fe	40	1.7	34	14

The slag formation onset temperature of the alumino-silicate components that have relatively higher melting temperatures was also computed under a reducing environment. Under this condition, the temperature of onset of slag formation was lowered to 910 °C and the amount of slag was 30-50 %, which was higher than under oxidizing conditions. Hence, in the presence of pockets of reducing gaseous conditions in the reactor, these components could cause propagation of agglomeration. FactSage analysis showed that the amount of slag-liquid formed from the alumino-silicate components was lower in case of Pittsburgh No. 8 coal rejects than the agglomerates obtained from Canada. Hence, the intermediate stage of the formation of a slag-coating on agglomerates, before bridging, which was seen for the agglomerates from Canada, was not seen in the Pittsburgh No. 8 rejects. Partially slag-covered agglomerates formed bridges in case of agglomerates from Pittsburgh No. 8 coal rejects.

5. Conclusions

In this work, a mechanism for agglomerate growth based on initiation at the particle-level followed by propagation in bed is proposed. SEM-EDX study of agglomerates from fluidized bed combustors substantiated this mechanism. It showed that iron-based minerals such as pyrite and also eutectics involving iron- and calcium-based minerals were identified as low-temperature initiators (≤ 850 °C) of agglomeration. The amount of slag formed at initiation is less than 10 %. As shown by the modeling work, this is enough only for sticking together of few smaller particles thereby initiating an increase in particle size. At that stage, the particles do not exceed the defluidization size. The alumino-silicates, which are present in larger amounts in bituminous coals, begin to form slag at higher temperatures (850 °C to 950 °C) and help propagation of the agglomeration process. Similarly, formation of a local reducing environment around char particles entrapped in agglomerates is likely to lead to larger amounts of slag at lower temperatures.

References

1. Tung, S.E., Williams, G.C., Atmospheric fluidized bed combustion—a technical source book, DE-AC21-80MC14536, U.S. Department of Energy, **1987,** 17-89–17-95.
2. Anthony, E.J., Talbot, R.E., Jia, L., Granatstein, D.L., Agglomeration and fouling in three industrial petroleum coke-fired CFBC boilers due to carbonation and sulfation, *Energy and Fuels*, 14, **2000,** 1021–1027.
3. Khadilkar, A. B.; Rozelle, P. L.; Pisupati, S. V., A study on initiation of ash agglomeration in fluidized bed gasification systems, *Fuel,* **2015**, 152, 48–57.
4. Khadilkar, A.B., Rozelle, P., Pisupati, S., Effect of heterogeneity in coal ash chemical composition on the onset of conditions favorable for agglomeration in fluid beds, *Energies* **2015**, *8* (11), 12530–12545.
5. Khadilkar, A.B., Rozelle, P., Pisupati, S., Incorporation of Hydrodynamics and Ash Chemistry into a Single Model for Prediction of Particle Size Increase in Fluidized Beds, *AIChE Annual Meeting,* **2015,** Utah, USA.
6. Khadilkar, A.B., Development of a fluidized bed agglomeration modeling methodology to include particle-level heterogeneities in ash chemistry and granular physics, Dissertation in Energy and Mineral Engineering, The Pennsylvania State University, 2016.
7. Atakül H., Hilmioğlu, B., Ekinci, E., The relationship between the tendency of lignites to agglomerate and their fusion characteristics in a fluidized bed combustor, *Fuel Processing Technology*, **2005,** 86, 12–13, 1369–1383.
8. Matuszewski, M, Detailed coal specifications, Final Report- DOE/NETL-401/012111, **2012**.
9. Brown, R.C.; Dawson, R.; Smeek, J. L. Bed material agglomeration during fluidized bed combustion, Final report for U. S. Department of Energy, No. DE-FG22-92PC92530, 1996.
10. Khadilkar, A., Rozelle, P. and Pisupati, S., Comparative study of fluidized bed agglomeration of high rank coal ash with consideration to particle-level heterogeneities, *Proceedings of the International Conference of Coal Science and Technology*, **2015,** Melbourne, Australia.
11. Visser, H. J. M., van Lith, S. C., Kiel, J. H. A., *Journal of Energy Resources Technology* **2008**, *130* (1), 011801-1- 011801-6.

12. Wen., C. Y., Krishnan, R., Kalyanaraman, R., Particle mixing near the grid region of fluidized beds, *Proceedings of the International Fluidization Conference* **1980**, 405.
13. Scala, F., Chirone, R., Characterization and Early Detection of Bed Agglomeration during the Fluidized Bed Combustion of Olive Husk, *Energy & Fuels* **2006**, 20, 120-132.
14. Bock, H.-J., Molerus, O., Influence of hydrodynamics on heat transfer in fluidized beds, *Proceedings of the International Fluidization Conference*, New Hampshire, U. S. A., **1980**, 217.
15. Basu, P., Combustion and Gasification in Fluidized beds, CRC Press, Florida, **2006.**
16. Scott, D. H., Ash behaviour during combustion and gasification, IEA Coal Research Report, **1999.**

453406 Numerical Simulation of the Effects of Size Change in a Coal Gasifier Using Method of Moments Approach

Abstract:

The population balance model (PBM) and the governing computational fluid dynamics (CFD) equations were used to simulate the effect of size change in a coal gasification process using the finite size domain complete set of trial functions method of moments (FCMOM) approach. The goal of this study was to demonstrate the effect of property (e.g. size and density) variation of coal char particles in the simulation, design, and scale-up of gasification processes. In this study, the process of coal char gasification in a fluidized bed at elevated temperature in the presence of hydrogen and steam as gasification media was simulated. The coupling of coal char size variations, hydrodynamics of the system, and heterogeneous gasification reaction rate were included in the simulation. The heterogeneous gasification reaction was modeled based on the available experimental data from Gas Technology Institute (GTI) using the shrinking core model. The simulation results showed that the effect of particle property variations (e.g., size) during a process such as gasification of coal char in a fluidized bed plays a very significant role in the modeling and design of such processes. Initially, as the coal char particles were gasified, the coal became smaller; therefore, the effect of mixing in the particulate phase became more pronounced (as a result of the formation of bubbles). Ultimately, the effects of mixing became less significant due to a reduction in the bed height, which, in turn, suppressed the formation of larger bubbles.

464063 Study of Structured Flow Pattern and Bubble Characteristics of a Pulsed Fluidized Bed Using Kinetic Theory and Turbulence Model

Abstract:

This work presentsÂ simulation results and analysis of the hydrodynamics and flow patterns of a pulsed fluidized bed using Eulerian-Eulerian two fluid modeling approach. The pulsed inflow of gas phase is in a form of sinusoidal function as $u_0/u_m = A + B \sin(2Ï€ft)$ where four pulsating frequencies of 3, 4, 5, 6 HZ are used. The kinetic theory and Johnson and Jackson Frictional Model for particles motions coupled with k-epsilon Turbulence Model for fluid phase, and Gidaspowâ€™s Drag Model for addressing interactions between two phases are used to close the system. The simulation successfully captured the structured and regular bubble patterns reported in other experimental papers. The hydrodynamics, bubble patterns, bubble sizes are obtained and analyzed for each frequency of inlet flow. The effects of inlet velocity, particle size and diameter on the hydrodynamics, flow patterns are also discussed. Discrete Fast Fourier Transform study is carried out on pressure drop, fluid velocities and volume fraction of a series of specific points in the bed to better understand the fluctuation and structure details of the fluid flow. A map of the structured pattern performance of particles with various density and size are marked on the Geldartâ€™s Chart for type B particles. Two Fluid Modeling results improve the understanding of the chaos and structured properties of the conventional and pulsed fluidization.

470285 Powder Flow Characterization at Low Consolidation: 1. Modeling and Experimental Values of Torque Estimation

Abstract:

The flow properties of fine powders at very low consolidation levels are relevant to small scale industrial application of powder flow, such as in small process hoppers and mixers of pharmaceutical powders. In addition, many process applications of powder technology involve gas-solids systems. The knowledge of flow properties of aerated powders is necessary for proper handling.

Characterization of powders and granular materials can be made with a variety of commercially available shear testing devices. However, in fact, sensitivity of these shear testers has been decreased greatly at low levels stress. Furthermore, materials which these testers find to be cohesion-less may have practical differences in flow behavior, especially in regions of low consolidation stress.

New testing devices, which use non-traditional methods (such as centrifugal force) to estimate cohesive forces at low levels of consolidation stress, are not well-established and their correlation to either traditional shear testers or practical problems is incomplete. The equipment used in this research project was the state of the art powder cell purposely developed by Anton Paar to fit on an Anton Paar rheometer and carry out torque measurements on both aerated and fluidized powders. In this instrument, the state of consolidation of the powder can be adjusted by changing the impeller depth and the aeration rate. Four free flowing powders, according to the Geldart classification, Sand, Glass beads, Alumina, and Calcium lactate powders were used in this research project. The theory developed by Bruni *et al.* 2007 and applied by Tomasetta *et al.* 2011 is used to evaluate powder flow properties of the powders. Two different models considering the state of stress, active and passive state of stress, during torque measurement were considered. The results indicated that the instrument allows to successfully exploring low consolidation ranges and well correlates with the expected powder flow properties.

REFERENCES

[1] Bruni, G., Barletta, D., Poletto, M., & Lettieri, P. 2007. A rheological model for the flowability of aerated fine powders. Chemical engineering science, 62(1), 397-407.

[2] Tomasetta, I., Barletta, D., Lettieri, P., & Poletto, M. 2012. The measurement of powder flow properties with a mechanically stirred aerated bed. Chemical Engineering Science, 69 (1), 373-381.

455145 CFD-DEM Simulation of Tube Erosion in a Bubbling Fluidized Bed with a Tube Bundle

Abstract:

CFD-DEM simulation of tube erosion in a bubbling fluidized bed with a tube bundle

Yongzhi Zhao (yzzhao@zju.edu.cn), Lei Xu

Department of Chemical and Biological Engineering, Zhejiang University, Hangzhou 310027, China

The erosion of the immersed tubes in a bubbling fluidized bed is studied numerically using CFD-DEM method and a particle-scale erosion model, SIEM (shear impact energy model), which is proposed based on the shear impact energy of particles in DEM simulations. By analyzing the simulation results, some characteristics of the tube erosion in the fluidized bed are obtained. The distributions of the erosion around the tubes located in different positions of the bed are different. For the tubes in the center of the fluidized bed, the profiles of erosion rate around the tubes have two peaks at the lower surface of the tubes. On the contrary, for the tubes near the sidewall of the fluidized bed, the maximum erosion usually occurs on the upper surface of the tubes. The values of the erosion rates on the tubes located in different positions of the bed are also different. With the increasing of the vertical position, the erosion rate of the tubes, including the maximum erosion rate and the average erosion rate, increases gradually. In the horizontal direction, the erosion rates of the tubes at the center of the fluidized bed are larger than that of the tubes on both sides of them in the same horizontal plane. For all tubes in the fluidized bed, the maximum erosion rates are about 1~3 times to the corresponding average erosion rates and most of the ratios are near to 2. The erosion model and the algorithm are proved to be effective for predicting the erosion of the tubes in the fluidized bed. The accuracy of the approach is good for predicting not only the distribution of the erosion around the tube, but also the absolute value of the erosion rate. The micro-scale behavior of particles around the tubes is also revealed and the linear relationship between the erosion and the shear impact energy is confirmed by the simulation results and experiment (Johansson et al., Chem. Eng. Sci., 2004, 59: 31-40). The agreement between simulation and experiment proves that the micro-scale approach proposed in this paper has high accuracy for predicting erosion of the tubes in the fluidized bed, and has potential to be applied to modeling the process in other chemical equipment facing solid particle erosion.

471444 Invited - Preventing Nanoparticle Plate-out in Heat Exchangers during Carrier Fluid Phase Change and Implications for Power Generation

Abstract:

For over a decade, a great deal of research has been conducted on the enhancement of heat transfer using nanofluids. Most exclusively, these results are rooted in the measured increase in thermal conductivity of the nanofluid when compared to the base fluid. Far less research points to negative drawbacks such as degradation of specific heat of the carrier fluid and viscosity increase with the addition of nanoparticles. We focus on the use of metal organic frameworks (MOFs) to enhance the energy transfer of a working fluid as it changes phase. The benefit of MOFs as a nanoparticle manifests itself within the energy equation in the form of the enthalpy of adsorption and desorption. We detail the many challenges faced over several years in working with nanofluids that use MOFs. Ultimately, using MOFs in power cycles can enhance the overall efficiency of the cycle with short payback period. As with most energy conversion cycles, working fluid phase change is a necessity, and resulted in nanoparticle plate-out on the various surfaces where phase change occurs. Our team has looked at several techniques for preventing this plate-out. One technique that shows promise in solving the plate-out problem involves forming a dispersion of the nanoparticles in a non-volatile low viscosity oil that is then added to the working fluid. Optical and IR spectroscopic techniques show that working fluid adsorption-desorption occurs unimpeded in the oil dispersion.

461379 Invited - Particle ALD Cost/Performance Advantages

Abstract:

The ability to place uniform nanoscale films on primary non-agglomerated particles, including nanoparticles, is a platform technology with many applications in the areas of electrochemistry, lighting, catalysis, micro-electronics, personal care, biotechnology, defense, construction and other areas - many yet to be discovered. ALD on particles is being successfully demonstrated on numerous particle systems including metals, ceramics, and polymers. Particle ALD allows the cost effective synthesis of unique materials that cannot be synthesized by conventional coating processes such as chemical (CVD) and physical (PVD) vapor deposition, or sol-gel methods. The process provides for a pathway to efficiently use expensive precursors with almost no waste product, hence, the low cost option for many functionalized particle applications. Although invented in the late 1990s, it has taken over a decade for commercial applications to be taken seriously and developed. Much of this has been the stigma attached to the higher cost of coating flat low surface area materials used in the micro-electronics industry. This presentation will review the development of Particle ALD on primary particles, their processing, scalability, and the opportunities presented for advanced applications.

453177 Invited - Thermal and Plasma Enhanced Atomic Layer Deposition on Powders and Particles

Abstract:

Surface engineering of micro- and nanoparticles is of great importance in fields such as catalysis, energy and sensing. For many of these applications particles are required with different bulk and surface properties. Atomic layer deposition (ALD) is known as a reliable technique for surface functionalization of powders, by conformally coating the individual particles with ultrathin layers. However, to perform ALD on powders, fluidization or agitation of the particles is required. During the past decades, both batch and spatial ALD reactor concepts have been developed, including fluidized bed, rotary reactors and pneumatic transport.[1] Although fluidized bed reactors are most often used for ALD on particles, this reactor concept is not compatible with plasma enhanced ALD, which is advantageous for e.g. coating on temperature sensitive polymer particles or deposition of metals and metal nitrides.

In this work, we report on a rotary reactor concept that was developed to enable both thermal and plasma-enhanced ALD on powders.[2] Particles ranging from nanometer size to millimeter size were successfully coated with layers of Al_2O_3, TiO_2, SiO_2, AlN and TiN. The ALD processes were characterized in-situ by means of mass spectroscopy (MS) and optical emission spectroscopy (OES). The composition and conformality of the coatings were evaluated by X-ray fluorescence (XRF), X-ray photoelectron spectroscopy (XPS), energy-dispersive X-ray spectroscopy (EDX), and transmission electron microscopy (TEM). These ultrathin conformal coatings may, for instance, be applied for encapsulation, inducing or enhancing catalytic activity, or altering electrical conductivity.

Our results prove that the developed rotary reactor enables conformal deposition on nano- and micropowders by thermal and plasma enhanced ALD. In this way, surface engineering of such particles can be achieved.

[1] D. Longrie et al., Journal of Vacuum Science and Technology A 32, 010802 (2014).

[2] D. Longrie et al., Surface & Coating Technology 213, 183-191 (2012).

460846 Employing Synergetic Effect of Doping and Thin Film Coating to Boost the Performance of Lithium-Ion Battery Cathode Particles

Abstract:

Atomic layer deposition (ALD) has evolved as an important technique to coat conformal protective thin films on cathode and anode particles of lithium ion batteries to enhance their electrochemical performance. Coating a conformal, conductive and optimal ultrathin film on cathode particles has significantly increased the capacity retention and cycle life as demonstrated in our previous work. In this work, we have unearthed the synergetic effect of electrochemically active iron oxide films coating and partial doping of iron on $LiMn_{1.5}Ni_{0.5}O_4$ (LMNO) particles. The ionic Fe penetrates into the lattice structure of LMNO during the ALD process. After the structural defects are saturated, the iron starts participating in formation of ultrathin oxide films on LMNO particle surface. Owing to the conductive nature of iron oxide films, with an optimal film thickness of ~0.6 nm, the initial capacity improved by ~25% at room temperature and by ~26% at an elevated temperature of 55 °C at a 1C cycling rate. The synergy of doping of LMNO with iron combined with the conductive and protective nature of the optimal iron oxide film leads to a high capacity retention (~93% at room temperature and ~91% at 55 °C) even after 1,000 cycles at a 1C cycling rate.

452329 Particle Atomic Layer Deposition for Synthesis of Fuel Cell Catalytic Material

Abstract:

Fluidized bed Atomic Layer Deposition (ALD) is a scalable method for nanoscale catalyst design, and has potential to address many of the persistent material challenges associated with PEM fuel cells, namely utilization and durability of the platinum catalyst. ALD was carried out to deposit platinum nanoparticles on functionalized and unmodified carbon black with both combustion and hydrogenation chemistries. Reactions throughout the ALD cycle were monitored using mass spectrometry and IR spectroscopy to clarify the deposition chemistry. The synthesized materials were evaluated in hydrogen pumping tests and as oxygen reduction reaction (ORR) catalysts in proton exchange membrane fuel cell experiments. Hydrogenation made finer, more dispersed Pt nanoparticles that performed similarly to their commercial equivalent when pumping hydrogen; oxygenation made a coarser catalyst that underperformed commercial Pt/C. The hydrogenation catalysts deposited on functionalized carbon were found to be ineffective in ORR testing due to poor water management in the fuel cell, which was attributed to reduced hydrophobicity within the catalyst layer.

472102 Nanostructured Powders Made from Flame Spray Pyrolysis for Li-Ion Cathode

Abstract:

Flame processes are the most economic ones to produce fine powders and have been used to produce pigment titanium oxide and fused silica in large volumes. In this work a flame spray pyrolysis (FSP) process is used to make positive electrode materials for Li-ion battery, which are made of complex metal oxide powders, such as $Li(Ni_{1/3}Mn_{1/3}Co_{1/3})O_2$. In particular, the work was focused on making the NMC materials with stoichiometric compositions and desired morphologies. During the pyrolysis process, different morphologies can be formed, involving powders often with primary nanoparticles. Such nanostructures were found to have affected the battery performance. Detailed results of the synthesis, characterization, and battery testing of the NMC powder materials will be presented.

465170 Dry Coating of Electrode Particle with Model Particle of Sulfide Solid Electrolytes for All-Solid-State Secondary Battery

Abstract:

We here present a dry coating of electrode particle with a model material of sulfide solid electrolytes for all-solid-state secondary batteries. LiNi1/3Co1/3Mn1/3O2 (NCM) was used as a host particle. Sodium sulfate (Na2SO4) was selected as a model material of sulfide solid electrolytes and used as a guest particle. Three different processes, which were vibration mixing, mortar-and-pestle mixing, and dry impact-blending process known as â€œHybridizerâ€, were used and compared to each other. FESEM and EDX images showed that the dry impact-blending process produced composite particles in which NCM was more fully covered with Na2SO4 than using the vibration mixing and mortar-and-pestle mixing. Cross sectional observations of composite particles showed that using the dry impact-blending process continuous layer of Na2SO4 were formed on NCM particle. In a compacted pellet prepared from the composite particles produced by the dry impact-blending process, following features were observed: better dispersion of NCM particles in Na2SO4 matrix without NCM agglomerates; less void space; and intimate NCM-Na2SO4 interfacial contacts. These features suggest that performance of all-solid-state batteries are likely to remarkably improve using the composite particles prepared from the dry coating process [1].

[1] T. Kawaguchi, H. Nakamura, S. Watano. Powder Technology, in press.

462966 Mixed-Oxide Based Redox Catalysts for Hydrocarbon Oxidation, Water-Splitting, and CO_2 Utilization

Abstract:

Using transition metal oxides as the reaction intermediate, the chemical looping strategy seeks to intensify fossil fuel conversion processes via a cyclic redox scheme. In such a process, carbonaceous molecules are oxidized by active lattice oxygen (O^{2-}) in the transition metal oxide, which functions as an oxygen carrier, a.k.a. redox catalyst. In a subsequent step, the O^{2-} deprived oxygen carrier is replenished by a gaseous oxidant. Although conventional chemical looping processes seeks to fully oxidize fossil fuels for carbon capture, tailored redox catalysts can be applied for a variety of applications including selective oxidation, water-splitting, and CO_2 reduction.

The present study exemplifies novel applications of iron and manganese containing mixed oxides for methane partial oxidation, oxidative dehydrogenation (ODH) of ethane, water-splitting, and CO_2 reduction. A number of redox catalysts composed of iron and/or manganese oxides are synthesized and characterized. Rational strategies to improve the redox activity, product selectivity, and/or water-splitting and CO_2 reduction conversions are investigated through the assistance of spectroscopy and *ab-initio* calculation tools. Compared to traditional heterogeneous catalysts, the redox catalysts in the current study show the potential to be more selective. They also possess excellent redox stability. Close to 100% water and CO_2 reduction efficiencies are shown to be achievable through rational selection of mixed metal oxide based redox catalysts.

455091 Development of a Resonant Acoustic Mixing Methodology for Particle Size Reduction

Abstract:

Particle size reduction plays an important role in various manufacturing industries, such as the chemical, pharmaceutical, and food and beverage industries. Decreasing particle size can be very beneficial for a process; materials can see an increased dissolution rate, an increased rate of reaction, a more-effective drug delivery, and a more consistent appearance and texture, among other things. Without these improved characteristics, many manufacturing processes would not succeed.

Nalas routinely operates a LabRAM resonant acoustic mixer to efficiently mix powder systems and to explore green methods of preparing different solid state products. We have developed a dry grinding method for the LabRAM that minimizes the time needed for particle size reduction and removes the need to suspend solids in a solvent typically employed in wet methods. This method has been found to be particularly suited to very dense solids and solids that tend to agglomerate upon reducing the particle size.

The development and reproducibility of this new methodology will be discussed as well as the impact of the grinding container size, milling media amount, amount of material to be milled, milling time, and the force of milling on the final particle size distribution. Finally, the method will be compared to existing milling methods, such as ball-milling and high-shear wet milling.

451360 A New Hybrid One-Dimensional Particle Attrition Model for Conveying Systems

Abstract:

Particle attrition during pneumatic conveying constitutes a major problem in a wide variety of industrial applications. When particles are transported pneumatically through a pipeline, they collide with each other and with the walls of the pipe, which may lead to particle breakage. In addition, particles may slide along the bottom of the pipeline, thereby being subjected to abrasive wear. These actions lead to particle degradation and hence to undesirable changes in material characteristics and product quality.

The attrition of particles during pneumatic conveying has been investigated experimentally in a large number of studies [1–5]. However, all are limited by the fact that the only way to experimentally examine particle attrition for a specific system is to compare the particle size distribution at the inlet to that at the outlet. Since particle attrition is governed by a vast number of parameters, it is extremely difficult to empirically correlate the particle size distribution at the end of the pipeline with that at the inlet, even for a large data set. For this reason, in the design of a new conveying line, there are no definitive 'rules' for obtaining the optimum operating conditions (pressure drop, air and particle mass flow rates, pipe diameter, number of bends, etc.) that will give the desired particle size distribution at the pipe outlet. One way to overcome this problem is to apply computational methods to predict the outcome product, but due to the complexity of the available models and the need for expensive computational resources, only a few researchers have implemented particle breakage in their simulations.

In the discrete element method (DEM) approach, particle breakage can be simulated by one of three different models [6]. Nonetheless, of these, only the 'fragments spawning' model of Brosh et al. [7] has been implemented to examine particle breakage in pneumatic conveying [8]. In this model, computational fluid dynamics (CFD)-DEM simulations are combined with the four following cumulative empirical comminution functions, which define the characteristics of the particles as a function of external loading: 1) the strength distribution to describe the compression force that would cause a particle to break [9]; 2) the breakage function, which is produced by horizontal impact system, to describe the sizes of the fragments after a breakage event [10]; 3) the equivalence function to describe the relationship between the impact velocity and the equivalent compression load that would cause the same breakage probability as that if the particles had collided with a particular impact velocity [11]; and 4) the fatigue function [12] to take into account the weakening of the particle strength due to repeated loading. By implementing these functions within DEM simulations it becomes possible to determine whether a discrete particle will or will not break in a particular collision. Therefore, in this method, the fatigue and breakage for every discrete particle are calculated within the 3D domain. This tool, while effective, has the major drawback of a high computational cost, since the number of particles increases as the particles break into finer segments.

An alternate approach is to adopt a macroscopic model for two-phase flow and to simultaneously predict particle breakage, albeit with reduced accuracy but with the advantage of far less computation time. Such a degradation model was proposed by Chapelle et al. [4] and implemented on a large-scale dilute-phase pneumatic conveying line [13]. In this degradation model, a breakage matrix was derived from single impact tests and then implemented on every impact event in the simulation. The zero-dimensional flow model used by Chapelle et al. [13] was based the assumption that all particles travel at the same velocity (and therefore impact at the same velocity). Furthermore, the collisions were assumed to take place perpendicular to the pipe wall and only inside bends, thereby neglecting particle-particle collisions, collisions in straight pipe sections, and fatigue phenomena. There is thus a need for a model that will address these drawbacks.

In the present study, a new method is presented for calculating particle attrition due to impact events under steady-state flow conditions in conveying pipeline systems. The model consists of two parts: a one-dimensional two-phase flow model for calculating the fluid dynamics [14], and the one-dimensional breakage algorithm (ODBA) developed in this study for determining particle collisions and breakage with respect to the flow field.

In the ODBA, the methodology of Kalman et al. [15] is adopted for combining probability comminution functions (strength distribution, breakage, equivalence and fatigue functions) as the particle characteristic tool with the dynamics of the particles inside the system, namely, the 'system behavior tool.' The system behavior tool in the previously mentioned fragments spawning model is a 3D CFD-DEM simulation that describes the explicit motion of every discrete particle in a 3D domain. In contrast, the ODBA utilizes 'machine functions' as the system behavior tool. The machine functions are the collision frequency function, which describes the distribution of time between each collision, and the collision velocity function, which defines the distribution of the collision impact velocity. These functions must be known or developed prior to applying the ODBA.

The impact velocity function is used as the input for the comminution functions that determine whether a particle will break or weaken, and the collision frequency function is used to locate the positions of the collisions. These functions were developed in the present study for a particular experimental system [3] for investigating dilute phase pneumatic conveying of potash particles at several superficial gas velocities. The new method facilitates the calculation of the particle size distribution at any location along the axial direction of the pipeline.

In CFD-DEM breakage simulations, every simulated case provides a standalone result for the particle size distribution, which is accurate only for those specific simulation conditions. In principle, this approach facilitates a better classification of the attrition phenomena than the ODBA, since it can characterize every particle's collision event in a 3D domain. Nevertheless, it is impossible to apply this method in practice for long-range conveying systems and for process optimization because of the high computational cost. To overcome this drawback, the ODBA was developed as a fast calculation method for predicting attrition in conveying pipelines. Note that for developing the machine functions required for the ODBA, CFD-DEM simulations were conducted without taking particle breakage into account, which speeds up the simulation time in comparison with full CFD-DEM breakage simulations. Thus, in the present study, a set of CFD-DEM simulations was conducted for different conditions, the collision data was recorded, and correlations for the velocity and collisions frequency distributions were developed. In addition, the newly developed model was validated by comparing the model predictions with experimental results for conveying potash [3].

[1] H. Kalman, Attrition of powders and granules at various bends during pneumatic conveying, Powder

Technol. 112 (3) (2000) 244-250.

[2] A.D. Salman, M.J. Hounslow, A. Verba, Particle fragmentation in dilute phase pneumatic conveying, Powder Technol. 126 (2) (2002) 109-115.

[3] H. Kalman, M. Hubert, E. Grant, Y. Petukhov, M. Haim, Fatigue behavior of impact comminution and attrition units. Powder Technol. 146 (1) (2004) 1-9.

[4] P. Chapelle, H. Abou-Chakra, N. Christakis, M. Patel, A. Abu-Nahar, U. Tüzün, M. Cross, Computational model for prediction of particle degradation during dilute-phase pneumatic conveying: the use of a laboratory-scale degradation tester for the determination of degradation propensity, Adv. Powder Technol.15 (1) (2004) 13-29.

[5] P. Rajniak, K. Dhanasekharan, C. Sinka, N. MacPhail, R. Chern, Modeling and measurement of granule attrition during pneumatic conveying in a laboratory scale system, Powder Technol. 185 (3) (2008) 202-210.

[6] Y. Guo, J.S. Curtis, Discrete element method simulations for complex granular flows, Ann. Review of Fluid Mech. 47 (2015) 21-46.

[7] T. Brosh, H. Kalman, A. Levy, Fragments spawning and interaction models for DEM breakage simulation, Granular Matter, 13 (6) (2011) 765-776.

[8] T. Brosh, H. Kalman, A. Levy, DEM simulation of particle attrition in dilute-phase pneumatic conveying, Granular Matter 13 (2) (2011) 175-181

[9] Y. Rozenblat, D. Portnikov, A. Levy, H. Kalman, S. Aman, J. Tomas, Strength distribution of particles under compression, Powder Technol. 208 (1) (2011) 215-224.

[10] Y. Rozenblat, E. Grant, A. Levy, H. Kalman, J. Tomas, Selection and breakage functions of particles under impact loads, Chem. Eng. Sci. 71 (2012) 56-66.

[11] Y. Rozenblat, A. Levy, H. Kalman, J. Tomas, Impact velocity and compression force relationship—Equivalence function, Powder Technol. 235 (2013) 756-763.

[12] Y. Rozenblat, A. Levy, H. Kalman, I. Peyron, F. Ricard, A model for particle fatigue due to impact loads, Powder Technol. 239 (2013) 199-207.

[13] P. Chapelle, N. Christakis, H. Abou-Chakra, I. Bridle, M.S.A. Bradley, M. Patel, M. Cross, Computational model for prediction of particle degradation during dilute-phase pneumatic conveying: modeling of dilute-phase pneumatic conveying, Advanced Powder Technol. 15(1) (2004) 31-49.

[14] A. Levy, D.J. Mason, D. Levy-Hevroni, I. Borde. Drying of wet solid particles in a steady-state one-dimensional flow, Powder Technol. 95 (1) (1998) 15-23.

[15] H. Kalman, V. Rodnianski, M. Haim, A new method to implement comminution functions into DEM simulation of a size reduction system due to particle-wall collisions, Granular Matter 11 (4) (2009) 253-266.

467799 A Combined Experimental and Computational Approach Using Discrete Element Method for the Development of a Mechanistically Motivated Breakage Kernel

Abstract:

Pharmaceutical manufacturing is gradually shifting from a batch to continuous process. Continuous manufacturing is known for its potential advantages in cost, efficiency and reduced â€˜time to marketâ€™ that increases the life span of the product. However, this emphasizes a greater need to understand the impact of material and process variability on product quality as disturbances canÂ propagate downstream [1]. In addition, the Quality by design (QbD) approach introduced by the US food and drug administration (FDA) requires that companies demonstrate the ability to predict the effect of material and process variability on product quality. Process modeling thus has a significant role to play in the transition to continuous manufacturing and the successful implementation of QbD.

Direct compaction, dry granulation (DG) and wet granulation (WG) routes for continuous manufacturing include numerous unit operations to process powder feed and make tablets as final product. The conical screen mill (Co-mill) is a widely used equipment in DG and WG routes for the purpose of breaking granules to a required size distribution. It is important for granules to adhere to a consistent and pre-determined size distribution as it impacts tablet compaction and drug bioavailability. The granule breakage mechanism in co-mill is a complex phenomenon where fracture and eventual breakage occur due to a rotating impellor and the smaller sized granules are discharged through a screen.

The population balance model (PBM) approach that has been used in the literature to simulate a co-mill process requires the formulation of a breakage kernel that represents the mill breakage phenomenon. Past work done by us has used a breakage kernel that is a function of impellor speed and granule size [2]. The model was calibrated and

validated using experimental data. However, material properties and milling conditions were not exclusively identified in the breakage function which limits its applicability. In addition, PBM does not capture particle-scale interactions that would render more accuracy to the model. Discrete element modeling (DEM) has potential to bridge this gap. Development of discrete element models (DEM) for breakage processes so far have been focused on ball mills, grinding mills, crushers etc. Despite its prevalence in pharmaceutical industry, work on DEM analysis of co-mill is very limited. This is possibly due to additional challenges posed such as, potential for continuous feed, discharge through screen, introduction of additional complexity in the particle-wall interactions from a rotating impellor, dearth of published experimental data etc.

Current work proposes a methodology to develop a breakage kernel that captures particle level dynamics through DEM and material properties through co-milling experiments. The DEM model takes into account threshold impact energy above which granules break. The impact energy distribution data for various size classes and impellor speeds is obtained from DEM. Co-mill experiments at various impellor speeds yield corresponding observed size distributions. The data from DEM and lab experiments are used to calibrate the breakage kernel through which material specific kernel parameters are determined. A multi-scale modeling framework utilizing capabilities from DEM, PBM as well as experimental data is thus established. This methodology to develop breakage kernel that identifies particle scale information, material and process specific parameters, and equipment geometry is implemented and qualitatively validated against experimental sensitivities.

References

[1] A. Rogers, A. Hashemi, M. Ierapetritou, Modeling of Particulate Processes for the Continuous Manufacture of Solid-Based Pharmaceutical Dosage Forms, Processes, 1 (2013) 67.

[2] D. Barrasso, S. Oka, A. Muliadi, J.D. Litster, C. Wassgren, R. Ramachandran, Population Balance Model Validation and Predictionof CQAs for Continuous Milling Processes: toward QbDin Pharmaceutical Drug Product Manufacturing, Journal of Pharmaceutical Innovation, 8 (2013) 147-162.

470438 Dynamic Process Behavior in Fluidized Bed Opposed Jet Mills

Abstract:

In fluidized bed opposed jet mills products can be ground into the submicron range. Therefore highly expanded gas jets are introduced through nozzles which are directed to a common focus point. Particles are entrained into the gas flow, accelerated and finally colliding with other particles. If the kinetic energy of the particles at the impact is high enough, breakage or abrasion occurs even for the hardest materials. Since no grinding tools are needed here, the autogenous comminution process allows grinding without contamination which is especially necessary for pharmaceutical products. Also thermosensitive materials like polymers can be processed in jet mills so that a wide scope of industrial applications is possible [1].

After the milling process, the particles are following the gas stream upwards to an integrated classifier wheel. Unground or oversized particles are rejected while product-sized particles can leave the machine through the classifier. The rejected material is circulating back to the bottom of the machine were it can be stressed again. Like in other classifier mills the processes in fluidized bed opposed jet mills can be divided in three unit operations: The comminution zone, where size reduction takes place, the classification zone and the pneumatic transport zone in between.

Classifier mills are characterized by a high solids holdup and therefore high solids concentrations appear in the unit operations. This leads to strong interaction between the zones whereas the gas flow is used for grinding, transport and classification process simultaneously. As a consequence, a higher gas flow might increase the comminution kinetics but would lead to a coarser product at the same time when the operating mode of the classifier is kept constant. Especially in dynamic processes, like a start-up, the analysis and the modelling of the mill behavior is challenging:

Until a steady state is reached the solids holdup and the particle size distribution inside the mill can vary strongly. This affects the fluid dynamics in all three unit operations and its interactions. Next to the internal parameters these changes cause a time dependency of external parameters, namely the product mass flow and its particle size distribution. This can cause at non-steady state conditions an unwanted product quality and should be avoided in the production of high-priced materials. In literature there is little written about the connection between external and internal parameters in classifier mills or the dynamic process behavior itself. For that reason, this work was carried out to contribute to a better understanding of comminution processes in classifier mills focusing the dynamic aspects.

To get excess to the complex mechanisms inside the mill, the presentation will show a strategic analysis and a modelling strategy for fluidized bed opposed jet mills. Therefore comminution experiments were carried out in a lab scale mill with an implemented online laser diffraction system on the product side. As test material well defined glass beads with narrow particle size distribution were used. They were chosen since they come directly from synthesis and hence have no prior load.

The comminution kinetics was determined in batch grinding experiments. Therefore the time evolution of the particle size distribution and solids holdup was measured. A model based on population balance, the so called Kapur function [2,3], was used to characterize the breakage mechanism. So the influences of parameters like feed particle size distribution, solids holdup, classifier speed etc. on the comminution zone and finally on the comminution process are describable. Additionally, the breakage probability of the used glass beads was determined in a single grain impact device to depict the impact event more precisely.

Since the flow around and in the classifier is inaccessible, well defined mixtures with product sized and non-product sized particles were used as feed stock. By analyzing the mass flow and the particle size distribution of the product a closer look into the recirculation and the rejecting process at the classifier wheel is possible. These experiments were intensified by using capacitance probes to measure the solids concentration in the transport zone with a high time resolution. So the flow to the classifier wheel and back to the comminution zone can be analyzed and lead to a better understanding of the local recirculating process.

At last a fed-batch mode was performed to simulate start-up and load change processes in classifier mills. These dynamic investigations show the decrease of particle size inside the mill just as the product mass flow and particle size distribution. In start-up processes first the mill contains a low solids holdup and therefore the breakage rate is limited caused by a low amount of particles which are entrained into the gas jets. Exceeds the solids stock a specific value the solids concentration in the jets is higher but on the other side the specific impact energy is decreased. For that reason experiments with variable, constant solids holdup were carried out to investigate the dynamic performance of the comminution, transport and classification processes.

The presented results of the comminution in the fluidized bed opposed jet mill contributes to a predictable dynamic model and therefore to material and energy saving process improvements.

The financial support of DFG (Deutsche Forschungsgemeinschaft) within the priority program SPP 1679 "Dynamic flowsheet simulation of interconnected solids processes" is gratefully acknowledged.

References:

[1] M. Benz, H. Herold, B. Ulfik, Performance of a fluidized bed jet mill as a function of operating parameters, Int. J. Miner. Process., 44-45:507-519, 1996

[2] P.C. Kapur,Â Kinetics of batch grinding, Part B. Transactions of the AIME 247:309â€"313, 1970

[3] H. Berthiaux, J.Dodds, Modelling fine grinding in a fluidized bed opposed jet mill Part I: Batch grinding kinetics, Powd. Techn., 106:78-87, 1999

471611 Population Balance Modeling of Paracetamol Crystal Breakage

Abstract:

Crystals exhibit various shapes including cubes and needles. While cubes have aspect ratios close to one, needle shaped crystals are characterized by larger aspect ratios. It is expected that higher aspect ratio crystals will break into shorter rods. That is, the aspect ratios of the smaller fragments would be lower than the aspect ratios of the parent crystals.

This work presents breakage distribution functions that are developed for modeling needle breakage. This includes functions where the child crystals are shorter rods with the same width as the parent particles. These functions can account for binary breakage and breakage into a greater number of child particles. This two-dimensional model includes the major axis length and the aspect ratio of the crystals. Simulation results are compared to experimental results of paracetamol crystal breakage in a stirred vessel, where the parent crystals have an average aspect ratio of approximately two. Because there are a variety of systems that produce needle-shaped crystals, this breakage model is expected to be applicable to a range of systems in addition to paracetamol.

472049 The Relationship Between Structure and Breakage Selectivity

Abstract:

Population balance modeling has been used to characterize the type of breakage in a given particle system. The basic approach to understanding particle breakage from a mechanistic point of view is to divide a given particle size distribution into bin sizes where adjacent bin sizes differ in particle size by a factor of two or possibly three. The next step is to expose the bulk material to a stimulus that will induce stress-strain or impact events on the particles in an increasing manner. The particle size distribution prior to any stress-strain event or impact events is measured as well as several particle size distributions after incremental application of stress-strain events or repeated impact events. Incremental mass balances as a function of stressâ€"strain rates or time at impact conditions are written for the breakage of any size particle bin into all particles of a smaller size bin. This set of equations is solved for the rate of breakage of any size particle, as well as the selectivity parameters that governs the size of new particles created from a broken particle. This can provide good information to easily fit breakage kernels with simple relationships between particle size, breakage type, and breakage stimulus when the particles are homogenous and are all a consistent shape. However, complex relationships exists when the original particles are of anisotropic structure or a non-random prescribed structure. This often causes breakage along preferred planes due to stress intensity functions that are structure dependent. The shape of the particle clusters and structured particles often determines the size of the broken particle pieces. However, a critical stress analysis of these structured particles can reveal points or planes of preferred breakage that can be coupled with selection coefficients to predict complex bimodal breakage models were the breakage is structure dependent. The paper provides a means of coupling these effects to generate a better estimate of the breakage mode for structured particles when exposed to stress-strain events.

449379 Modelling of the Combined Milling Circuit for the Maximum Production of a Desired Particle Size Range

Abstract:

Abstract

Ball milling is mainly conducted in closed circuits in order to avoid material overgrinding and thus maximize the production of the desired product size. Modelling the desired milling circuit is initially the first step taken to optimise its operation. To that end, a mathematical simulation of the combined milling circuit, consisting of classifiers and grinding models was applied to optimize the operation of the circuit, that is, to determine the optimal conditions in terms of energy consumption under which production of the floatable size range of platinum-bearing ore can be achieved. The results from the simulation showed useful insights fundamental to the enhancement of the operation of the milling efficiency. Regions showing the best combination of the design and operational factors that enable the production of the highest amount of the floatable size class with the minimum energy consumption were shown.

Keywords: *Combined milling circuit, Floatable size range, Milling parameters, Population balance model, Attainable region*

School of Chemical Engineering

Analysis of fine particle grinding process on vertically stirred media mills

Yang Yang, Neil Rowson,

IMERYS

UNIVERSITY OF BIRMINGHAM

Introduction

The research mainly aims to analyse the performance of vertically stirred media mill with wet grinding method through investigating particle size reduction, flow pattern and power draw by varying some operating parameters such as viscosity, tip speed and design of geometry(Altun et al., 2013). Positron emission particle tracking (PEPT) is used to observe flow characteristic inside the grinding chamber(Conway-Baker et al., 2002). Extensive experiments are conducted on a bench-scale mill and a pilot mill.

Material and methods

Fig.1. Experimental rig of bench-scale mill

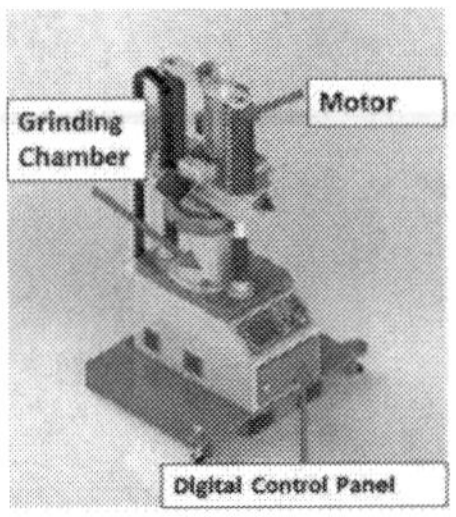

Fig.2. Photography of pilot mill

Fig.3. Photograph of feeding GCC under SEM

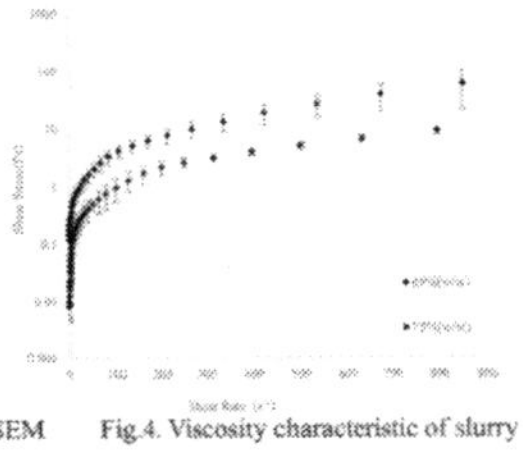

Fig.4. Viscosity characteristic of slurry

Ground Calcium Carbonate (GCC), a main product of IMERYS Mineral Limited (formerly known as English China Clay) commonly used as paper filler, is chosen as feeding material($D_{80} \approx 26\mu m$). Two types of ceramic beads are used as grinding media in the study. One referred as Carbolite is supplied by CARBO ceramic grinding media company. The bead size is +0.8 to -1.2mm with an apparent specific gravity 2.7±0.1. The other is named as ZirPro, a zirconium-based materials supplied by Saint-Gobain. The range of size is +0.7 to -1.4mm, with an apparent specific gravity 3.2±01. Obviously, Carbolite is finer and lighter than ZirPro, but surface is not as smooth as ZirPro.

Results and discussion

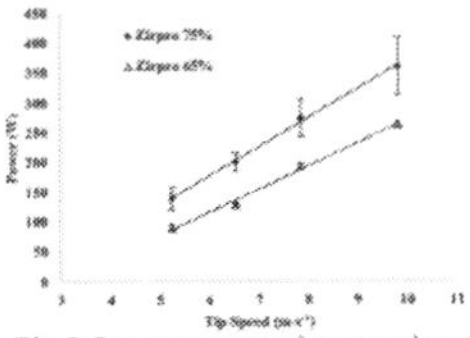

Fig.5. Power consumption at various tip speed and solids concentration

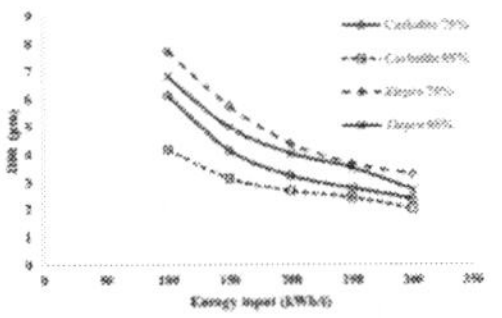

Fig.6. D_{80} against energy input by grinding media Carbolite and Zirpro

Fig.7.Impeller A-1

Fig.8 Impeller A-2

Fig.8.Impeller A-3

Fig.9.Impeller A-4

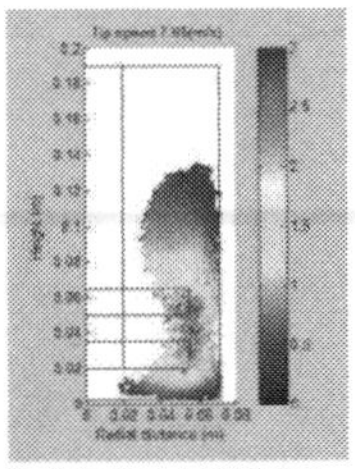

Fig.10. Impeller A-1 at Vt=7.85m/s

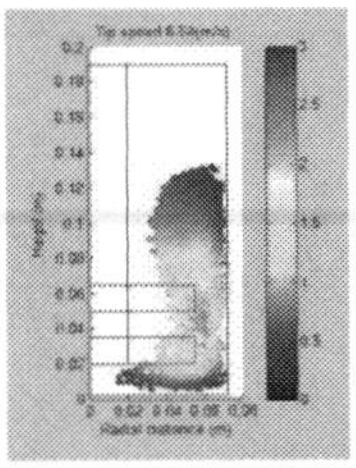

Fig.11. Impeller A-1 at Vt=6.54m/s

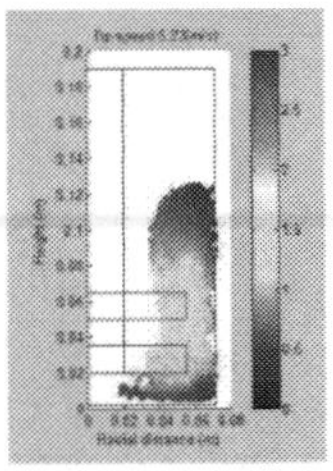

Fig.12. Impeller A-1 at Vt=5.23m/s

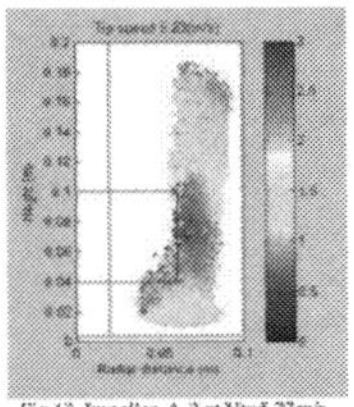

Fig.13. Impeller A-2 at Vt=5.23m/s

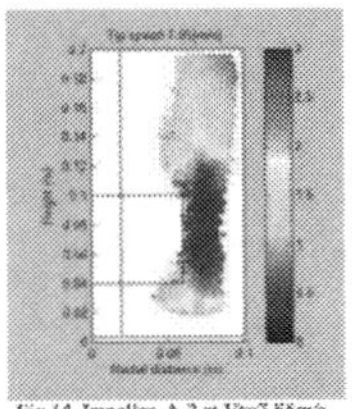

Fig.14. Impeller A-2 at Vt=7.85m/s

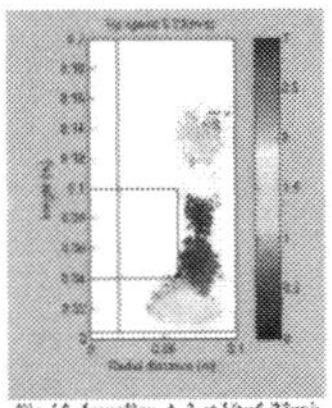

Fig.15. Impeller A-3 at Vt=5.23m/s

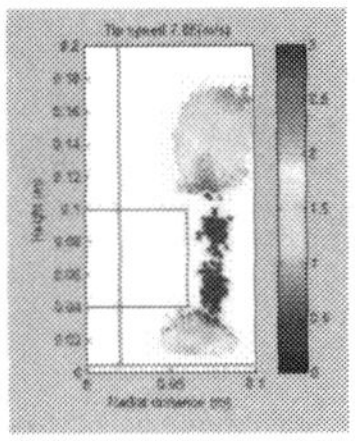

Fig.14. Impeller A-3 at Vt=7.85m/s

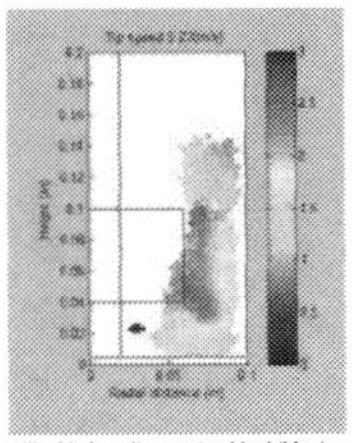

Fig.15. Impeller A-4 at Vt=5.23m/s

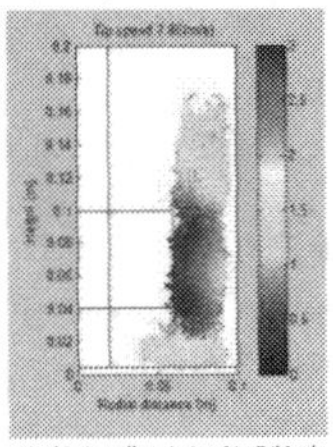

Fig.16. Impeller A-4 at Vt=7.85m/s

Conclusions

- Mineral slurry shows typical Non-Newtonian characteristic (shear thinning) at both 65% and 75% solids concentration.
- Tip speed is liner proportional against power draw within Vt=5.23~9.82 m/s on bench scale.
- The velocity map and flow pattern with novel design of impeller is demonstrated by PEPT.

REFERENCE

Altun, O., Benzer, H., Enderle, U., 2013. Effects of operating parameters on the efficiency of dry stirred milling. Minerals Engineering 43-44, 58-66.

Conway-Baker, J. Barley, R.W., Williams, R.A., Jia, X., Kostuch, J., McLoughlin, B., Paker, D.J., 2002. Measurement of the motion of grinding media in a vertically stirred mill using position emission particle tracking (PEPT). Minerals Engineering 15, 53-59.

CONTACT INFORMATION

Yang Yang
School of Chemical Enineering
University of Birmingham
B15 2TT, UK
Yxy357@bham.ac.uk

455219 One-Pot Peptide Synthesis Assisted By Hydrophobic Magnetic Nanoparticles

Abstract:

[Introduction]

In addition to traditional small molecules and antibodies, peptides have recently been recognized as promising pharmaceutical candidates. Typically, solid-phase liquid phase method are widely used for peptide syntheses. Due to the easy separation process, the solid-phase method is widely used for peptide syntheses, while high reaction efficiency and real-time monitoring are achieved in the liquid-phase method, also enabling facile scale-up. Recently, a new peptide synthesis method has been reported that uses hydrophobic benzyl alcohols as soluble supports[1] with hydrophobic magnetic nanoparticles.[2] In this method, the supported hydrophobic products are precipitated by the addition of poor solvent and rapid magnetic collection of the desired products can be realized with hydrophobic magnetic nanoparticles. However, the examples reported so far are limited and the detached hydrophobic modifiers from the particles cause undesired side reactions since the modifiers are attached through weak ionic interactions. The purpose of this study is the preparation of hydrophobic magnetic nanoparticles which are tightly interacted between particles and modifiers. And these particles are applied in the liquid-phase peptide syntheses. By using new modifiers which has strong bond such as chemical bond on the particles, it will be refrain from undesired side reactions. In this research, octadecyl isocyanate is used as hydrophobic moiety. It has been reported that isocyanate forms carbamate bond with hydroxy groups on the particles.[3] Then, the modified particles are applied for synthesis of typical repeating sequence of elastin as model peptide. It is known that the synthesis of elastin is difficlut because of the gelation property.

[Experimental method]

Iron (III) oxide particles were purchased from Sigma-Aldrich and were characterized several analyses,

including SEM, IR, and BET before use. The iron (III) oxide particles and octadecyl isocyanate were added in dry toluene and the mixture was sonicated. Disperse particles were collected by magnet, dried in vacuum, and characterized by FT-IR, FE-SEM and elemental analysis. By using these modified particles, the model peptide was synthesized. The peptide was elongated by repeating couplings and deprotections over 9 steps using valine, glycine and proline. All the reactions were monitored by TLC and yields were determined by the recovered weights.

[Results and discussion]

From FT-IR analysis, the corresponding peaks to the modifiers were observed compared to iron (III) oxide particles before treatment. The particles are formed fiber-like structure composed of iron (III) oxide nanoparticles and loading amount was estimated to be 4 μmol/m^2 from elemental analysis. The peptide was synthesized using these modified particles and monitored by TLC analysis. Based on TLC analysis, no significant side reactions were observed in all steps and the desired peptide was obtained in 87% over 9 steps in one pot. By adding this nanoparticles, all steps are completed without gelation.

[Conclusion]

Iron (III) oxide particles were hydrophobically modified using octadecyl isocyanate and these particles were applied for synthesis of typical repeating sequence of elastin. By using this method, all the reaction were proceeded with no significant side reactions and the desired product was prepared in 87% yield.

[References]

[1] (a) Okada, Y. *et. al. Org. Lett.* **2015**, *17*, 4264–4267. (b) Okada, Y. *et. al. J. Org. Chem.* **2013**, *78*, 320–327. (c) Tana, G. *et. al. Chem. Commun.* **2010**, *46*, 8219–8221.

[2] Iijima, M. *et. al, Langmuir* **2009**, *25*, 11043–11047.

[3] Carrara, C. *et al., Org. Lett.* **2014**, *16*, 460-463.

455857 Numerical Simulations of the Effect of Ionic Surfactant/Polymer on Oil-Water Interface Using Dissipative Particle Dynamics

Abstract:

The dissipative particle dynamics (DPD) is an ideal method to simulate such a coarse-grained system. DPD method can reproduce the forming process of the water-oil microstructure. In term of the effect of surfactants on emulsion structures, it is of fundamental importance to study the properties of emulsions. In this study, effects of the ionic surfactant (CTAB) and polymer (SMAc) on the oil/water interface are studied using dissipative particle dynamics (DPD) at the mesoscopic scale. Effects of the water content on equilibrated morphology of the oil/water emulsion are studied with the help of ionic surfactant. In addition, oil/water emulsion has the patterns of the compound emulsifier at the very low and very high water content. The interfacial tension increases, reaches maximum, and then decreases with an increase of water content. Both experiments and simulations show that the inorganic salts can improve the interfacial efficiency of the ionic surfactants, and lower the interfacial tension. The influence of polymer number is studied on their ability to reduce interfacial tension. The distributions of interfacial tension and mean interfacial density are predicted. The lower the interfacial tension and faster oil coalescence are, the more polymers is. In addition, the influence of the polymer on the surfactant aggregation behavior of the oil-water interface is discussed. The molecules of polymer (SMAc) are mainly gathering at the interface, where the surfactant molecules associate on the polymer chains in the composite oil/water/surfactant/polymer systems. The surfactant beads both located at the interface and associated on the polymer chains are in more upright and orderly arrangement at the interface with an increase of polymer number, as well as, the peak of polymer interfacial density become higher and broader. The interfacial tension impairment of the composite system is higher than that of the single system. The lower the interfacial tension and faster oil coalescence are, the more the polymer is. The model predictions reveal the basic mechanisms of ionic surfactant in the oil-water system and the coalescence process of oil and water droplets at the different ratio of water and oil. From these results, the conclusion can be made that the coarse grained model is a powerful and flexible tool to study emulsion processes in the oil-water systems, and further determines the properties surfactant and polymer in the emulsion processes.

457099 Uneven Distribution of Particle Flow in the Riser of Rfcc: Effect of the Number of Feed Injector

Abstract:

Two typical troubles associated with reactors in commercial Residue Fluidized Catalytic Cracking (RFCC) process are uneven distribution of particle flow in the riser and formation of carbonaceous deposits in the secondary reactor cyclone. The uneven distribution of particle flow means each outlet of the riser has a different particle mass flow rate and the formation of carbonaceous deposits cause serious catalyst carryover into the fractionators. These two problems are found by the number and location of the completely blocked cyclone with deposits in the commercial RFCC process. This study analyzes particle-fluid flow patterns in the riser using computational particle fluid dynamics (CPFD) method to troubleshoot the real industrial problem. The CPFD method, which has been developed to complement the Eulerian–Eulerian and the Eulerian–Lagrangian methods, applies the Navier–Stokes equation for fluid phase and multi-phase-particle-in-cell (MP-PIC) model for particle phase. Particle flow distribution is different each of outlets from 15.5% to 18.7% in the 1-injector case in which solid loading ratio of each of cyclones is different affecting the cyclone separation efficiency and the formation of carbonaceous deposits. Through this analysis, the uneven distribution of particle flow is diagnosed, and the standard deviations of particle flow distributions were investigated for the cases of 1, 2, 4, 6, and 8 injectors and evaluated to be 1.12, 0.58, 0.36, 0.24, and 0.39, respectively. From the results, the effect of the number of feed injectors was investigated and the 6-injector case is found to have a largely even particle flow distribution.

455469 Multicomponent Particle Granulation in Spray Drying

Abstract:

One of the granulating methods for multicomponent particle is spray drying. It is generally known the spray drying produces the homogeneous structure in the composite. In some cases, it is desirable the segregation in the composite takes place. We focused on the surface tension during drying process. Experimental approach and simple simulation are carried out to confirm the effect of the primary particle size on the composite structure during drying. Some experimental and simulation results will be presented.

457373 Synthesis of Fe_3O_4 and Au-Fe_3O_4 Dumbbell Nanoparticles By Thermal Decomposition Using Different Reducing Agents

Abstract:

Over the last decade, magnetic nanomaterials have gained a lot of interest in the fields of data storage, hyperthermia therapy, magnetic resonance imaging (MRI) as contrast agent, etc. Au nanoparticles are available for various applications due to their optical and catalytic properties. Synthesis of multifunctional nanomaterials such as core/shell, patchy, janus or flower-like nanoparticles have attracted much more attention in recent years. Especially dumbbell-shaped nanoparticles facilitate the attachment of different functionalization of each particle due to its asymmetric surface. These structures usually have significant properties such as optical, magnetic, plasmonic, catalytic, etc. and these outstanding properties provide usage in many diverse platforms [1]. Also, biocompatible Au-Fe_3O_4 dumbbell nanoparticles are advantageous to be used in biomedical applications. Furthermore, easy modification of iron oxide and gold sides of Au-Fe_3O_4dumbbell nanoparticles allows them to obtain bifunctional nanostructures.

In the literature, stabilization and synthesis of iron oxide nanoparticles with narrow size distribution have been studied in detail [2-4]. Preparation of Fe_3O_4 and Au-Fe_3O_4 nanoparticles was achieved at high-temperatures and reduction of $Fe(acac)_3$ (acac= acetylcetonate) was performed by using 1, 2-hexadecanediol or 1, 2-tetradecanediol [3, 4]. In this study, Fe_3O_4 nanoparticles have been synthesized via thermal decomposition of iron precursor and Au-Fe_3O_4 dumbbell nanoparticles were obtained with nucleation and growth of iron oxide nanoparticles over different sized AuNPs. In order to synthesize monodisperse Fe_3O_4 and Au-Fe_3O_4dumbbell nanoparticles via thermal decomposition method, we used an alternative reducing agent which has not been used before. In order to analyze the morphology, size and size distribution of the particles, High Resolution Transmission Electron Microscopy (HR-TEM) analysis was

performed. Also, Dynamic Light Scattering (DLS), Vibrating Sample Magnetometer (VSM), X-ray diffraction (XRD) and Fourier Transform Infrared Spectroscopy (FT-IR) were used to implement a detailed characterization of nanoparticles.

To obtain monodisperse Fe_3O_4 NPs, $Fe(acac)_3$ was used as precursor, oleylamine and oleic acid as stabilizing agents and a reducing agent, which contains one hydroxyl and a carboxyl group. According to TEM images, 6 nm sized magnetite nanoparticles have narrow size distribution and high crystallinity. Monodisperse Au nanoparticles were synthesized according to previous reports [5]. Synthesis of 5 and 9 nm sized gold nanoparticles was achieved by the reduction reaction of $HAuCl_4$ solution. Afterwards, these oleylamine capped gold nanoparticles were used to obtain different sized Au-Fe_3O_4 dumbbell nanoparticles. To perform the synthesis of dumbbell nanoparticles, reaction solution in 1-octadecene was heated at 200°C under a flow of N_2. Then under a blanket of N_2, AuNPs was added to the solution and the solution was heated to reflux. After cooled down to room temperature, the particles were washed with iso-propanol, centrifuged and redispersed into 1-octadecene or hexane.

TEM images of Fe_3O_4 NPs showed that the particle size was 6 nm and Au-Fe_3O_4 dumbbell nanoparticles were 5-15 nm and 9-15 nm. From DLS measurements, it was seen that hydrodynamic radii of Fe_3O_4 NPs and dumbbell nanoparticles were 7.4 nm and 11 nm respectively in 1-octadecene solvent. It can conclude that the results of TEM and DLS analyses were consistent with each other due to attraction of solvent and ligands on the surface of particles. Five XRD peaks at 30.1°, 35.5°, 42.5°, 52.5°, and 62.1° were observed which can be assigned as the fcc Fe_3O_4. Moreover, characteristic peaks for Au at 38.12°, 44.4°, 64.6° and 77.5° were obtained for dumbbell NPs.

We have investigated the influence of reducing agents in the synthesis of iron oxide nanoparticles. Results showed that high-temperature synthesis of Fe_3O_4 and Au-Fe_3O_4nanoparticles can be performed successfully.

REFERENCES

[1] H. Yu, M. Chen, P. M. Rice, S. X. Wang, R. L. White, S. Sun, "Dumbbell-like Bifunctional Au –Fe_3O_4Nanoparticles" Nano Letters, 5, 379-382, 2005.

[2] Z. Xu, C. Shen, Y. Hou, H. Gao, S. Sun, "Oleylamine as Both Reducing Agent and Stabilizer in a Facile Synthesis of Magnetite Nanoparticles", Chem. Mater., 21, 1778–1780, 2009.

[3] S. Sun, H. Zeng, "Size-Controlled Synthesis of Magnetite Nanoparticles" J. Am. Chem. Soc., 124, 8204-8205, 2002.

[4] M. Lattuada, T. A. Hatton, "Functionalization of Monodisperse Magnetic Nanoparticles", Langmuir, 23, 2158-2168, 2007.

[5] B. Wu, H. Zhang, C. Chen, S. Lin, and N. Zheng, "Interfacial Activation of Catalytically Inert Au (6.7 nm)-Fe_3O_4 Dumbbell Nanoparticles for CO Oxidation", Nano Res, 2, 975-983, 2009.

459119 Cyberinfrastructure Enabled Parallelization of Population Balance Models for Efficient Simulation of Granulation Processes

Abstract:

Cyberinfrastructure Enabled Parallelization of Population Balance Models for Efficient Simulation of Granulation Processes

Anik Chaturbedi, Franklin Bettencourt, Srinivas Mushnoori, Subhodh Karkala, Shantenu Jha, Marianthi Ierapetritou, Rohit Ramachandran

Rutgers, the State University of New Jersey, Piscataway, NJ, USA 08854

Particulate processes, in general, are complex to understand and control efficiently. Granulation is one of the most widely used particulate process and is used to produce commercial products such as pharmaceuticals, catalysts, fertilizers etc. In granulation, multiple rate processes, such as aggregation, breakage, liquid addition, nucleation, consolidation, layering take place simultaneously. The incorporation of these multi-dimensional physics is a challenge in any model formulation. Previous research has shown that a population balance model (PBM) can be used to simulate granulation processes and to predict the evolution of macroscopic properties of the system such as particle size distribution (PSD), average composition and porosity. Due to the multitude of dimensions present and the associated integrals, population balance models can become computationally intensive. Moreover, for modeling continuous granulation, additionally, spatial variation of particle properties has to be incorporated in already complex population balance models [1]. This increases the computational cost even more. To tune these models for accurately simulating real systems and eventually move towards real-time prediction and control of these processes we will need to massively speed up these simulations. Some previous research has been done on the parallelization of PBMs using either the MATLAB Distributed Computing Toolbox [2] or Message Passing Interface (MPI) [3].

In this work, a 3D PBM (distributed solid volume, lumped liquid and gas volumes) and two additional spatial dimensions has been parallelized using a simpler Message Passing Interface (MPI) approach and a more complex hybrid MPI+OpenMP approach. The speed up and scale up of these parallelization techniques have been studied. Also, a pilot job system, RADICAL-Pilot has been used to efficiently parallelize and distribute computational work among different computational units in a National Science Foundation-supported distributed computing resource, Stampede and Rutgers University School of Engineering High Performance Computer (SOE HPC). Parallel simulation of PBMs through RADICAL-Pilot will not only aid in more efficient and user friendly implementation of these parallel PBMs but also enable coupling of PBMs with micro-physics models such as discrete element models (DEM) to increase the accuracy of PBMs and accurately capture the multi-scale dynamics of complex particulate processes. All codes were compiled with the '–ofast' option. Initial runs show a 40 times speedup for the hybrid code compared to the serial one on 128 cores on 8 nodes. Also the hybrid code resulted in 80% less memory usage compared to the MPI code, which presents an advantage in simulating larger systems with same computing resources. This work will eventually lead to the incorporation of these fast and more mechanistic population balances in real-time applications such as control and optimization.

References

[1] D. Barrasso, S. Walia and R. Ramachandran, "Multi-component population balance modeling of continuous granulation processes: A parameteric study and comparison with experimental trends," *Powder Technology*, pp. 85-97, 2013.

[2] A. V. Prakash, A. Chaudhury, D. Barrasso and R. Ramachandran, "Simulation of population balance model-based particulate processes via parallel and distributed computing," *Chemical Engineering Research and Design*, pp. 1259-1271, 2013.

[3] R. Gunawan, I. Fusman and R. D. Braatz, "Parallel High-Resolution Finite VolumeSimulation of Particulate Processes," *AIChE Journal*, pp. 1449-1458, 2008.

459511 Scale-up of Fluidized Reactors Based on Choas Analysis Approch and New Scale-up Methodology

Abstract:

The fluidized bed is consider one of importance solid-gas reaction systems due to its widely using in different industrial application such as catalyst regeneration, drying, gas adsorption, and gas-solid polymerization. Although fluidized bed reactors are simple in construction, a proper understating of their hydrodynamics, design, scale-up, and performance are still challenging due to the complexity of the flow structure and complex interaction between the phases. In the present work, the methods of chaos analysis and new scale-up methodology have been implemented and proposed to study the scale-up of two fluidized bed (0.44 m and 0.14 m inside diameter). The chaos analysis is based on the degree of chaos that is assessed by Komlogorov entropy, which is a measure of the rate of information loss in the system. The basic concept of the chaos scale-up approch is that the rate of information loss should kept simiale when scaling up a fluidized bed from small sacle to the large scale, in order to ensure hydrodynamic similarty betweed the two scaled beds. While, the basic idea of the new methodology of sacle-up that we proposed is that Radial profile or cross sectional distribution of gas holdup (or solids holdup) should be the same or closer for two beds to be hydrodynamically similar or closer. CFD was used as an enabling tool to identify the conditions of matching and mis-matching cases. It was found that the cases with same radial gas holdup or closer (maching casaes) are also possess the same radial Komlogorov entropy or closer and the cases with different radial gas holdup (mis-maching cases) are also possess different radial Komlogorov entropy.

459715 CFD Simulation of Gas and Solids Mixing in FCC Packing Strippers

Abstract:

The gas and solid mixing in fluid catalytic cracking (FCC) stripper with and without packing were investigated using computational fluid dynamics (CFD). The Eulerian-Eulerian two-fluid model coupled with the modified Gidaspow drag model was used to simulate the gas-solid flow behavior. The grid independency study and the comparison between 2D and 3D simulations were carried out. Many parameters may affect the stripping efficiency, including backing-mixing degree, the gas and solids mixing degree and available vessel area. The residence time distribution model (RTD) and axial dispersion model were utilized to obtain the parameters indicating the back-mixing degree, such as mean residence time t_m, dimensionless variance and Peclet number of gas and solids. The time-averaged bed density at various superficial gas velocities, the bubble size and gas/solid distribution indicated the gas and solids mixing degree. The dead zone, plug and well-mixed volume indicated the available vessel area. It was found that stripping efficiency depends strongly on the operating conditions as well as the internal configurations of the strippers. Higher superficial gas velocity increased stripping efficiency. Internal packing could improve break-up and redistribution of bubbles, reduce particle volume fraction and enhancing the gas-solids contacting remarkably. Compared with the empty stripper, the packing stripper exhibited a smaller , a shorter t_m and a bigger Pe, standing for a reduced back-mixing degree of gas and solids. In addition, internal packing could reduce the dead zone and well-mixed volume, increase plug volume, thus improving the utilization of the vessel area.

THE APPLICATION OF POWDER RHEOLOGY FOR EVALUATING THE FLOW BEHAVIOUR OF POWDERS FOR ADDITIVE MANUFACTURING APPLICATIONS®

Jamie Clayton[1], Tim Freeman[1], Katrina Brockbank[1], Robert Deffley[2]
[1]Freeman Technology, 1 Miller Court, Severn Drive, Tewkesbury, Gloucestershire, GL20 8DN.
[2]LPW Technology, 16 Berkeley Court, Manor Park, Runcorn, Cheshire, WA7 1TQ, UK USA

freemantechnology

INTRODUCTION

Many Additive Manufacturing (AM) techniques require powder to be distributed in a uniform manner and are therefore sensitive to minor variations in the properties of feedstocks. Particle size alone is often used to define the suitability of a particular feedstock, however, to suggest that any single parameter or test method, e.g. particle size analysis, Hall Flowmeter or Hausner Ratio, can reliably describe a powder's suitability for a particular process is a simplified approach to a complex issue. The ability of a powder to perform in a desired manner in a specific piece of equipment is not an inherent material property. Successful production demands that a powder is compatible with the process and it is not uncommon for the same powder to perform well in one unit operation, or machine, but poorly in another. This means that multiple characterisation methods are required, the results from which can be correlated with process ranking to produce a design space of parameters that correspond to acceptable process behaviour.

The increasing application of AM processes in an expanding number of industrial sectors is due to the ability to manufacture complex components quickly and precisely. While the main application has historically been prototyping, typically for biomedical, aerospace and automotive industries, there is an increasing emphasis on higher volume manufacturing. In order to manage the costs associated with larger production scales, manufacturers may look to evaluate alternative suppliers of feedstock materials as well as consider the re-use of powder that did not become part of previous components.

This study investigates the flow behaviour of samples of a commonly used titanium alloy sourced from different suppliers and produced using different manufacturing methods. The data show that despite similar physical specifications, the flow behaviour of the samples varies significantly resulting in inconsistent final build quality and requiring adjustments to be made to machine settings in order to achieve the necessary throughput and desired properties of the finished part. A further study demonstrates how the measurement of dynamic flow properties can be applied to optimise the re-use of metal powders.

The results illustrate how a multi-variate approach to powder characterisation can help optimise AM processes by identifying powders that are conducive to successful manufacture, enabling machine settings that handle variations in feedstock to be defined, as well as enhancing efficiency through the re-use of unused material.

RESULTS: MANUFACTURING METHODS

Three Titanium Alloy feedstock powders with the same D50 and PSD (two from different suppliers made using the same method, and two from the same supplier made using different manufacturing methods) were evaluated using the FT4, to evaluate whether the different suppliers and manufacturing methods can influence in-process performance.

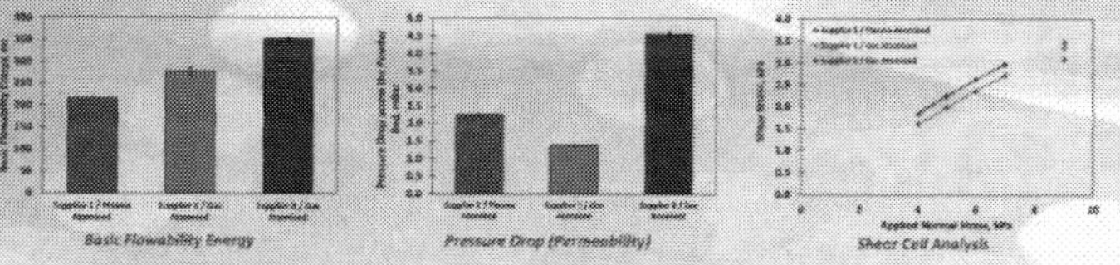

Dynamic tests not only identified differences caused by the change in manufacturing method, but also identified differences in the gas atomized samples from different suppliers. The sample from Supplier 2 has a higher Basic Flowability Energy and Specific Energy than the sample from Supplier 1, indicating more cohesive behaviour in dynamic operations such as filling and spreading. Marked variations were also demonstrated by the Permeability test with the sample from Supplier 1 being significantly more permeable than the sample from Supplier 2. Changing suppliers may therefore have a significant impact on process performance which must be considered alongside any financial or logistical benefits of making the change.

Interestingly, while shear cell tests were able to differentiate between the gas and plasma atomized samples, this technique was unable to distinguish between the gas atomized samples from different manufacturers, potentially overlooking fundamental differences in properties that can influence process performance.

When considering the data from these tests, it is evident that without a comprehensive evaluation of a powder's behaviour, substantial variations in performance across a wide range of processes may not be identified which could result in significant process interruptions and/or degradation in product quality, ultimately impacting profitability.

METHOD

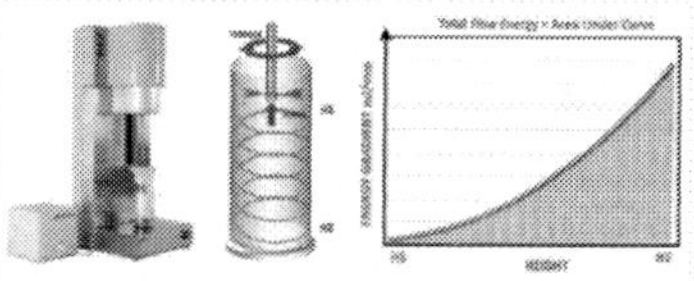

An FT4 Powder Rheometer® (Freeman Technology) was used to quantify dynamic flow, bulk and shear properties. The dynamic flow results are obtained by means of a patented measurement principle that evaluates the resistance to the motion of a specially shaped blade passing along a prescribed path through a precise sample volume. The required torque and force are recorded and converted into a flow energy.

The repeatability of all the measurements was enhanced by the use of a conditioning cycle prior to testing which removes packing history and operator induced variability.

RESULTS: EVALUATION OF FRESH AND USED FEEDSTOCK

A range of different feedstocks containing varying proportions of fresh and used material were evaluated using the FT4's dynamic flow methodologies to determine if critical characteristics of the used powder differed from those of the virgin material, and if so, what strategies might be successful in returning the powder to a condition that would enable its re-use.

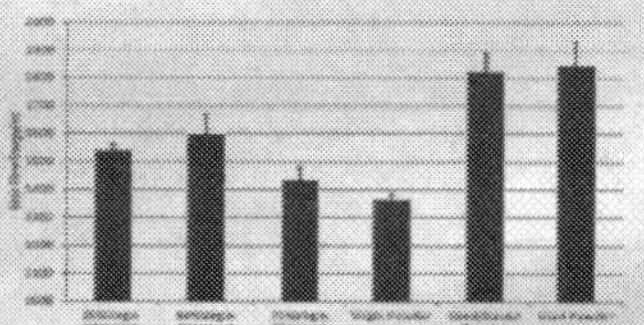

The results for Virgin and Used powder show that processing has significantly increased the flow energy of the powder. This indicates that the Used powder may not flow as readily as the Virgin material and consequently may be less likely to perform as well in process.

Powder exiting an AM machine may contain splatter from the melt pool, in the form of larger particles, or may have changed chemically, from picking up contaminants on the powder surface. Experiments were therefore undertaken to determine whether sieving the Used powder would return it to a state where its flow energy was acceptable. Sieving slightly improved flowability but did not return it to the original flow energy values measured for the Virgin material.

Further experiments were then conducted to see if Used and Virgin powders could be blended together to form an acceptable feed for subsequent processing. A ratio of 75% Virgin to 25% Used powder produced a flowability most similar to that of the fresh material. Interestingly, the 50:50 blend generated the highest flow energy of the blended samples, suggesting that flowability does not change linearly with respect to Virgin powder content.

These results highlight the ability of dynamic flow testing to detect subtle changes in powders that are of direct relevance to their performance in AM machines. As a result, dynamic testing can support successful optimisation and lifecycle management of metal powders for AM.

CONCLUSION

These case studies demonstrate how powder rheology can be employed to identify and quantify subtle differences in powders that directly impact performance in AM processes. These case studies also illustrate how a multivariate approach is required to fully describe powders that are compatible with a given process.

Powder flowability is not an inherent material property, but is more about the ability of powder to flow in a desired manner in a specific piece of equipment. Successful processing demands that the powder and the process are well-matched. It is not uncommon for the same powder to perform well in one unit operation but poorly in another. This means that multiple characterization methods are required, the results from which can be correlated with process ranking to produce a design space of parameters that correspond to acceptable process behavior.

References
[1] S. Mellor, L. Hao, D. Zhang, Int. J. Prod. Econ., (2014) 149 194-201.
[2] B.P. Conner, G.P. Manogharan, A.N. Martof, L.M. Rodomsky, C.M. Rodomsky, D.C. Jordan, J.W. Limperos, Add. Man., 1–4 64-76.
[3] M.K. Taylor, J. Ginsburg, A. Hickey, F. Gheyas, AAPS PharmSciTech, (2000) 1 20-30.
[4] M. Leturia, M. Benali, S. Lagarde, I. Ronga, K. Saleh, Powder Technol., (2014) 253 406-423.
[5] R. Freeman, Powder Technol., (2007) 174 25-33.

Process Analytical Technology for High Shear Wet Granulation: Wet Mass Consistency Reported by In-Line Drag Flow Force Sensor Is Consistent With Powder Rheology Measured by At-Line FT4 Powder Rheometer®

Tim Freeman[1*], Ajit S. Narang[2], Valery Sheverev[3], Douglas Both[4], Vadim Stapaniuk[3], Doug Millington-Smith[1], Kevin Macias[4], Ganeshkumar Subramanian[4], Michael Delancy[5]
[1], Freeman Technology, 1 Miller Court, Severn Drive, Tewkesbury, GL20 8DN, UK
[2], Current affiliation – Genentech, 1 DNA Way, South San Francisco, CA 94080-4990, USA
[3], Lenterra Inc. 105 Lock Street, Suite 301, Newark, NJ 07103, USA
[4], Bristol Myers Squibb, 1 Squibb Drive, New Brunswick, NJ 08903, USA
[5], Current affiliation – Shimadzu, 19 Schoolhouse Rd # 107, Somerset, NJ 08873, USA

freemantechnology

INTRODUCTION

High shear wet granulation (HSWG) is commonly used in the pharmaceutical industry to tailor powder properties to the requirements of tableting. The effect of changes in the equipment, scale, or formulation are typically assessed on the attributes of dried granules or compressed tablets.[1] Consequently, quality-by-design drug product development paradigms rely primarily on end product critical quality attributes (CQA),[2] with limited information on in-process material attributes that can be measured in-line and real-time using appropriate process analytical technology (PAT) that can enable proactive and flexible control of unit operations.

Optimization of wet granulation process parameters is usually an empirical process that relies on statistical design-of-experiment studies to correlate critical process parameters with CQAs.[1,3-6] Development of high (temporal) resolution in-line PATs that correlate well with end product CQAs is crucial to improve process understanding and monitor quality in-line. The primary technologies investigated for HSWG monitoring include power consumption, near-infrared spectroscopy, Raman spectroscopy, capacitance measurements, microwave measurements, imaging, focused beam reflectance measurements, spatial filter velocimetry, stress and vibration measurements, as well as acoustic emissions.[7,8] All these technologies have potential but have considerations that hinder their widespread acceptance.[5]

The application of a drag flow force (DFF) sensor as a monitoring tool, which overcomes the limitations of other HSWG PATs while providing high frequency and high (temporal) resolution in-line granule densification data that correlates well with the off-line granule densification and tablet dissolution, was recently reported.[5,9,10] The DFF sensor is a thin, hollow cylindrical pin, whose deflection in the flow is measured by two optical strain gauges affixed at the inner surface of the pin.[5] The sensor is calibrated to measure the DFF that is related to fundamental parameters of the material such as density and shear viscosity. In addition to force, the DFF sensor outputs temperature. The DFF sensor was shown to be sensitive to changes in formulation composition and process parameters using both a placebo and a brivanib alaninate formulation.[5]

In this study, the in-line response of the DFF sensor is correlated with at-line measurements of flow properties of the wet granules collected at different time points during processing and measured using an FT4 Powder Rheometer®. This correlation allows better understanding of the DFF sensor response as representing fundamental properties of the granules which are known to directly impact tablet CQAs[11].

METHODS

Components	% wt/wt		
	Mixture 1	Mixture 2	Mixture 3
MCC	61	59	57
Lactose Anhydrous	37	37	37
Sodium Croscarmellose	1	1	1
HPC	1	3	5

Three mixtures of MCC (PH102, FMC Biopolymer), anhydrous lactose (Sheffield Bioscience), sodium croscarmellose (AcDiSol, FMC Biopolymer) and hydroxypropyl cellulose (HPC, Klucel EXF, Ashland Speciality Ingredients) were wet granulated with 40% wt/wt water in a GEA PharmaConnect™ high shear wet granulator

For each formulation, changes in DFF during the granulation step were directly monitored in-line using an in-line DFF sensor (Lenterra, USA) and at-line using an FT4 Powder Rheometer (Freeman Technology, UK), by withdrawing aliquots from the granulator and measuring their Basic Flowability Energy (BFE), to investigate the correlation between changes in DFF signal and BFE as a function of the granulation process. The BFE is a measure of resistance to flow as a precision blade is rotated along a helical path through a fixed volume of conditioned powder and is expressed as a flow energy.

RESULTS

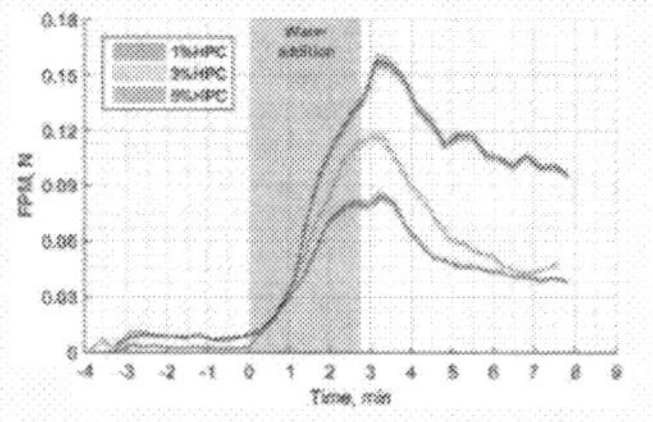

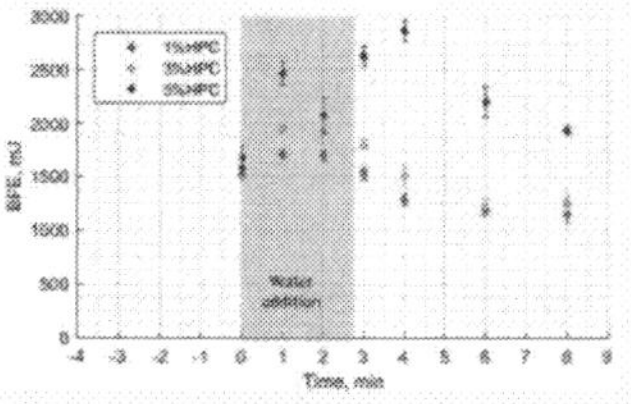

Force Pulse Magnitude (FPM) as a function of time was relatively consistent for the three formulations until the point of water addition when an increase in FPM was observed as the granules developed and began to increase in size. A maximum FPM value occurred soon after the end of the water addition period. This is consistent with conventional understanding that wet granulation end point is achieved shortly after the end of water addition. The peak position suggests that whilst a higher HPC binder content increases the time taken to achieve granulation end point it also potentially results in stronger granules.

Evolution of BFE presented a similar trend to the DFF with an increase and subsequent decrease in BFE as a function of time after the start of water addition. Higher levels of binder also result in higher BFE values, indicating stronger, denser and larger granules.

The relatively high BFE of the end product also suggested that granules with a higher binder content are more robust whereas granules with lower binder contents show a significant reduction in BFE during the same period, probably due to granulate degeneration.

CONCLUSIONS

Wet mass consistency measured in-line and real-time by the DFF sensor is a good indicator of powder rheology, and correlates well with BFE measurements from the FT4 Powder Rheometer. The DFF sensor provides data rich process fingerprinting and can be a valuable tool for wet granulation formulation and process development, as well as for routine monitoring and control during manufacturing. The results generated by both the DFF sensor and the FT4 can be correlated to granulation end point and can help with experimental studies of the effect of process parameters on drug product CQAs.

In order to fully understand the relationship between material properties, process parameters, and product CQAs, process-relevant characterisation methodologies are required, the results from which can be correlated with CQAs to produce a design space of parameters that correspond to optimal process configuration. Rather than relying on classical techniques that produce one value to describe behavior across all processes, the FT4 and DFF approach measures granule properties in conditions relevant to the wet granulation operation, or directly within it, allowing for the direct investigation of a powder's response to various process and environmental conditions.

1. Badawy SI, Narang AS, Lamarche K, Subramanian G, Varia SA. Mechanistic basis for the effects of process parameters on quality attributes in high shear wet granulation. Int J Pharm. 2012;439(1-2):324-333.
2. Badawy SIF, Narang AS, LaMarche KR, et al. Integrated application of quality-by-design principles to drug product development: case study of brivanib alaninate film coated tablets. J Pharm Sci. 2015; submitted.
3. Badawy S, Menning MM, Gorko MA, Gilbert DL. Effect of process parameters on compressibility of granulation manufactured in a high-shear mixer. Int J Pharm. 2000;198:51-61.
4. Vemavarapu C, Surapaneni M, Hussain M, Badawy S. Role of drug substance material properties in the processibility and performance of a wet granulated product. Int J Pharm. 2009;374:96-105.
5. Narang AS, Sheverev VA, Stepaniuk V, et al. Real-time assessment of granule densification in high shear wet granulation and application to scale-up of a placebo and a brivanib alaninate formulation. J Pharm Sci. 2015;104(3):1019-1034.
6. Zhao Z, Tao L, Li J, Bindra D, Narang Ajit S. Granules Consolidation Kinetics during High Shear Wet Granulation: Effect of Binder Type, Molecular Weight, and Method of Addition. 2013. American Association of Pharmaceutical Sciences (AAPS), ed.,San Antonio, TX.
7. Faure A, York P, Rowe RC. Process control and scale-up of pharmaceutical wet granulation processes: a review. Eur J Pharm Biopharm. 2001;52(3):269-277
8. Hansuld EM, Briens L. A review of monitoring methods for pharmaceutical wet granulation. Int J Pharm. 2014;472(1-2):192-201.
9. Narang AS, Breza R, Macias K, et al. Real-Time Measurement of Granule Densification and Size in High Shear Wet Granulation: Combined Use of Focused Beam Reflectance Measurement with Drag Force Sensor. 2013. AAPS Annual Meeting, ed., San Antonio, TX.
10. Narang AS, Sheverev V, Stepaniuk V, et al. American Association of Pharmaceutical Sciences Annual Meeting, San Diego, CA. 2014.
11. Freeman T, Birkmire A, Armstrong B. A QbD approach to continuous tablet manufacture. Procedia Engineering 2015;102:443-449

460632 Incorporating the Effect of Fluidized-Bed Temperature in CFD Simulation through Particle-Particle Interaction Coefficient

Abstract:

Abstract

Experimental studies on the hydrodynamics of dense gas-solid fluidized beds at elevated temperatures are very limited. This lack of studies is due to difficulties associated with measuring techniques under these conditions. However, numerous industrial applications of fluidized beds are at high temperatures, e.g., combustion and gasification, and it is desired to better understand the effects of temperature on hydrodynamics. The first-principles based computational fluid dynamics (CFD) is an effective tool to explore the complex hydrodynamics behavior in gas-solid fluidized bed at elevated temperature. It is well accepted that the changes of fluid physical properties, such as density and viscosity are not sufficient to describe the hydrodynamic behavior of fluidized beds at high temperature. Particle-particle interactions play a vital role at high temperature, which in turn has a substantial impact on the hydrodynamics of a fluidized bed. In this study, a series of CFD simulations was performed using glass beads particles (Geldart-B) in a two dimensional bubbling fluidized bed at different temperatures . Effects of particle-particle interactions were studied by changing the restitution coefficient, and its effect on the hydrodynamic properties (i.e., bubble size, bubble shape, bubble rise velocity, pressure drop fluctuations, and bed expansion) were investigated. The coefficient of restitution has an effect on bed hydrodynamics which is most pronounced when near the point of inelasticity. Our CFD results show that decreasing the coefficient of restitution results in larger bubbles. A similar result is experimentally reported for high-temperature fluidized beds (Sanaei et al. 2007. "Effect of Temperature on Hydrodynamics of Fluidized Beds"; Velarde et al. 2016. "Development Of An Endoscopic-Laser Piv/Dia Technique For High-Temperature Gas-Solid Fluidized Beds." *Chemical Engineering Science*.)

We explain the effect of particle-particle interactions on bubble size through the two-phase theory of fluidization.

When particles experience inelastic collisions, i.e. no loss of energy, there is no change in total particle momentum. However, elastic collisions result in a loss of kinetic energy and reduced particle velocity. , As the fluid passes by particles with reduced velocity, the bed experiences increased relative fluid velocity. In two-phase theory, any fluid flow exceeding the minimum fluidization velocity passes through the bed as bubbles. Thus, decreased coefficient of restitution can result in increased bubble size.

The interpenetrating two-fluid model (TFM) using Eulerian-Eulerian flow field method employed in the open source MFIX software is used for all simulations studied here. The complete set of mass and momentum balance equations are solved through numerical simulation to obtain gas and solid flow fields at high temperatures. The simulation data analyzed and results were then compared with recent experimental results. Velarde et al. (among others) showed that high temperature beds exhibit relatively large bubbles, with an increased possibility of slugging behavior. We believe the effect of increased bed temperature can be accounted for in TFM simulations by decreasing the coefficient of restitution.

460475 Platelet Aggregation at Sites of Vascular Injury

Abstract:

Platelet aggregation at sites of vascular injury is necessary for hemostasis and arterial thrombosis and occurs via plateletâ€“platelet adhesion, tethering and rolling on the injured endothelium, a critical initial step in blood clot formation. Smaller platelet size yielded less mural thrombus mass and smaller emboli, either with or without RBCs. Aggregation of platelets is accelerated by shear rate during blood flow. Adhesion of platelets to each other and platelets to RBC depend on collisions of elastic platelets to deformable RBC in a hydrodynamic flow with Johnson-Kendall-Roberts (JKR) models. In this way the cohesive impulse, required for the hard-sphere model, is calculated together with other parameters, namely the collision duration and the coefficient of restitution. Aggregate morphology, elastic modulus and surface energy are critical in determining adhesion to straight and tortuous capillary walls and to Red Blood Cells for required optimizing coagulation. Model predictions on platelet to platelet collisions and of platelets to the capillary wall showed that platelet size, platelet Youngâ€™s modulus and shear rate are most significant variables controlling capillary aggregation and thrombus. Results are compared with available experimental results.

461343 CFD Simulation of a Laboratory-Scale Spray Dryer Fitted with Ultrasonic Nozzle for Production of Skim Milk Powder

Abstract:

Spray drying is a well-established method used in the food, chemical, and pharmaceutical industries for the production of various powder products. The basic principle of spray drying is based on contact between the droplets and the drying air, where the drying air provides the energy for the rapid evaporation of the solvent in the feed and absorbs the solvent vapour. The process of spray drying involves multiphase flow with heat, mass and momentum transfer between the continuous phase of three-dimensional swirling drying air flow and the discrete phase of polydispersed droplets/particles. Therefore, a good understanding of the dynamic flow behaviour in the drying chamber is essential to optimise the process of spray drying and ensure product quality.

Recent studies have shown that Computational Fluid Dynamics (CFD) is a powerful tool for the optimisation of spray dryer operational conditions. CFD is the science of predicting fluid flow, heat and mass transfer, and related phenomena. To predict these phenomena, CFD solves equations for conservation of mass, momentum, energy etc. One important advantage of this approach is the ability to evaluate the effect of the actual air flow pattern, particularly at the air inlet region near the atomiser, on the drying behaviour. CFD simulation can also be used to evaluate regions of high particle deposition flux of food powder which can aid the positioning of air hammers or manipulation of the operating conditions. Advanced modelling approaches based on CFD thereof plays an increasingly important role in developing next-generation drying technology. However, the modelling of the spray drying of dairy products has been relatively neglected. Furthermore, spray drying using the conventional pressure or rotary atomisers have drawbacks of broad droplet size distributions and risk of clogging. These disadvantages can be reduced using ultrasonic nozzles generating uniform droplets and low velocity.

Therefore, the aim of this study is to develop and validate a three-dimensional (3D) CFD model for predicting the air flow fields (i.e., velocity and temperature fields) and particle drying behaviour of skim milk powder produced by a laboratory-scale spray dryer fitted with an ultrasonic nozzle. In the present study, simulations are conducted using commercial CFD solver ANSYS Fluent (version 17). The 3D volume geometry of drying chamber of laboratory-scale spray dryer is sketched with the CAD System (SOLIDWORKS) and inserted into ANSYS Fluent. The Lagrangian-Eulerian approach is implemented to simulate multi-phase flows in sprays and atomisers. Continuous-phase, i.e., air, conservation equations are formulated in the Eulerian model while the droplet or particle equations are set up in the Lagrangian model. Two-way coupling between the continuous and dispersed phases is taken into account in the governing equations. The 3D simulation results give a better understanding of several phenomena that appear during the spray drying process. As such this approach has the potential to considerably reduce the optimisation time for spray drying processes and also assist in finding a solution to the fouling problem.

460859 Enhanced Performance of Lithium-Ion Batteries By Atomic Layer Deposition

Abstract:

Lithium-ion batteries (LIBs) are one of the most promising candidates for energy storage because of their exceptional characteristics, such as high energy density and light weight. Thin film coatings on LIB electrode materials have proven to be an effective way to improve the capacity retention, rate capability, and thermal stability of electrode materials. Atomic layer deposition (ALD) is a layer-by-layer thin film coating technique. It consists of a set of sequential reactions that occur only between the gaseous precursors administered to the reactor, and the functional groups present on the surfaces of primary particles. ALD is believed to be an important technique for LIB applications. Currently, most of ALD-LIB research focuses on insulating ALD materials. Clearly, there is a trade-off between the species transport (increasing capacity) and protection of the particles (expecting long cycle life) using insulating materials. We reported a breakthrough to overcome this trade-off by coating an ultra-thin conformal cerium dioxide (CeO_2) film on the surfaces of $LiMn_2O_4$ particles. The optimized CeO_2 film (~3 nm) coated particles exhibited a significant improvement in capacity and cycling performance compared to uncoated, Al_2O_3 ALD coated, and ZrO_2 ALD coated samples at room temperature and 55 °C for long cycling numbers. The initial capacity of the 3 nm CeO_2 coated sample showed 24% increment compared to the capacity of the uncoated one, and 96% and 95% of the initial capacity was retained after 1,000 cycles with 1 C rate at room temperature and 55 °C, respectively. Iron oxide films were coated on $LiMn_{1.5}Ni_{0.5}O_4$ particles by ALD for the synergetic effect of performance enhancing by iron doping and conformal iron oxide film coating. With an optimal film thickness of ~0.6 nm, the initial capacity improved by 25% at room temperature and by ~26% at 55 °C at 1C cycling rate. The synergy of doping of $LiMn_{1.5}Ni_{0.5}O_4$ with Fe near surface combined with the conductive and protective nature of the optimal iron oxide film led to high capacity retention (~93% at room temperature and ~91% at 55 Â°C) even after 1,000 cycles at 1C cycling rate. This work has

the potential to produce the foundational knowledge necessary to solve the â€œcapacity-protectionâ€ dilemma.

462777 Comparison and Validation of Coupled CFD-DEM and CFD-TFM Simulation Predictions for a Laboratory-Scale Wurster-Coating Process

Abstract:

Fluidized-bed Wurster coaters are commonly employed in the pharmaceutical industry to coat beads containing drug products. The Wurster-coating process is comprised of an internally-circulating fluidized bed —an internal cylindrical riser tube (Wurster insert) is used to channel the air and bead flow to establish the desired circulating bead-flow pattern. Two computational fluid dynamics (CFD) based approaches to simulate a lab-scale Wurster coater were compared: CFD-DEM (discrete-element method) and CFD-TFM (two-fluid model). The CFD-DEM model provides better particle trajectories but is computationally more expensive, whereas the CFD-TFM simulations are numerically faster but sacrifice the resolution of particle trajectories and collisions.

Predictions from both approaches are validated against the experimental measurements reported by Li et al. [AIChE J. (2015), 61(3), 756-768]. The CFD-DEM simulations were performed using the XPS code (eXtended Particle System, developed internally at RCPE) coupled with commercial CFD software AVL Fire™ (AVL List GmbH Graz, Austria). The CFD-TFM simulations used Ansys Fluent™ (ANSYS Inc. Canonsburg, USA). The objective of this study to determine which modeling methodology is best-suited to simulate the Wurster coating process. More specifically, the ability of CFD-TFM simulations to generate the desired residence-time-distribution data in various regions of the coater is contrasted to the more-expensive CFD-DEM method.

For both CFD-DEM and CFD-TFM approaches, the bead flow velocities, residence time in various zones, and the bead-cycle-time distributions were measured to allow direct comparisons to Li et al. The present results, using both CFD-DEM as well as CFD-TFM, agree quantitatively and qualitatively with the experimental

results given by Li et al. However, a comparison of the clustering and bubbling dynamics reveal significant differences between the CFD-DEM and CFD-TFM models. Although it is difficult to precisely identify the source of these differences, it can be stated that the CFD-DEM predictions are more visually similar to the flow of actual beads.

Simulations using half (180°) and quarter (90°) models, taking advantage of azimuthal symmetry, are also performed—such a strategy would be important to enable simulation of larger-scale Wurster coaters. The CFD-DEM method shows better agreement between the full- (360°), half-, and quarter-sized systems, whereas the CFD-TFM approach was not amenable to this type of geometric idealization. Therefore, CFD-DEM has been demonstrated to be better suited to simulate the current Wurster-coating

16AIChE
Annual Meeting, San Francisco, CA

Synthesis and Characterization of Nickel-Cobalt Oxide for Ethanol Decomposition

Authors: Anchu Ashok[1], **Anand Kumar**[2], Faris Tarlochan[1]

1. Department of Mechanical and Industrial Engineering, Qatar University, Doha, P O Box 2713, Qatar,
2. Department of Chemical Engineering, Qatar University, Doha, P O Box 2713, Qatar

Corresponding Author: Anand Kumar, Department of Chemical Engineering, Qatar University, Doha, P O Box 2713, Qatar,

Email: akumar@qu.edu.qa

Phone: (974) 4403-4164

Fax: (974) 4403-4131

Synthesis and Characterization of Nickel-Cobalt Oxide for Ethanol Decomposition

Anchu Ashok[1], **Anand Kumar[2]**, Faris Tarlochan[1]

1. Department of Mechanical and Industrial Engineering, Qatar University, Doha, Qatar,
2. Department of Chemical Engineering, Qatar University, Doha, Qatar

Abstract

In this work we use combustion synthesis method to prepare cobalt catalysts which have been reported to be active for ethanol-hydrogen production. This work focuses primarily on understanding the reaction mechanism leading to various products using in situ DRIFTS studies. Bimetallic NiCo were synthesized from the aqueous solution of nickel nitrate ($Ni(NO_3)_2 \cdot 6H_2O$), cobalt nitrate ($Co(NO_3)_2 \cdot 6H_2O$) and glycine ($C_2H_5NO_2$) in solution combustion synthesis (SCS) method using various fuel to oxidizer ratio. Time temperature profile of the synthesis was recorded using high speed data logger during combustion. A correlation between combustion temperature profile and fuel to oxidizer ratio is established to understand the effect of synthesis conditions to the expected nanomaterials properties. These materials are tested for ethanol hydrogen production, primarily focusing on the decomposition mechanism on Ni, Co and NiCo bimetal catalysts using an in-situ DRIFTS analysis. Other characterizations include XRD, SEM, TEM and XPS for bulk and surface analysis of the nanomaterials synthesized.

Keywords: Catalysis, Energy and Nanomaterials

Introduction

Solution combustion synthesis (SCS) being a simple, economic and energy efficient method for the synthesis nanomaterials has gained a considerable interest over the past few years as indicated by the exponential growth in SCS publications relating its application in a variety of industries owing to the high surface area and uniform properties in the synthesized nanomaterials [1-12]. The underlying principle in SCS utilizes the exothermic energy generated during redox reaction of its constituent components to synthesize high purity crystalline materials in a single step without requiring further post-synthesis treatments such as separation, purification or calcination. The material produced is ready for catalytic application. The redox reaction involving metal nitrate precursor and glycine as fuel is believed to be based on following between stoichiometric reaction [1, 5]:

$$M^{v}(NO_3)_v + \left(\frac{5}{9}v\varphi\right) NH_2CH_2COOH + v\frac{5}{4}(\varphi-1)O_2$$

$$\downarrow$$

$$M^{v}O_{v/2}(s) + \left(\frac{10}{9}v\varphi\right)CO_2(g) + \frac{25}{18}v\varphi H_2O(g) + v\left(\frac{5\varphi+9}{18}\right)N_2(g) \qquad (1)$$

Where M is the metal with valency υ. The value φ known as fuel/oxidizer ratio determine the oxidation state of the metal in the output $\varphi > 1$ is fuel rich condition, $\varphi < 1$ is fuel lean condition and $\varphi = 1$ is the stoichiometric condition with no need of atmospheric oxygen for fuel oxidation.

The SCS synthesized materials are being used for hydrogen generation, solar fuel production, solar oxidation, wastewater treatment etc. using various reactions is our lab [4-8, 10, 11, 13-21], along with being used for validation of theoretical predictions[22]. Currently, the work focuses more of reaction mechanism study leading to hydrogen production from ethanol on metal/bimetal surfaces. The ethanol decomposition/dehydrogenation reactions are represented as following:

Ethanol decomposition (ED):

$$C_2H_5OH \longrightarrow CH_3CHO+H_2 \tag{2}$$

$$C_2H_5OH \longrightarrow CH_4CO+H_2 \tag{3}$$

The purpose of this study is to investigate the production of hydrogen at lower temperature over nickel and cobalt nanoparticles synthesized using SCS technique. At this stage only synthesis part of the work is completed, the detailed ethanol hydrogen production studies are currently underway and expected to be ready by the time of annual meeting.

Experimental

Nanoparticles of cobalt and nickel, were synthesized using cobalt nitrate hydrate ($Co(NO_3)_2{\cdot}6H_2O$), nickel nitrate hydrate ($Ni(NO_3)_2{\cdot}6H_2O$) and glycine as fuel. These precursors were mixed together and dissolved in deionized water and kept these for 1 hr to obtain a uniform mixture of solution. The quantity of precursors used satisfies the stoichiometric equation with φ values of 0.5, 1 and 1.75. A flow chart describing the synthesis procedure is presented in the chart below (chart-1). The desired amounts of precursors were calculated to obtain an output product of 1.5 g and those were mixed in 75 ml of water in 200 ml capacity beaker. Thus prepared solution was heated over a hot plate heater at 200°C in air at atmospheric pressure until all the water gets evaporated and the mixture reaches to its ignition temperature. Once ignition starts, the combustion wave moves from one end to other end producing the nano-powders in the beaker. For the synthesis of monometallic nickel and cobalt catalysts, same procedure was followed except only one individual metal nitrate precursors were used instead of both. The obtained particles were further characterized using SEM, TEM, XPS etc. before being used for catalytic application.

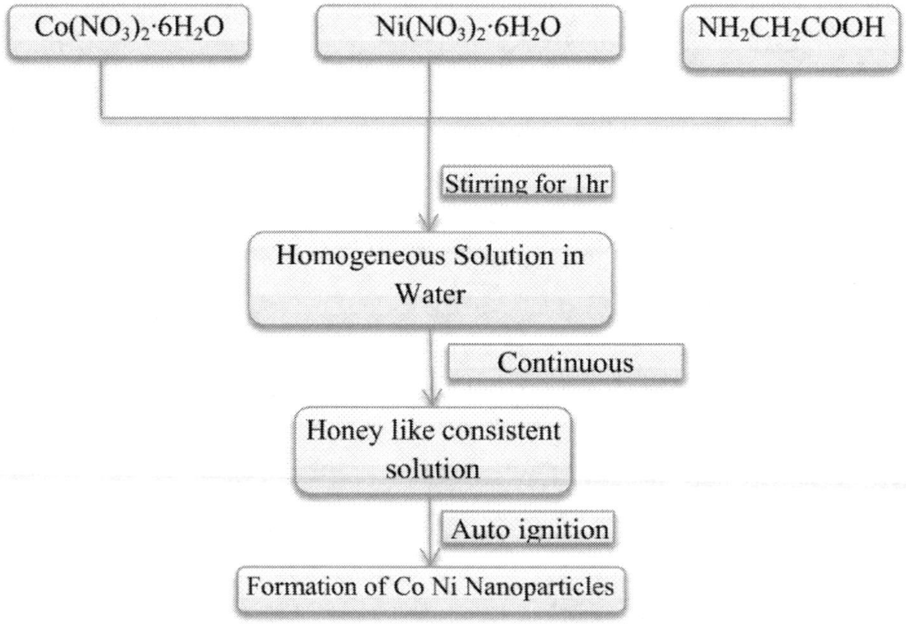

Chart 1: Steps for the synthesis of Copper nanoparticles.

Crystalline phase of the particles were calculated with XRD analysis using Rigaku MiniFlexII Desktop X-ray powder diffractometer with a wavelength of 0.154056 nm (Cu-Kα radiation) at a scan range of 10 – 90 degrees. Morphology study was conducted with Scanning Electron Microscope (Nova Nano 450, FEI, 200 kx magnification), along with EDS for the elemental analysis of the samples. A high resolution transmission electron microscopy (HRTEM, TECNAI G^2 F20, FEI) has been used for the particle size identification of the derived nanopowders. The ethanol decomposition property of prepared catalyst was conducted with in situ Diffused Reflectance Infrared Fourier Transform Spectroscopy (DRIFTS) technique. Thermo Nicolet 6700 FTIR with a reaction chamber (Harrick, HVC) along with an optic assembly of praying mantis was used to carry out the DRIFTS study. The sample was reduced first by placing in the chamber and heated the temperature at 300°C in H_2 (100 sccm) atmosphere. Once the reduction is over, approx. after one hour, N_2 was purged over the chamber to collect the background spectrum. Ethanol was passed over the catalyst with N_2 as carrier gas through an ethanol bubbler at two different temperatures of 50°C and 400°C, and the spectra was collected with 256 scan between 700-4000 cm^{-1} at a resolution of $4cm^{-1}$.

Results and Discussion

Time-temperature profiles of the samples during synthesis of NiCo nanopowders were recorded to measure the maximum combustion temperature for each fuel to oxidizer ratio that can be used to correlate the particle size wrt φ values. Although the results are not included in this report, our preliminary analysis indicates the maximum temperature corresponding to $\varphi = 1$ and displaying a volcano trend as the peak temperature decreases for both $\varphi < 1$ and $\varphi > 1$ as the φ values move farther from $\varphi = 1$. The XRD shows the presence of nickel-cobalt component in their oxidized states at lower fuel to oxidizer ratio.

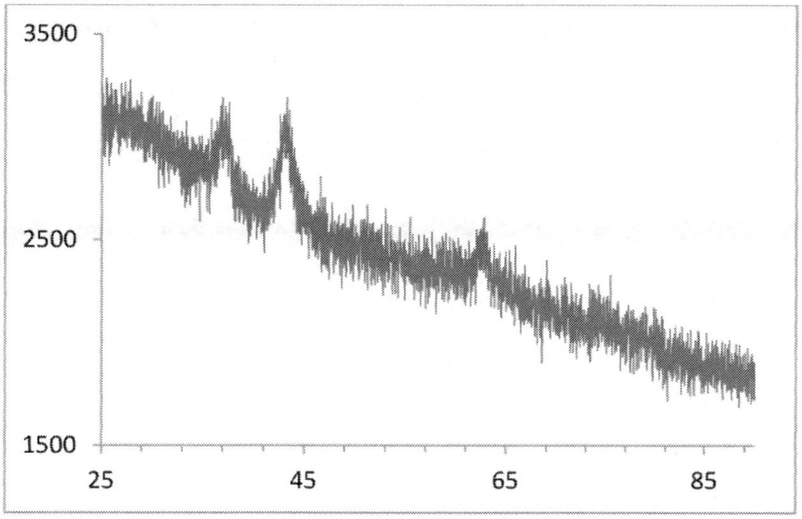

Fig-1: XRD of SCS synthesized NiCo nanopowders for φ=0.5

Fig-1 presents the NiCo XRD for $\varphi = 0.5$. In our initial analysis, the peaks do indicate the oxides of Ni and Co, although detailed analysis is pending at this stage. Based on increasing φ values for monometallic Cu, Ni and bimetallic CuNi, it is anticipated that CoNi may also show a more reducing trend producing alloys of CoNi metals rather than a mixture of metal oxides [1, 5, 12, 13].

Fig-2: SEM microstructure SCS synthesized CoNi nanopowders for φ=0.5

SEM of as-synthesized nanoparticles for $\varphi = 0.5$ in SCS mode shows the nanoparticles of high porosity that are randomly distributes as well as agglomerated. Mostly this high porosity is due to the escaping of excess gases during combustion process. Also the EDS results (not shown, planned to be included during annual meeting presentation) are in consistent with the XRD results showing higher amount of oxide in the synthesized NiCo compound. The details of TEM and other characterization will be presented in the annual meeting; however a summary is presented here excluding the actual data and figures to save space and avoid repetition. TEM image show agglomeration that is common in solution combustion mode with non-uniform sized particles. The particle size increases with temperature and in the range of 10-40nm. Prior to conducting catalytic evaluations, nickel-cobalt oxides were reduced to metallic form by passing hydrogen in the reaction chamber at 300°C. An FTIR spectrum at 50 and 100°C shows the presence of adsorbed ethanol and ethoxy species over the reduced catalyst. IR band at 3669 cm^{-1} indicates the presence of OH from adsorbed ethanol on the catalyst surface. At higher temperature the molecularly adsorbed ethoxy species is converted into acetate species along with the presence of carbonate species. After increasing the temperature from 200°C the intensity CO_2 band at 2335-2367 cm^{-1} becomes evident. At this temperature the acetate species are dehydrogenated to acetaldehyde and other intermediate species. Strong acetate band of 1760 cm^{-1} above 200°C could be due to the transformation of acetaldehyde formed by the dehydrogenation of ethanol to either ethyl acetate or acetic acid though the nucleophilic reaction of ethoxy or hydroxyl species with the surface aldehyde. These results indicate that, in a single step, nickel-cobalt and their oxides nanoparticles can be formed using SCS method, that are active for ethanol hydrogen production.

Conclusion

NiCo nanoparticles were synthesized using solution combustion synthesis technique with a fuel ratio of 0.5. The nanoparticles were reduced to metallic form and used as catalyst for ethanol decomposition reaction. The synthesized and spend catalyst were characterized with XRD, SEM and TEM. Initial DRIFTS indicate the formation of ethoxy species which undergo aldehyde and acetate formation and subsequently decomposes to release carbon dioxide. Further studies are underway to identify the reaction pathway followed on Co, Ni and NiCo bimetallic surface. These studies will be reported in future publications.

Acknowledgement:

This publication was made possible by the NPRP grant (NPRP8-145-2-066) from the Qatar National Research Fund (a member of Qatar Foundation). The statements made herein are solely the responsibility of author(s).
The authors also wish to acknowledge the technical services granted by the Central Laboratory Unit (CLU) and the Gas Processing Centre (GPC) team in Qatar University.

References:

[1] A. Kumar, E.E. Wolf and A.S. Mukasyan, "Solution Combustion Synthesis of Metal Nanopowders: Nickel- Reaction Pathways " AIChE Journal 57 (2010) 2207-2214.
[2] A. Kumar, A.S. Mukasyan and E.E. Wolf, "Impregnated layer combustion synthesis method for preparation of multicomponent catalysts for the production of hydrogen from oxidative reforming of methanol" Applied catalysis.A, General 372 (2010) 175-183.
[3] A. Kumar A.S. Mukasyan, E.E. Wolf, "Modeling Impregnated Layer Combustion Synthesis of Catalyts for Hydrogen Generation from Oxidative Reforming of Methanol" Industrial engineering chemistry research 49 (2010) 11001-11008.
[4] A. Kumar, Combustion synthesis of novel catalysts for hydrogen production from light alcohols, University of Notre Dame, 2011.
[5] A. Kumar, E.E. Wolf and A.S. Mukasyan, "Solution combustion synthesis of metal nanopowders: Copper and copper/nickel alloys" AIChE Journal 57 (2011) 3473-3479.
[6] A. Kumar, A. Mukasyan and E. Wolf, "Combustion synthesis of Ni, Fe and Cu multi-component catalysts for hydrogen production from ethanol reforming" Applied Catalysis A: General 401 (2011) 20-28.
[7] A. Cross, A. Kumar, E.E. Wolf and A.S. Mukasyan, "Combustion Synthesis of a Nickel Supported Catalyst: Effect of Metal Distribution on the Activity during Ethanol Decomposition **2012**, *51* (37), pp 12004–12008" Ind Eng Chem Res 51 (2012) 12004-12008.
[8] A. Kumar, J.T. Miller, A.S. Mukasyan and E.E. Wolf, "In situ XAS and FTIR studies of a multi-component Ni/Fe/Cu catalyst for hydrogen production from ethanol" Applied Catalysis A: General 467 (2013) 593-603.
[9] G. Carotenuto, A. Kumar, J. Miller, A. Mukasyan, E. Santacesaria and E. Wolf, "Hydrogen production by ethanol decomposition and partial oxidation over copper/copper-chromite based catalysts prepared by combustion synthesis" Catalysis Today 203 (2013) 163-173.
[10] R.R. Bhosale, I. Alxneit, L.L.P. van den Broeke, A. Kumar, M. Jilani, S.S. Gharbia, J. Folady and D.Z. Dardor, "Sol-Gel Synthesis of Nanocrystalline Ni-Ferrite and Co-Ferrite Redox Materials for Thermochemical Production of Solar Fuels" MRS Online Proceedings Library 1675 (2014) 203-208.
[11] A. Kumar, A. Cross, K. Manukyan, R.R. Bhosale, L.J.P. van den Broeke, J.T. Miller, A.S. Mukasyan and E.E. Wolf, "Combustion synthesis of copper–nickel catalysts for hydrogen production from ethanol" Chem. Eng. J. 278 (2015) 46-54.
[12] A. Ashok, A. Kumar, R.R. Bhosale, M.A.H. Saleh and van den Broeke, Leo JP, "Cellulose assisted combustion synthesis of porous Cu–Ni nanopowders" RSC Advances 5 (2015) 28703-28712.
[13] A. Kumar, A. Ashok, R.R. Bhosale, M.A.H. Saleh, F.A. Almomani, M. Al-Marri, M.M. Khader and F. Tarlochan, "In situ DRIFTS Studies on Cu, Ni and CuNi catalysts for Ethanol Decomposition Reaction" Catal. Lett. 146 (2016) 778-787.

[14] R.R. Bhosale, A. Kumar, L.J.P. van den Broeke, S. Gharbia, D. Dardor, M. Jilani, J. Folady, M.S. Al-Fakih and M.A. Tarsad, "Solar hydrogen production via thermochemical iron oxide–iron sulfate water splitting cycle" Int J Hydrogen Energy 40 (2015) 1639-1650.
[15] D. Dardor, R.R. Bhosale, S. Gharbia, A. Kumar and F. Al Momani, Journal of Emerging Trends in Engineering and Applied Sciences 6 (2015) 129-135.
[16] F.A. Almomani, R.R. Bhosale, A. Kumar and C. Kennes, "Removal of volatile sulfur compounds by solar advanced oxidation technologies and bioprocesses" Solar Energy 135 (2016) 348-358.
[17] R.R. Bhosale, A. Kumar, F. AlMomani, U. Ghosh, S. Al-Muhtaseb, R. Gupta and I. Alxneit, "Assessment of CexZryHfzO2 based oxides as potential solar thermochemical CO2 splitting materials" Ceram. Int. 42 (2016) 9354-9362.
[18] R.R. Bhosale, A. Kumar, F. AlMomani and I. Alxneit, "Sol–gel derived CeO2–Fe2O3 nanoparticles: Synthesis, characterization and solar thermochemical application" Ceram. Int. 42 (2016) 6728-6737.
[19] R.R. Bhosale, A. Kumar, F. AlMomani, U. Ghosh, D. Dardor, Z. Bouabidi, M. Ali, S. Yousefi, A. AlNouss, M.S. Anis, M.H. Usmani, M.H. Ali, R.S. Azzam and A. Banu, "Solar co-production of samarium and syngas via methanothermal reduction of samarium sesquioxide" Energy Conversion and Management 112 (2016) 413-422.
[20] R.R. Bhosale, A. Kumar, F. AlMomani and I. Alxneit, "Propylene oxide assisted sol–gel synthesis of zinc ferrite nanoparticles for solar fuel production" Ceram. Int. 42 (2016) 2431-2438.
[21] F.A. Almomani, M. Shawaqfah, R.R. Bhosale and A. Kumar, "Removal of emerging pharmaceuticals from wastewater by ozone-based advanced oxidation processes" Environmental Progress & Sustainable Energy 35 (2016) 982-995.
[22] A. Kumar, R.R. Bhosale, S.S. Malik, A.E. Abusrafa, M.A.H. Saleh, U.K. Ghosh, M.J. Al-Marri, F.A. Almomani, M.M. Khader and I.M. Abu-Reesh, "Thermodynamic investigation of hydrogen enrichment and carbon suppression using chemical additives in ethanol dry reforming" Int J Hydrogen Energy 41 (2016) 15149-15157.

464977 Mechanical Characterization of Pollen Grains By Atomic Force Microscopy

Abstract:

Plant pollen grains are gaining interest as a unique adhesive micro-structured particle and a promising filler for polymer composites, due to their light-weight, high strength and sustainable nature. Sporopollenin, the major component of pollen shell, is one of the toughest naturally-occurring polymers with extreme chemical and thermal stability. Mechanical properties of pollen shells have a significant influence on the performance of composites, as well as the adhesion of pollen grains. However, there exist few literature reports of mechanical characterization of individual pollen grains. Here, the mechanical properties of three pollen species were also characterized by atomic force microscopy (AFM), including ragweed, pecan and Kentucky blue grass.

In the mechanical characterization, the elastic moduli of pollen shells were mapped quantitatively on 5Ã—5 Î¼m^2 areas. The values of modulus were determined to be between 7~9Â GPa for ragweed, 3~7 GPa for pecan, and 8~18 GPa for Kentucky blue grass pollen particles. Saturation at a relative humidity of 90% reduces the Kentucky blue grass pollen modulus to 3~6 GPa after absorbing water. These values are used to place in context recent reports of pollen behavior as a filler. To the best of our knowledge, this is the first time that the interactions between complex spiny microstructured particles have been directly measured by AFM. Moreover, pollen grains have excellent mechanical strength as an environmentally-friendly filler for high-performance composites.

465396 Microfluid-Aassisted Hollow Silica Nanoparticle Synthesis for Inkjet-Printed Antireflective Coating

Abstract:

Hollow spherical silica nanoparticles (HSNPs) are a promising material for anti-reflective coating (ARC) due to their void space and cage-structured shells. Tuning the void space, diameter and wall thickness of HSNPs could reach various refractive indices (RI) to produce an optimal ARC for different substrates. To prepare the HSNPs, template method is a facile and straightforward approach. Hard templates such as polystyrene particles [1] and hydroxyapatite nanoparticles [2], were successfully applied to produce the uniform HSNPs in tunable void space and shell thickness. Nevertheless, the high temperature (around 500â‚ƒ) was required to calcinate out these hard templates thoroughly after the formation of silica shell. Oppositely, soft templates made from emulsion micelles [3] and vesicles [4] can be easily washed away by a selective solvent at room temperature. However, to prepare the soft templates in a uniform shape usually involved at least two surfactants and tedious procedures, especially to obtain and control the uniform HSNPs under 100 nm which have higher surface area is still a challenge. The groups of Du et al. [6] and Wan et al. [7] proposed a simplified and environment-friendly approach by applying a single polyelectrolyte - poly(acrylic acid) (PAA) as the soft-template. However, the rapid aggregation of the PAA templates in the batch reaction only can produce the HSNPs in a broad size distribution which results in an uneven ARC thin film and finally diminishes its anti-reflective capability.

By our microfluidic system with the hydrodynamic focusing micromixer (HFM), the uniform HSNPs in 50 nm were successfully synthesized in a short-time scale. The fast and homogeneous mixing of the microfluidic system enable the PAA globules monomers protected instantly by evenly dispersed silica precursor. Along the further attachment of silica precursor and the rapid growth of silica shell, the aggregation of PAA monomers are efficiently avoided due to the hydroxide bond on the particle surface. A micro-channel following the HFM was applied to accomplish the growth of silica shell. By controlling the residence time of the microfluidic flow, different thickness and density of the silica shell can be obtained. In this study, a numerical model was established by using COMSOL Multiphysics to investigate the mixing profile under different flow condition. Incorporated with this computational model, we describe the experimental operation of the microfluidic chemical reactor to control the continuous synthesis of HSNPs. The effects of the mixer geometry and the operating variables on the particle structure and size distribution are discussed.

As coating the uniform HSNPs in 50 nm on glass substrates by the inkjet printer, the enhanced smoothness and transmittance has been detected. Compare to the thin film of the HSNPs in broad size distribution, the thin film of uniform HSNPs raised the increment of the transmittance from 3.6% to 3.8%. It successfully helped to increase the efficiency of the monocrystalline solar panels by 1.82%. Reference

1. Ding, Xuefeng, Kaifeng Yu, Yanqiu Jiang, Hari-Bala, Hengbin Zhang, and Zichen Wang. "A Novel Approach to the Synthesis of Hollow Silica Nanoparticles." Materials Letters 58.27-28 (2004): 3618-621.
2. Williamson, Peter A., Philip J. Blower, and Mark A. Green. "Synthesis of Porous Hollow Silica Nanostructures Using Hydroxyapatite Nanoparticle Templates." Chem. Commun. 47.5 (2011): 1568-570.
3. Yildirim, Adem, and Mehmet Bayindir. "A Porosity Difference Based Selective Dissolution Strategy to Prepare Shape-tailored Hollow Mesoporous Silica Nanoparticles." J. Mater. Chem. A 3.7 (2015): 3839-846.
4. Wu, Xue-Jun, and Dongsheng Xu. "Soft Template Synthesis of Yolk/Silica Shell Particles." Adv. Mater. Advanced Materials 22.13 (2010): 1516-520.
5. Li, Yunqi, Bishnu Prasad Bastakoti, Masataka Imura, Jing Tang, Ali Aldalbahi, Nagy L. Torad, and Yusuke Yamauchi. "Dual Soft-Template System Based on Colloidal Chemistry for the Synthesis of Hollow Mesoporous Silica Nanoparticles." Chemistry - A European Journal Chem. Eur. J. 21.17 (2015): 6375-380.
6. Du, Yi, Lunet E. Luna, Wui Siew Tan, Michael F. Rubner, and Robert E. Cohen. "Hollow Silica Nanoparticles in UVâ€"Visible Antireflection Coatings for Poly(methyl Methacrylate) Substrates." ACS Nano 4.7 (2010): 4308-316.

465709

Insights from Dem Simulation in Comparative Studies of Granular Material and Its Liquid Counterpart

Abstract:

DEM (discrete element method) simulation can offer more detailed location, velocity and interaction information for each particle compared with lab experiment, thus enable detai led analysis of the fundamental mechanisms. For liquid-solid systems we have developed and applied the coupled method of CFD (Computationa l Fluid Dynamics) and DEM simulations for similar purposes. Many experimental setups for studying liquid exist, which provide a rich source of inspiration and quantitative data f or comparative studies of granular materials. Two examples are presented in detail. A dri ven cavity is filled with granular material to create a granular driven cavity, which displa ys unique behaviors compared with the classic liquid driven cavity under similar conditio ns. The flow pattern is measured experimentally to calibrate simulation parameters, whic h is then used to calculate the detailed force distribution of the granular flow field and ex plain the cause of the difference from the liquid cavity. The second application example i s study of a granular jet impinge on a flat, horizontal plate, which has many liquid counte rparts as well. We focus on the impinging region between the jet and the plate, as well as the hydraulic jump formed as the granular material spreads over the plate. Different granu lar and liquid material properties are used to find their effects on the flow pattern and dyn amics.

465319 Understanding Ligand-Nanoparticle Interactions for Silica, Ceria, and Titania Nanopowders

Abstract:

We perform a study on aqueous suspensions of three types of nanoparticles (TiO2-NPs, CeO2-NPs,SiO2-NPs) and investigate the material properties and the surface reaction with surfactant-type ligands to their subsequent colloidal stability, the key factors to their impact on the environmental health and safety (EHS). Bovine serum albumin, sodium dodecyl sulfate, and cetyltrimethylammonium bromide were used as representative surfactant ligands. Electrospray differential mobility analysis, dynamic light scattering, zeta potential measurement, transmission electron microscopy, inductively coupled plasma mass spectrometry, x-ray diffraction spectrometry were employed orthogonally to obtain particle size, morphology, surface corona, concentration and composition, and the change in colloidal stability. Results show that colloidal stability was strongly affected by the surface properties of nanoparticles and the subsequent interactions with ligands under different environmental conditions. Aggregation was shown to be the dominant process to colloidal instability. Our work demonstrates a prototype study to investigate the change of surface charges and the fate of nanomaterialâ€™s colloidal stability after interacting with ligands and their subsequent impact in EHS.

466703 Simulations of Flow Behavior of Particles in a Liquid-Solid Fluidized Bed Using a Second-Order Moments Model

Abstract:

The flow behavior of particles is simulated using the Eulerian-Eulerian two-fluid model approach with a second-order moments method for solids phase in the liquid-solid fluidized bed. Transport equations for the velocity moments and closure equations for the particle phase are used on the basis of the second-order moments method and the kinetic theory of granular model, respectively. The anisotropic characteristic of fluctuation is predicted based on second-order moments of particles. Simulated results are compared with the experimental data of Razzak et al. in a liquid-solid fluidized bed. Simulated results indicate that the axial second-order moments M_{zz} in the direction of flow are much larger than the radial second-order moments M_{xx}. The components of second-order moments in all direction are increased with increasing liquid velocity. The simulated granular temperature is increased, reaches a maximum, and then decreases with the increase of solid volume fraction. We also conduct the simulations to verify the numerical model for the downcomer, and the same conclusion as the riser is obtained from the two different models. Through comparison of various simulation models, it was found that the velocity fluctuation of particles is a key factor in influencing the hydrodynamics characteristics.

466739 Measurement of Permeability of Powder with Reference to Flow Behaviour

Abstract:

Recent research has highlighted the importance of particle-air interactions in powder processing. For example, air pressure increase during powder discharge into closed cavities opposes powder flow and affects powder dosing processes such as die fill on rotary powder presses. The permeability of the powders is a property which influences flow performance under such conditions.

In this research a system is designed to measure the permeability of a range of powders used as excipients in pharmaceutical formulations. Starting from bulk density, the system is used to measure the permeability of the powders at different solid fractions by applying a small degree of compression.

The results are discussed in relation to flow rate measurements during bin discharge in an enclosure where the air pressure (or vacuum) is controlled.

468291 Multi-Scale Approach to Developing Non-Spherical Particle Models for Realistic Granular Flow Simulations

Abstract:

The simulation of granular flows can be performed with either a Lagrangian or Eulerian approach. The Lagrangian approach, i.e. discrete element method (DEM) simulations, has the ability to simulate realistic particle shapes, but only a limited number of particles can be simulated due to the high computational cost. The Eulerian approach, i.e. two-fluid simulations, can simulate large-scale granular flows, but the models in this approach typically assume that the particles are spherical. The spherical-particle assumption hampers the accuracy of two-fluid simulations, due to the fact that particle shape has a significant impact on flow behavior. As a result, better Eulerian models should be developed to account for the shape effect. This work focuses on the development of a workflow for the numerical modeling of non-spherical particle flow at a large scale. In this workflow, the DEM simulations are used to determine the terms in the granular kinetic theory for the description of non-spherical particle flow. The developed non-spherical particle models are then plugged into the two-fluid model for the large-scale simulations.

Granular kinetic theory terms that are important for two-fluid simulations include: solid stress, collisional dissipation rate, and granular conductivity. The former two terms are dominant for dense granular flows. DEM simulations with the glued-sphere particle model and true cylinder model have been utilized to develop the solid stress and collisional dissipation rate models for elongated, rod-like particles. The developed models are implemented into a two-fluid model code for the simulations of rod-like particle flows. A validation study is presented that utilizes an experiment with monodisperse beds of cylindrical particles. The experiment involves the impingement of an air jet onto a particle bed, which produces a crater. Varying flow rates and particle sizes with aspect ratios from 2 to 8 have been used. The depth of the crater over time is measured for each case. The air jet-induced particle bed cratering is simulated using the two-fluid code and the comparison of the experimental and simulated data is presented.

468350 Intra- and Inter-Particle Resolved Simulations and Experiments on Thermal Transport in Confined Particle Beds

Abstract:

Currently, numerical studies on thermal transport in dense gas-particle suspension flows often rely on a number of simplifications, e.g., (i) a uniform intra-particle temperature profile [1], or (ii) a two dimensional formulation [2]. This clearly limits the applicability of these previous studies, often leaving open questions related to the relative importance of intra-particular transport phenomena. It is the goal of our present work to close this gap by establishing an open-source simulation environment, and demonstrate the validity of the employed models and tools within an array of canonical flow situations.

In the present contribution we experimentally and numerically investigate the energy transport in dense particle beds that are confined by two walls. Therefore, we use the well-established Eulerian-Lagrangian simulator CFDEM® [3]. Also, we make use of the novel coupling capabilities of this simulator with the recently developed intra-particle solver ParScale [4]. Our numerical studies focus on two methods: first, we simulate the details of flow and temperature fields in the fluid phase by means of Direct Numerical Simulation (DNS). Therefore, we use a novel Hybrid Fictitious-Domain/Immersed-Boundary (HFDIBM) approach to accurately impose the Dirichlet boundary condition at the particle surfaces. Coupling of the HFDIBM with ParScale enables us (i) to make direct predictions of intra- and inter-particle heat fluxes, as well as (ii) to derive particle-based Nusselt numbers. The latter are compared to correlations available from literature in order to identify opportunities for the improvement of these correlations. Second, we perform unresolved CFD-DEM-based simulations, which benefit from the understanding gained from our DNS in terms of closures for the heat transfer rate. By coupling the CFD-DEM simulations with ParScale, we are now in the position to evaluate the effect of intra-particular heat fluxes on the overall transferred amount of heat with relative ease (see Figure). Finally, we compare our results with experiments of fixed and fluidized beds, as well as simple 1D models.

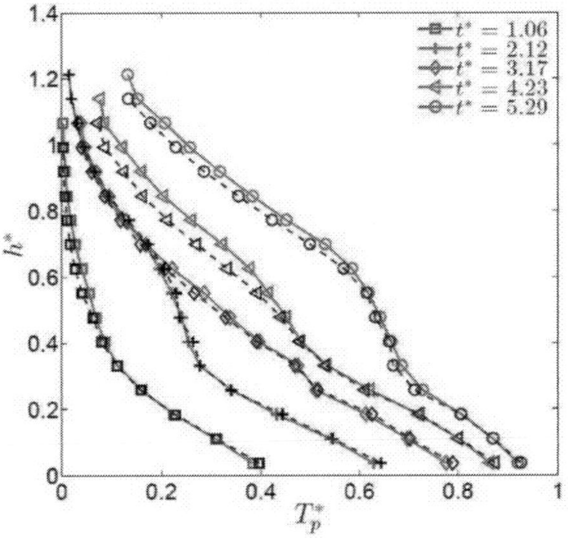

Figure: Temporal progression of the mean particle temperature vs. bed height in a fluidized bed based on CFD-DEM simulations with ParScale (solid lines), and without intra-particle heat transfer (dashed lines; in both simulations the Biot number is ca. 0.8).

References

[1] Z. Feng, S. Musong. Direct numerical simulation of heat and mass transfer of spheres in a fluidized bed. Powder Technology, Vol. 262, pp. 62-70, 2014.

[2] R. Schmidt, P. Nikrityuk. Numerical simulation of the transient temperature distribution inside moving particles, The Canadian Journal of Chemical Engineering, Vol. 90, pp. 246-262, 2012.

[3] C. Kloss, C. Goniva, A. Hager, S. Amberger, S. Pirker (2012) "Models, algorithms and validation for opensource DEM and CFD-DEM", Progress in Computational Fluid Dynamics, Vol. 12, Nos. 2/3, pp.140–152, 2012.

[4] S. Radl, T. Forgber, A. Aigner, C. Kloss, ParScale - An Open-Source Library for the Simulation of Intra-Particle Heat and Mass Transport Processes in Coupled Simulations, in: E. Onate, M. Bischoff, D.R.J. Owen, P. Wriggers, T. Zhodi (Eds.), IV Int. Conf. Part. Methods – Fundam. Appl. (PARTICLES 2015), ECCOMAS, Barcelona, Spain, 2015: pp. 1–9.

469421 Experimental Study on the Velocity Distribution in a Hydrocyclone with a Highly Viscous Fluid

Abstract:

Hydrocyclones are widely used in chemical or mineral industry to separate solid particles from liquid media. Therefore, several studies in the past six decades were dealing with the velocities inside hydrocyclones [1-4]. These investigations mainly focused on water, which is the most common liquid media by using hydrocyclones. In the field of purification technology such as wastewater and sludge treatment, the rheological characteristics of the liquid media are significantly different from water [5]. As a result, most hydrocyclone calculation models [6-8] are not suitable for these applications. For the modelling of the rheological characteristics of sludges in the wastewater sector glycerin and water with different mixing ratios are used. A hydrocyclone with a diameter of 100 mm and a length of 500 mm was manufactured of transparent material to allow non-invasive optical measurements by using PIV (Particle Image Velocimetry). The light source for the measurement is carried out by a high-power LED, which is mounted on a linear-guiding unit what allows measuring all areas of interest. Since the viscosity of the water-glycerin mixture is strongly depending on the shear forces, the local velocities in the hydrocyclone will have a strong influence on the efficiency and pressure loss of the separator system, respectively. By measuring the velocity distribution, these investigations should help to define operation parameters for the application of hydrocyclones in wastewater treatment plants in order to minimize the pressure drop and maximize separation efficiency.

[1] Kelsall, D.F. A study of the motion of solid particles in a hydraulic cyclone, Transactions of the Institution of Chemical Engineers, 30, 1952.

[2] Bradley, D., Pulling, D.J., Flow patterns in a hydraulic cyclone and their interpretation in terms of performance, Transactions of the Institution of Chemical Engineers, 37, 1959.

[3] Bergström, J., Vomhoff, H., Experimental hydrocyclone flow field studies, Separation and Purification Technology, 53, 2007.

[4] Wang, C., Ji, C., Zou, J., Simulation and experiment on transitional behaviours of multiphase flow in a hydrocyclone, The Canadian Journal of Chemical Engineering, 93, 2015.

[5] Baudez, J.C., Markis, F., Eshtiaghi, N., Slatter, P., The rheological behaviour of anaerobic digested sludge, Water Research, 45, 2011.

[6] Mitzmager, A., Mizrahi, J., Correlation of the pressure drop through small cyclones operation with dilute pulps of various liquids, Transactions of the Institution of Chemical Engineers, 42, 1964.

[7] Bohnet, M., Neuere Untersuchungen über die Trennwirkung und den Druckverlust von Hydrozyklonen, Verfahrenstechnik, 3, 376-381, 1968.

[8] Braun, T., Theoretische und experimentelle Untersuchungen des Einflusses der Feststoff-konzentration und der Partikelgrößenverteilung auf das Trennverhalten von Hydrozyklonen, Dissertation, TU Carolo-Wilhelmina, Braunschweig, 1989.

469582 Investigation of the Effects of Humidity on Powder Flow By Experimentally Validated Computational Methods

Abstract:

The flow properties of powder are governed by its intrinsic materials properties (cohesion, surface energies, elastic moduli, coefficient of friction and yield stress) and particle properties (size distribution, morphology, porosity and roughness). Environmental variables that affect the powder flow properties include relative humidity, temperature, and electrostatic charging. This large number of variables and their mutual interactions, along with unknown material properties, make it difficult to predict the flow properties for a new entity (API) or excipient or their mixture without very detailed experimental investigation (based on Design of Experiments-DOE) requiring significant amount of time and resources.

Properties such as flow, compaction, disintegration, dissolution, hardness and chemical stability are all influenced by moisture. The effect of moisture has been introduced in a Discrete Elementary Method based model by incorporating Granular Bond numbers. The experimental appproach involves the flow discharge study of two different hopper angles at 20%, 40% and 60% RH of different grades of Lactose and Avicel. The DEM models for those hopper models, are validated by predicting the accurate granular bond numbers for different RH conditions and comparing the flow properties against the two hopper angles. After systemica compareison of the hopper flow experiments of different individual powders, further study will involve their binary mixtures , to determine the appropriate DEM parameters (bond number, friction coefficients etc).

The preliminary results suggest improved flow properties with the decrease of relative humidity.

472046 A Population Balance Based Model to Describe the Rheology of Thixotropic Suspensions with a Yield Stress

Abstract:

A systematic development of an evolution equation to describe the structural dynamics in a thixotropic dispersion is presented. Ideal thixotropy is defined as the continuous decrease of viscosity with time when flow is applied to a sample that has been previously at rest and the subsequent recovery when flow is discontinued [1]. Yield stress is also frequently observed in thixotropic suspensions [2] presenting additional challenges in modeling these systems. At the heart of these modeling efforts is the complex effects of flow on the microstructure such as aggregation and breakage that give rise to thixotropic behavior. In this work, population balance modeling is adopted as a framework to describe the microstructure evolution that underlies thixotropy. A key advantage of using population balances is the incorporation of parameters in the aggregation and breakage kernels with a clear physical meaning that can be connected to specific properties of the colloidal aggregates through independent experiments.

Starting from the population balance equation for colloidal dispersions, a monodisperse closure rule is used to develop a coarse grained structure evolution equation. Moving beyond previous efforts, important modifications are made to account for dynamic arrest of aggregation kinetics at the onset of the yield stress and to enforce a minimum particle size below which breakage is not feasible. The resulting population balance model is coupled to a simple constitutive equation to complete the model allowing us to predict the effects of aggregation and breakage processes on the rheology. The overall model provides a reasonable representation of experimental data for a model thixotropic suspension available in the literature [3]. The resulting model is able to capture the dominant thixotropic timescales for unidirectional shear flows in step-up and step-down transient experiments. In addition, predictions of rheology under large amplitude oscillatory shear (LAOS) and flow reversal experiments are also compared to experimental measurements. The coarse grained model equations are also compared to prevalent structure kinetics models and shown to be distinct, emphasizing the novelty of using a population based model as a basis for thixotropic suspension

modeling.

References

[1] Mewis J,Wagner NJ. Colloidal suspension rheology. Cambridge University Press 2012.

[2] Hoffmann H, Rauscher A. Aggregating systems with a yield stress value Colloid and Polymer Science. 1993; 271:390â€“395

[3] Armstrong MJ, Beris AN, Rogers SA, Wagner NJ. Dynamic shear rheology of a thixotropic suspension: Comparison of an improved structure-based model with large amplitude oscillatory shear experiments Journal of Rheology. 2016; 60:433-450.

476542 Solids mixing in bubbling fluidized beds: CFD-based analysis of BubbleDynamics and Time Scales

Abstract:

Bubbling fluidized bed reactors are commonly employed for combustion and gasification applications. In these reactors, mixing of the bed material is critical because it directly affects fuel distribution/segregation and residence time as well as wall-to-surface heat transfer. Although solids mixing has been the focus of substantial effort over the past few decades, there continues to be a lack of understanding, especially at commercial-scales, because (a) most diagnostic techniques are only feasible in lab-scale setups, where the hydrodynamics are substantially different [1], and (b) the dynamics are significantly sensitive to the operating conditions (such as superficial gas velocity and particle characteristics). Thus, it is not surprising that quantitative estimates of mixing (e.g., dispersion coefficient, mixing indices) and their dependence on the operating conditions often span orders of magnitude. Nevertheless, it is well accepted that the micro-mixing and gross circulation of solid particles is driven by bubble motion through wake lift, emulsion drift and bubble eruptions. Quantifying this dependence is the focus of this study.

Solids mixing is investigated using fine-grid 3D CFD simulations. The fluidization of two distinct Geldart B particles (0.5 mm glass and 1.15 mm LLDPE) is simulated in an intermediate size 50 cm diameter fluidized bed since hydrodynamics at this scale are wall-independent and, therefore, scalable [1]. Detailed diagnostics of the computed flowfield data are performed using MS3DATA [2], a tool that we developed to detect and track bubbles in 3D, and the solids axial and lateral motion are correlated with bubble rise. Subsequently, the mixing time-scales in the fluidized bed are estimated using the spatial and size distribution of bubbles. This is one of the first studies investigating solids mixing using 3D simulations an intermediate size bubbling fluidized bed. It provides valuable insights for estimating mixing time scales in larger beds, as well as for the reactor network modeling of bubbling fluidized beds (e.g. [3]). All simulations are based on the Two-Fluid Model (TFM) framework and performed using MFiX.

References

[1] A. Bakshi, C. Altantzis, R.B. Bates and A.F. Ghoniem, Study of the effect of reactor scale on fluidization hydrodynamics using fine-grid CFD simulations based on the two-fluid model, Powder Technology, 299: 185-198, 2016

[2] A. Bakshi, C. Altantzis, R.B. Bates and A.F. Ghoniem, Multiphase-flow Statistics using 3D Detection and Tracking Algorithm (MS3DATA): Methodology and application to large-scale fluidized beds, Chemical Engineering Journal 293: 355-364, 2016

[3] A.K. Stark, C. Altantzis, R.B. Bates and A.F. Ghoniem, Towards an advanced reactor network modeling framework for fluidized bed biomass gasification: Incorporating information from detailed CFD simulations. Chemical Engineering Journal. 303: 409-424, 2016

476543 Steam-Blown Biomass Gasification in Fluidized Beds:Gas-Flow Distribution for Advanced Reactor Network Models

Abstract:

Bubbling fluidized beds are used extensively in the energy and chemical industries because of their excellent heat and mass transfer characteristics. Computational Fluid Dynamics (CFD) is an extermely useful tool for reactor design, analysis and optimal operation. Application to large scale gasifiers continues to be challenging because of limitations on computational resources. Given that the hydrodynamics can largely be characterized by bubbles rising through the bed, a more feasible approach for investigating large-scale reactors is to quantify bubble dynamics and specifically, reactive gas distribution in different phases- visible bubble flow, bubblethrough flow and dense-phase flow.

In this study, reactive 3D CFD simulations are conducted for steam-blown biomass gasification in bubbling fluidized beds of varying diameters and pressures. The hydrodynamics are coupled with a global devolatalization mechanism for biomass conversion [1], while the steady state char concentration is predicted using a detailed char particle gasification and combustion model [2]. The physical model and numerical tool were developed and validated in previous studies [3], while 3D Bubble statistics are computed using MS3DATA (Multiphase Statistics using 3D Detection and Tracking Algorithm) [4]. Simulation data is analyzed and the distribution of both the oxidant and devolatalized gases, and their residence times, are computed. Accurate description of the gas-flow is critical for large-scale reactor design since quantifying the gas distribution can determine the fuel rich zones (in the emulsion) as well as the oxidant bypass through the bubbles (through-flow) leading to inefficient performance, and the formation of recalcitrant tar compounds. Additionally, insights from this study will be valuable for reactor network modeling of fuel conversion systems [1].

REFERENCES

[1] A.K. Stark, C. Altantzis, R.B. Bates and A.F. Ghoniem, Towards an advanced reactor network modeling framework for fluidized bed biomass gasification: Incorporating information from detailed CFD simulations. Chemical Engineering Journal. 303: 409-424, 2016.

[2] R. Bates, C. Altantzis, and A.F. Ghoniem, Modeling of Biomass Char Gasification, Combustion, and Attrition Kinetics in Fluidized Beds. Energy Fuels 30(1):360–376, 2016

[3] A. Bakshi, C. Altantzis, R.B. Bates and A.F. Ghoniem, Study of the effect of reactor scale on fluidization hydrodynamics using fine-grid CFD simulations based on the two-fluid model, Powder Technology, 299: 185-198, 2016

[4] A. Bakshi, C. Altantzis, R.B. Bates and A.F. Ghoniem, Multiphase-flow Statistics using 3D Detection and Tracking Algorithm (MS3DATA): Methodology and application to large-scale fluidized beds. Chemical Engineering Journal, 293: 355-364, 2016

465981 Computational Modeling of Large Aspect Ratio, Fibrous Particles

Abstract:

In this poster, we present a computational study of mechanical behavior of fibrous particles. A bonded sphero-cylinder chain method is developed for the modeling of large aspect ratio, flexible fibers. This flexible fiber model is verified by examining a single fiberâ€™s deformations (i.e., bending, twisting, and stretching) subject to external loads. The simulation results of the single fiberâ€™s mechanical responses are in good agreement with the analytical solutions. To validate the numerical method for the modeling of a collection of fibrous particles with strong particle-particle interactions, the stress-controlled shear flows of fibers are simulated using the developed method, and the yield shear stresses obtained from the simulations are compared with those measured from the Schulze shear cell experiments.

COMPARING THE FLOW BEHAVIOUR OF FCC CATALYSTS USING THE FT4 POWDER RHEOMETER®

Tim Freeman[1], Jamie Clayton[1], Ray Cocco [2], S.B. Reddy Karri[2] Katrina Brockbank[1]
[1]Freeman Technology, 1 Miller Court, Severn Drive, Tewkesbury, Gloucestershire, GL20 8DN.
[2]Particulate Solid Research Inc., 4201 West 36th Street, Building A, Chicago, Illinois,

freemantechnology

PSRI

INTRODUCTION

The Fluid Catalytic Cracking (FCC) unit remains the primary hydrocarbon conversion unit in the modern petroleum refinery. It is widely used to convert the high-boiling, high-molecular weight hydrocarbon fractions of petroleum crude oils, for example straight-run atmospheric gas oil, vacuum gas oils and heavy stocks recovered from other operations to more valuable gasoline, olefin gases, and other products [1]. There are currently around 350 FCC units operating worldwide with a total processing capacity of over 14.7 million barrels per day [2]. The profitability of an FCC unit depends largely on the type of feed being processed and the FCC catalyst employed, with a typical unit processing up to 75,000 barrels per day and circulating over 55,000 tonnes of catalyst per day [3].

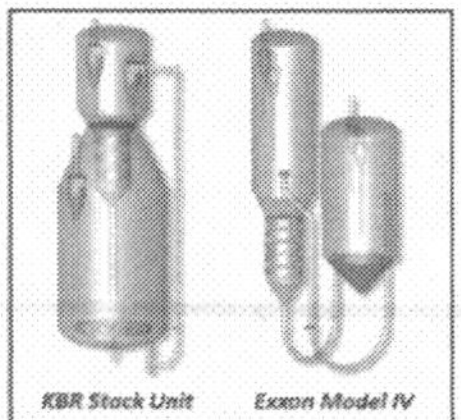

KBR Stack Unit *Exxon Model IV*

During the cracking process, the hydrocarbon vapours "fluidize" the powdered catalyst and the mixture of vapours and catalyst flows upwards entering the reactor at a temperature around 700°C and a pressure of approximately 1.72bar. The regenerated catalyst is then circulated back to the riser reactor. This is performed continuously at rates ranging from 5 to 25 tons per minute. It is therefore essential that the catalyst exhibits the desired fluidization behavior. This is especially true in the catalyst regenerator were poor fluidization can result in higher temperatures, lower productivity and higher NOx emissions. Furthermore, as the catalyst will undergo a large number of collisions during the process, the particles must also be resistant to attrition [4].

Modern FCC catalysts are fine powders with typical bulk densities of 0.80 to 0.96 g/cm³ and a particle size distribution ranging from 10 to 150 µm with an average of 60 to 100µm. The main active component is usually a stabilized form of zeolite Y but other components may include clay, alumina and silica sources as well as compounds for trapping metals [2, 4]. The components are typically mixed in aqueous slurry, and then spray-dried to form near uniform spherical particles.

In this study, FCC catalysts with differing fines content, were characterized using an FT4 Powder Rheometer® [5] with the aim of understanding the observed variation in process behavior, in particular, the propensity of some powders to from weakly bound clusters, which influence the entrainment rate due to differences in the aerodynamic properties compared with independent particles. Measurements included evaluating dynamic flow, bulk and shear properties of the powders including an assessment of fluidization behavior and permeability.

METHOD

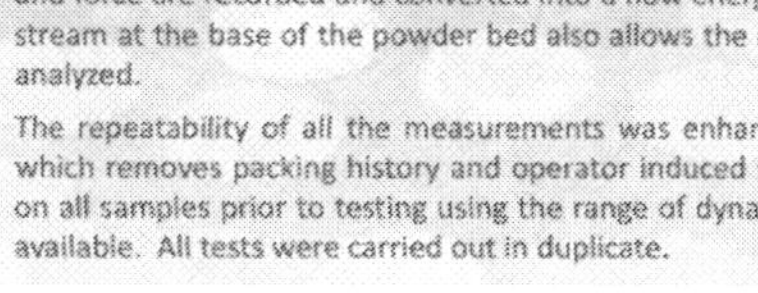

Three samples were evaluated which consisted of FCC catalyst powder with varying levels of fines concentration (<44 microns). The samples were Low Fines (3% fines), Medium Fines (12% fines) and High Fines (100% fines). Low Fines had been found to form weakly bound clusters which impacted on entrainment in fluidized beds.

An FT4 Powder Rheometer (Freeman Technology) was used to quantify the dynamic flow, bulk and shear behavior. The dynamic flow results are obtained by means of a patented measurement principle that evaluates the resistance to the motion of a specially shaped blade passing through a precise volume of the sample along a prescribed path. The required torque and force are recorded and converted into a flow energy. The introduction of a controllable gas stream at the base of the powder bed also allows the aeration behaviour of the samples to be analyzed.

The repeatability of all the measurements was enhanced by the use of a conditioning cycle which removes packing history and operator induced variability. Conditioning was undertaken on all samples prior to testing using the range of dynamic flow, bulk and shear measurements available. All tests were carried out in duplicate.

RESULTS

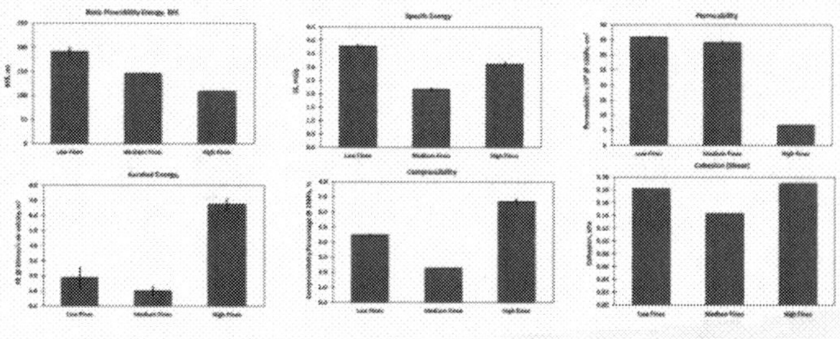

Fines/smaller particles typically have a larger specific surface area, resulting in increased inter-particular forces due to the increased contact area. As such, it would be expected that powders with a higher fines content might exhibit more cohesive behavior.

For High Fines, this was shown to be the case with this sample presenting the highest Aerated Energy (AE) and Cohesion (as derived from Shear Cell tests).

This sample also exhibited the least efficient particle packing structure as demonstrated by the higher Consolidation Index (CI) and Compressibility values, and lower Permeability and Conditioned Bulk Density (CBD) values. This would also explain the lower Basic Flowability Energy (BFE) as air voids in the bed offer minimal resistance to the passage of the blade.

Based on this hypothesis it would also be expected that Low Fines would be the most free-flowing. However, Medium Fines was demonstrated to have more efficient particle packing with the lowest Compressibility. This may be due to reduced mechanical interlocking and friction as indicated by the lower Specific Energy (SE) and Cohesion values, allowing particles to move more readily with respect to each other in order to occupy the available space

In optimal quantities, fines can act as a flow aid/lubricant, possibly explaining the improved behavior of Medium Fines. By coating the particles, fines can "smooth" the surface texture, reducing friction, as well as break cohesive bonds. This may inhibit the formation of stable agglomerates/clusters as the coated particles are less likely to interlock with each other, explaining why Medium Fines formed fewer clusters compared with Low Fines.

CONCLUSIONS

Without understanding the underlying nature of the formation of particle clustering it is difficult to capture the hydrodynamics of a fluidized bed. These results show that specific concentrations the fine particle fraction can actually improve flowability, inhibiting the formation of weakly bound clusters which influence the entrainment rate. This is most likely due to the fine particle fraction acting as flow aid/lubricant, reducing mechanical interlocking and cohesion.

This demonstrates the complex relationship between the physical powder properties and flow behavior, making powder behavior extremely difficult to predict. As such it is necessary to characterize powder behavior using a method(s) which reflects the conditions to which the powder is subjected to in process.

The wide range of test methodologies provided by the FT4 allows powders to be tested under a range of stress and flow regimes, enabling clear and repeatable differences to be quantified, and rationalising the variations in process behavior.

REFERENCES
[1] Coker, A.K., Ludwig's applied process design for chemical and petrochemical plants. (Elsevier Science), 2014.
[2] Sadeghbeigi, R., Fluid catalytic cracking handbook: An expert guide to the practical operation, design, and optimization of fcc units. (Butterworth-Heinemann), 2012.
[3] Gary, J.H. Handwerk, G.E., Petroleum refining. (Taylor & Francis), 2001.
[4] Vogt, E.T.C. Weckhuysen, B.M., Fluid catalytic cracking: Recent developments on the grand old lady of zeolite catalysis. 2015.44.7342-7370.
[5] Freeman, R., Measuring the flow properties of consolidated, conditioned and aerated powders — a comparative study using a powder rheometer and a rotational shear cell. Powder Technol.2007;174:25-33.

469484 Ink Formulation and Direct Ink Writing of Thermites

Abstract:

3D printing is a rapidly emerging technology involving the layer-by-layer deposition of material to form a final part. This technique has great potential for reactive materials, in that dynamic behavior may be controlled through the strategic placement of material into features within a part. Direct ink writing (DIW) is an extrusion-based method of 3D printing, involving the room temperature extrusion of custom inks through micronozzles. Since heating is not required, this technique is suitable to print reactive materials such as pyrotechnics or thermites. Here we investigate the formulation of both aluminum and copper oxide based inks for the direct printing of thermite. Choices of volume loading, surfactant concentration, and matrix solvent are all critical with regard to establishing an ink capable of shear thinning, but also rapidly setting upon extrusion. Without the latter behavior, the ink would sag upon printing and would not produce a quality part. The formulation and rheological properties of the inks are reported here. We demonstrate the ability of each material to be extruded for 3D printing and discuss several parameters important to the process. Additionally, the components are mixed with a custom mixing head for direct printing of the thermite composite. This work was performed under the auspices of the U.S. Department of Energy by Lawrence Livermore National Laboratory under Contract DE-AC52-07NA27344. LLNL-ABS-691141.

460263 Production of Nano-Sized Aluminum through Cryogenic Milling

Abstract:

There is tremendous interest in nano-sized aluminum powder for numerous energetic applications such as liquid and solid propellant additives. The challenge however, is the selection of a scalable process that produces nano-sized aluminum powder that is stable for storage, while energetically favorable in desired uses. Thus, protection or passivation of the surface of nano-aluminum particles without substantially reducing their energy density is another significant problem. Cryogenic mechanical milling was investigated here as a novel method for the production of aluminum nanopowders from conventional micron-sized aluminum feedstock. Aluminum is milled in solid hydrocarbons at cryogenic temperatures, and remains in the organic liquid after recovering at ambient temperatures. The nanopowder is never exposed to any atmosphere. If it is used as an additive for liquid fuel formulations, it does not need to be specifically passivated. The prepared materials are characterized using SEM, ICP-MS, UV-VIS, laser scattering and other techniques. Combustion and ignition studies are conducted and results from experiments are discussed in detail.

454051 Effect of Polar Environments on the Aluminum Oxide Shell Surrounding Aluminum Particles:

Abstract:

Density functional theory (DFT) calculations were performed to understand molecular alterations of a polar environment on an alumina surface. The analysis has strong implications for the reactivity of aluminum (Al) particles passivated by an alumina shell. Recent studies have shown a link between the solvation media used for Al powder intermixing and the reactivity of Al with fluorine containing reactive mixtures. Flame speeds show a threefold increase when polar liquids are used as the carrier fluid when intermixing aluminum and fluoropolymer powder mixtures. It was hypothesized that the alumina lattice structure could be transformed due to the hydrogen bonding forces exerted by these solvents that induce modified bond distances and charges that influence reactivity. In this study, the alumina surface was analyzed by using DFT calculations and model clusters as isolated systems embedded in polar environments (acetone and water). Five defect models for specific active â€"OH sites were investigated in terms of structures and vibrational â€"OH stretching frequencies. The observed changes of the surface OH sites invoked by the polar environment were compared to the bare surface. The calculations revealed a strong connection between the impact of carrier fluid polarity on the hydrogen bonding forces between the surface OH sites and surrounding species. Changes were observed in the OH characteristic properties such as OH distances (increase), atomic charges (increase), OH stretching frequencies (decrease); and these consequently improve OH surface reactivity. The difference between medium (acetone) and strong (water) polar environments was minimal.

469720 Silicon Particles for on-Chip Energetic Material Applications

Abstract:

Silicon has found increased use as a fuel in energetic materials, largely due to the ability to porosify crystalline silicon and form a high surface area structure. The porous structure, which can also be tuned to higher or lower porosity to tune the reaction, allows for an intimate mixture between fuel and oxidizer allowing for (at times) a very rapid reaction. With several emerging methods for on-chip energetic material deposition including printing, electrophoretic deposition, or evaporation driven assembly, the ability to integrate energetic particle formulations on-chip is becoming more and more intriguing. This study investigates the integration of silicon particles on-chip, with the end goal of creating unique on-chip porous formations. Multiple fabrication methods were utilized, including deposition of porous particles, or deposition of crystalline particles with a secondary porosification step. Interest also lies in creating silicon energetics with metal oxide nanoparticle oxidizers, rather perchlorates which are often used. Analysis includes BET porosimetry, SEM, DSC, and high speed imaging.

EVAPORATION DRIVEN ASSEMBLY OF ON-CHIP NANOTHERMITE DEVICES

Matthew H. Ervin, Sarah S. Bedair, Nicholas W. Piekiel, Cory R. Knick, Brian Isaacson and Harvey Tsang
US Army Research Laboratory, Adelphi, MD

Introduction

Energetic nanomaterial systems such as nanoporous silicon with sodium perchlorate as well as nanoparticulate aluminium/metal-oxide thermites have been developed for many applications. In general, these applications are utilizing the mechanical force, heat, light, gases or reaction products that result from the energetic reactions. These nano-energetics have been demonstrated for explosives initiation (primers[1,2] & airbag initiators[3]), thermoelectric pulse power generation,[4] microfluidic micropump actuation,[5] intercellular biomolecule delivery,[6] microthrusters,[7,8] emission spectroscopy,[9] and other[10] applications. Porous silicon devices have useful features such as being easily integrated on-chip using standard microfabrication techniques[11] and being tailorable to exhibit few m/s to few km/s burn rates.[12] However, the fabrication of these devices is dependent on the conductivity and doping of the silicon substrate used and require the exposure of the wafer to a very corrosive etchant.[13] Nanothermites, on the other hand, may be deposited onto any substrate using a number of methods such as: drop casting,[14] electrophoretic deposition,[15] electrospray,[16] or printing.[17] Here we have used evaporative driven assembly (EDA) to fabricate on-chip nanothermite devices.[18,19]

In EDA, a volume of a solution/suspension is deposited into a well connected to a micromachined trench or directly onto the trench itself. The solution gets drawn into the trench by capillary action and the solution meniscus in the trench results in locally enhanced evaporation of the solution.[20] The evaporation of the solvent in the trench deposits the suspended nanoparticles in the trench, while drawing more solution into the trench. This process continues until the trench has been filled with the particles, which in our case are nanoparticles of aluminium (Al) and copper oxide (CuO). EDA is attractive as it is an additive process that allows very thick features to be fabricated while using/wasting very little material.

Experimental

The thermite suspension used in this work are made using 70 nm Al (Novacentrix) and <50 nm CuO (Sigma Aldrich) particles suspended in dimethyl formamide (DMF) from Sigma Aldrich. The CuO:Al mass ratio, assuming a 70% activity for the Al (surface oxide), should be 3.1:1 according to the reaction stoichiometry. We have used a slightly Al-rich mixture of 2.5:1 here. The thermite suspension is made with a mass loading of 50 mg of thermite particles in 0.5 ml of DMF with probe sonication for 20 minutes immediately before use.

The substrates used were silicon wafers with trench structures etched into them using a standard deep reactive ion etch performed with a PlasmaTherm VLN Deep Silicon Etcher. The trenches were 15, 25, or 50 μm wide, 45 or 100 μm deep, and 10 or 25 mm long. Some trenches had attached wells while others had no associated well. Suspended trenches, for thermal isolation, were fabricated by DRIE etching 100 μm deep trenches and isolation structures from the topside, and then continuing etching the isolation structures 400 μm deep from the backside resulting in released trenches. In some cases, the released trenches were fabricated with 100 μm thick, 3 x 3 mm paddles attached to them in order to have a large enough area for accurate temperature measurement using the IR camera. Fig. 1 shows an example EDA trench structure with an associated solution well. The trench to be filled is surrounded by another trench used as a reference for the optical profilometry measurements after nanoparticle deposition. Typically, a few μl of thermite suspension are hand pipetted at the mouth or onto the end of the trench, with additional depositions as needed to fill the trench. The longer trenches do not necessarily fill to their complete length, with the fill length determined by the solution properties and trench geometry.

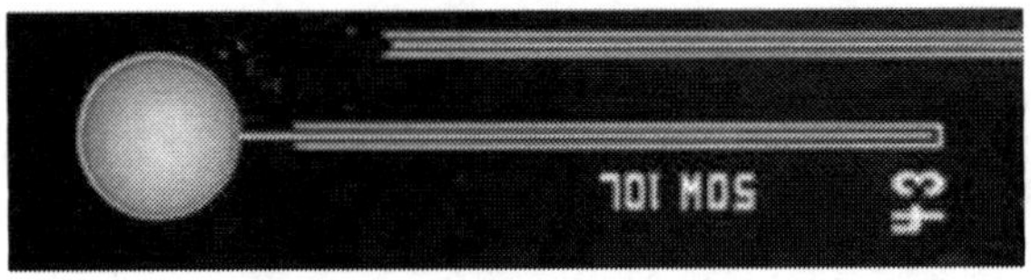

Figure 1: EDA structure with a well on left and an attached 50 wide, 45 μm deep, and 10 mm long trench (surrounded by another trench for a depth reference).

The nanothermite devices were imaged using optical and scanning electron microscopy (SEM), the later performed with a Hitachi S4500 SEM. Chemical analysis of the nanothermite was done using an FEI Quanta 200 F SEM equipped with an Oxford INCA X-sight energy dispersive x-ray (EDX) detector. The depth of the trench filling was measured using a Wyko NT1100 optical profilometer. Micro x-ray computer aided tomography (CT) was performed using a Zeiss Xradia 510 Versa system. The burn rates of the thermite devices were characterized using a Photron FAST-CAM SA5 high-speed camera, and a FLIR model A40 mid-IR camera was used to measure temperature changes. These thermite devices were ignited using a battery operated barbeque ignitor wired to two probe tips, one of which contacted the substrate while the other probe tip was located over the thermite filled trench.

Results and Discussion

When the nanothermite suspension is deposited in the well of the device shown in Fig. 1, or on the end of a similar device that does not have a well, the suspension proceeds to wet the device channel and deposit thermite in the channel from the tip back to the well. The deposited thermite looks as shown in the SEM images in Fig 2. Fig 2A shows the end of the trench filled with thermite and the surrounding empty trench used as a reference for the thermite thickness measurement. Fig 2B shows a middle section of the device that reveals some cracking of the thermite deposit. At this point, cracking of the deposit is not unusual since a binder material was not used in the suspension. Fig 2C is a close-up of the thermite deposit showing the structure and residual porosity of the deposited thermite.

The spherical particles are the Al while the irregularly shaped particles are the CuO. EDX analysis performed at 20 keV sampling 10 x 30 μm areas along the length of a 10 mm long device showed significant variation of the Al to CuO ratio. In a localized area, the atomic percent compositions vary by only a couple percent, but over the length of the device the Al:Cu ratio may vary from 1.5, near the well, to 0.57 at the end of the trench. This could be the result of the Al and CuO NPs not forming a stable suspension, and that they are settling out at different rates either in the vial or during the EDA. More work needs to be done to determine if EDA results in any separation of components during the deposition process and if this can be prevented with more stable particle dispersion.

The deposited thermite is tightly packed enough that a 3 mm segment of one device was observed to be self-supporting after having, in a rare occurrence, pulled out of the device channel as shown in Fig 3. Optical profilometry shows how thick the deposited thermite is when using a 25 μm wide and 45 μm deep trench flanked on either side by empty channels. The thermite deposited in this device is found to be 40 μm thick, as shown in Fig 4. Using an additive manufacturing method such as EDA greatly reduces the waste, not to mention the hazard, associated with making such a thick device. Such device thicknesses are required as devices that are too thin will succumb to thermal quenching that will not allow the reaction to propagate along the device.

Micro-CT images shown in Fig 5 reveal that the deposited thermite contains no voids (empty included volumes significantly larger than the particle size) although it may have cracks that form as the deposit shrinks during drying. Frequently, the thermite either pulls away from one trench sidewall or it cracks down the middle as it dries. The micro-CT also shows that there are some larger particles along the bottom of the deposit which are presumed to be settling out of the suspension during the deposition process due to their larger size.

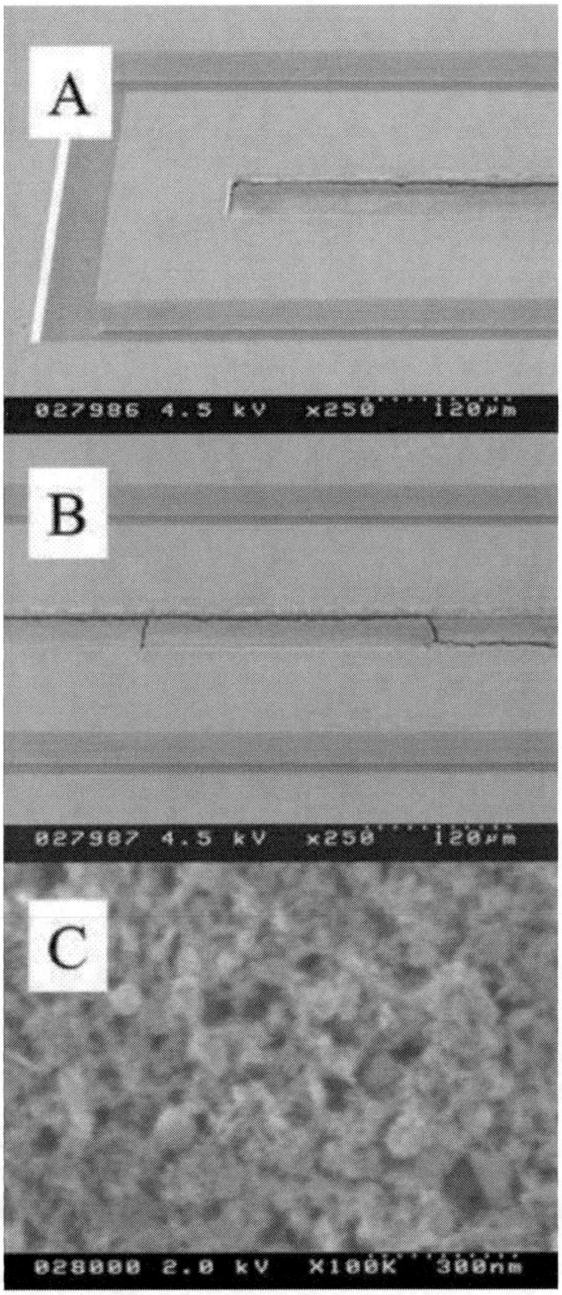

Figure 2. A) Thermite filled 45 µm deep trench end, B) portion of filled trench showing cracks in the thermite, C) high magnification image of deposited thermite.

While filming the burning of the thermite in these devices, burn rates of 1-15 m/s have been observed. Fig 5 is a still image from an ignition video. The probe tip on the left was used to ignite the device. The small flame 10 mm to the right of the probe tip is the thermite in the trench burning away from the ignition point. The burn rates measured with the high speed camera not only show variation between devices, but within devices as well. While some sections burn at a constant rate over mm's, other sections may burn more erratically. Some of this erratic behaviour is likely due to the cracks in the thermite. An appropriate binder that will eliminate the

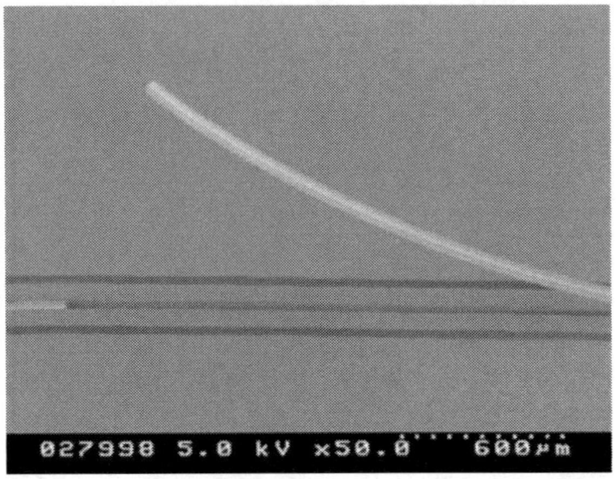

Figure 3. Delaminated portion of deposited thermite showing that it is well packed enough to be self-supporting.

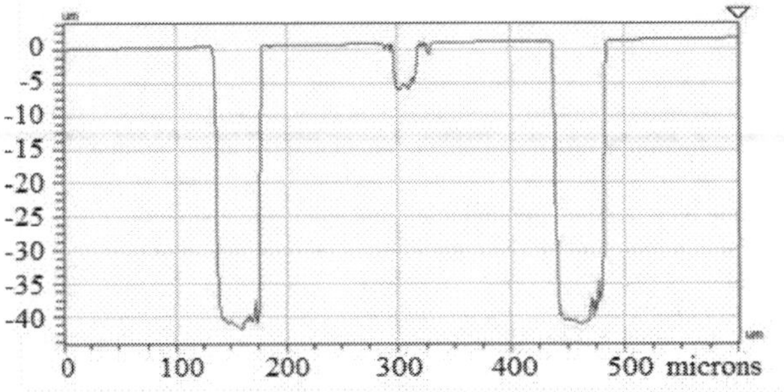

Figure 4. Optical profilometry plot showing a cross-section of the thermite-filled trench in between the surrounding empty trenches. The thermite is ~35 µm thick in this location.

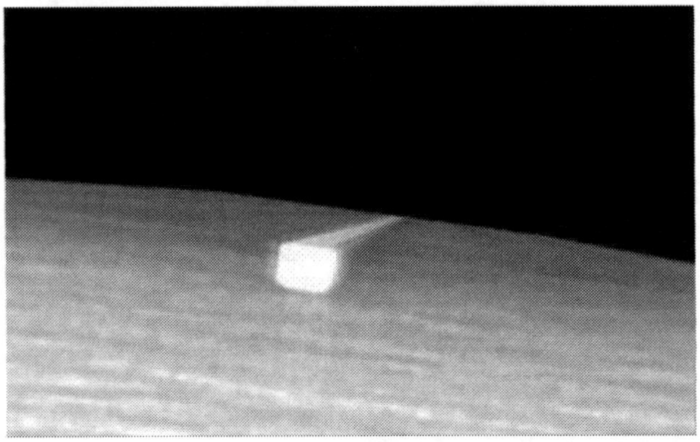

Figure 5. Micro-CT showing a cross-section of the thermite-filled trench.

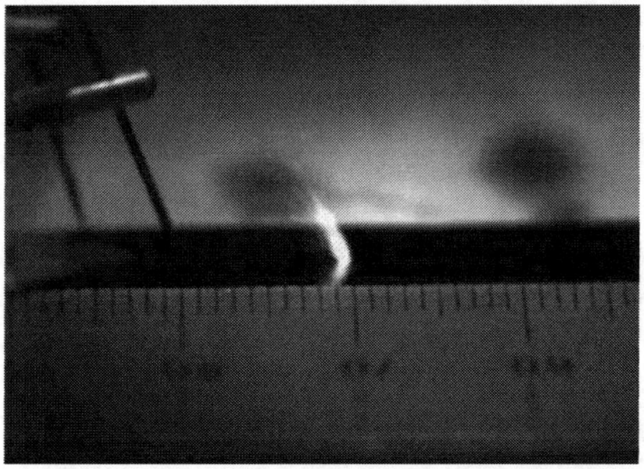

Figure 6. Still image from high-speed video showing the thermite burning along the trench away from the ignition point at the probe tip.

cracking is needed. Other factors affecting the burn rate along a device may include variations in the thermite thickness[15] as well as the local variations in the Al to CuO ratios. More homogeneous mixing of these materials may improve the burn consistency. The slow burn rate may be indicative of the packing density of the nanothermite as it has been found that at higher packing density nanothermite burns at slower rates due to being in a thermal conduction burn regime compared to the faster burn rates achieved with lower packing densities that result in convective flame propagation.[21,22]

In order to determine how high a device may be heated by these thermite devices, an IR camera was used to film the backsides of burning devices. The backside is filmed to minimize measuring the thermal effects of the fireball above the sample surface. Fig 7 is a still frame from one such video. In order to maximize the temperature rise, suspended trench structures were fabricated which thermally isolate the trench from the rest of the substrate. It is apparent that the temperature rise of the device is significantly lower than would be predicted from the expected energy release from the thermite. This is in large part due to the energy carried away from the device by the gaseous reaction products. In order to improve on this, devices that confine the hot gases next to the devices for longer times are being investigated as are reactants that produce fewer gaseous products. The highest temperature rise measured to date is 25.3 C which was for a suspended 200 μm wide 100 μm deep trench. The actual temperature rise is certainly larger since these suspended cantilevers were not wide enough for the camera areal measurement resolution, so that the cold background on either side of the cantilever was averaged into the measurement.

Figure 7: IR image of a suspended silicon structure heated by burning the themite in a trench running across it.

Conclusions

On-chip nanothermite devices yield some of the same functionalities as porous silicon, but with greater substrate flexibility. EDA has been demonstrated as a suitable method for making devices 10 μm thick, 10's of μm wide, and mm's long while using a minimum of material. Few m/s burn rates have been demonstrated. These initial devices have shown regions of constant burn rates while other areas are more erratic which is believed to be due to cracking in the deposits. Binders need to be investigated for eliminating these cracks. Heating of devices in the 10's of degree C have been achieved. In order to achieve even greater device heating, designs for increasing the time the hot gases are in contact with the devices are being investigated as are reactants that produce fewer gaseous products.

References

1. Glavier, L; et al. (2015) "Nanoenergetics as pressure generator for nontoxic impact primers: comparison of Al/Bi_2O_3, Al/CuO, Al/MoO_3 nanothermites and Al/PTFE," Combustion and Flame, 162, p. 1813.

2. du Plessis, M. and Conradie, C. (2006) "Nano-explosions in porous silicon," in Device and Process Technologies for Microelectronicsk, MEMS, and Photonics IV, J. C. Chiao et al. eds, Proc SPIE, 6037, p. 1.

3. Cle´ment, D.; et al. (2005) "Highly explosive nanosilicon-based composite materials," Phys Status Solidi A, 202, p. 1357.

4. Puszynski, J. A.; et al. (2012) "Thermoelectric pulse power generation using self-sustaining gasless nanoreactant systems," Intl J Energetic Materials and Chemical Propulsion, 11, p. 389.

5. Rodriguez, G. A. A.; et al. (2009) "A microactuator based on the decomposition of an energetic material for disposable lab-on-chip applications: fabrication and test," J Micromechanics and Microengineering, 19, p. 8.

6. Korampally, M.; et al. (2012) "Transient pressure mediated intranuclear delivery of FITC-Dextran into chicken cardiomyocytes by MEMS-based nanothermite reaction actuator," Sensors and Actuators B-Chemical, 171, p. 1292.

7. Churaman, W. A.; et al. (2012) "The first launch of an autonomous thrust-driven microrobot using nanoporous energetic silicon," IEEE J Microelectromech Syst, 21, p. 198.

8. Staley, C. S.; et al. (2015) "Fast-Impulse Nanothermite Solid-Propellant Miniaturized Thrusters," J Propulsion and Power, 29, p. 1400.

9. Mikulec, F. V.; Kirtland, J. D. and Sailor, M. J. (2002) "Explosive nanocrystalline porous silicon and its use in atomic emission spectroscopy," Advanced Materials, 14, p. 38.

10. Zhou, X.; et al. (2014) "Nanostructured Energetic Composites: Synthesis, Ignition/Combustion Modeling, and Applications," ACS Applied Materials & Interfaces, 6, p. 3058.

11. Ervin, M. H.; Isaacson, B. and Levine, L. B. (2015) "Process for integration of energetic porous silicon devices," in Proc PowerMEMS 2015, J of Phys Conf Ser, 660, p. 012067.

12. Piekiel, N. W.; et al. (2015) "Combustion and material characterization of highly tunable on-chip energetic porous silicon," J Propellants Explosives Pyrotechnics, 40, p. 16.

13. Korotcenkov, G. and Cho, B. K. (2010) "Silicon porosification: state of the art," Critical Reviews in Solid State and Material Science, 35, p. 153.

14. Nellums, R. R.; et al. (2014) "Preparation and Characterization of Aqueous Nanothermite Inks for Direct Deposition on SCB Initiators," Propellants Explosives Pyrotechnics, 39, p. 463.

15. Sullivan, K.; et al. (2012) "Electrophoretic deposition and mechanistic studies of nano-Al/CuO thermites," J Applied Physics, 112, p. 024316.

16. Wang, H. Y.; et al. (2014) "Assembly and reactive properties of Al/CuO based nanothermite microparticles," Combustion and Flame, 161, p. 2203.

17. Tappan, A. S.; Cesarano, J. and Stuecker, J. N. (2011) "Thermite ink useful in inkjet, screen and gravure printing, comprises aluminum particles, oxidizer particles, a dispersant and a non-aqueous dispersing medium," US Pat. US8048242-B1.

18. Bedair, S. S.; et al. (2010) "Magnetic Nanoparticle Dosing of MEMS Structures by Solvent Evaporation in Capillaries," IEEE Transactions on Magnetics, 46, p. 2198.

19. Smith, G. L. and Bedair, S. S. (2014) "Apparatus and method for suspension wicking of nanoparticles into microchannels" US Pat. 8728409.

20. Bedair, S. S. and Fedder, G. K. (2009) "Picogram material dosing of microstructures," J of Applied Physics, 106, p. 104913.

21. Apperson, S. J.; et al. (2009) "Characterization of Nanothermite Material for Solid-Fuel Microthruster Applications," J. Propulsion and Power, 25, p. 1086.

22. Pantoya, M. L.; et al. (2009) "Effect of Bulk Density on Reaction Propagation in Nanothermites and Micron Thermites," J. Propulsion and Power, 25, p. 465.

471480 On the Fly Mixing and 3D Printing of Al/CuO Thermite

Abstract:

The ability to spatially control the behavior of reactive materials *within* a part is now a reality with advances in 3D printing. This vastly opens up the design space for rapidly deflagrating materials, such as pyrotechnics or thermites, to yield a precise property or dynamic performance. In order achieve this goal, a mixing print head was used to mix an aluminum and a copper oxide ink on the fly. The mixing and printing parameters were first investigated for a stoichiometric mix of fuel and oxidizer to determine at what point the material can be assumed well-mixed. The equivalence ratio was then changed, and the critical mixing parameters established. The reactivity was characterized by printing a strip of material, then initiating the thermite and measuring the propagation velocity with a high-speed camera. Once the velocity reached a plateau, we considered the system well mixed. 3D printing was then used to make parts where the local stoichiometry, which corresponds to performance, is spatially varied. Collective effects of having incorporated features with differing reactivates were investigated.

This work was performed under the auspices of the U.S. Department of Energy by Lawrence Livermore National Laboratory under Contract DE-AC52-07NA27344. LLNL-ABS-691142

Emanuel Marsis, PhD.; Tudor Ionescu, PhD.; Paul Stead; Scott Lehrer; Sunder Ramachandran, PhD.

Outline

- Motivation
- Introduction
  - H_2S Scavenger
  - Computational Fluid Dynamics
- Test Results
- CFD Model
- Discussion & Next Steps
- Conclusion

Motivation

- Accurate prediction of H_2S concentration during the scavenging process is important to highlight any risk associated with high localized H_2S quantities

- Factors affecting the dynamics of H_2S scavenging:
  - Fluid phases
  - Fluid dynamics
  - Injection rate
  - Scavenger particle diameter
  - Reaction rates
  - Mass transfer

Introduction (H_2S Scavenger)

- Risk of having H_2S in a production system
- It can be present in either gas or liquid
- H_2S Scavengers are application sensitive
- An accurate model of H_2S scavengers is needed to minimize the risk on structure integrity and environmental impact due to H_2S production
- Triazine is a widely used H_2S scavenger, but it can be slow acting
- A fast acting H_2S scavenger was developed, tested and modeled

Introduction (Computational Fluid Dynamics)

- CFD Applications
  - Fluid Dynamics
  - Multiphase Flow
  - Chemical Reactions
  - Moving Geometry (impellers)
  - Thermodynamics
  - 2D & 3D
  - Steady State & Transient

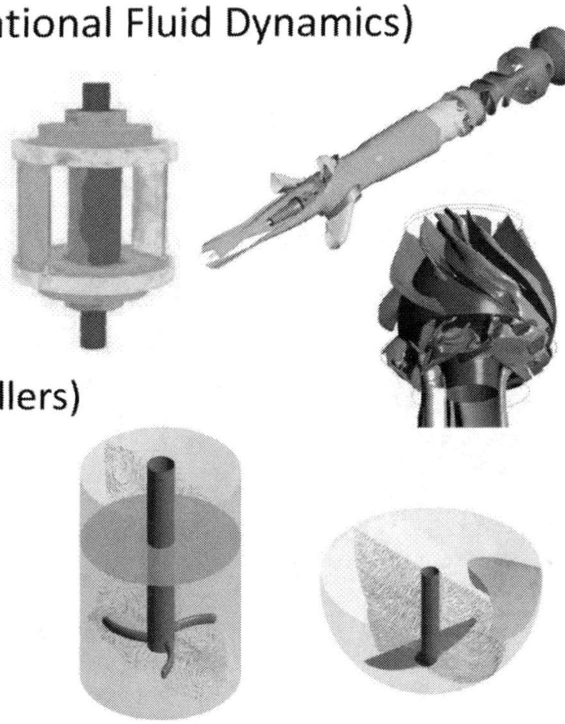

Test Setup

- **Features**
  - Oil / brine/ mixed phase
  - Continuous mixed gas flow
    - N_2 (dilution gas), CO_2, H_2S
  - Stirred
  - Continuous H_2S measurement
  - Heated (atmospheric pressure)
- **Benefits**
  - Relatively time efficient
  - Measures scavenger efficiency & kinetics

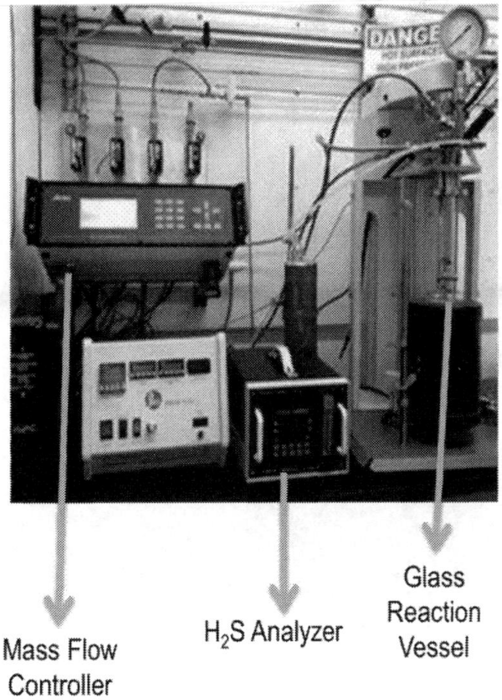

Test Setup

Test Setup

Reactor

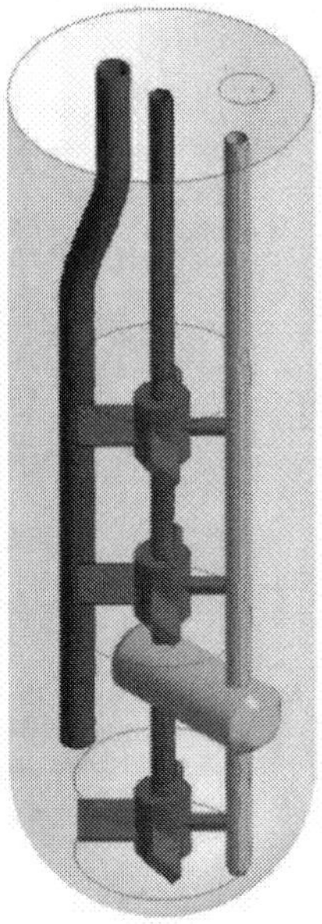

3D Computer Model

Test Results

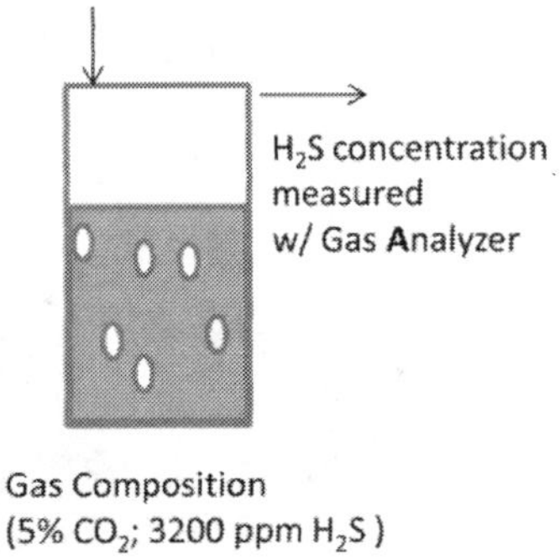

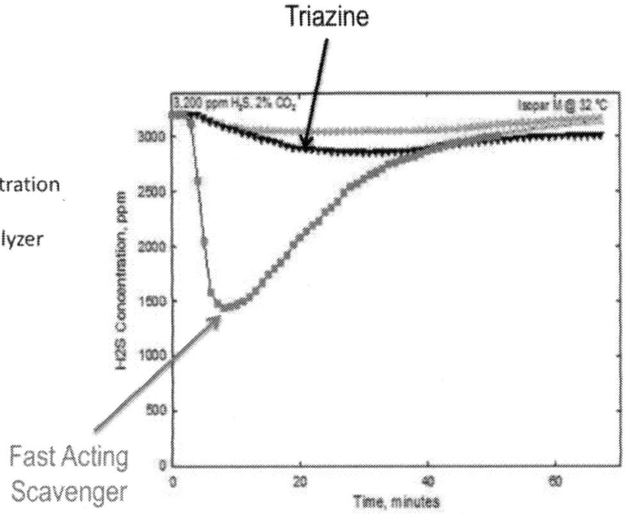

CFD Model

- Full 3D Transient
- Moving Mesh (Rotating Impellers)
- Eulerian Multiphase
- Species Model
- Mass transfer of H_2S from liquid to gas phase
- Chemical reaction takes place in the liquid phase only
- Polyhedral mesh

CFD Model

- All the species were defined separately for each phase
- Mass transfer rate of H_2S was obtained from the literature
- The slope of H_2S Concentration with time was obtained from the CFD model
- The reaction rate was determined in the CFD model by matching experimental data with CFD data

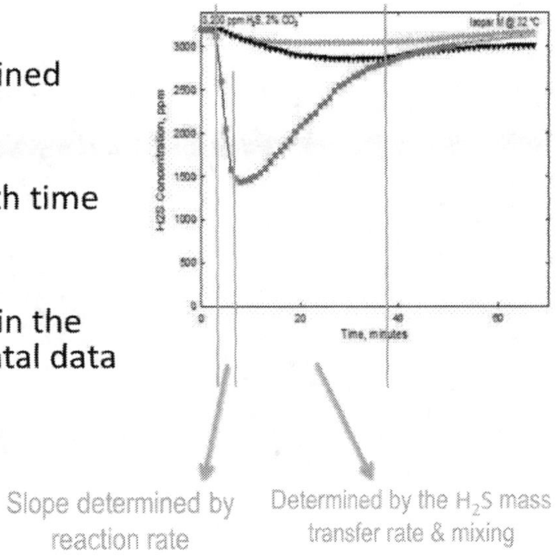

CFD Model

- Gas Volume Fraction

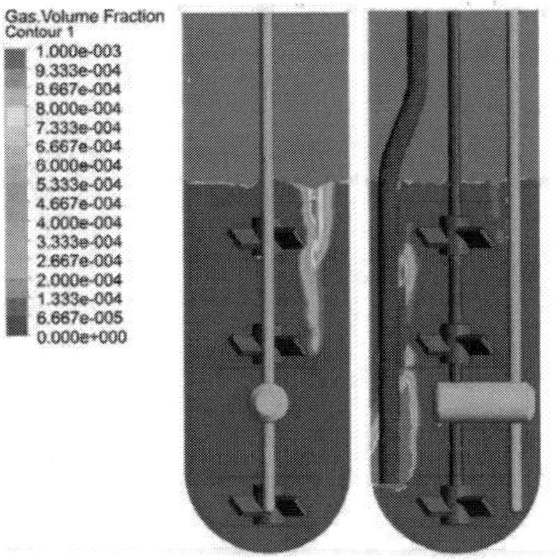

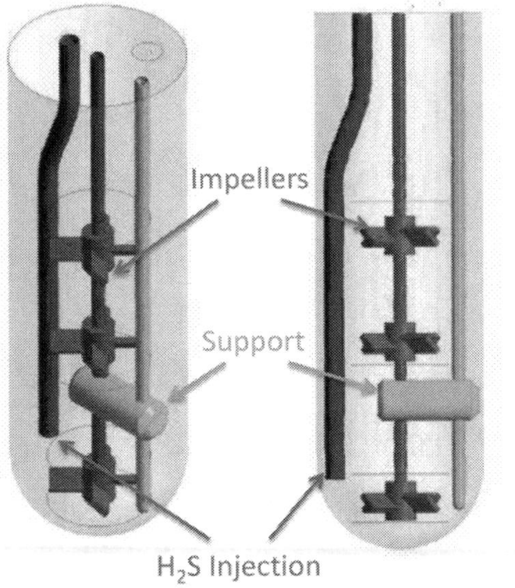

CFD Model

- H_2S Injection

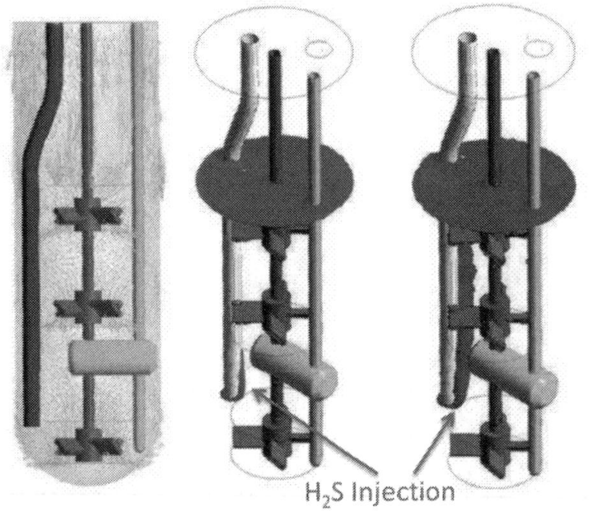

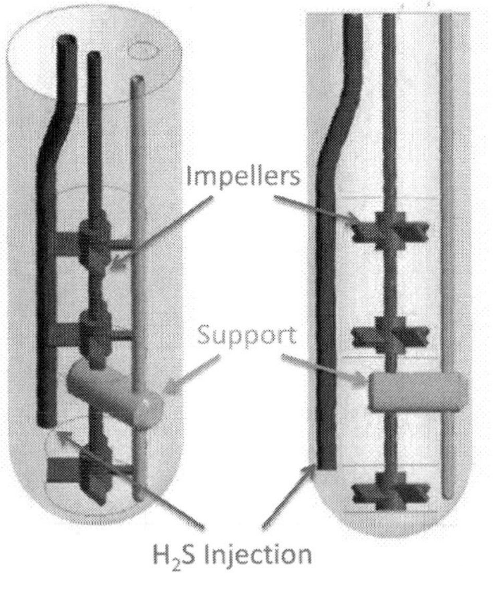

CFD Model

- Injecting the Fast Acting H_2S Scavenger

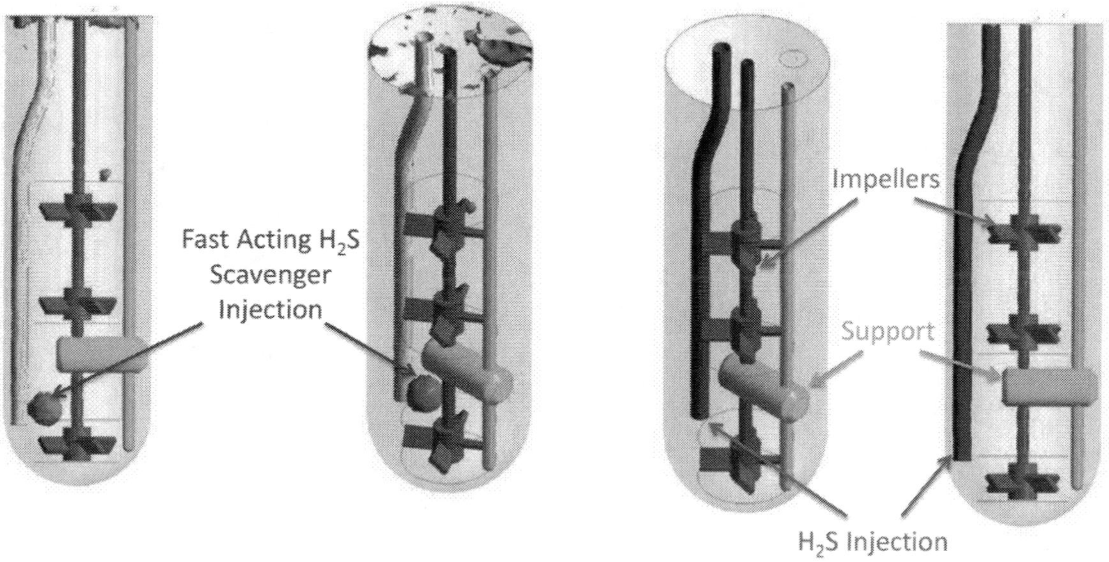

CFD Model

- Monitoring the H_2S Concentration vs Time
- Different CFD models represent different kinetic rates
- Kinetic rate in CFD Model #3, shown in the graph, matches the experimental results
- Limitations on the model
  - Time step size

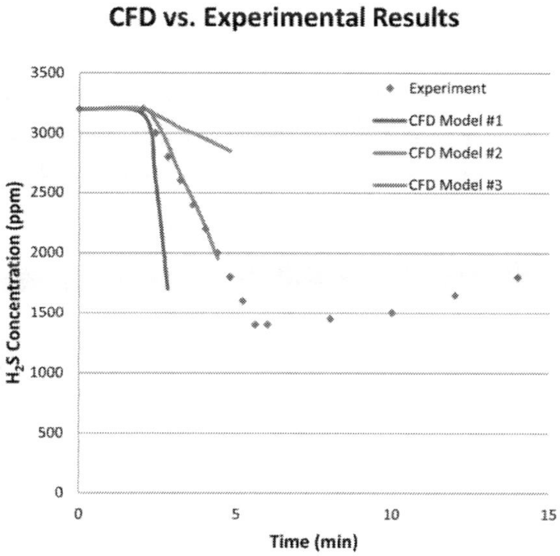

Discussion & Next Steps

- H_2S Concentration is dependent on:
  - Reaction rates
  - Fluid dynamics (rpm of the impeller, gas injection rate, mass transfer between the phases)
- By modeling the whole setup, an accurate prediction of the kinetic rate was carried out
- The validated model can be used in complicated field applications; like valves, wellheads, flowlines, or pipelines to predict the concentrations of H_2S at different locations
- Next Step is to use the validated model to predict H_2S Scavenging in a gas line

Discussion & Next Steps

- A production tubing was modeled using a species model
- The reaction rates computed earlier using the lab test were used to show the localized concentration of hydrogen sulfide in the production tubing

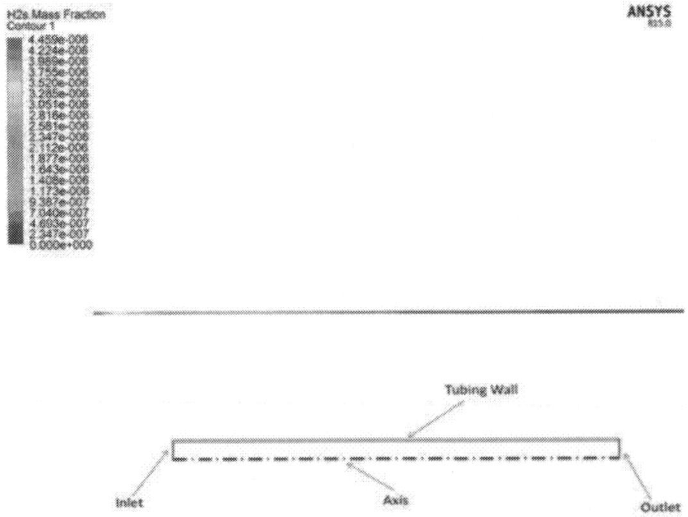

Discussion & Next Steps

- CFD allows the use of Discrete Phase Models (DPM), which is a Lagrangian particle tracking model to simulate H_2S Scavenger particles flowing in a gas line
- The mass transfer from the gas to the liquid particle was identified
- The reaction is taking place only in the liquid phase

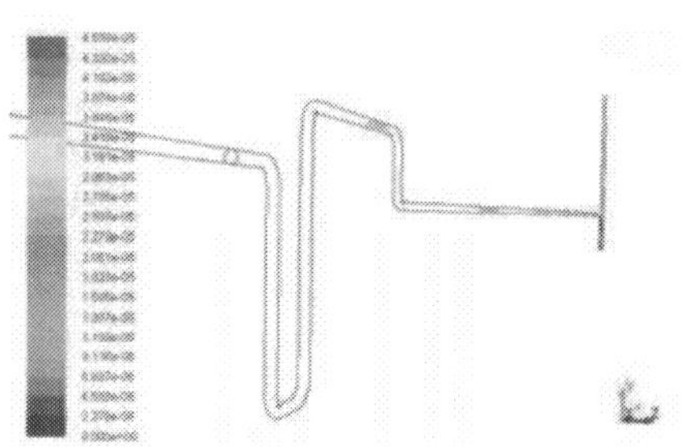

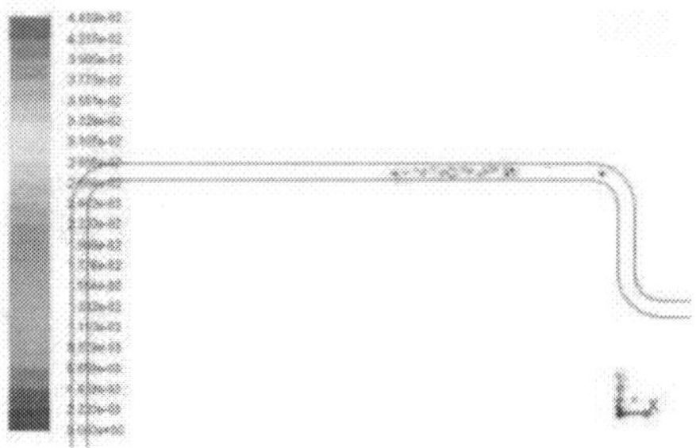

Conclusion

- Fast Acting H_2S Scavenger was tested in the lab and compared to Triazine Scavengers at the same concentrations
- The Fast Acting H_2S Scavenger has much higher kinetic rate compared to Triazine
- CFD was used to build a 3D model of the test setup and predict the reaction rates including the effect of fluid dynamics
- CFD is capable of modeling multiphase flow with chemical reactions and multiple species including mass transfer between the phases
- CFD modeling can be used to simulate complicated production problems and predict H_2S concentration in a given geometry using the fast acting H_2S Scavenger

QUESTIONS

Computational study of the bubbling-to-slugging transition in a laboratory-scale fluidized bed

Emilio Ramirez[1,2], C. Stuart Daw[1], Charles E. A. Finney[1], Sreekanth Pannala[3], Jack Halow[4], Qingang Xiong[5]

[1] Oak Ridge National Laboratory, Oak Ridge TN 37831 USA
[2] University of Tennessee, Knoxville TN 37996 USA
[3] SABIC Americas, Sugar Land TX 77478 USA
[4] Separation Design, Garards Fort, PA 15334 USA
[5] FCA US LLC, Auburn Hills, MI 48326 USA

Presented at the 2016—AIChE Meeting
Group/Topical: Particle Technology Forum

San Francisco, California
November 16, 2016

OAK RIDGE National Laboratory | NATIONAL TRANSPORTATION RESEARCH CENTER

Acknowledgements

This research was supported by the U.S. Department of Energy Bioenergy Technologies Office (BETO). The authors would like to thank BETO sponsors Jeremy Leong, Cynthia Tyler, and Kevin Craig for their guidance and support.

NETL – Tingwen Li, Rupen Panday, William Rogers

ORNL - Gavin Wiggins, Maggie Connatser, Mike Kass, Jae Soon Choi, John Turner

MIT - Akhilesh Bakshi, Christos Atlantzis

NREL reactor experimentalists
Rick French, Danny Carpenter, Kristiina Ilsa, Mark Jarvis, Esther Wilcox, Kristin Smith, Katherine Gaston

University of Tennessee - Bredesen Center for Interdisciplinary Research and Graduate Education

Why are we interested in slugging?

"The arrival time of a space probe traveling to Saturn can be predicted more accurately than the behavior of a fluidized bed reactor!"
- Derek Geldart

"The **art** of engineering science is the art of seeing which steps should be taken to erase from real life (hopefully) irrelevant details which make it so complex, to look at the essential aspects of it, or, in other words, to decide what is the appropriate meta-problem to look at: to conceive the appropriate chain of idealization."
- Gianni Astarita and Julio Ottino

Are there challenges with slugging correlations?

Literature slugging correlations vary between studies
Possibly due to:
- qualitative methods used to identify slugging

Quantitative methods can provide a robust method to determine:
- onset of bubbling-to-slugging transition
- fully developed slugging

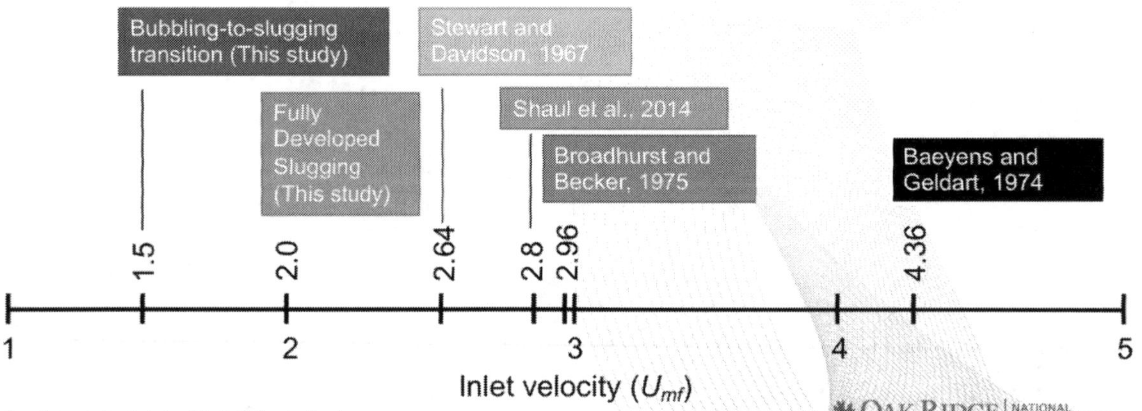

How does slugging affect biomass pyrolysis oil yield & composition?

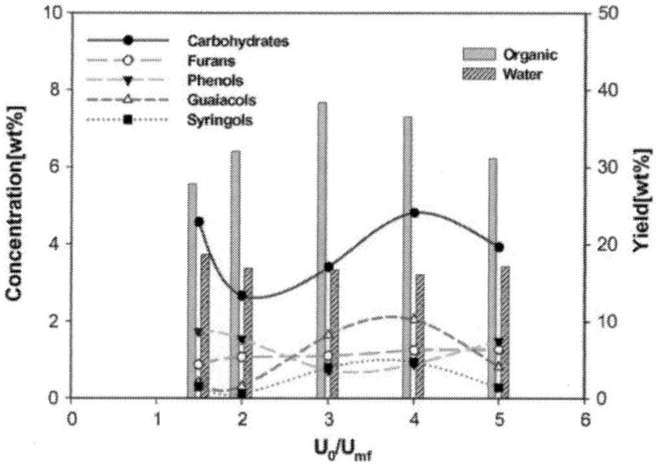

Figure: S.-H. Lee, M.-S. Eom, K.-S. Yoo, N.-C. Kim, J.-K. Jeon, Y.-K. Park, B.-H. Song, S.-H. Lee, The yields and composition of bio-oil produced from *quercus acutissima* in a bubbling fluidized bed pyrolyzer, J. Anal. Appl. Pyrolysis 83 (2008) 110-114. http://dx.doi.org/10.1016/j.jaap.2008.06.006

Hydrodynamics directly impact:

1. Particle residence time
2. Gas residence time
3. Particle heating rate
4. Particle attrition/fragmentation
5. Particle and ash elutriation
6. Particle segregation

All the above significantly impact raw oil yield and composition.

Particle heating rate

Affects rate of chemical kinetic yield and composition

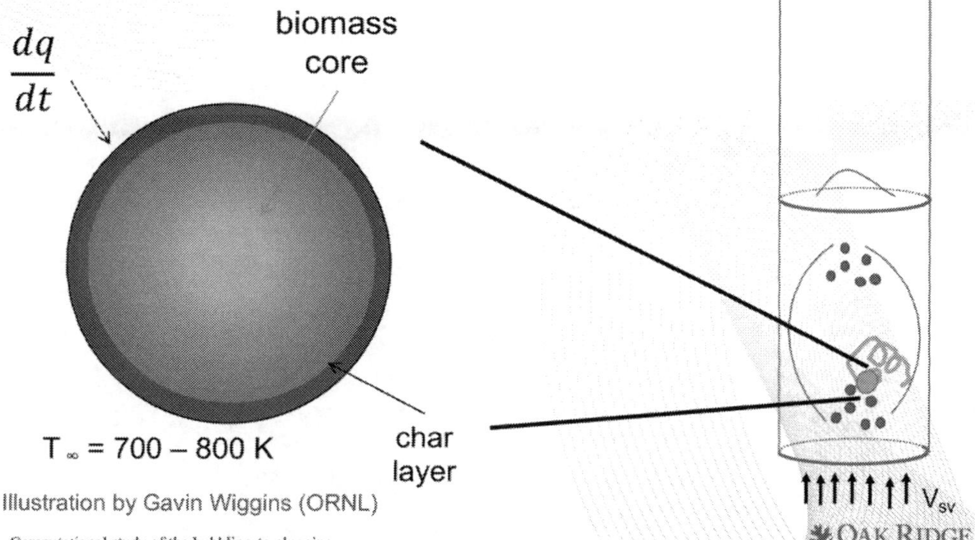

Illustration by Gavin Wiggins (ORNL)

Residence time, elutriation, and segregation

Particle and gas residence time, $\boldsymbol{\tau}$, in the bed and freeboard

$$\tau_{freeboard} = \frac{H_{reactor} - H_{exp\,bed}}{V}$$

$$\tau_{\exp bed} = \frac{H_{\exp bed}}{V / \varepsilon_{\exp bed}}$$

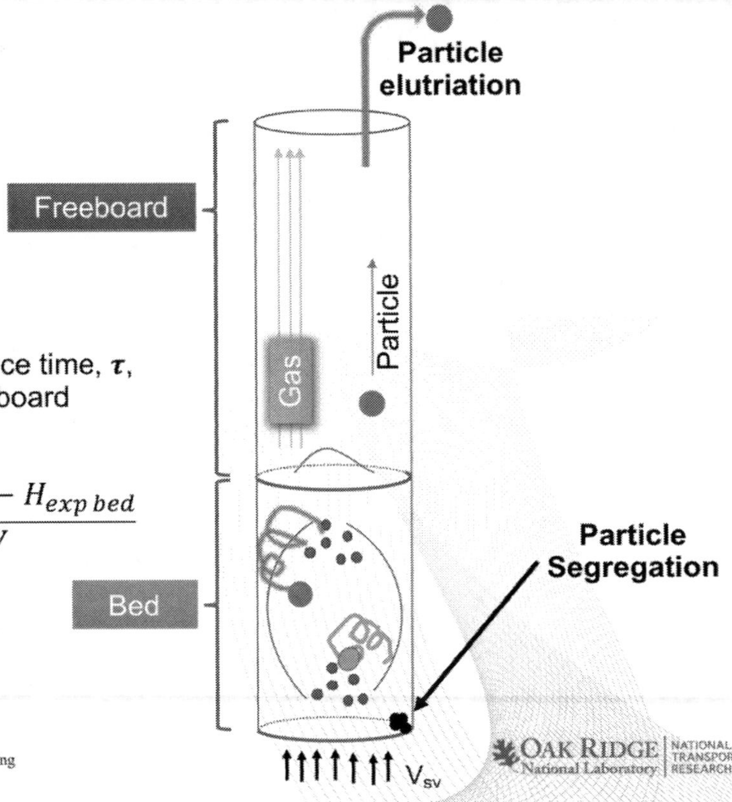

Particle attrition/fragmentation

Caused though mechanical impact, agitation, shearing, and internal pore structure

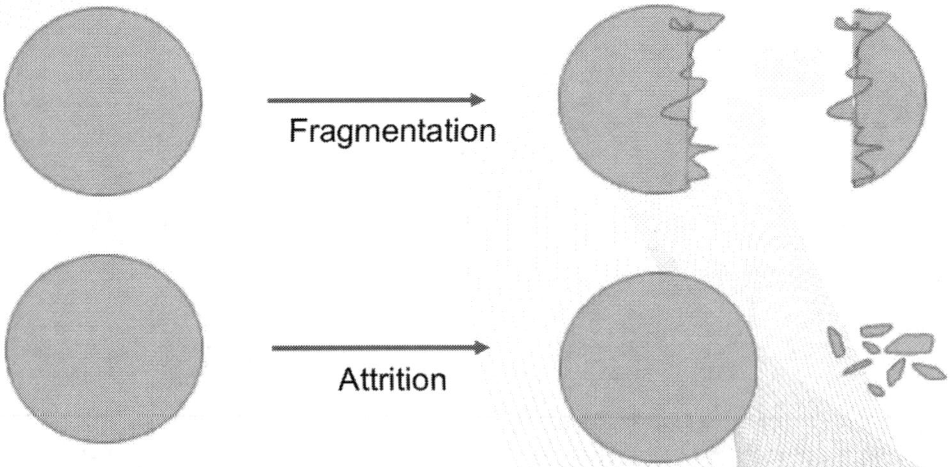

Objectives

Develop clearer understanding of slugging physics and develop possible on-line diagnostic
- What measurement can be used?
- Can we be quantitative?

Can a numerical model be used as a tool to understand the physics involved in slugging beds?
- How can an MFiX simulation be used to validate slugging studies?
- What quantitative numerical tools can we use?

- Scope of this study:
  - Bubbling/slugging beds
  - Bench scale reactor ~2 inch
  - Geldart type B particles
  - Hydrodynamics
  - Bubble analysis
  - Pressure analysis

Motivation: Slugging transition appears to be associated with a critical performance limit

Heat up time of biomass particles depends on particle size, shape, thermal properties, and reactor conditions.

Low tar yield and incomplete particle conversion if $\tau_{particle} < \tau_{tar}$

Tar yield decreases due to secondary reactions if $\tau_{particle} > \tau_{tar}$

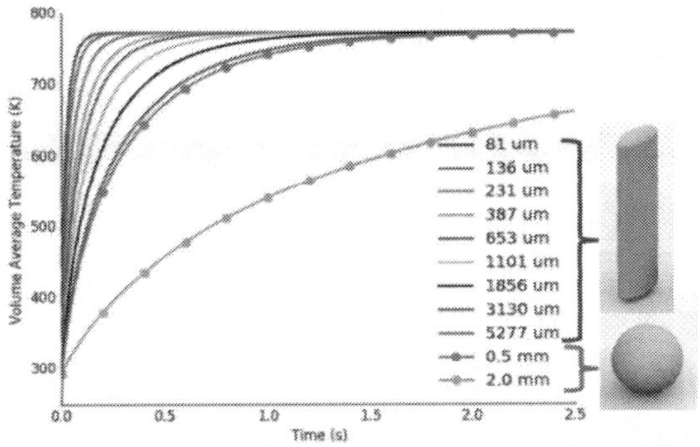

Temperature profiles from loblolly pine low-order particle model for D_F = 81 – 5277 um and single sphere with D = 0.5 and 2 mm.

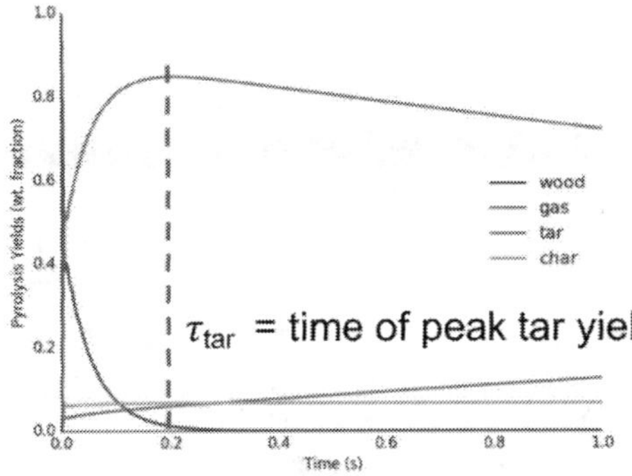

Estimated pyrolysis yields from Miller and Bellan 1997 kinetic scheme for beech wood at 500°C.

See the following AIChE 2016 session for details of this work:
458989 Modeling the Impact of Biomass Particle Residence Time on Fast Pyrolysis Yield and Composition

Background: Experiment of interest

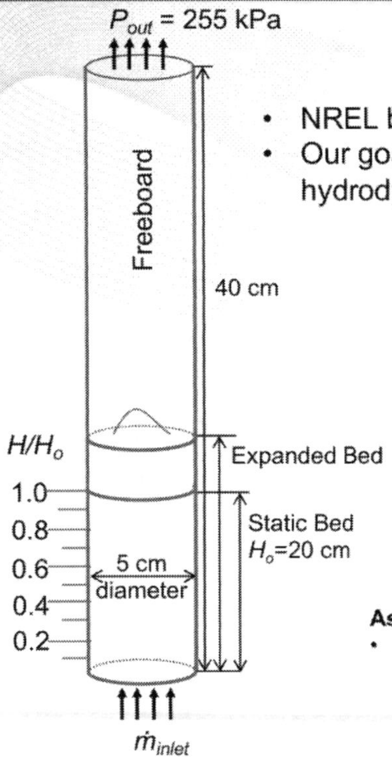

NREL

- NREL bubbling bed reactor used to study biomass fast pyrolysis
- Our goal is to help improve control and interpretation of the hydrodynamic aspects of this experiment

definition	Units	experiment	simulation
Particle diameter	m	$250x10^{-6}$	$250x10^{-6}$
Particle density	kg/m^3	2484	2484
Bulk density	kg/m^3	1552	
Temperature	K	773	773
Pressure	kPa	289	289
Fluidizing N_2 (range)	kg/s	$8.46\text{-}18.6\times10^{-5}$	$7.70\text{—}23.1\times10^{-5}$
Minimum fluidization	kg/s	-	7.70×10^{-5}
Minimum fluidization	m/s	-	0.03
coefficient of restitution	-	-	0.8
angle of repose	°	-	30
friction coefficient	-	-	0.1

Assumptions:

- Minimum fluidization – Richardson correlation
  - N_2 density – ideal gas temperature and pressure corrected
  - N_2 viscosity – NASA 7-coefficient polynomial, temperature corrected

Technical approach: computational

MFiX TFM

MFiX simulation parameters:

- Eulerian-Eulerian (Two-Fluid Model)
- Syamlal-O'Brien drag-model
- Kinetic theory of granular flow
  - Schaeffer frictional stress tensor formulation
  - Sigmoidal blending stress function
- Second order discretization
- Modified SIMPLE with variable time stepping
- Jackson and Johnson partial-slip wall boundary condition
- Cylindrical Mesh

Technical approach: Identify spatio-temporal transition metrics in MFIX simulation results

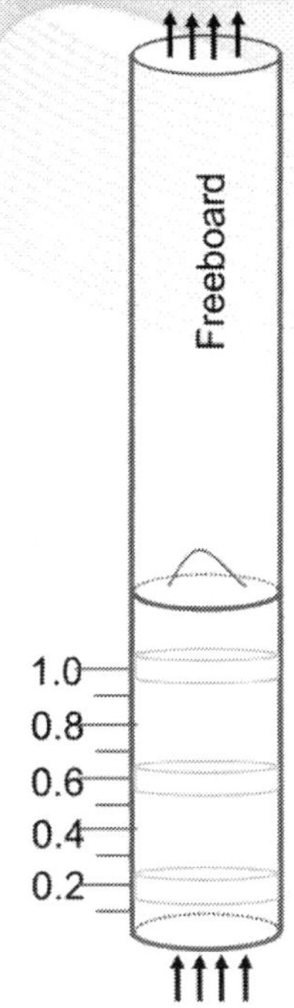

MFiX simulated pressure and bubble probes utilized data from locations relative to static bed height (H/H_o)

Pressure Statistics

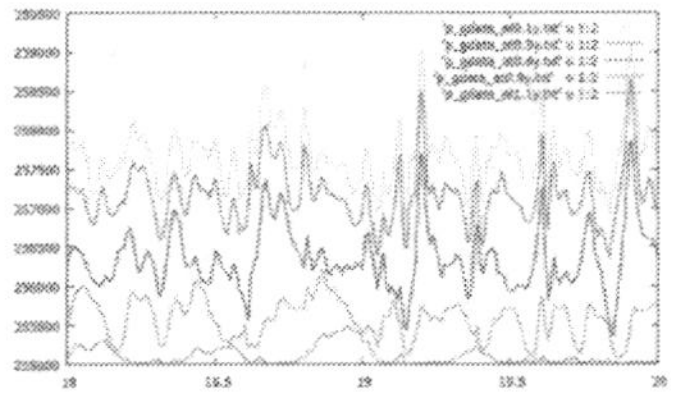

Analysis of PostMFiX pressure time series
- fluctuation distributions and statistics
- time-scale metrics

This work

Bubble Statistics

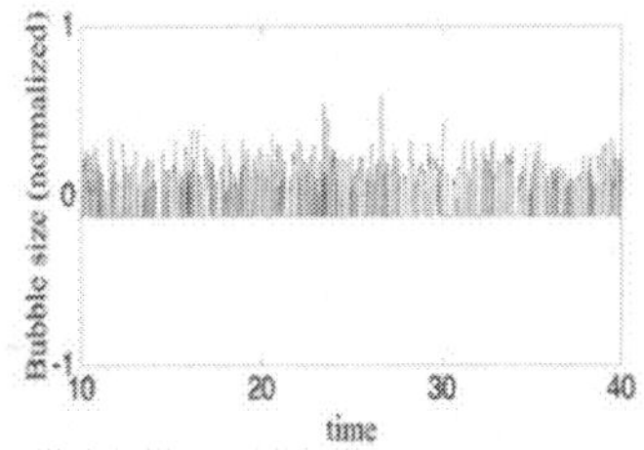

Analysis of PostMFIX void fraction time series processed with MS3DATA
- bubble size distributions and statistics
- time-scale metrics

A. Bakshi, C. Altantzis, R.B. Bates, A.F. Ghoniem, Multiphase-flow Statistics using 3D Detection and Tracking Algorithm (MS3DATA): Methodology and application to large-scale fluidized beds, CEJ, Volume 293, 1 June 2016, Pages 355-364.

OAK RIDGE National Laboratory | NATIONAL TRANSPORTATION RESEARCH CENTER

13 Computational study of the bubbling-to-slugging transition in a laboratory-scale fluidized bed

Results (1): Bubble statistics indicate the slugging transition occurs gradually, not abruptly

Max and min bubble frequency indicates beginning and end of bubbling to slugging transition in upper part of bed

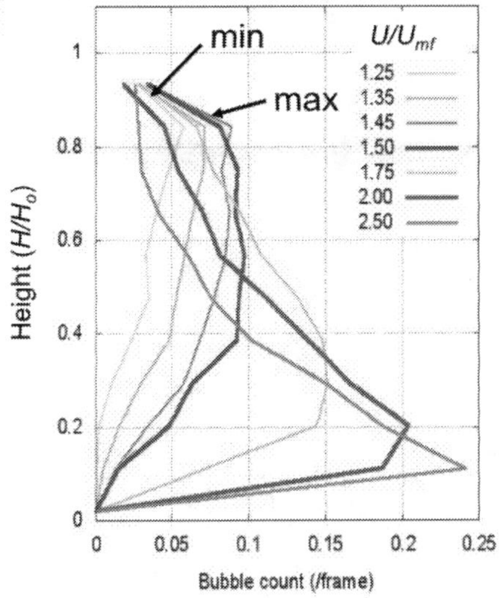

Bimodal distribution appears at bubbling to slugging transition which varies with inlet velocity and height

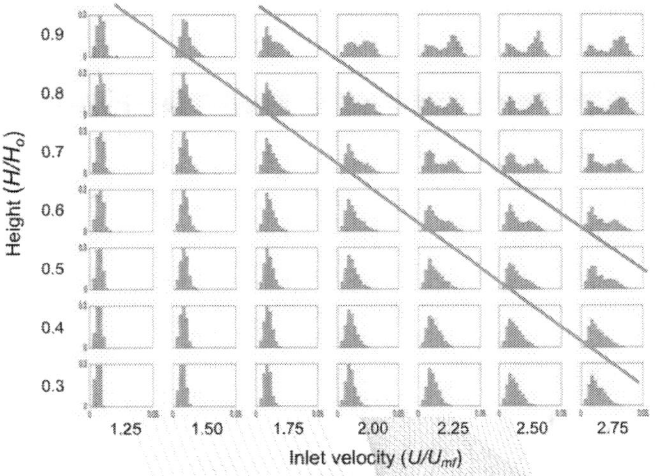

OAK RIDGE National Laboratory | NATIONAL TRANSPORTATION RESEARCH CENTER

14 Computational study of the bubbling-to-slugging transition in a laboratory-scale fluidized bed

Results (2): Pressure variations appear to track with bubble statistics but the relationship is complex

Pressure distribution tails reveal different asymmetries at different heights

1.25 U_{mf} – Free Bubbling

Pressure variations at different heights look similar

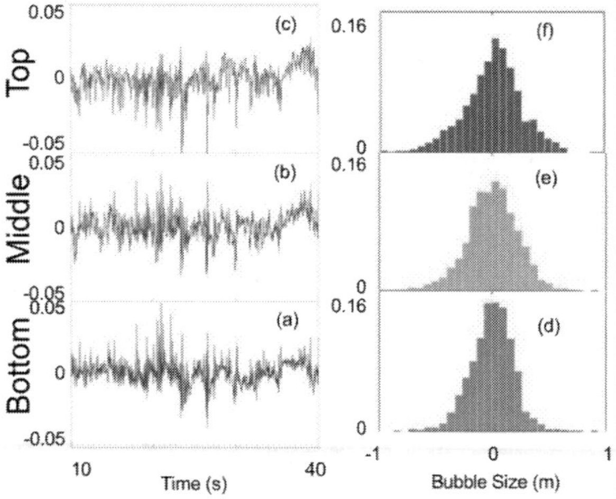

2.75 U_{mf} – Fully developed slugging

Bimodal patterns emerge

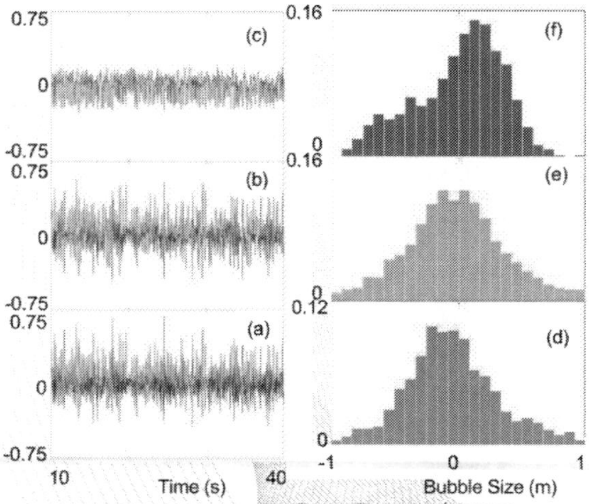

Results (3): Shifts in statistical metrics for pressure fluctuations change with location

1.25 U_{mf} – Free Bubbling

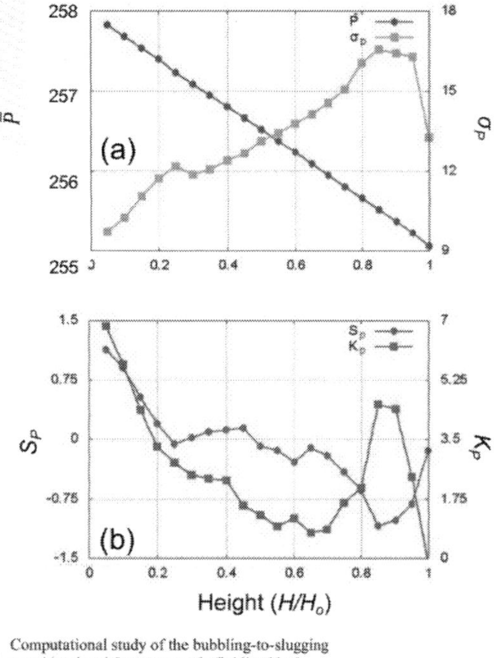

2.75 U_{mf} – Fully developed slugging

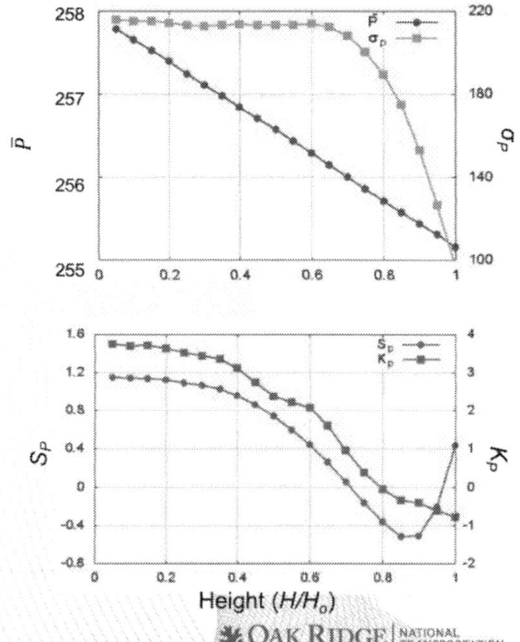

Results (4): MFIX predicted pressure trends appear consistent with experimental studies

Example: Predicted P fluctuation trends at 0.7-0.9 H/H_o appear to agree with Lee et al. 2002 experimental results in upper bed region (0.75 H/H_o).

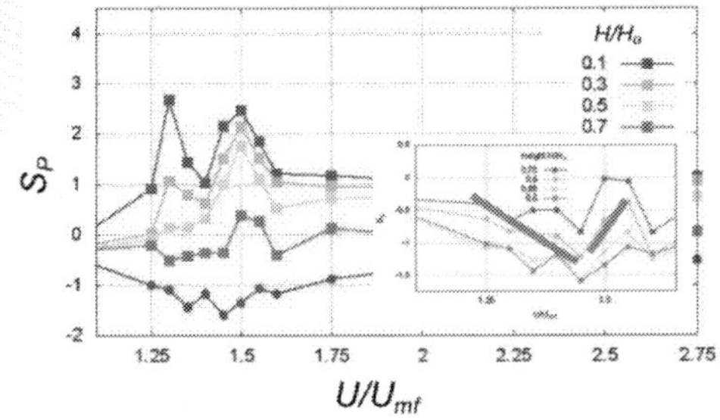

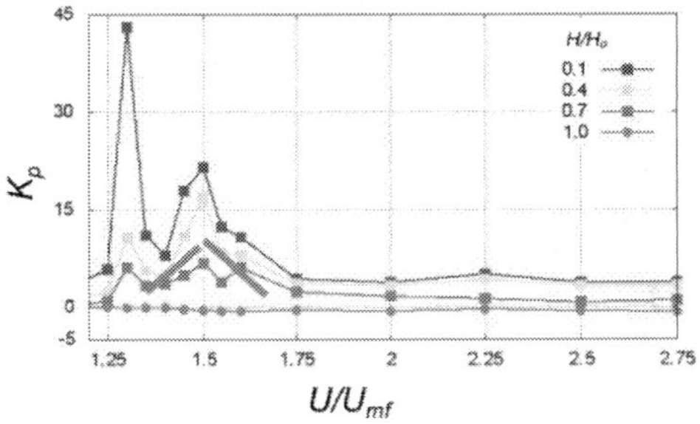

Skewness at 0.75 H/H_o has a downward spike

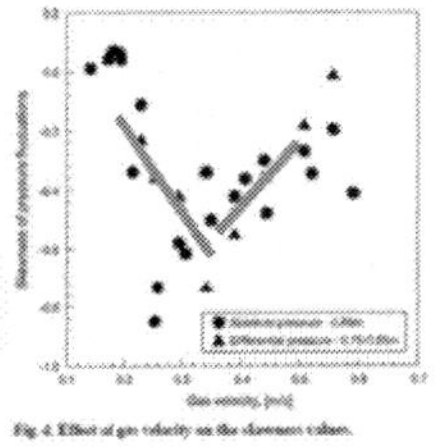

Kurtosis at 0.75 H/H_o has a upward spike

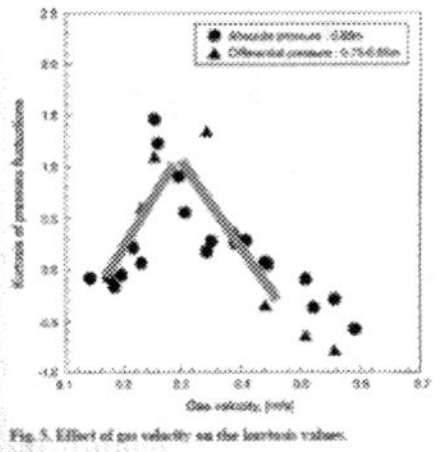

Statistical and Deterministic Chaos Analysis of Pressure Fluctuations in a Fluidized Bed of Polymer Powders, 2002
See Hoon Lee, Sang Done Kim† and Soung Hee Park

Results (5): MFIX results imply temporal metrics can be used to relate bubbles and pressure measurements

- Pressure and void fraction signal appear related
- Cross correlation magnitude increases with inlet velocity
- Upper bed section appears the most useful for slugging diagnostic

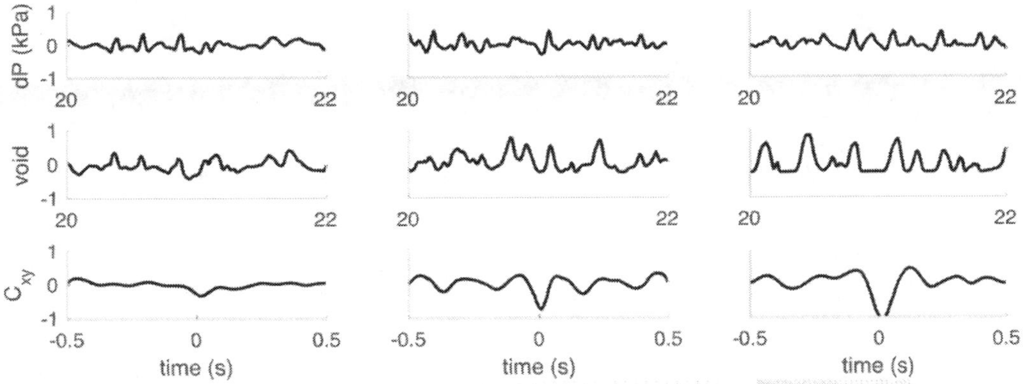

1.25 1.75 2.50

Inlet velocity (U_{mf})

OAK RIDGE National Laboratory | NATIONAL TRANSPORTATION RESEARCH CENTER

Summary and Conclusions to Date

- The bubbling-to-slugging transition is a gradual process, not an abrupt event
  - The bubbling-to-slugging transition occurs over a range of flows
  - Bubble coalescence begins near top of the bed and progresses downward as gas flow increases
  - Slugging becomes fully developed when the first major bubble coalescence occurs neat the distributor
- It may be easiest to detect the initial transition toward slugging at higher levels in the bed
- Statistical and temporal analyses of local pressure and void fluctuations can be used to develop quantitative metrics to indicate/diagnose the slugging transition
- MFIX and MS3DATA are useful tools for understanding the complex relationship between dynamic pressure and void fluctuations (bubble behavior)
- The optimal degree of slugging for fast pyrolysis of biomass has not yet been resolved

Future work (1): Relate simulated pyrolysis yields and bubble patterns

Preliminary 2D simulation

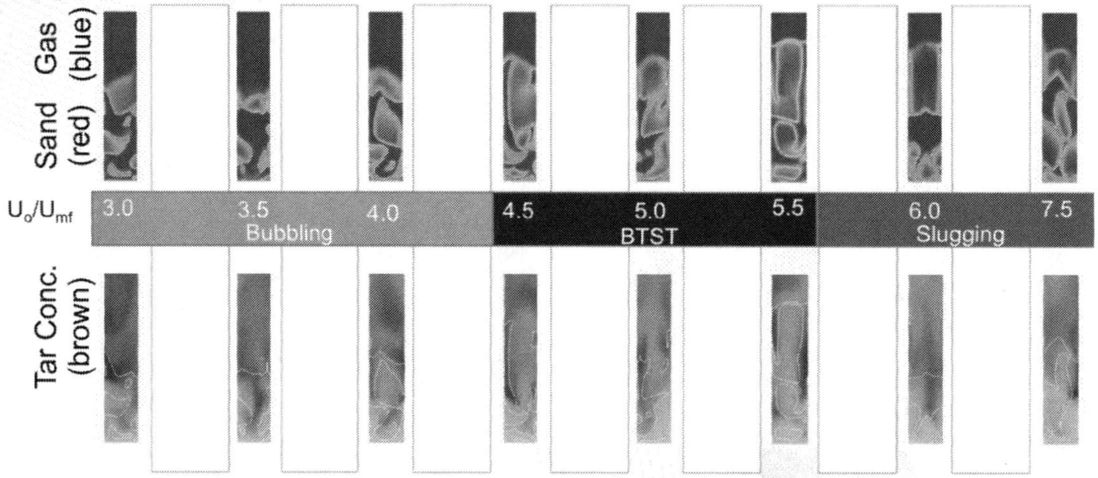

Bubble contours at 0.7 void cut off

Gas/sand emulsion and tar (oil) generation are highly correlated

Future work (2): Improve MFIX kinetic and elutriation parameters

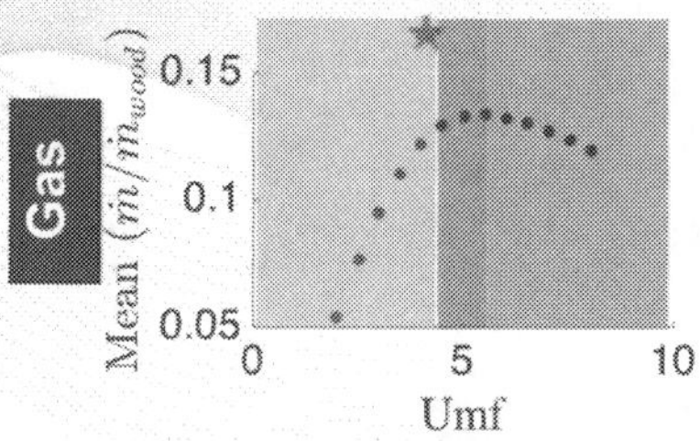

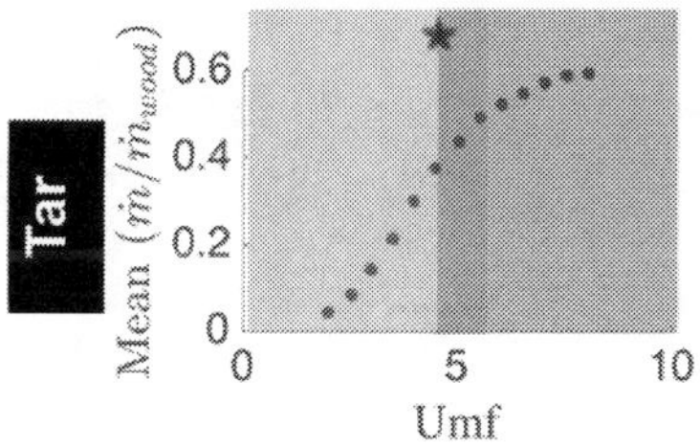

Bubbling = 3.50 U_o/U_{mf}
Transition (BTST) = 5.00 U_o/U_{mf}
Slugging = 7.50 U_o/U_{mf}

Preliminary Kinetic yield trends appear to be affected by the bubbling-to-slugging transition.

Table: Yield at ~4.4 U_{mf}

	Experiment	Simulation
Tar	0.661	0.377
Gas	0.162	0.129
Char	0.119	0.494

Mean yields did not match experiment

Uncertainty not accounted in results

2D simulation does not capture the correct physics in the fluidized bed

Future work (3): Do exit pyrolyzer fluctuations impact downstream processes?

Amplitude of tar oscillations increase with increased superficial velocity

Oscillations have potential to affect down stream equipment operation, such as VPU or LPU reactors.

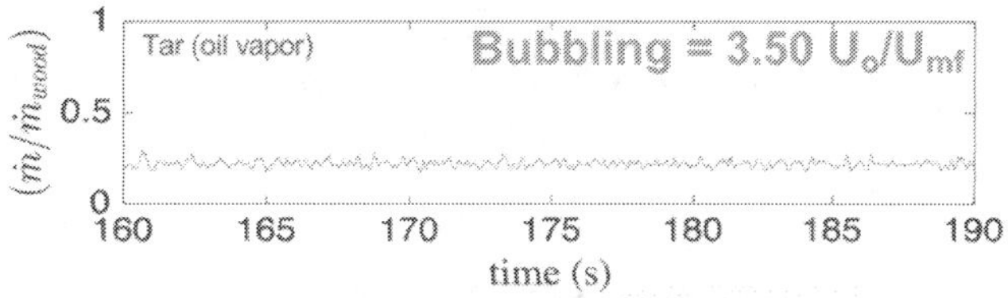

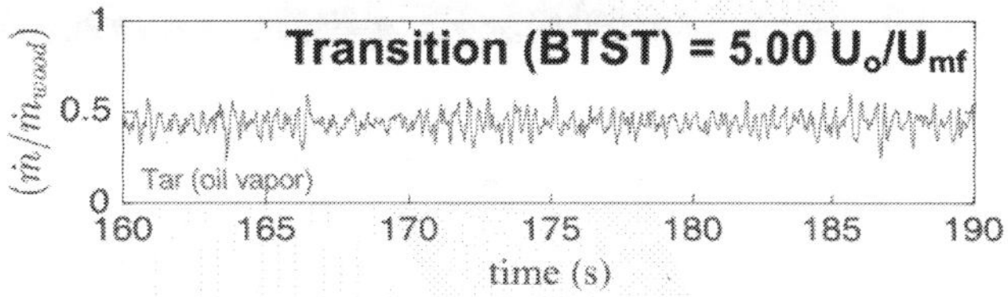

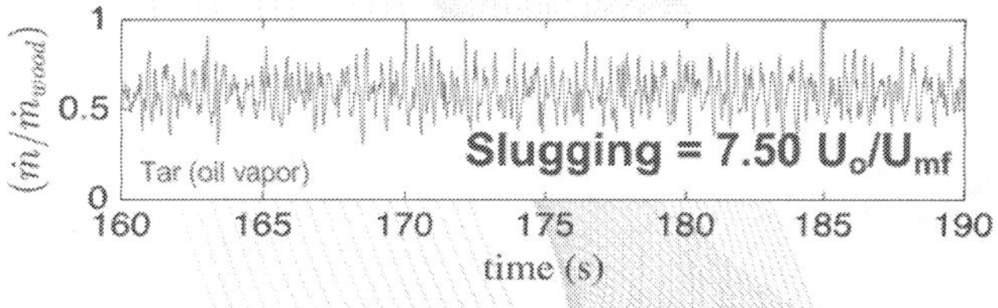

Future work (4): Experimental

Experimental validation

NETL
- ambient reactor experimental trends vs. MFIX predictions

NREL
- Analysis of existing reactor data
- Added instrumentation and diagnostics on the reactor

Questions ?

Emilio Ramirez – eramire2@vols.utk.edu

E. Ramirez, C.E.A. Finney, S. Pannala, C.S. Daw, J. Halow, Q. Xiong, Computational study of the bubbling-to-slugging transition in a laboratory-scale fluidized bed, Chemical Engineering Journal 308 (2017) 544-556.http://dx.doi.org/10.1016/j.cej.2016.08.113

CPC Computational Pyrolysis Consortium

http://cpcbiomass.org

Transitioning to Consortium for Computational Physics and Chemistry (CCPC)

OAK RIDGE National Laboratory · NREL National Renewable Energy Laboratory · Pacific Northwest NATIONAL LABORATORY · NETL · Argonne NATIONAL LABORATORY

469537 Particle and Gas Velocities in a Vortex Driven Externally Recirculated Fluidized Bed

Abstract:

The ever demanding need for lower capital costs requires the development of new contactors that can potentially utilize some hydrodynamic feature to enhance heat and mass transfer and thereby achieve process intensification. One such concept being explored at NETL is for a vortexing circulating fluidized bed. In this reactor concept, the gas enters tangentially, the rotating gas flow establishes a pressure gradient between the wall and center of the reactor. The geometry of the riser design exploits this gradient to generate a recirculating flow of solids from the top outer wall of the reactor to the bottom center. Barracuda computational findings and experimental results from a test reactor will be compared to assess the capabilities, and potential of the system. Specifically, gas and solids velocities will be compared between the simulation and the experimental results as well as comparing the pressure profiles.

CFD DEM COUPLING VALIDATION FOR PNEUMATIC SEPARATION OF SUGARCANE BAGASSE

Eduardo de Almeida, Nicolas Spogis and Maria A. Silva, University of Campinas, Campinas/SP, Brazil

Introduction

Sugarcane bagasse can be defined as the byproduct of the sugarcane milling process consisting of the sugarcane's fibrous portion after the extraction of its juice. Currently many countries ferment and then distill the sugarcane juice to produce ethanol; an important renewable fuel used as fossil fuel substitute aiming, among other factors, to reduce the global warming caused by fossil fuel burning.

The bagasse is nowadays used as a burning fuel in boilers in sugarcane mills, as animal feed, and in paper production. However, if the cellulose present in this biomass was converted to ethanol, by hydrolysis process, called second generation ethanol the yield per hectare of planted sugarcane could rise from current 5500 L ethanol/ha to around 8000 L ethanol/ha without increase the planted area in Brazil's case (1,2).

However, the hydrolysis process still presents some important bottlenecks that need to be overcome like high cost for example. The high cost is affected by many factors highlighting one important step of the process called physical pretreatment responsible for homogenize the material and decrease the bagasse particle's size allowing good hydrolysis yields but consuming high energy because the bagasse should be submitted to milling, grinding and chipping or a combination of these.

In order to propose alternatives to the physical pretreatment, consuming less energy than other ways of obtaining particles of different sizes it was analyzed the performance of a device, called pneumatic classifier equipment, developed at the School of Agricultural Engineering (Feagri / Unicamp). Based its operation principle on fluid dynamic theories, the pneumatic classifier equipment makes the separation of bagasse in an air column by dragging the small particles to the top and collecting the coarse particles at the bottom. The results from experiments showed the pneumatic classifier equipment allowed the continuous separation of sugarcane bagasse particles wherein the small particles, around 37% of total bagasse at dry basis, reached excellent yields when submitted to enzymatic hydrolysis without any other extra pretreatment thus avoiding energy consumption costs (3).

Inspired on previous results, it has started a theoretical study for bagasse separation process where the scope was to obtain a simulated model validated by experimental findings in order to be applied for example in different equipment structures aiming to evaluate the best one for separation of particles with a desired range of sizes or to help the simulation of other studies involving pneumatic transport of bagasse like drying, pyrolysis, gasification and others.

Pneumatic conveying and particle separation process are a complex phenomenum that requires a tighter coupling between air/fluid flow and particle-particle interaction. In order to better characterize these interactions, this work was performed coupling ANSYS FLUENT (16.2) and ROCKY (3.11). ANSYS FLUENT is a computational fluid dynamics (CFD) software used to help analyses of systems involving fluid flow, heat transfer, combustion, etc., by means of computer simulation based on finite volume to solve the continuum equations of motion. ROCKY uses a numerical technique called discrete element method (DEM) for predicting the behavior of polydispersed solids. The coupling between the simulated air flow at FLUENT and the particles motion at ROCKY allows a realistic approach of the separation process.

To validate the simulation results it was used a PhD thesis entitled "Characterization of sugarcane bagasse for energy generation" from School of Mechanical Engineering (FEM/Unicamp) where the author used a pneumatic separator device to evaluate bagasse particle's fluid dynamics characteristics besides to determine bagasse's physical properties from experimental data. As these results were not published in scientific journals, they are described in the next session (4).

Materials and Methods

The next paragraphs describe the equipment used to obtain the bagasse experimental data by Paulo Lenço in a PhD thesis entitled "Characterization of sugarcane bagasse for energy generation" and the models used by the softwares ANSYS FLUENT and ROCKY.

Operation Principle of the Separator Equipment

In Lenço's (4) work, the first step to start the bagasse fractions separation into the equipment is to turn on the centrifugal fan to cause a pipeline pressure differential, forcing a flow of ambient air through the opening at the bottom of the separation column to balance this difference. After the ascending airflow stabilization is reached in the air column, the bagasse is fed by an endless worm and falls by gravity into an opening near to the outlet of the column [see Figure 1].

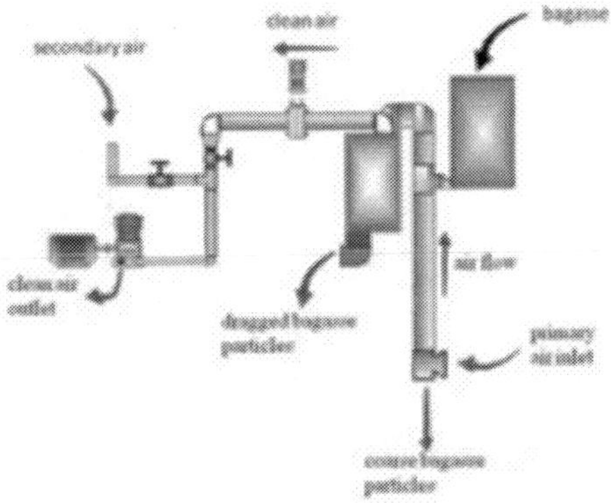

Figure 1. Equipment scheme used to separate bagasse particles (4)

Thus, the ambient air in contact with the bagasse, in a countercurrent flow, promotes the entrainment of the fine particles in the ascending flow until their collection in a particle precipitator at the top of the equipment. The coarse particles heavy enough to not be either dragged or put into lifting, fall by gravity action into an opening at the bottom of the air column. These coarse particles collected at the bottom are fed again and a higher velocity was used and a new fraction was separated. This process was repeated until all the remaining particles are dragged. In such study 7 fractions were found.The properties of the bagasse fractions as mass separated, particle size and apparent density were evaluated according the different velocities of separation. The basic dimensions of the riser are height 1.5m and internal diameter 0.1m.

Physical Properties of the Air and Bagasse

The air physical properties followed the default FLUENT properties for air, dynamic viscosity 1.7894×10^{-5} kg.m^{-1}s^{-1} and density 1.225 kg.m^{-3}.The bagasse properties were that obtained by Lenços' [9] studying the biomass fluid dynamics characteristics according to the experiments in the separator equipment. Table 1 summarizes all the properties applied on simulations [mass, d_c [particle diameter], L [particle length] and density] and the calculated terminal velocity using the Ganser's drag coefficient (5) that was a parameter to define the velocities simulated according the selected group to be dragged: 1.5 m/s; 1.7 m/s; 2.0 m/s; 3.0 m/s; 3.5 m/s and 4.0 m/s.

Table 1. Values Obtained by Practical Experiments (4) and calculated terminal velocity

Particle group	**Air velocity experimenal (m/s)**	**Mass (kg)**	**d_c (m)**	**L (m)**	**Density (kg/m^3)**	**Terminal velocity calculated (m/s)**
1	0.501	0.00924	0.00028	0.00061	545.3	0.74
2	0.710	0.00579	0.00091	0.00286	143.2	0.99
3	1.062	0.01091	0.00164	0.00587	164.3	1.64
4	1.578	0.01301	0.00191	0.00736	287.0	2.39
5	1.963	0.00609	0.00317	0.01279	248.0	2.81
6	2.277	0.00337	0.00318	0.01317	273.7	2.94
7	>2.277	0.01174	0.00329	0.01318	471.5	3.90

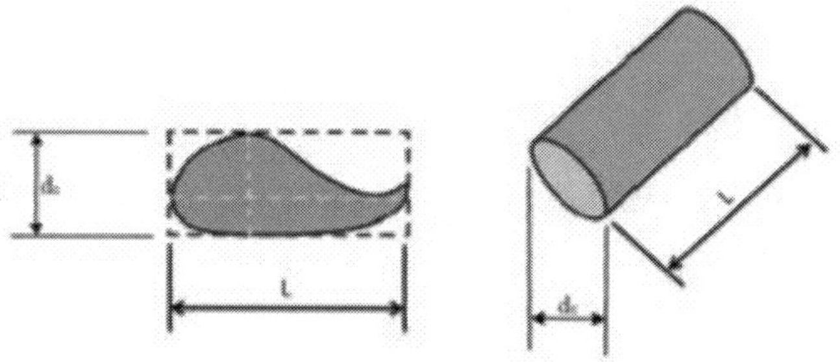

Figure 2. Particle dimensions (4)

Computational Fluid Dynamics (CFD)

The differential equations that describe the air flow into the riser are determined from Navier-Stokes' Equations of mass and momentum conservation.

Mass conservation equation (6):

$$\nabla.\left(f_g \rho_g V_g\right) = \frac{\partial(f_g \rho_g v_{g,x})}{\partial x} + \frac{\partial(f_g \rho_g v_{g,y})}{\partial y} + \frac{\partial(f_g \rho_g v_{g,z})}{\partial z} = S_g^{\rho} \qquad [01]$$

Where, f_g represents the volumetric fraction of air

ρ is the air density [kg.m^{-3}];

V is the velocity vector that can be decomposed into $v\hat{x}$, $v\hat{y}$ e $v\hat{z}$ [m/s];

S_g^p is the source term, in this case, $S_g^p = 0$

Momentum conservation equations to directions x, y and z:

$$\sum_{\xi=x,y,z} \frac{\partial}{\partial \xi}\left[f_g \rho_g v_{g,x} v_{g,\xi} - f_g \mu_g \left(\frac{\partial v_x}{\partial \xi} + \frac{\partial v_\xi}{\partial x}\right)\right] = S_{g,x}^m \quad [02]$$

$$\sum_{\xi=x,y,z} \frac{\partial}{\partial \xi}\left[f_g \rho_g v_{g,y} v_{g,\xi} - f_g \mu_g \left(\frac{\partial v_y}{\partial \xi} + \frac{\partial v_\xi}{\partial y}\right)\right] = S_{g,y}^m \quad [03]$$

$$\sum_{\xi=x,y,z} \frac{\partial}{\partial \xi}\left[f_g \rho_g v_{g,z} v_{g,\xi} - f_g \mu_g \left(\frac{\partial v_z}{\partial \xi} + \frac{\partial v_\xi}{\partial z}\right)\right] = S_{g,z}^m \quad [04]$$

Where, μ_g is the air dynamic viscosity [kg.m^{-1}.s^{-1}];
S_g^m represents the momentum transformation in the air.

These differential equations are used to describe the fluid motion, establishing that changes in momentum and acceleration of a fluid particle are resulting from changes in the pressure and viscous dissipative forces, originated from the molecular interaction, acting within the fluid.

The air turbulence was taken into account in the model using *k*-ε model. This model deal with two separated transport equations obtaining the turbulent kinetic energy, *k*, and the kinetic energy dissipation, ε, whereas the Reynolds Stress Models (RSM) are proportional to the average velocity gradients.

The turbulence kinetic energy, *k*, and its rate of dissipation, ε, are obtained from the following transport equations (6):

$$\nabla \cdot \left(f_g \rho_g \bar{v}_g k\right) = \nabla \cdot \left[\left(\mu_{g+} \frac{\mu_{g,t}}{\sigma_k}\right) \nabla k\right] + P_k - f_g \rho_g \varepsilon \quad [05]$$

$$\nabla \cdot \left(f_g \rho_g \bar{v}_g \varepsilon\right) = \nabla \cdot \left[\left(\mu_{g+} \frac{\mu_{g,t}}{\sigma_\varepsilon}\right) \nabla \varepsilon\right] + \frac{\varepsilon}{k}\left(C_{\varepsilon 1} P_k - C_{\varepsilon 2} f_g \rho_g \varepsilon\right) \quad [06]$$

In which the constants take the values: $C_{\varepsilon 1} = 1{,}44$; $C_{\varepsilon 2} = 1{,}92$; $\sigma_k = 1{,}0$ e $\sigma_\varepsilon = 1{,}3$.

P_k is the turbulent product due to viscous and buoyancy forces, expressed by equation (4):

$$P_k = \mu_{g,t} \nabla v_g \cdot \left(\nabla \bar{v}_g + \nabla \bar{v}_g^T\right) - \frac{2}{3} \nabla \cdot \bar{v}_g \left(3\mu_{g,t} \nabla \cdot \bar{v}_g + f_g \rho_g k\right) + P_{kb} \quad [07]$$

Considering the buoyancy production term (P_{kb}) null because there is no turbulence generation by it.

Discrete Element Method (DEM)

Discrete Element Method (DEM) models the total force experienced by individual particles or grains of a granular system and subsequent lists the accelerations, velocities and positions in a time period. The total force is the sum of the contact forces (particle-particle and particle-boundary) and non contact forces (such as gravity, magnetism and electrostatic forces). DEM is a mesh-free method and does not solve the continuum equations of motion instead of it solves the equations of motion for every individual particle that are numerically integrated with time.

DEM can be coupled with CFD to study fluid-particulate systems in which both solid and fluid phases contributes for dynamic flow, such as pneumatic transport and fluidized beds. In this approach, commonly referred as Lagrangian-Eulerian method, the solid phase is first modeled discretely using DEM, and these results are coupled with the CFD model.

The accomplished work by ROCKY DEM and ANSYS allows ANSYS FLUENT to work with Rocky in parallel CPU (Central Processing Unit) and GPU (Graphics Processing Unit) capabilities providing very fast DEM-CFD results. This processing methodology also allows to simulate large multiphase flow problems with millions of particles presenting different sizes and shapes.

For this study it was used the one-way coupling simulation because the particle loading is low and coupling parameters are almost null, thus, CFD solution provides information to the DEM solution, i.e., the fluid flow affects the particle trajectory but the particles do not affect the fluid flow.

The contact forces used by DEM, including in ROCKY software, consist of two parts, normal-to-contact plane and tangential-to-contact plane. The contact plane is a plane perpendicular to the line connecting closest points of the particles and boundaries. So, the algorithm for determining the contact plane for non-round particles, like those simulated in this study, takes account the calculations for one of any the following options: the closest points of two particles; or the closest points of a particle and triangle; or the two points with the maximum overlap distance in the case of a physical contact, depending on the contact type for each pair of particles.

The model that has been chosen to predict the normal contact forces in this work was an elastic-plastic model. The elastic-plastic models follow two major requirements that are: the force has to be a repulsive one; the normal contact force model has to allow significant energy dissipation since a granular medium is an extremely dissipative system. For the tangential contact prediction it was used an elastic-frictional force model.

One of the elastic-plastic models, the linear hysteresis model was adopted by this study because it presents many advantages as: to avoid the overhead of long simulation time when simulating the plastic energy dissipation on a contact; to make the energy dissipation insensitive to other contacts, since no viscous damping term is used and the energy dissipation is not dependent on the relative velocities of neighboring particles; and to allow modeling compressible materials accurately due to the fact that the contact forces can be almost zero even at residual overlaps.

The following equations describe the linear hysteresis model used by ROCKY (7):

$$F_n^t = min(F_n^{t-dt} + K_{nu} \cdot ds_n, K_{nl} \cdot S_n) \quad \text{if } ds_n \geq 0 \qquad [08]$$

$$F_n^t = min(F_n^{t-dt} + K_{nu} \cdot ds_n, 0.001 \cdot K_{nl} \cdot S_n) \quad \text{if } ds_n < 0 \qquad [09]$$

$$ds_n = s_n^t - s_n^{t-dt} \qquad [10]$$

Where, F_n^t is the normal elastic-plastic contact force at the current time, *t*

F_n^{t-dt} is the force at the previous time, *t-dt*, where *dt* is the timestep

ds_n is the change in the contact normal overlap during the current time (it is assumed to be positive as particles approach each other and negative when they separate)

s_n^t and s_n^{t-dt} are the normal overlap values at the previous and the current time
K_{nl} and K_{nu} are the values of loading and unloading contact stiffness respectively

The particle's loading stiffness is calculated in the following manner (7):

$$K_{npl} = E_p \cdot L \quad [11]$$

$$K_{nbl} = E_b \cdot L \quad [12]$$

Where, E_p is the particle material bulk Young's Modulus or Elastic Modulus
E_b is the boundary material Young's Modulus
L is the particle size.

For the contact of two particles, or of a particle with a boundary, the loading and unloading stiffness are defined respectively as (7):

$$K_{nl} = \frac{K_{npl\_1} \cdot K_{npl\_2}}{(K_{npl\_1} + K_{npl\_2})} \quad [13]$$

$$K_{nl} = \frac{K_{npl} \cdot K_{nbl}}{(K_{npl} + K_{nbl})} \quad [14]$$

$$K_{nu} = \frac{K_{nl}}{\varepsilon^2} \quad [15]$$

Where, ε is the coefficient of restitution, in ROCKY it is a measure of energy dissipation for the contacting pair of materials; and subscripts *_1* and *_2* refer to particle 1 and particle 2 of two contacting particles.

Considering the tangential force model it is an elastic-frictional force model described by the equations (7):

$F_t^t = min(F_t^{t-dt} + K_{nl} \cdot ds_t, \mu_s \cdot F_n^t)$, if no sliding is taking place for the contact [16]

$F_t^t = min(F_t^{t-dt} + K_{nl} \cdot ds_t, \mu_d \cdot F_n^t)$, if sliding is taking place for the contact [17]

Where, F_t^t is the tangential contact force at the current time;
F_t^{t-dt} is the same force at the previous time;
ds_t is the tangential relative particles' displacement during the current time;
μ_s and μ_d are the static and dynamic friction coefficients;

If the tangential force exceeds the limit of $\mu_s \cdot F_n^t$ the sliding is considered to be taking place for the contact, otherwise if the tangential force falls below value of $\mu_d \cdot F_n^t$, the contact is considered non-sliding.

Results

The group 1 of particles, around 450,000, weren't simulated because they would all certainly be dragged by the air due the chosen velocities and the simulation time would increase a lot considering the computational power available making the simulations infeasible. Nevertheless, the simulation with the other groups of particles, representing 82% of the true bagasse mass, proved to be very predictive as shown hereafter.

Figures

The next figures show the visual results obtained by Rocky software with zoom of particles near the bottom and top of the equipment.

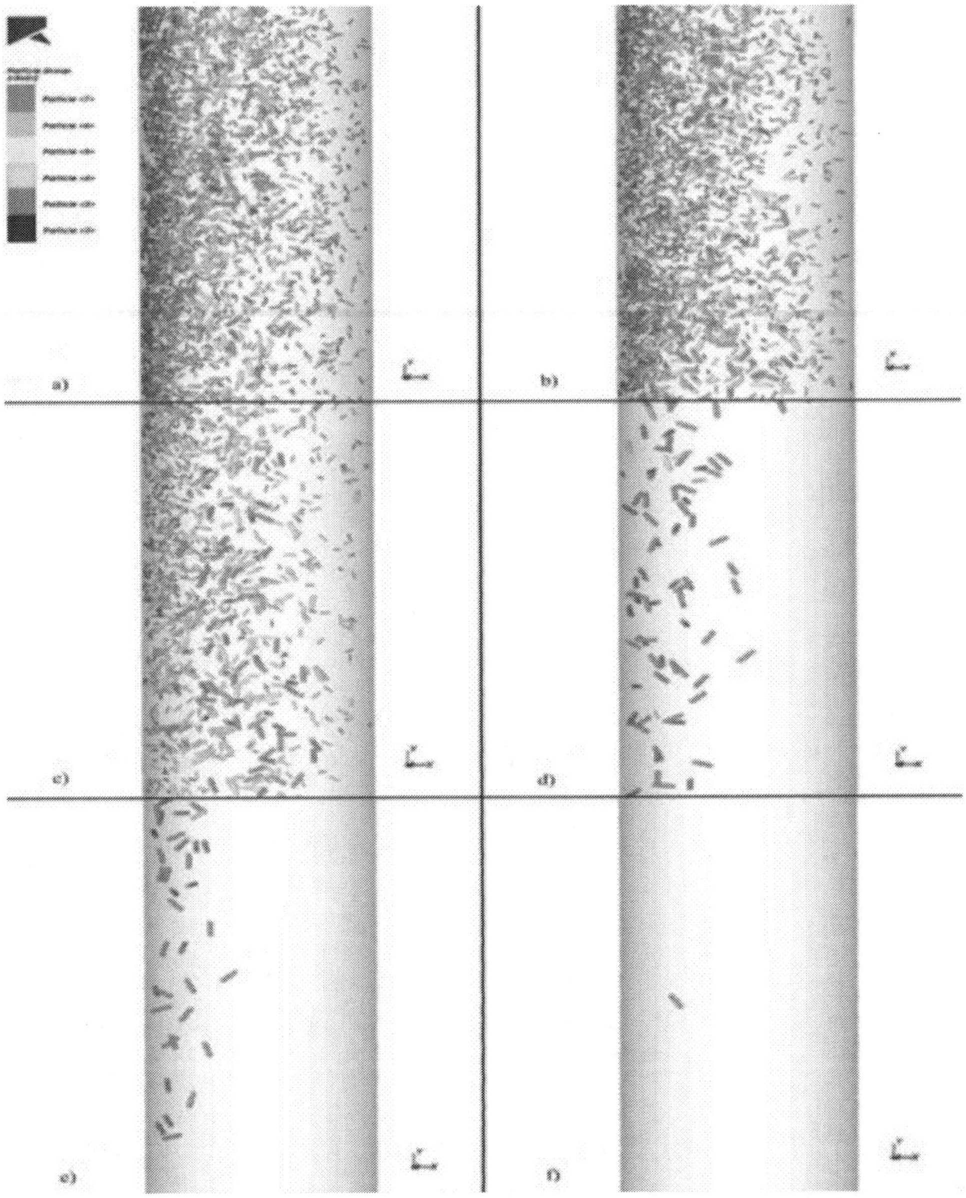

Figure 3. Zoom in bagasse separation simulation near the bottom at same time using air average velocities of a) 1.5m/s b) 1.7 m/s c) 2.0 m/s d) 3.0 m/s e) 3.5 m/s and f) 4.0 m/s

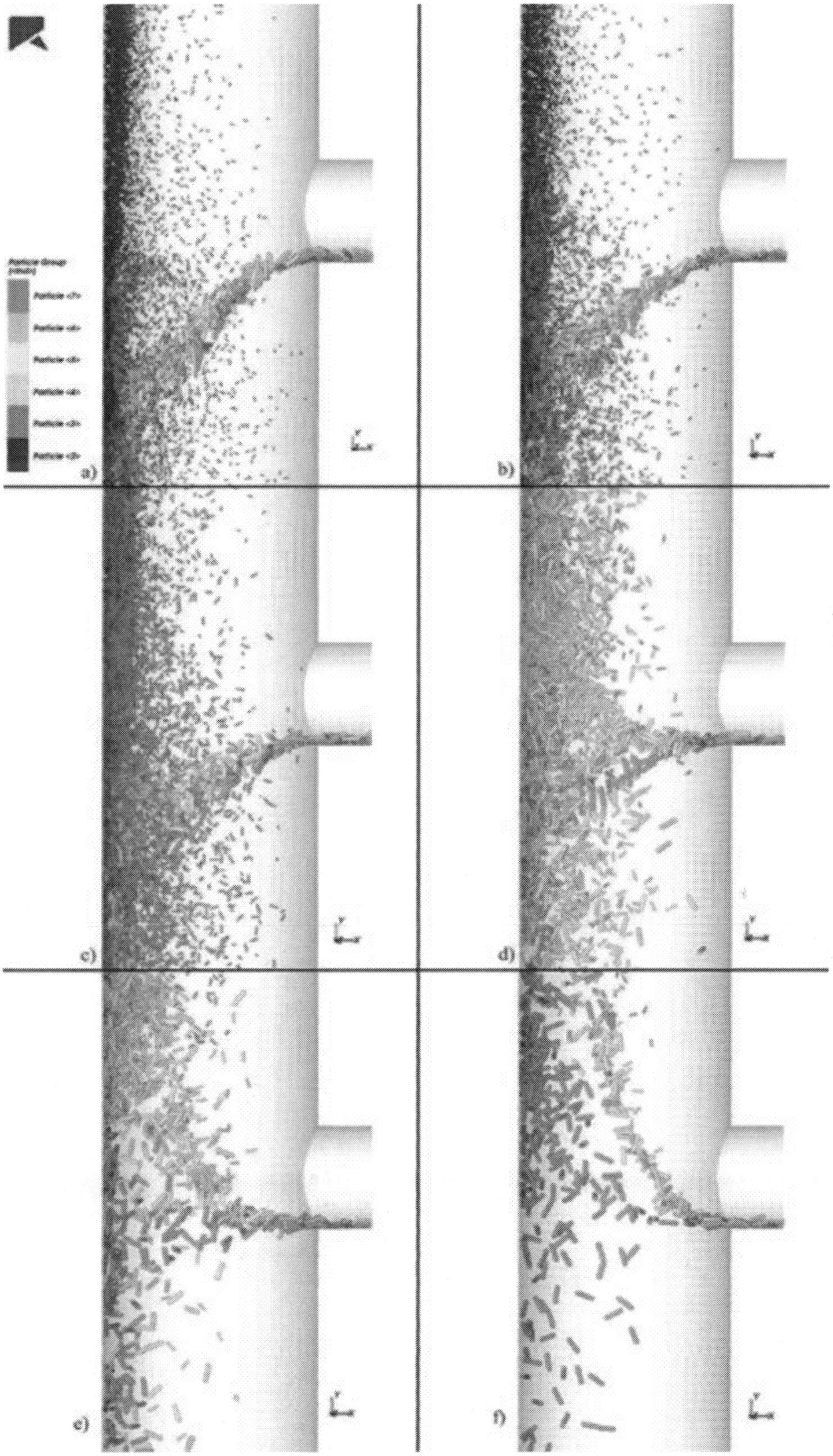

Figure 4. Zoom in bagasse separation simulation near the top at same time using air average velocities of a) 1.5m/s b) 1.7 m/s c) 2.0 m/s d) 3.0 m/s e) 3.5 m/s and f) 4.0 m/s

Graphics

The graphics show de particle distribution of groups collected on the top and bottom of the equipment varying the air velocity of separation.

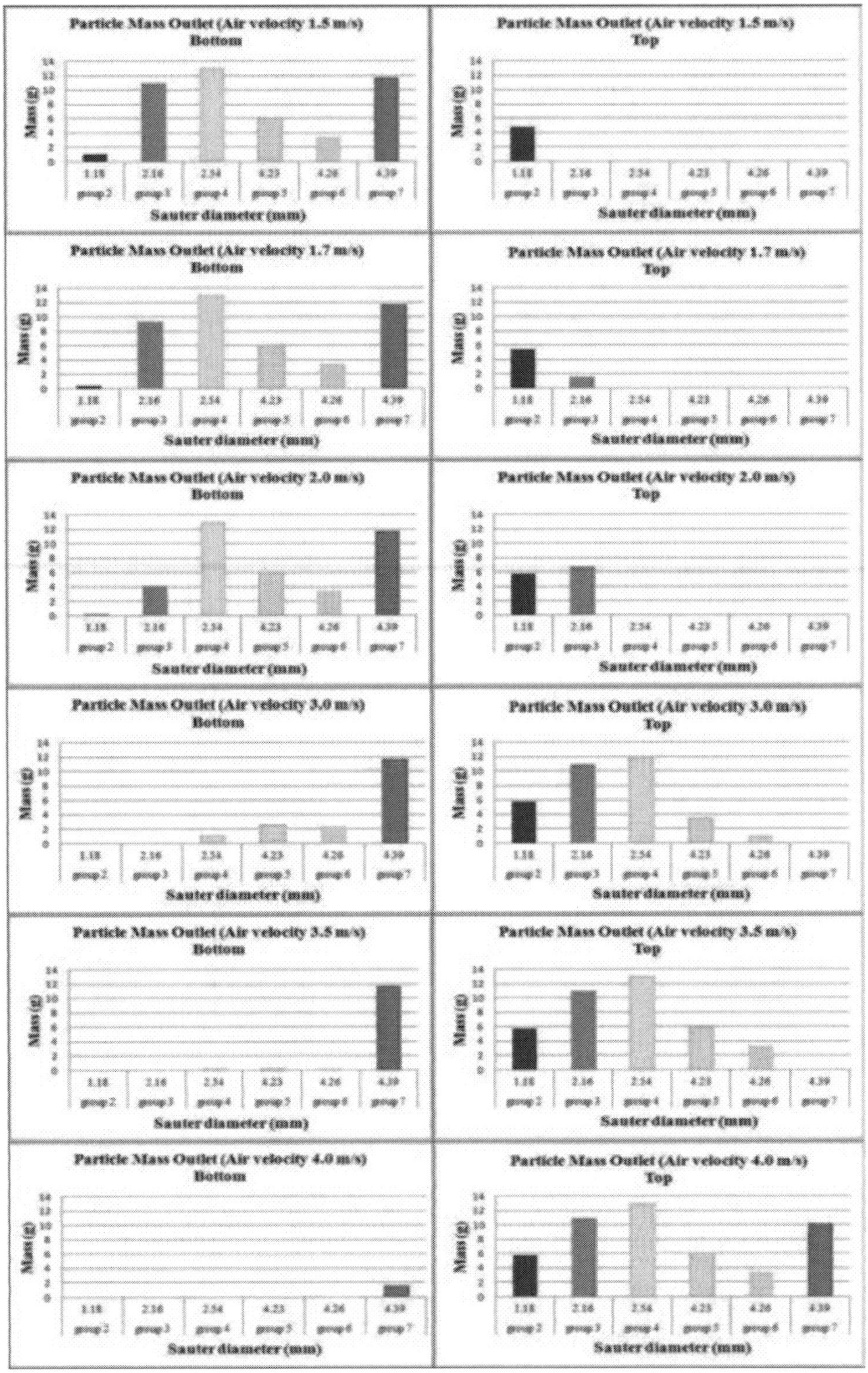

Figure 5. Mass distribution of particles on top and bottom of the equipment varying the simulated air velocities

Analyzing the results both by images and mass distribution was possible to understand that at lower velocities even particles that should be dragged to top are collected by bottom possibly because with these velocities the contacts particle/particle and particle/wall are higher. Thus, the particle contact with the riser wall decrease the drag force over it, prevailing the weight force on the forces balance, causing the particle fall by gravity action and besides, the falling larger particles can also drag the smaller to the bottom due to contact during separation.On the other hand, using the air average velocity of 3.5 m/s into the riser, almost all the particles were dragged to the top just separating group 7, as expected, when compared the air velocity with the particles` terminal velocities in Table 1. Simulating the air velocity 4.0 m/s almost all the particles were completely dragged separating just few grams of particles from group 7 at the bottom outlet.

Conclusions

CFD/DEM techniques were used in this paper to simulate the bagasse particles` separation that showed to be a powerful tool since the adopted model described with good accuracy the particles movements and their interactions with other particles and boundaries. When the air velocity was enough to separate intermediate group of particles it was observed that the particle/particle and particle/wall contacts were larger than when high air velocities were adopted, thus the separation was more efficient when the objective was to separate the coarsest particles.The parameters of simulation studied can be extended to other processes that use sugarcane bagasse fluidization as drying, pyrolysis, gasification and others, allowing the study and development of equipment close to an ideal configuration saving time and money for it.

Acknowledgments:

CNPq - National Council for Scientific and Technological Development (Brazil)
ESSS - Engineering Simulation and Scientific Software

References

1. Companhia Nacional de Abastecimento: Acompanhamento da safra brasileira: cana-de-açúcar. http://www.conab.gov.br/OlalaCMS/uploads/arquivos/12_04_10_09_19_04_boletim_de_cana.pdf (2012) Accessed 28 April 2016
2. Leite, R.C.C., Leal, M.R.L.V., Cortez, L.A.B., Griffin, W.M., Scandiffio, M.I.G. (2009), "Can Brazil replace 5% of the 2025 gasoline world demand with ethanol?", *Energy 34*, pp. 655-661.
3. Almeida, E., Cortez, L.A.B., Silva, M.A.(2013), "Sugarcane bagasse pneumatic classification as a technology for reducing costs on enzymatic hydrolysis process. " *In: International Society of Sugar Cane Technologists*, Brazil, Abstract, Brazil: STAB & The XXVIIIth ISSCT Organising Committee, p.358.
4. Lenço, Paulo A. (2010), "Characterization of sugarcane bagasse for energy generation," PhD Thesis, University of Campinas, p.181.
5. Ganser, Gary H.(1993),"A rational approach to drag prediction of spherical and non-spherical particles," *Powder Technology 77*, p.p. 143-152.
6. Ansys Fluent 16.2 (2011), *Theory Guide*
7. Rocky 3.11 (2015), "A brief introduction to DEM and an Overview of Contact Force models in Rocky," *Theory Guide*

464140 CFD-DEM Simulation of Hydrodynamics in Wet Gas-Solid Fluidized Beds

Abstract:

It is common in many industries, including energy, pharmaceuticals and polymer production, for small amounts of liquid to be injected into gas-solid fluidized beds. This liquid can serve as a reactant or facilitate agglomeration or heat transfer. The liquid often covers the surface of particles and proceeds to form cohesive liquid bridges between particles, ultimately leading to the agglomeration. The cohesive force caused by small amounts of liquid can alter hydrodynamics dramatically, since agglomerates, rather than individual particles, need to be fluidized. In extreme cases, liquid bridging and agglomeration can lead to local or device-scale defluidization. Despite this industrial importance, the effect of wetness on hydrodynamics is not fully understood. A limited number of experimental studies show differences in bed fluidity [1,2], particle velocities [3–6], bed height [1], bubble size [6,7] and minimum fluidization velocity [6] caused by liquid bridging in gas-solid fluidized beds.

Various efforts have been undertaken to model the effect of wetness on fluidization computationally. Most work has involved an Euler-Lagrange approach often referred to as the Computational Fluid Dynamics – Discrete Element Method (CFD-DEM) [8] in which each individual particle is modeled, gas flow is simulated on a CFD grid and the two phases are linked via a drag law. In this approach, contacts between particles are modeled, often with a coefficient of restitution dictating the level of inelasticity in the collisions. A number of studies have accounted for wetness by lowering this coefficient of restitution [9,10], since the cohesive force provided by liquid bridges will act to slow the particles from rebounding off of one another after a collision. A fundamental issue with approach is that liquid bridging does not merely make collisions more inelastic; liquid bridges provide cohesion necessary to keep particles together for long periods of time and form agglomerates. To address this issue with a modified coefficient of restitution approach, a few more detailed studies have directly modeled forces arising from liquid bridges [11,12]. One issue with these studies is that they have assumed liquid bridges to form instantaneously. In fact, the surface tension and viscosity of the liquid will affect the rate at which liquid pours from the surface of a particle into a liquid bridge [13], and in certain cases, particles can rebound off of one another before a significant liquid bridge has had time to form.

In this study, we model the effects of liquid on fluidization hydrodynamics using a CFD-DEM model in which liquid bridges are modeled directly and liquid bridges form at a finite rate. We simulate laboratory-sized fluidized beds and investigate the effects of liquid content, viscosity and surface tension on bed height, particle velocity, bubble behavior and minimum fluidization velocity for comparison with effects seen in the literature.

References:

[1] S. L. McDougall, M. Saberian, C. Briens, F. Berruti, and E. W. Chan, Int. J. Chem. React. Eng. **2**, (2004).

[2] S. McDougall, M. Saberian, C. Briens, F. Berruti, and E. Chan, Chem. Eng. Process. Process Intensif. **44**, 701 (2005).

[3] V. S. Sutkar, N. G. Deen, A. V. Patil, E. A. J. F. Peters, J. a. m. Kuipers, V. Salikov, S. Antonyuk, and S. Heinrich, AIChE J. **61**, 1146 (2015).

[4] Q. Zhang, Y. Zhou, J. Wang, B. Jiang, Y. Yang, S. Stapf, C. Mattea, and Q. Gong, Chem. Eng. Technol. n/a (2015).

[5] M. L. Passos and A. S. Mujumdar, Powder Technol. **110**, 222 (2000).

[6] Y. Zhou, C. Ren, J. Wang, and Y. Yang, AIChE J. **59**, 1056 (2013).

[7] J. Ma, D. Liu, and X. Chen, Ind. Eng. Chem. Res. **55**, 624 (2016).

[8] Y. Tsuji, T. Kawaguchi, and T. Tanaka, Powder Technol. **77**, 79 (1993).

[9] P. Darabi, K. Pougatch, M. Salcudean, and D. Grecov, Powder Technol. **214**, 365 (2011).

[10] V. S. Sutkar, N. G. Deen, J. T. Padding, J. a. m. Kuipers, V. Salikov, B. Crüger, S. Antonyuk, and S. Heinrich, AIChE J. **61**, 769 (2015).

[11] T. Mikami, H. Kamiya, and M. Horio, Chem. Eng. Sci. **53**, 1927 (1998).

[12] M. Girardi, S. Radl, and S. Sundaresan, Chem. Eng. Sci. **144**, 224 (2016).

[13] M. Wu, J. G. Khinast, and S. Radl, in (Barcelona, 2014).

CIRCULATION AND SEPARATION OF BINARY SOLIDS IN CONNECTED FLUIDIZED BEDS

[1]*Yusumi Nagahashi,* [1]*Hideki Takeuchi,* [1]*Tomohiro Kawamura*
[1]*National Institute of Technology, Kochi College, Nankoku, JAPAN*
[2]*John R. Grace*
[2]*University of British Columbia, Vancouver, CANADA*
[3]*Yuataka Asako*
[3]*MJIIT, University Technology Malaysia, Kuala Lumpur, MALAYSIA*

Introduction

In many fluidization and particle processing systems, mixing and segregation of multi-particle mixtures occur, sometimes being problematic. On the other hand, segregation is useful in some applications for separating one type of particle from other particles. An understanding of the mixing and segregation of particles is therefore of considerable importance in the operation of fluidized bed processes, e.g. in fluidized bed reactors for fast pyrolysis producing bio-oil, both for pyrolysis and char removal.

Although a fluidized bed pyrolyzer has some advantages for processing capacity and fast residence time compared with other types of pyrolysis reactor, it also has disadvantages such as the difficulty of char removal from the bed material heated medium. Because wood-char acts as a vapor cracking catalyst, rapid and effective separation and removal of char particle from the bed material is needed for high quality and favorable oil-yields [(1)], and it is also essential for continuous operation of bio-oil production.

There have been many investigations related to particle mixing and segregation in binary solids fluidized beds, e.g. those by Rowe et. al [(2), (3)]. Mixing and segregation depend on particle density, size and gas flow rate in binary solids mixture gas-fluidized bed. In general, large low-density particles such as wood char gather in the upper region as flotsam in gentle large-density particle beds, whereas high-density small particles of bed material like silica sand behave as jetsam. In this work we utilize Rowe's segregation terminology (char: flotsam, bed material: jetsam).

Working on char removal from the fluidized bed pyrolysis, Berruti's ICFAR group has reported unique results of char removal using movable partition plates which divide the bed into two sections for separation and mixing [(4), (5)]. Their results indicate that separation efficiency depends on biomass feed rate. The driving force transporting particles into the separation section in their configuration could only be hydraulic head difference (bed pressure difference) between two sections. Therefore it needs a high feed rate for faster separation, but excessive feed rates lead to choking when transporting char particles to the separation section.

In this work, particle circulation and separation of binary solids have been investigated in a novel fluidized bed system as a fast pyrolysis reactor for bio-oil production. The system was constructed with two sections connected through a slit installed on the partition wall immersed in the bed. This configuration is similar to ICFAR's facility, but with a unique feature, an inclined gas distributor [(6)] which assists particle transportation into the separate section. One side contains a vigorously fluidized bed with an inclined distributor for biomass pyrolysis, while the other provides gentle fluidization to facilitate particle separation.

Particle exchange through the slit is observed and analyzed by PIV (Particle imaging velocimetry) to explain the driving force transporting particles between sections. Influences of particle properties of modeled chars and slit locations and also gas velocities of each section on separation performance are investigated in the bed with a conventional flat gas distributor and an inclined gas distributor. The system with the inclined distributor is shown to perform well for char-particle separation and removal.

Experimental Apparatus and Procedure

The experimental apparatus is shown in Fig. 1. Experiments were carried out using an acrylic two-dimensional fluidization column of rectangular cross-section 23 mm × 160 mm (D_1 = 56 mm for the particle separation section, D_1 = 100 mm for the mixing section), and height 270 mm. The inclination angle of the inclined distributor was set at 20° with respect to the horizontal. Silica sand of mean diameter 0.11 mm (minimum fluidization velocity U_{mf} = 0.00295 m/s) was the bed material and six types of particles represented char particles. Table 1 shows the properties of the modeled char (d_p: particle diameter, ρ_0: particle density, ρ_e: particle bulk density). Most experiments were carried out with L_0/D = 1.0, where L_0 is the static bed height and D is the column width of the mixing section.

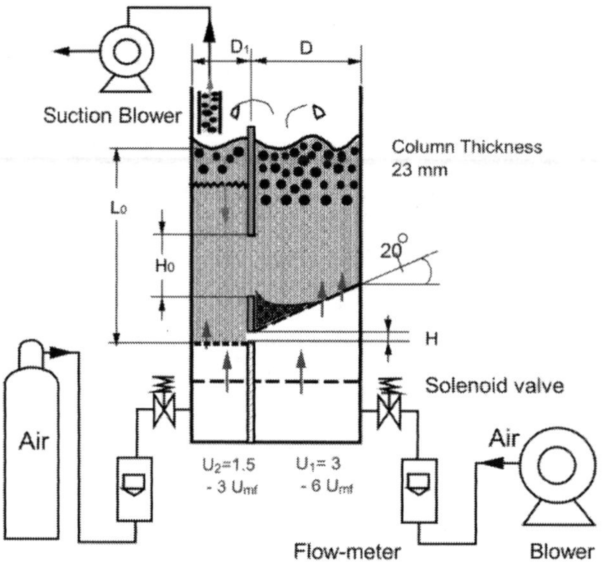

Fig. 1. Experimental Set-up

Table 1. Particle Properties of Modeled Char in Air at Atmospheric Temperature and Pressure

	d_p [mm]	ρ_0 [kg/m^3]	ρ_e [kg/m^3]	U_{mf} [m/s]
Glass Beads	1.0	2450	1410	0.42
Actv. Carbon 1 (sphere)	0.78	850	560	0.19
Actv. Carbon 2 (broken block)	0.80	-	450	0.25
Actv. Carbon 3 (broken block)	1.1	-	450	0.39
Actv. Carbon 4 (broken block)	1.45	-	450	0.48
Polystyrene	3.2	1010	510	0.70

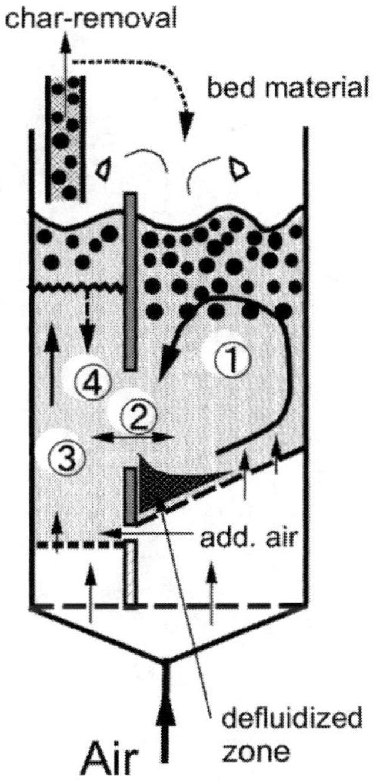

Fig. 2. Pattern and Mechanisms of Particle Circulation and Particle Exchange

Figure 2 shows the pattern and mechanisms of particle circulation and particle exchange. Based on our observations, the following mechanism is presented. (1) Particle circulation occurs along the distributor plate. (2) Particles in the separation section are entrained into the mixing (pyrolysis) section. (3) Mixed char-inert particles transfer into the separation section via the slit. (4) Char is removed as segregated flotsam from the top of the bed, as shown by Rowe [(2), (3)], in the gentle fluidized bed.

Results and Discussion

Observation of Particle Exchange

Figures 3 and 4 indicate particle transportation and return between the two compartments. Enhanced serial pictures (Fig. 3), when model char particles were placed the bed surface initially to the same level as the bed surface, illustrate particle movement due to surface-level-difference (hydraulic head difference). For batch operation, the same phenomena occur, but movement was less severe. Particle exchange occurs by entrainment and local pressure differences, even if the levels are the same in the two compartments, as shown in Fig. 4, which shows particle return from the separation section to the mixing section.

Particle Exchange (transfer) through a Slit

In order to confirm particle exchange through a slit, PIV analysis was carried out at the center of the slit. Figure 5 shows the particle oscillatory movement (velocity vector chart) at the center of slit. In this figure, the positive velocity means that there was transportation from the separation section to the mixing section. The negative value of the velocity means that particles from the mixing section moved to the separation section. The results indicate high-frequency oscillating movement, with a frequency of a few Hz. From these results, it can be confirmed that the particle exchange between two sections is possible, even when they have the same bed level.

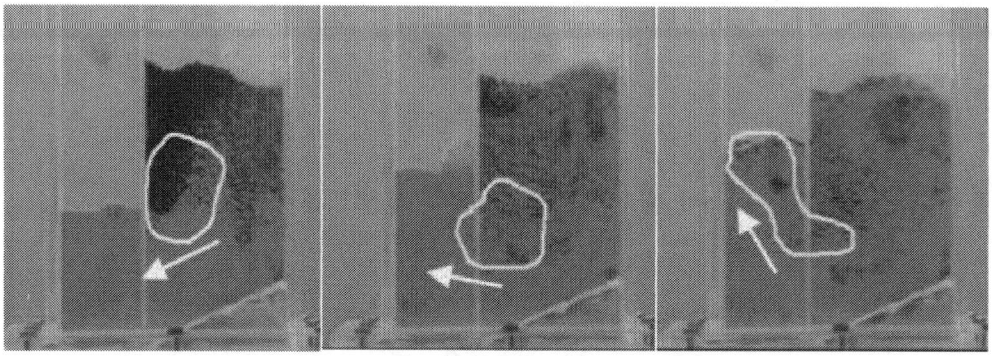

Fig. 3. Particle Transportation to Separation Section

Fig. 4. Particles Return to Mixing (pyrolysis) Section

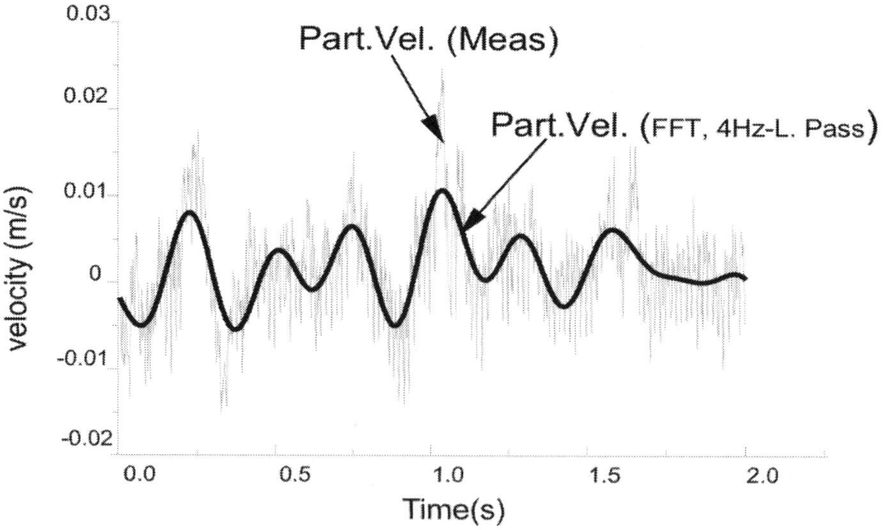

Fig. 5. Particle Oscillatory Movement at Center of Slit (vector chart)
(+) Separation Section → Mixing Section, (−) Separation Sect ← Mixing Sect

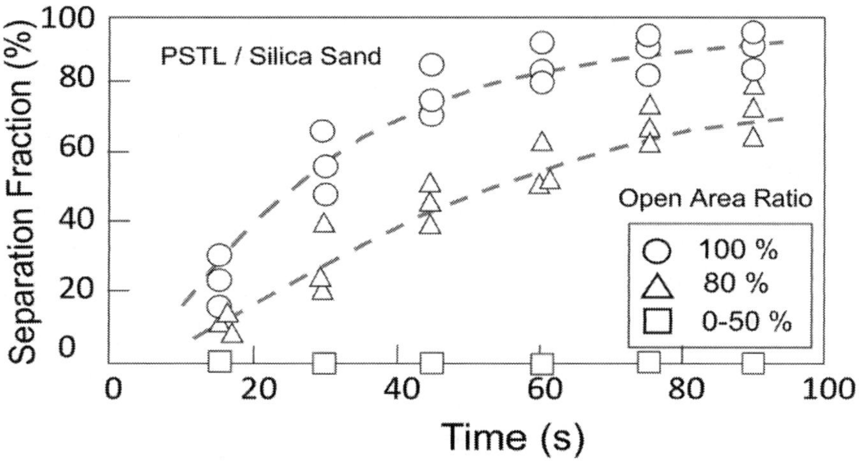

Fig. 6. Effect of Added Air on Separation Fraction
(Air flow rate depends on open ratio: open height 0-3 mm)

Effect of Added-Air on Particle Separation

Fluidization with the inclined distributor was found to become poor because of the high pressure drop at the bottom of the inclined distributor. Earlier work [(7)] on reducing a defluidized zone with an inclined gas distributor, indicates that particle circulation can be improved by controlling the local gas velocities, with an optimum ratio of surrounding gas velocity to core area gas velocity of $u_{0s}/u_{0c} = 2$. In this work added air is also supplied to improve the mobilization of particles around the defluidized zone as shown in Fig. 2. Figure 6 shows the effect of added air on the separation of polystyrene particles. The particle separation fraction is defined by m_s/m_0, where m_s is the mass of particles suctioned from the separation section and m_0 is the initial mass of the input particles. It was found that the particle separation fraction increased with increasing open area ratio which results in changing the air flow rate. As the particle separation fraction for the open area ratio of 100% (3 mm / 3 mm) was > 90% after 90 s, the 10%-slit height (= 3 mm, slit height is 30 mm as shown in Fig. 1) for the open height gives separation performance comparable to a flat plate distributor. This means about 10% height clearance of the slit height gives sufficient air flow to improve the mobilization around the low end of the inclined distributor plate.

Particle Separation Fraction and Particle Transport Fraction

Figure 7 indicates the particle separation fractions for all five types of particle for modeled char particles (Actv. Carbon 1-4 and Polystyrene in Table 1) obtained with batch operation. The suction and separation were carried out for 15 s, and then measurements were taken, and only bed material was returned to the mixing bed. In general, the separation fraction for binary solid fluidized beds strongly depends on the density ratio of these particles. Larger particles give better separation performance for particles of the same material (density). This tendency is also seen in Fig. 7, with the separation fraction improving with increasing particle diameter. The separation performance was worst for the smallest spherical particles.

Figure 8 shows the particle separation fraction for the flat and inclined distributors. These results are also from repetition of 15 s batch operation. After 90 s of operation, the results for the inclined and flat distributors were almost the same, but the inclined distributor performed better in short-time operation (up to 75 s). Since faster char particle removal resulted in better oil quality and much larger yields, the inclined system is desirable for reactor design. These results suggest that the particle circulation forced by the inclined gas distributor enables rapid transportation into the separation bed.

The particle transport fractions for the flat and inclined distributors are shown in Fig. 9. The particle transport fraction is defined by m_t/m_0, where m_t is the mass of particles transported from the mixing section to the separation section. For convenience and to obtain the results under the same bed level conditions, the particle transport fraction (strongly correlated with separation fraction) was also measured. In these experiments, the mass of char particles in the whole separation section was measured during 60 s of continuous operation. It is seen that a higher slit location gave a larger transport fraction. The inclined distributor system also resulted in a good transport fraction for the whole slit location. The transport fraction (obtained in this continuous operation) was less than the separation fraction obtained in a batch system because of no-head-difference.

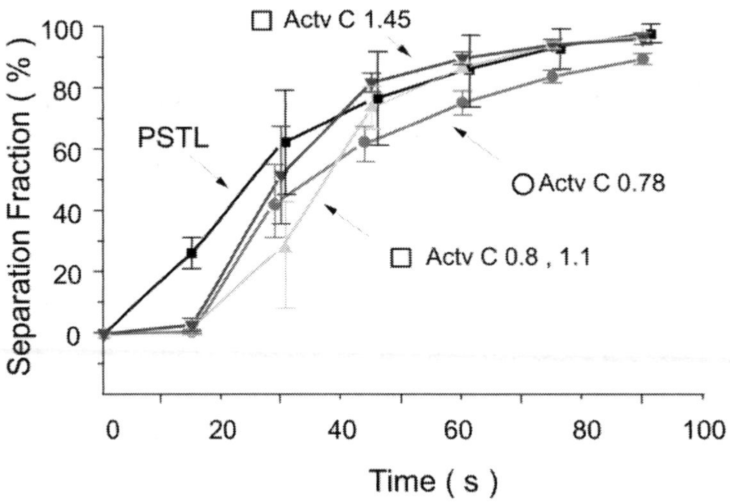

Fig. 7. Particle Separation Fraction (5 char models)

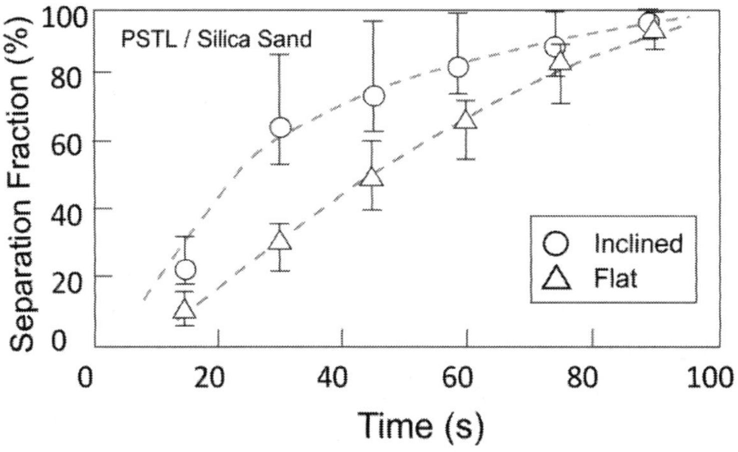

Fig. 8. Particle Separation Fraction for Flat and Inclined Distributors

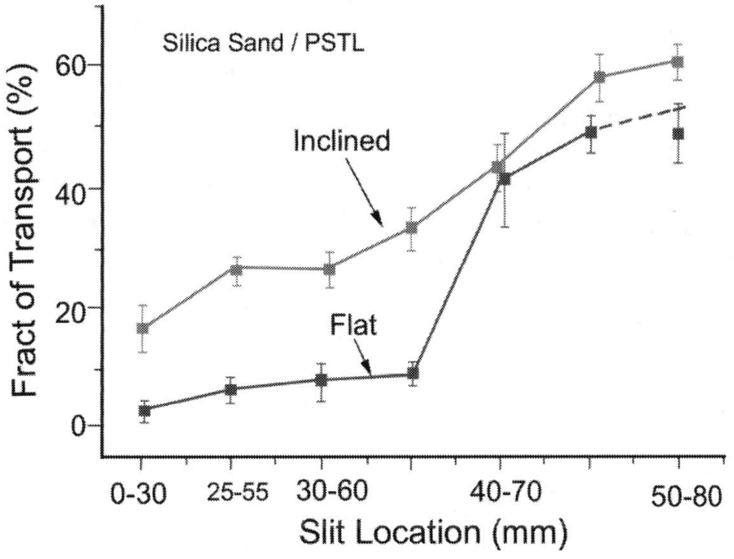

Fig. 9. Particle Transport Fraction for Flat and Inclined Distributor

Conclusions

A fast pyrolysis reactor, with char separation and char removal from bed material, was investigated in a novel connected fluidized bed system under intermittent operation. This investigation led to the following conclusions:

1. Particle exchange through a slit occurs, probably due to pressure difference around the slit zone. It is enhanced by pressure difference when char particles are continuously withdrawn from the bed.
2. The inclined gas distributor performed well in transporting particles from the bed to the other connected bed.
3. The system with the inclined distributor for particle separation and removal worked well, with its hydrodynamics elucidated by PIV analysis around the slit region.

These results are based on intermittent operation. Continuous operation (biomass feed and char removal) will be studied in the near future.

Acknowledgments

The authors are grateful for support from both the Natural Sciences and Engineering Research Council of Canada (RGPIN 7111-11) and JSPS KAKENHI Grant Number JP24560248.

References

1. V. Bridgewater (2003), "Renewable Fuels and Chemicals by Thermal Processing of Biomass", *Chemical Journal*, **91**, pp.87-102.
2. N. Rowe, A.W. Nienow and A.J. Agbim (1972), "The Mechanisms by which Particles Segregate in Gas-Fluidized Beds - Binary Systems of Near Spherical Particles", *Trans. Inst. Chem. Eng.*,**50**, pp.310-323.
3. N. Rowe and A.W. Nienow (1976), "Particle Mixing and Segregation in Gas Fluidized Beds. A Review", *Powder Technol.*, **15**, pp.141-147.

4. Mara, L. Ferrante, F. Berruti and C. Briens (2011), "New System for the Continuous Removal of Biochar from Bubbling Bed Pyrolyzers", *Bioenergy 3*, Lanzarote, Canary Islands, Spain.
5. M. Adegboye (2014), Master Thesis of Eng. Sci., Univ. Western Ontario, (Electronic Thesis and Dissertation Repository).
6. Y. Nagahashi, J.R. Grace, N. Epstein, Y. Asako and A. Yokogawa (2007), "Flow Characteristics of Fully Wetted Binary Solid Mixtures in Gas Fluidized Beds with Inclined Gas Distributors", *Fluidization 12*, ed. X, Bi, F. Berruti and T. Pugsley, ECI, Brooklyn , NY, pp.337-344.
7. T. Kita, H. Kamiya and M. Horio (1998), "Mixing and Segregation of Solid Wastes in a Fluidized Bed with an Inclined Distributor", *Kagaku Kogaku Ronbunshu*, **24-6**, pp. 901-906.

463998 Numerical Study on the Hydrodynamics Behaviour in Bubbling Fluidized Bed for Sewage Sludge Gasification

Abstract:

Sewage sludge is high moisture bio-solid produced during the sewage treatment. Gasifying sewage sludge based on the fluidization technology becomes a promising method of waste management to achieve energy recovery for sustainability and environmental protection. The hydrodynamics behaviour of sewage sludge in the fluidized bed is crucial to obtain an effective gasification process. In this numerical study, sewage sludge and saw dust are adopted as the feedstock for gasification, and air is continuously injected from the bottom of the fluidized bed. The multiphase flows of sewage sludge, saw dust and air are studied and compared under different operating conditions based on the Euler-Euler model. The results are capable of serving as reference for the design and optimisation of the sewage sludge gasification to improve the effectiveness.

460783 Realization, Control and Stability Analysis of Multiple Temperature Zones in Liquid-Containing Gas-Solid Fluidized Bed Reactor

Abstract:

Multiple temperature zones fluidized bed reactors (MTZFBRs) can overcome the limitations of uniform temperature in traditional gas-solid fluidized beds, and thus are widely used in numerous chemical engineering processes, such as fluid catalytic cracking, coal gasification, drying and granulation to mention a few. Investigations of the common scientific problems from the perspective of multiple temperature zones and multiple fluidization patterns coexistence, will greatly enhance the understanding of the fluidization characteristics and control mechanism of MTZFBR. Various mature technologies have shown that spraying liquid into a fluidized bed is one of the most effective means to realize MTZFBR. Therefore, this thesis proposes a novel MTZFBR by means of liquid-spraying within a fluidized bed reactor based on gas-phase fluidized bed polyethylene condensed mode operation process. Due to its optimization and improvement in the fluidized bed reactor process, high performance polyethylene products can be expected, which provides a new research direction and realization for product optimization and reactor process development. However, liquid-containing operation significantly affects hydrodynamic behavior and fluidization stability of fluidized beds. Realization and control of multiple temperature zones in liquid-containing gas-solid fluidized beds on the premise of stable fluidization states, is an extremely challenging research task which possesses great theoretical significance and industrial value. This thesis focuses on realization, control and stability of the MTZFBR and thus the research work has been carried out from the following five aspects:

1. A new research methodology, with combination of time scale analysis and force balance analysis, has been proposed in this thesis and thus the stability analysis model for liquid-bridge induced particle agglomerations is acquired. Firstly, the judgment of the effective particle agglomerations among liquid-coating particles is made based on time scale analysis of liquid-related key processes (such as droplet evaporation, droplet-particle collision and liquid-coating particles collision). To be specific, the particle agglomerations are induced by the liquid bridge when the droplet evaporation time scale is greater than the collision time scale among liquid-coating particles. Furthermore, the accurate judgment of fluidization instability caused by particle agglomerations is made according to the proposed force balance analysis and important criterion. Finally, based on results of stability analysis model, the spectrums of gas-liquid-solid (G-L-S) three phases fluidization patterns under different liquid-solid contact states are obtained as well as the particle agglomeration behaviors and the dominant mechanisms during the liquid-containing fluidization processes.
2. Multi-scale characterization method has been established with multiple measurement techniques including acoustic emission, pressure fluctuation and camera, which has been used for the first time to reveal the variation law of particle, bubble behavior and overall fluidization states simultaneously in the liquid-containing gas-solid fluidized bed reactor. Results demonstrate that when particle size reflected by acoustic signal varies significantly, the bubble scale (from pressure fluctuation) and overall fluidization state scale (from camera) behaviors show regular variation trends, such as formation of bubble shrinkage and gas channeling. Moreover, through Hurst and V-statistics analysis of the acoustic signal, the cyclic behavior characteristics induced by increased liquid in the fluidized bed are distinguished for the first time. The frequency of the characteristic cycle component is 400 Hz and thus the cycle time is 2.5 ms, which indicates the newly formed characteristic behavior is correlated to motion behavior of the meso-scale particle agglomeration. There are verification and complementarity among three kinds of measurements results, and all the three techniques can reflect the unstable fluidization states such as agglomerations and gas channeling during the liquid addition process.
3. With a self-designed hot mode fluidized bed apparatus, systematic experiments have been performed to realize MTZFBR. Results demonstrate that compared with the bottom liquid-spraying scheme, the upper liquid-spraying scheme shows larger temperature differences between upper zone and bottom zone, as well as higher fluidization stability, and thus the upper liquid-spraying scheme is preferable in the realization of MTZFBR. Based on upper liquid-spraying scheme, liquid flow-rate, bed height and gas inlet temperature are found to be the most significant operating parameters to affect multiple temperature zones. Moreover, studies have shown that liquid evaporation and liquid bridge are the two competitive factors during liquid-containing fluidization and thus the relative intensity of these two factors will significantly affect the temperature differences and fluidization stability. Furthermore, based on the stability analysis method proposed in the Chapter 4, two different three phases fluidization patterns are found to coexist in the stable MTZFBR. One is G-L-S three phases fluidization pattern with dynamic particle agglomerations mechanism dominated by a balanced action between liquid evaporation and liquid bridge, and the other is G-L-S three phases fluidization pattern with agglomeration breakup mechanism dominated by liquid evaporation action.
4. In the MTZFBR, the action law of particle, bubble and particle circulation pattern induced by increased liquid flow-rate have been revealed. The study shows that both increases in liquid flow-rate and decreases in gas velocity cause lower gas-liquid relative action intensity, and thus the liquid-containing fluidization process dominant mechanism shifts from gas flow (or liquid evaporation) controlling mechanism to liquid bridge controlling mechanism. As a result, the particle circulation pattern changes from one mode (particles move upward in the center and particles move downward in the wall) to another mode (downward motion of agglomerations has been enhanced significantly), meanwhile the particle and bubble motion behaviors also change with regular trends. Besides, the minimum fluidization velocity of particles is found to increase with liquid content increase. Based on temperature profile measurements, one feasible method is proposed to reflect particle circulation modes in the fluidized beds, which helps to provide a new means to study particle circulation modes.
5. For industrial multiple temperature zones fluidized bed polymerization reactor, based on the concept of G-L-S and G-S fluidization patterns coexistence, an improved emulsion-bubble two phase fluidized bed model has been proposed with help of the liquid evaporation model. The model can simulate accurately the temperature profile of industrial reactor, which proved that the industrial MTZFBR can be studied by means of two fluidization patterns coexistence. Finally, the process simulation results show that high performance products can be realized by using the reactor with composite fluidization patterns.

456821 Chiroptical Activity Enhanced Nanoparticles

Abstract:

Chirality can be considered as a ubiquitous relationship between structure and function in nature. Thus it is one of the cornerstones of todayâ€™s fundamental and applied research for the investigation of the intimate structure and property relationships of matter. Since matter structured on a length scale comparable to or smaller than the wavelength of light can exhibit unusual optical properties, chiral nanoparticles (NPs) have received a great deal of attention as promising candidates for exotic optical functions as in non-linear optics or negative refraction and so forth. The interdisciplinary character of NPs and the ease of their integration into composite materials also provide excellent opportunities to explore new dissymmetric materials. To achieve this goal, the high chiroptical activity of individual NPs is in demand as building blocks.

Here we demonstratedÂ a simple one-pot synthesis method which yields NPs having highly enhanced optical activity.Â The obtained NPs showed anisotropy factors of g=0.02 which is comparable to naturally chiral proteins, ~10x higher than conventionally synthesized NPs. Due to the high optical activity and simple synthesis procedure, this strategy offers wide range of applications as chiral sensors, catalysts and as metamaterials in advanced optical devices.

457533 Polymer Coated Nanoparticle As Additives for Interfacial Modification

Abstract:

Additives that can tailor oil-water interfacial properties are potentially useful for enhanced oil recovery processes. Here, we investigate polymer coated nanoparticles as additives for reducing oil-water interfacial tension and stabilizing oil/water emulsions. First, we investigate polymer-coated nanoparticles comprised of oxidized carbon black and covalently grafted amphiphilic polymeric chains. These polymer-coated nanoparticles can be dispersed in the aqueous phase, and when blended with surfactants the nanoparticles migrate to bicontinuous microemulsion phases. The domain size and characteristics of bicontinuous phases are characterized by interfacial tension measurements and small-angle X-ray scattering measurements. Second, we synthesize silica nanoparticles grafted with pH-responsive poly((dimethylamino)ethyl methacrylate) (PDMAEMA) chains. pH-responsive polymeric nanopoarticles can be used to stabilize oil-in-water emulsions under neutral conditions, and at low-pH triggers de-mulsificaiton of heavy oil emulsions. A combination of surface energy and absorbance measurements are used to analyze phase behavior of these emulsions. This work demonstrates the potential of stimuli-responsive polymer-coated nanoparticles for enhanced oil recovery processes.

458604 Multifunctional Magnetic Nanoparticles By Surface Initiated Atom Transfer Radical Polymerization

Abstract:

Superparamagnetic and ferromagnetic nanoparticles are of huge interest in various areas of science (*e.g.* chemistry, biology, bio-chemistry, medicine, sensor technology)[1]. The requirements for particles are as multifaceted as the areas, in which they are applied. Nanoparticles enjoy great popularity as particles at the nanoscale exhibit a favorable surface to volume ration in comparison to the same bulk material. However, this property is not that valuable, if other requirements do not fit.

For chemical applications, the particles most often have to be quickly separable, inert and stable against harsh environment like temperature, challenging pH-values and solvents.[2,3] In contrast to this, medical and biological applications usually afford dispersion stability in aqueous solutions and compatibility with biological relevant media (*e.g.* phosphate buffered saline (PBS) or Dulbeccoâ€™s modified eagleâ€™s medium (DMEM))[4]. For sensor application, the requirements to the magnetic nanoparticles contain huge dispersion stability, a high saturation magnetization and the possibility for further modification (*e.g.* attachment of a target substance *etc.*)[4,5,6]. Requirements to the magnetic nanoparticles which are highly demanded, often contain multiple combinations of different properties. This afford smart materials which contains amphiphilic character. Such smart materials can for example change the properties by an external trigger (temperature, pH, gas)[7].

Most of the different requirements can be achieved using carbon coated ferromagnetic nanoparticles as a platform. The graphene-like layers protect the magnetic metal core from oxidation and allows the covalent attachment of chemical functionalities *via* diazonium chemistry[8]. There are numerous functionalities which can be implemented by this approach. Amongst them, alcohol-, amine-, carboxy-, azide- or an initiator moiety for atom transfer radical polymerization (ATRP) can be listed, just to mention a few[7,9,10]. Adsorption of a polymer can usually not satisfy the requirements, as adsorbed polymers tend to form only single layers and desorption frequently occurs.

In contrast to this, surface initiated ATRP (SI-ATRP: polymer-growth from the surface) is suitable with a plethora of different monomers (desired functionality) and the polymerization can be well controlled (layer thickness and block-co-polymerization)[4,11]. Thus SI-ATRP gives the toolbox for various modification of the surface that allows achievement of several desired requirements.

Using charged monomers (*e.g.* 3-Sulfopropyl methacrylate potassium salt (SPM) or 3-(Methacryloylamino)propyl]trimethylammonium chloride (MAPTAC)) leads to highly stable dispersions in aqueous solutions. The stability can be tuned by the number of repeating units[4,6]. Using poly(ethylene glycol) methacrylate (PEGMA) as monomer the highly lipophilic graphene layers can be turned into hydrophilic and antifouling surfaces[12]. The use of glycidyl methacrylate (GMA) gives access to post-modification by epoxide ring opening.[4] Most challenging modifications require amphiphilic properties (dispersion stability vs. fast separation or lipophilicity vs. hydrophilicity). Such requirements need a coating with smart polymers, which can change their properties upon external trigger (temperature, pH *etc.*). By using N-isopropylacrylamine (NIPAM) as building block, the particles can reversibly change between hydrophilic and hydrophobic upon temperature change[7]. Here the promising potential of SI-ATRP on magnetic nanoparticles in terms of designing a desired nanomaterial for chemical, biological and medical applications is shown[13].

[1] A.H. Lu, E.L. Salabas, F. Schueth, *Angew. Chem. Int. Ed*, **2007**, *46*, 1222.

[2] A. Schaetz, O. Reiser, W.J. Stark, ***Chem. Eur. J.***, **2010,** *16*, 8950.

[3] Elia M. Schneider, Martin Zeltner, Niklaus Kranzlin, Robert N. Grass, and Wendelin J. Stark, *Chem. Commun.,* **2015**, *51*, 10695.

[4] C.J. Hofer, V. Zlateski, P.R. Stoessel, D. Paunescu, E. M. Schneider, R.N. Grass, M.Zeltner, W.J.Stark, *Chem. Comm*., **2015**, *51*, 1825.

[5] Q. A. Pankhurst, J. Connolly, S. K. Jones and J. Dobson, *J. Phys. D. Appl. Phys.*, **2003**, *36*, R167.

[6] M. Zeltner, R.N. Grass, A. Schaetz, S.B. Bubenhofer, N.A. Luechinger, W.J. Stark, ***J. Mater. Chem.,*** **2012**, *22*, 12064.

[7] M. Zeltner, A. Schaetz, M.L. Hefti, W.J. Stark, ***J. Mater. Chem.***, **2011**, *21*, 2991.

[8] Robert N. Grass, Evagelos K. Athanassiou, and Wendelin J. Stark, *Angew. Chem. Int. ed.,* **2007**, 4909.

[9] Alexander Schatz, Robert N. Grass, Wendelin J. Stark, and Oliver Reiser, **Chem. Eur. J., 2008,27,8262.**

[10] Chun Ghee Tan, and Robert N. Grass, *Chem.Commun.,* **2008**, *36*, 4297.

[11] R. Barbey, L.Lavanant, D. Paripovic, N. Schuwer, C. Sugnaux, S. Tugulu, H.A. Klok, *Chem. Rev.*, **2009**, *109*, 5437.

[12] Elia M. Schneider, Martin Zeltner, Vladimir Zlateski, Robert N. Grass, and Wendelin J. Stark, *Chem. Commun.,* **2016**, *52*,938.

[13] Tong Bu, Tamotsu Zako, Martin Zeltner, Karin M. Sörgjerd, Christoph M. Schumacher, Corinne J. Hofer, Wendelin J. Stark, and Mizuo Maeda, *J. Mater. Chem.B*, **2015**, *16*, 3351.

465272 Gas-Phase Synthesis of Functional Nanoparticles for Energy Applicationsin

Abstract:

In this study we establish an aerosol-based spray pyrolysis system to generate high purity metal oxide nanoparticles (NPs) with tunable particle size and composition. Copper oxide-based nanoparticle (CuOx-NP) and copper-aluminum oxide nanoparticle (CuAlOx-NP) are chosen as the representative material, and the CO oxidation is used as the model reaction. Differential mobility analysis (DMA) was employed to in-situ monitor the secondary particle size of the synthesized NPs in the gas phase, and the scanning electron microscopy was used for providing the particle imagery of the electrostatically-deposited NPs and the primary size distributions of NPs. Thermo-gravimetric analysis was employed to determine the required temperature of the pyrolysis process, and x-ray diffractometry was used to correlate the crystallite size of NP with the pyrolysis temperature. Temperature-programmed reduction system is employed to measure the ability in the oxidation-reduction of CuOx-NPs and CuAlOx-NP, which can be correlated to the ignition temperature and the conversion rate of catalytic reaction. Result shows that we can successfully synthesize CuOx-NPs and CuAlOx-NP with tunable size and composition with ability of in-situ characterization of physical size in gas phase (i.e., by DMA). By tuning the concentration and the composition of Cu and Al precursor, we can control the particle size and the morphology of nanoparticle. The reduction temperature of CuOx and the corresponding light-off temperature of the CO oxidation reaction can be effectively improved. Addition of Al precursor was found to affect the ability of reduction and catalytic performance. Our work provides a prototype study to fabricate CuOx-NPs and CuAlOx-NP with an optimal performance in catalysis. The results can be used to develop a correlation of material properties versus the oxidation-reduction ability and the catalytic activity, which have shown to be a strong indicator to the performance in the energy release and the ability in catalysis.

465894 Monolayer Assembly of Monodisperse Nanoparticles for Magnetic and Catalytic Applications

Abstract:

With unique properties different from their bulk materials, nanoparticles (NPs) with controlled size, shape and composition have been extensively studied not only for scientific interests but also for technological developments. Organic solution-phase synthesis has been demonstrated to be a versatile and robust method for synthesizing these NPs. Over the years, assembling these well-defined NPs into functional devices through "bottom-up" approach has being actively pursued. In this talk, I will give two examples of synthesis and assembly of monodisperse NPs for magnetic and catalytic studies.

Specifically, I will discuss the controlled synthesis of monodisperse cobalt ferrite NPs with tunable magnetic properties and the large-area monolayer assembly of these magnetic NPs through the water-air interfacial self-assembly approach.[1] The monolayer arrays can be used as magnetic recording media, demonstrating great potentials in magnetic data storage application. This self-assembly approach can be extended to uniform cobalt NPs, which are of great interests for catalyzing OER in water splitting.[2]In this case, highly stable cobalt NPs are demonstrated to be more active and stable than the state-of-the-art iridium catalyst. More importantly, the well-defined monolayer catalyst of the cobalt NPs enables more detailed study of the intrinsic OER activity of cobalt NPs, showing 15 times higher turnover frequency and mass activity than the NPs deposited on conventional carbon black. These two examples demonstrate the importance of assembling well-defined NPs into ordered monolayer arrays for important technological applications and fundamental studies.

References

1. Wu, L.; Jubert, P. O.; Berman, D.; Imaino, W.; Nelson, A.; Zhu, H.; Zhang, S.; Sun, S., Monolayer Assembly of Ferrimagnetic $Co_xFe_{3-x}O_4$ Nanocubes for Magnetic Recording. *Nano Lett.* **2014,** *14*, 3395-3399.

2. Wu, L.; Li, Q.; Wu, C. H.; Zhu, H.; Mendoza-Garcia, A.; Shen, B.; Guo, J.; Sun, S., Stable Cobalt Nanoparticles and Their Monolayer Array as an Efficient Electrocatalyst for Oxygen Evolution Reaction. *J. Am. Chem. Soc.* **2015,** *137*, 7071-7074.

466107 Design, Synthesis, and Structure-Function Relationship of Amphiphilic Organic Ligands for Stable Nanoparticle Dispersions in Various Solvents

Abstract:

【Introduction】

Inorganic nanoparticles such as metals, semiconductors, and oxides have recently been recognized as promising functional materials which have physicochemical properties, including electrical, optical and mechanical properties. For example, polymer nanocomposites are one of the most attractive applications using nanoparticles. In order to take advantage of the unique property of nanoparticles, it is necessary to obtain stable dispersion of nanoparticles in several solvents, hopefully in high concentration. In most cases, the colloidal stability is achieved by modifying nanoparticles with a monolayer of organic ligands. However, the choice of suitable organic ligands for each solvent and particle generally requires trial-and-error approaches and can be problematic. In this context, we found that TiO_2 nanoparticles modified with a commercial phosphoric acid surfactant can be dispersed in both polar and less-polar organic solvents and applied this surfactant to prepare several stable colloidal dispersions. The surfactant has hydrophobic alkyl chain and hydrophilic ethylene glycol chain, which might be a key for the dispersion stabilities. However, the relationship between molecular structure and colloidal stability is unclear and the structure is not optimized for better dispersion. In this study, we designed and synthesized a series of novel amphiphilic organic ligands and used them to modify TiO_2 nanoparticles as a model. The resulting modified TiO_2nanoparticles were then dispersed in several organic solvents to study the relationship molecular structure and colloidal stability.

【Experimental method】

We designed the amphiphilic organic ligands with different lengths of alkyl and ethylene glycol chains and a phosphoric acid group as an anchoring moiety. A series of the ligands were synthesized in 3 steps from commercially available alkyl bromides. The ligands structures and purities were characterized by NMR analysis. TiO_2nanoparticle aqueous dispersion was purchased from Ishihara Sangyo Kaisha, Ltd. (STS-100, primary particle size is 5 nm) and were modified with the synthesized ligands by mechanical stirring. The modified nanoparticles were aggregated in the aqueous solution, which were then collected by centrifugation and washed with water to remove free ligands and dried in vacuum. The resulting modified nanoparticles were characterized by elemental analysis and were used for dispersion study in a wide variety of organic solvents using dynamic light scattering measurement.

【Result and Discussion】

We found that the loading amount of all synthesized ligands were approximately 5 μmol/m^2, suggesting that the stable monolayers were formed through the phosphoric acid-TiO_2 bonding. It should be noted that the excess addition (10 μmol/m^2) of the ligands caused clear aqueous solutions, while white aggregations were formed when lower amounts were used. This could be explained that the bilayers were constructed at the surface of TiO_2nanoparticles to show ionic phosphoric acid group, rendering the modified particles hydrophilic. The recovered modified particles were then dispersed into several organic solvents at 3.0 wt% using sonication. We confirmed that the dispersions were stable and the mean particle sizes were estimated to be 10-20 nm by DLS measurements in most organic solvents used. We found that the slight structural differences significantly affected the colloidal dispersion stabilities, which will be discussed in details in the presentation.

【Conclusion】

The novel amphiphilic organic ligands were designed and the synthesis routes were established. We found that the slight structural differences significantly affected the colloidal dispersion stabilities, leading to the optimization of the structure. The modified TiO_2 nanoparticles were nicely dispersed in various organic solvents at useful concentrations of 3.0 wt%.

469119 Hierarchical Nickel Carbide "Dandelion" Nanostructure: Controlled Synthesis and Potential Applications

Abstract:

Transition metal carbides (TMCs), *e.g.*, Ni_3C, Mo_2C, and WC, can exhibit high electrical and thermal conductivity combined with excellent thermal and mechanical stability due to their mixed covalent-ionic bonding. TMCs can be excellent catalysts with performance comparable to noble metals for some reactions, and can also serve as supports for noble metal catalysts. Rhombohedral nickel carbide, Ni_3C, is a common product of nickel carburization during catalysis using Ni. Research on the transformation of Ni to Ni_3C helps elucidate its formation mechanism. Recently we reported a hierarchical nickel carbide (Ni_3C) nanostructure that resembles a dandelion flower (Qiao *et al. Angew. Chemie*, 2016). A typical hierarchical nanocrystal (HNC) consists of one core and many radially-arranged branches, most of which are topped by a hexagonal cap. To understand the formation of the Ni_3C HNCs, we conducted time-resolved experiments. Quasi-spherical aggregates of primary particles, referred to as "cores", were observed after 75 min. The Ni content at this point exceeds that of stoichiometric Ni_3C. After 80 min, the radial arms have grown in one dimension, and the platelets have begun to grow in the perpendicular direction. From the 75th min to the 80th min, the cores blossomed into midsize HNCs, while also densifying to become non-porous. Further growth of both the arms and the platelets proceeds through the 90th min. The atomic composition of the HNCs matched the stoichiometry of Ni_3C, *i.e.*, 75% Ni and 25% C. During evolution from cores to dandelions, the specific surface area (SSA) increased from about 3 $m^2\ g^{-1}$ to over 60 $m^2\ g^{-1}$. We are presently testing this novel nanomaterial in electrocatalytic applications and will report these results along with the synthesis in this talk. In short, we developed a hierarchical Ni_3C "dandelion" nanostructures with specific surface area much larger than that of dense particles of comparable size. Insights into their growth mechanism can enlighten further research on the controlled synthesis of hierarchical nanostructures in

solution. In addition, we believe that our current efforts will demonstrate promising applications of this material.

461832 High Density Reactive Composite Powders

Abstract:

Tungsten is used by military and civilian organizations as a penetrator and densifier due to its high density and strength. Tungsten components may be even more attractive in kinetic projectiles if their rapid oxidation could be triggered by impact. However, igniting tungsten is difficult, probably because of its high heat capacity and relatively high ignition temperature. This study is aimed to prepare and characterize composite powders comprising tungsten, titanium, and boron. The targeted density of the composite material was that of steel, 8050 kg/m3. Both boron and titanium have high heats of combustion; more importantly, they are also capable of a highly exothermic intermetallic reaction producing TiB and TiB2. This reaction is expected to be very rapid, and cause fast temperature increase of composite particles, assisting ignition of tungsten. Further, it is expected that at high temperatures all metals will proceed to oxidize producing thermodynamically favorable oxides as the final combustion products. In this study, composite materials were prepared by mechanical milling. In order to achieve fine mixing, tungsten was first balled milled with boron. Both metals are hard and produce a finely mixed composite. This composite was then milled with titanium to produce the targeted ternary compound. SEM imaging showed that the produced powder includes equiaxial fine particles with all three components mixed on the submicron scale. Particle size distributions were measured using low-angle laser light scattering. The mean particle size was 11.2 μm and the standard deviation was 1.7 μm. Differential Thermal Analysis was performed for milled binary B·Ti composite as a reference, and for the ternary W·B·Ti powders. Results showed an exothermic reaction between boron and titanium occurring in the temperature range of 700-1400 K. XRD analyses of powders heated to 1400 K confirmed formation of multiple new phases including all three elements. Powders were fed into a CO2 laser beam and burned in air. Burn times of the particles and their combustion temperatures were measured. Results will be presented and discussed in this talk.

462123 Metal-Rich Aluminum-PTFE Reactive Composite Material

Abstract:

Reactive materials combining aluminum with polytetrafluoroethylene (PTFE) with different compositions have been manufactured earlier. Expected advantages of such materials include high reactivity and greater volatility of high-temperature combustion products leading to reduced condensation and two-phase losses in propulsion systems. However, previously synthesized materials included substantial amounts of PTFE varied from 35 to 70 wt %, and thus have low density and low energy density. Materials with greater aluminum concentrations would have more attractive characteristics, but are difficult to prepare. In particular, mixing aluminum and PTFE is challenging because of substantial difference in their mechanical properties. In this study, composite Al·PTFE powders with 90 wt% Al are prepared using mechanical milling at 77 K, the liquid nitrogen temperature (cryomilling). Reference samples of the same composition are also prepared by room-temperature milling. Prepared powders are characterized using thermo-analytical measurements in both oxidizing and inert gas environments. The volatile species released upon heating are identified using mass spectrometry. Prepared and reacted materials are analyzed using electron microscopy and x-ray diffraction. Finally, the powders are ignited using different ignition stimuli. Stability and reactivity of the prepared powders are compared to those reported earlier for the mixtures of nano-aluminum and PTFE and for ball-milled Al-PTFE composites with 70 wt% Al. The mixing of PTFE is more homogeneous in the cryomilled than in room-temperature milled materials. PTFE is retained in both cryomilled and reference room-temperature milled materials upon their heating up to 820 °C in inert environment. Conversely, neat PTFE or PTFE included in the Al·PTFE composites prepared elsewhere decomposed at approximately 490 °C. In an oxidizing environment, cryomilled materials exhibit greater reaction rates as compared to room-temperature milled powders. The oxidation for all materials started well below the aluminum melting point. Finally, cryomilled samples can be ignited by electrostatic discharge (spark), whereas reference room-temperature milled materials cannot. However, when placed on a burning paper above a metal mesh, room-temperature milled materials ignited but their cryomilled counterparts did not.

467502 Preparation and Characterization of Nanostructured Composition B

Abstract:

Explosive compositions employing nanoscale crystals of high explosives (HEs) have demonstrated reduced sensitivities to external stimuli. Up to date, the investigated compositions are limited to plastic bonded explosives (PBX). It is also desirable to use nanoscale HEs in explosives that are conventionally produced by melt cast to address their sensitivity issues. However, the integration of nanoscale HEs into the melt cast process is challenging due to their high surface area. In this work, we explored the preparation of nanoscale HEs based Composition B (Comp B), a widely used high performance explosive commonly prepared by melt cast process, by spray drying and sequential pressed formulation. The composite particles obtained after spray drying were characterized by scanning electron microscopy (SEM), X-ray diffraction (XRD), and differential scanning calorimetry (DSC). The particles were also tested in impact, friction, and electrostatic discharge tests before pressed formulation. The formulated Comp B explosive was analyzed using focused ion beam- (FIB-) SEM, illustrating a novel nanostructure characterized by a large amount but small voids (~ 200 nm). The nanoscale HEs based Comp B demonstrated improved shock sensitivity in small scale gap test (SSGT). This work is consistent with previous observations that the use of nanoscale HEs can reduce the shock sensitivity of explosives. It also opens an avenue for producing nanoscale HEs based compositions unachievable using the conventional melt cast formulation process.

469508 Microwave Enhancement of Energetic Materials Combustion

Abstract:

The development of a new class of smart energetic materials with on-command controllable performance is essential to meeting future energetic material mission needs. In previous work, we have shown that alkali earth metal doping of an ammonium perchlorate composite propellant can enable strong electromagnetic flame coupling, leading to enhancement of flame structure (flame temperature), and ~70% increase in atmospheric burning rate due to the formation of microwave-supported plasmas within propellant flames. In this overview presentation, we report on the use of 1 kW, 2.45 GHz RF energy interaction with a variety of propellant and pyrotechnic energetic materials within an impedance matched, tuned waveguide cavity. We specifically discuss the use of alternative alkali dopants with improved (lower) ionization energies for use in further improving gas phase enhancement of compositions as RF-condensed phase considerations in gas phase enhancement as well as techniques to engineer the condensed phase of pyrotechnic formulations to enable switchable burning rates through RF-mediated reconfiguration of condensed phase constituents. For these techniques, we study the enhancements to burning rate, flame temperature, and emission using high speed video, two-color high speed video pyrometry, and UV-VIS-IR spectroscopy, respectively.

462593 Nanocomposite Thermites with Calcium Iodate As an Oxidizer

Abstract:

Iodine bearing reactive materials and fuel additives are being developed to inactivate aerosolized harmful microorganisms by combined thermal and chemical effects. Composite thermite powders were prepared by mechanical milling with 80 wt% of calcium iodate, $Ca(IO_3)_2$, as an oxidizer and with 20 wt % of aluminum or boron as fuels. Different milling protocols have been explored for synthesizing different materials. Morphology and particle sizes of the prepared materials were characterized using scanning electron microscopy (SEM). Aluminum containing materials formed composites with particles sized from 1 to 10 μm. After material was stored in air for two weeks, particle surface morphologies changed indicating substantial aging. In particular, smooth surface areas, indicative of formation of aluminum hydroxide, were detected under SEM. Boron containing materials produce smaller size composite particles with calcium iodate well mixed with boron. Exposure to air for two weeks caused no changes in the material morphology. Both materials were ignited as thin coatings on an electrically heated filament. The ignition temperatures of $B{\cdot}Ca(IO_3)_2$ were close to 700 K, much lower than approximately 1100 K at which ignition of $Al{\cdot}Ca(IO_3)_2$ was observed. Particles were also burned in the combustion products of an air-acetylene flame. Particle burn times were recorded optically and correlated with their sizes. The burn times for $B{\cdot}Ca(IO_3)_2$ were longer than for $Al{\cdot}Ca(IO_3)_2$. Both materials were also ignited by electro-static discharge. $Al{\cdot}Ca(IO_3)_2$ composite was difficult to ignite, while $B{\cdot}Ca(IO_3)_2$ composites ignited readily. Finally, the composite powders were pressed into pellets and ignited with a CO_2 laser. For the same size pellets, the aluminum-based materials required more energy for ignition than boron-based materials. Two Hamamatsu R3896-03 photo-multiplier tubes were equipped with 700 and 800-nm interference filters to record the pellet combustion emission. At the same time, a high speed camera and StellarNet BW16 spectrometer were employed to observe the combustion events and obtain integrated emission spectra. Bulk burn rate for $Al{\cdot}Ca(IO_3)_2$ pellet was 0.8 g/s and it was doubled for the pellet of $B{\cdot}Ca(IO_3)_2$. Results will be discussed in this talk.

468781 Boron Carbide As a Fuel for Printable Biocidal Inks

Abstract:

Boron carbide is typically considered a cheap and readily available abrasive; however, it can also be used as a ceramic fuel in pyrotechnic applications. For example, Shaw et al. recently demonstrated its use in a time delay formulation in combination with sodium periodate. This system is also interesting as a potential biocide where the limitations of aluminum can be overcome (more favorable aging and chemistry). In this work, we build upon our previous efforts focused on printable biocides and examine the role of boron carbide on the combustion and total iodine generation when combined with a suitable fluoropolymer binder and as fuel/oxidizer. It was observed that polymer content as low as 10% will result in printable formulations. Iodine generation was found to be higher than Al based formulations and enhanced with the presence of the binder. Bomb Calorimetry under both inert (heat of reaction) and air (heat of combustion) atmosphere are investigated to shed light on its effect on the end products. Simultaneous differential scanning calorimetry and thermo-gravimetric analyses (DSC/TGA) were performed to explain the interactions between the polymer, iodate, and boron carbide. Rheological properties of the biocidal ink and combustion characteristics of the printed formulations will also be presented in detail.

465257 Synthesis and Characterization of Energetic Borides

Abstract:

Borides are advanced materials for use as fuel additives for energetic systems. The potential of use of metal powders (Al, Mg and others) is almost exhausted today. New additives with higher energy characteristics are required.

The use of potentially high boron exothermicity becomes possible if it is used in the form of metal borides which also have high values of the heat of combustion. It was found that borides of Al and Mg as well as Ti and Zr are the most promising ones.

This paper describes a laboratory technology of production of borides of Al, Mg etc. including double and mixed compounds. Elemental boron, boron carbide, iron boride were used as boron source. The experimental results of particle size analysis, XRD-analysis, TGA for synthesized powders in compare with pure B and Al are given.

Ignition delay and burning rate for AlB2 are determined. The same research for model mixture of borides with binder also are analyzed. Preliminary studies have indicated that it is possible to produce borides with control and high content of target phase for applications in high energy materials..

Borides are advanced materials for use as fuel additives for energetic systems. The potential of use of metal powders (Al, Mg and others) is almost exhausted today. New additives with higher energy characteristics are required.

The use of potentially high boron exothermicity becomes possible if it is used in the form of metal borides which also have high values of the heat of combustion. It was found that borides of Al and Mg as well as Ti and Zr are the most promising ones.

This paper describes a laboratory technology of production of borides of Al, Mg etc. including double and mixed compounds. Elemental boron, boron carbide, iron boride were used as boron source. The experimental results of particle size analysis, XRD-analysis, TGA for synthesized powders in compare with pure B and Al are given.

Ignition delay and burning rate for AlB2 are determined. The same research for model mixture of borides with binder also are analyzed. Preliminary studies have indicated that it is possible to produce borides with control and high content of target phase for applications in high energy materials..

Borides are advanced materials for use as fuel additives for energetic systems. The potential of use of metal powders (Al, Mg and others) is almost exhausted today. New additives with higher energy characteristics are required.

The use of potentially high boron exothermicity becomes possible if it is used in the form of metal borides which also have high values of the heat of combustion. It was found that borides of Al and Mg as well as Ti and Zr are the most promising ones.

This paper describes a laboratory technology of production of borides of Al, Mg etc. including double and mixed compounds. Elemental boron, boron carbide, iron boride were used as boron source. The experimental results of particle size analysis, XRD-analysis, TGA for synthesized powders in compare with pure B and Al are given.

Ignition delay and burning rate for AlB2 are determined. The same research for model mixture of borides with binder also are analyzed. Preliminary studies have indicated that it is possible to produce borides with control and high content of target phase for applications in high energy materials.

470781 Combustion Performance of Laser Pyrolysis-Synthesized Boron Nanoparticles with Polytetrafluoroethylene

Abstract:

Boron has been studied as an energetic material for future use in hybrid rocket motors, as theoretical energy release of boron combustion with oxygen is much higher than that of aluminum. However, the two-stage combustion of boron and the physics/phase behavior of its oxide layer slow its reaction and hamper ignition and combustion. Polytetrafluoroethylene (PTFE) is believed to enhance the ignition behavior of boron by gasification of the oxide layer. Fluorine species can react with B_2O_3 at a much lower temperature which accelerates the boron ignition process. In this effort, the reactivity, ignition, and combustion performance of compositions of boron with nanoscale PTFE were investigated in 20/80 wt.% B/PTFE compositions. Boron powders consisted of laser-pyrolysis synthesized nano-size boron (10-15 nm) and commercially available boron powders were used. To assess the ignitability of compositions, DSC/TGA experiments indicate that nanoscale boron based compositions oxidize / react at temperatures ~490 °C which is well below those of commercial boron based compositions. T-jump experiments were conducted at heating rates of 1×10^5 °C/s and show that at high heating rate, nano Boron compositions have much lower ignition temperature than micron boron compositions. Reaction and combustion rate of compositions is assessed with closed vessel combustion experiments. Measured pressurization rates of synthesized and commercial boron/PTFE mixtures indicate the pressurization rate of nano-B /PTFE mixture is about 10 times higher than commercial boron/PTFE mixtures in both high pressure air and argon. We also report that nano boron is sensitive to photoflash ignition and can be ignited via broadband light exposure. Overall, mixtures of laser pyrolysis-synthesized boron with PTFE show much better performance than commercial boron compositions in respect to ignition temperature, combustion rate (pressurization rate), and unique photo-flash ignition sensitivity.

465512 Capillary-Driven Clustering of Semiconductor Nanorods into End-to-End Network and Vortex Structures

Abstract:

Semiconductor nanorods in the strong quantum confinement regime exhibit both size-dependent energy gap and anisotropy-derived optical properties, such as linearly polarized emission. Self-assembly of these elongated nanomaterials into long range order has been pursued for long in the past, only to garner relatively marginal progress. In this presentation, the capillary force between nanorods at the assembly interface will be discussed in the context of long-range clustering of of CdSe nanorods into two-dimensioanl end-to-end network. In addition, nanorods that cluster into upright assembly often exhibit vortex structures, which will ultimately hamper perfect arrangement of nanorods into long range. The vortex formation has been discussed in previous literature; however, in most cases, the mechanism with which the vortices form has been left to speculation. In this regard, we investigated the effect of solvent polarity on the formation or prevention of vortex structures. In this presentation, we will discuss the implication of the vortexes from the viewpoint that the nanorod-to-nanorod distance will affect carrier dynamics via inter-nanorod energy transfer.

References

[1] Kim D, Kim W D, Kang M S, Kim S-H, Lee D C 2015 *Nano Lett.* **15** 714

[2] Kim W D, Chae W.-S., Bae W K, Lee D C 2015 *Chem. Mater.* **27** 2797

463010 Three-Phase Self-Assembly of Gold Nanoparticle Monolayers – Overcoming Ligand Size Limitations

Abstract:

A simple yet effective technique was developed to prepare centimeter-scale monolayer gold nanoparticle (Au NP) films of long-range order with hydrophobic ligands. Aqueous Au NPs were entrapped at an oil/water interface where the Au NP surface could be modified with various types of amine-terminated ligands, including alkanes and polystyrene. The Au NPs then spontaneously translocated to an air/water interface to form a monolayer film. The spontaneous formation of an Au NP film at an air/water interface was due to the minimization of the system Helmholtz free energy. The self-assembled Au NP films are of a hexagonal close packed structure. The interparticle center-to-center space was dictated by the amine ligand length. These Au NP monolayers exhibit tunable surface plasma resonance and excellent spacial homogeneity for surface-enhanced Raman-scattering. The "air/water/oil" self-assembly method developed here not only benefits the fundamental understanding of NP ligand conformations, but is also applicable to scale up the manufacture of plasmonic nanoparticle devices with precisely designed optical properties.

449692 Self-Assembly of Directionally Interacting Spheres and Rods

Abstract:

There has been considerable recent interest in understanding the self-assembly of mixtures of differently shaped objects. Non-directional entropic and energetic effects are known in the specific case of rods mixed with spheres to exclusively yield layered phases. [1,2] However, here we use computer simulations and theory to demonstrate that the introduction of directional energetic attractions, between pairs of rod ends and between rod ends and isotropically interacting spherical nanoparticles (NPs), can be used to intelligently stabilize specific crystal morphologies through the interplay of energetic and entropic effects. To illustrate, we rationally design rods and spheres to obtain either face-centered cubic (FCC) or hexagonal close-packed (HCP) NP crystals at will. By tuning the relative size of the rods and spheres we control the interaction between non-nearest neighbor NPs in these crystals allowing us to control their relative stability with respect to each other, and with respect to an amorphous phase. Experiments with mixtures of gold spheres isotropically decorated with single stranded DNA (ssDNA) and rods with complementary ssDNA ends, unequivocally verify this behavior. We therefore propose that directionally specific attractions, which reflect only some features of the building block anisotropies, offer an entrée into completely different classes of self-assembly behavior.

References

[1] Adams, Dogic, Keller, and Fraden, Nature 393, 349-352 (1998).
[2] Dogic, Frenkel, and Fraden, Phys. Rev. E. 62, 3925-3933 (2000).

473985 Recovering Dynamic Parameters of Nanoparticle Assembly from Disjointed Images of Nanoparticle-Polymer Composites

Abstract:

The incorporation of nanoparticles (NPs) into polymers constitutes a powerful strategy for enhancing their thermomechanical properties and for introducing new optical, electrical, and magnetic functionalities into the polymers. Here we will describe an approach for inferring dynamic parameters of NP assembly from spatially and temporally disjointed images of NP-polymer composites [1]. The approach involves adjustment of the parameters of a kinetic model of assembly until the size statistics of NP clusters computed from the model match those obtained from high-throughput analysis of the experimental images [2]. Application to shaped, metal NPs in polymer films reveals that NP structures grow via a cluster-cluster aggregation mechanism, where NPs and their clusters diffuse in a Stokes-Einstein manner and stick with a probability that depends strongly on the size and shape of the NPs and the molecular weight of the polymer.

[1] C. R. Murthy, B. Gao, A. R. Tao, and G. Arya, "Dynamics of nanoparticle assembly from disjointed images of nanoparticle-polymer composites," *Physical Review E*, 93, 022501 (2016).

[2] C. R. Murthy, B. Gao, A. R. Tao, and G. Arya, "Automated quantitative image analysis of nanoparticle assembly," Nanoscale, 7, 9793-805 (2015)

467512 Generalized Mechanistic Model for the Chemical Vapor Deposition of 2D Transition Metal Dichalcogenide Monolayers

Abstract:

Transition metal dichalcogenides (TMDs) like molybdenum disulfide (MoS_2) and tungsten disulfide (WS_2) are layered materials capable of growth to one monolayer thickness *via* chemical vapor deposition (CVD). Such CVD methods, while powerful, are notoriously difficult to extend across different reactor types and conditions, with subtle variations often confounding reproducibility, particularly for 2D TMD growth. In this work, we formulate the first generalized TMD synthetic theory by constructing a thermodynamic and kinetic growth mechanism linked to CVD reactor parameters that is predictive of specific geometric shape, size, and aspect ratio from triangular to hexagonal growth, depending on specific CVD reactor conditions. We validate our model using experimental data from Wang *et al.* (*Chem. Mater.*, 2014, 26 (22), 6371-6379) that demonstrate the systemic evolution of MoS_2 morphology down the length of a flow CVD reactor where variations in gas phase concentrations can be accurately estimated using a transport model (C_{Sulfur} = 9–965 μmol/m^3; C_{MoO3} = 15–16 mmol/m^3) under otherwise isothermal conditions (700 °C). A stochastic model which utilizes a site-dependent activation energy barrier based on the intrinsic TMD bond energies and a series of Evans-Polanyi relations leads to remarkable, quantitative agreement with both shape and size evolution along the reactor. The model is shown to extend to the growth of WS_2 at 800 °C and MoS_2 under varied process conditions. Finally, a simplified theory is developed to translate the model into a "kinetic phase diagram" of the growth process. The predictive capability of this model and its extension to other TMD systems promise to significantly increase the controlled synthesis of such materials.

467605 Structure-Dependent Stability of Magic-Number Thiolated Metal Nanoparticles

Abstract:

Magic numbers in the stability of $M_n(SR)_m$ nanoparticles (NPs) in Brust-Schiffrin-type syntheses have been recently identified with advancements in synthesis and characterization techniques.[1] The atomically-precise nature of these particles makes them attractive candidates for understanding the thermodynamic and colloidal stability of NPs in solution. Traditionally, this magic stability has been explained by electron shell closure methods such as the "superatom model".[2] However, this model has not held in rationalizing the stability of several recently-identified NP structures, such as the $Au_{20}SR_{16}$ and the $Au_{36}SR$ introduce a theory developed using density functional theory calculations relating the exceptional stability of thiolated-metal NPs to structure-dependent, thermodynamic descriptors. Using the identified descriptors we rationalize the stability of various monometallic and multimetallic atomically-precise nanoparticles both for charged and neutral systems. Our theory is able to capture a series of experimentally observed stability trends in nanoscale thiolate-protected metal NPs. The thermodynamic implications of our findings help rationalize the observed structures of NPs and help to inform thiolate ligand selection in NP synthesis. Our findings aid in accelerating the discovery of atomically precise, stable metal nanoparticles.

1. Jin, R. Atomically precise metal nanoclusters: stable sizes and optical properties. *Nanoscale* **15,** 1549–1565 (2015).

2. Walter, M. *et al.* A unified view of ligand-protected gold clusters as superatom complexes. *Proc. Natl. Acad. Sci. U. S. A.* **105,** 9157–9162 (2008).

3. Zeng, C., Liu, C., Chen, Y., Rosi, N. L. & Jin, R. Gold-Thiolate Ring as a Protecting Motif in the Au20(SR)16 Nanocluster and Implications. *J. Am. Chem. Soc.* **20,** 16–19 (2014).

466716 Rapid Microwave Assisted Synthesis of Zinc Oxide Nano-Forest for Solar Cell Applications

Abstract:

Nanomaterials based photo-electro-chemical solar cells require high surface area as well as high electron transport quality to achieve good efficiency. Nanowires or hierarchical nanostructured arrays fulfill these two requirements. Synthesis of such complicated structures are conducted either using vapor phase deposition technique or lengthy wet-chemical route. We have synthesized Zinc oxide nano-forest structures on FTO glass using microwave assisted wet-chemical route. FESEM images revealed formation of highly ordered nanoforests structure. The main stem of zinc oxide nanowires have diameter around 100-150 nm and length about 10 μm, whereas the branch nanorods have diameter less than 50 nm and length about 1 μm. Instead of days for preparation using conventional wet-chemical routes, such nano-forest structures were prepared within 4 h using microwave assisted synthesis. XRD study revealed such nano-structures are highly crystalline c-axis oriented materials. Photo-sensitizer dye molecules were loaded on such semiconductor nanostructures to fabricate photo-anode for dye-sensitized solar cells. ZnO nanoforests based such solar cells showed enhanced photovoltaic efficiency.

470005 Iron OXIDE Nanoparticle-Graphene Patterned Interfaces

Abstract:

IRON OXIDE NANOPARTICLE-GRAPHENE PATTERNED INTERFACES

Abhilasha Dehankar[1], Justin Young[2], Josh Goldberger[3], Ezekiel Johnston-Halperin[2], Jessica O. Winter[1,4]

[1]William G. Lowrie Department of Chemical and Biomolecular Engineering, [2]Department of Physics, [3]Department of Chemistry and Biochemistry, [4]Department of Biomedical Engineering

Nanoparticles, such as superparamagnetic iron oxide nanoparticles (SPIONs), exhibit unique properties because of their small size; and therefore, have significant potential to improve optical, electronic, and magnetic devices. Significant research has occurred in the synthesis of these particles; however, integration of these particles with higher order constructs, as required for device development, remains a challenge. Further, such integration could lead to the development of synergistic structures with novel properties. To harness the full potential of these composites, nanoparticles should be capable of being assembled in a specific, user-defined, and scalable manner on larger film-based materials. Therefore, the broad aim of this research is to develop facile and controlled techniques for generation of nanoparticle-film composites and to study the emergent properties of their interfaces.

As a model system, we are investigating graphene-SPION interfaces. Graphene is a diamagnetic single layer of carbon with excellent electronic properties. However, it lacks intrinsic magnetic ordering, which could be exploited for magneto electronic or spintronic functionality. Existing strategies for generating magnetism in graphene have disrupted native electronic properties. However, recent investigations have shown induced magnetism in graphene without impairing its electronic properties when magnetic materials are placed in close proximity with graphene films[1], [2]. Thus, we deposited SPIONs, which can locally modify the magnetic field felt by charge carriers in the graphene, with

the goal of creating ordered assemblies with emergent magnetic properties. As a first approach, we deposited materials through direct deposition and determined that solvent properties significantly influence aggregation state. To create ordered structures, we first encapsulated SPIONs in micelles, that can in turn self-assemble on thin films following Langmuir-Blodgett[3] and spin coating deposition[4]. These materials were characterized using Atomic Force Microscopy (AFM) to evaluate composite surface topography. Magnetic and electronic properties of these composites would be evaluated. Knowledge gained from this research could enable novel 0D-2D interfaces. Similar techniques could be applied to a broad array of nanoparticles and surfaces to generate novel emergent behaviors.

1. Cavalcanti-Adam, E.A., et al., *Cell spreading and focal adhesion dynamics are regulated by spacing of integrin ligands.* Biophys J, 2007. **92**(8): p. 2964-74.

2. Wang, Z., et al., *Proximity-induced ferromagnetism in graphene revealed by the anomalous Hall effect.* Phys Rev Lett, 2015. **114**(1): p. 016603.

3. Spatz, J.P., et al., *Ordered Deposition of Inorganic Clusters from Micellar Block Copolymer Films.* Langmuir, 2000. **16**(2): p. 407-415.

4. Liou, J.-Y. and Y.-S. Sun, *Monolayers of Diblock Copolymer Micelles by Spin-Coating fromo-Xylene on SiOx/Si Studied in Real and Reciprocal Space.* Macromolecules, 2012. **45**(4): p. 1963-1971.

470691 Alloy-Core@Shell Nanoparticles for Catalysis

Abstract:

Alloy-core@shell nanoparticles are designed to combine the strengths of the alloy and core@shell particles, and avoid their flaws. The noble shell protects the particle core during the catalytic processes, and the alloy-core composition allows for fine tuning of the catalytic properties.[1,2] The Dendrimer-encapsulated nanopartilces (DENs) as a model catalyst is sufficiently small and well-characterized that its function can be directly predicted by theory. Specifically, our work seeks to develop a fundamental and detailed understanding of the relationship between the structure of nanoscopic electrocatalysts and their function. Two categories of stories will be given in the talk:

1) **tailoring catalytic function by tuning compositions**: Trends in reaction descriptor (binding energies of key reactants) were calculated with density functional theory (DFT) to probe the catalytic activity of two types of alloy nanoparticles structures: random alloy (X/Y)[3] and alloy-core@shell (X/Y@Z)[1,2] with various compositions. Establishing the general principal of correlation between compositions and activities provides guidelines for designing novel catalysts.[4] First-principal theory prediction of PdAu@Pt DENs for ORR were examined by experiment and leads to a great agreement with experiment results.[2]

2) **enhanced stability by alloying core:** $Pd_xAu_{140\_x}$@Pt DEN electro catalysts were tested for CO oxidation at wider alloying range.[5] Both experiments and DFT calculations suggest that this unusual behavior is caused primarily by structural changes of the DENs at high and low values of x. The alloy PdAu cores stabilize the core@shell structures by preventing Au and Pd from escaping the core.[5] These findings illustrate the importance of controlling both the stability and reactivity of the catalysts, and they provide guidance as to how core composition can be used to do that.

Overall, we demonstrate that iteration between theory and experiment can facilitate an understanding of nanoparticle catalysts and reduce the time and effort involved in the design of new catalysts.[6,7]

[1] L. Zhang, G. Henkelman, J. Phys. Chem. C **116** 20860-20865 (2012).

[2] L. Zhang, R. Iyyamperumal, D. F. Yancey, R. M. Crooks, and G. Henkelman, *ACS Nano* **7**, 9168-9172 (2013).

[3] W. Tang, L. Zhang, G. Henkelman *J. Phys. Chem. Lett.* 2011, **2**, 1328-1331.

[4] L. Zhang and G. Henkelman, *ACS Catal.* **5**, 655-660 (2015)

[5] L. Luo, L. Zhang, G. Henkelman, and R. M. Crooks, *J. Phys. Chem. Lett.* **6**, 2562-2568 (2015)

[6] L. Zhang, R. M. Anderson, R. M. Crooks, and G. Henkelman, *Surf. Sci.* **640** 65-72 (2015)

[7] R. M. Anderson, D. F. Yancey, L. Zhang, S. T. Chill, G. Henkelman, and R. M. Crooks, Acc. Chem. Res. **48** 1351-1357 (2015).

451367 Thermal Stability of Shape-Controlled Pd-Pt Core-Shell Nanoparticles

Abstract:

Thermal Stability of Shape-Controlled Pd-Pt Core-Shell Nanoparticles

Luke T. Roling[a], Ahmed O. Elnabawy[a], Madeline Vara[b], Xue Wang[b], Younan Xia[b], and Manos Mavrikakis[a]

[a]Department of Chemical & Biological Engineering, University of Wisconsin-Madison, Madison, WI
[b]Department of Biomedical Engineering, Georgia Institute of Technology, Atlanta, GA

We recently reported the synthesis and electrocatalytic performance of shape-controlled Pd-Pt core-shell nanoparticles that demonstrated exceptional activity and stability toward the oxygen reduction reaction.[1,2] In this presentation, we investigate and explain the durability of these catalysts in high-temperature environments typical of a broader range of catalytic applications. We observe that the cubic and octahedral nanocatalysts are stable to at least 500 °C; interestingly, the core-shell structure and overall particle shape are each deformed at different temperatures in the two catalyst geometries.

Detailed DFT calculations rationalize the observed behaviors, attributing the differences in reconstruction to lower activation energy barriers for metal atom migration from the nanoparticle edges. Calculations also indicate that different concentrations of subsurface vacancies are seeded during nanoparticle synthesis, and that these are responsible for differences in Pd-Pt interdiffusion, which causes disruption of the core-shell structure at elevated temperatures.

1. S. Xie, S.-I. Choi, N. Lu, L. T. Roling, J. A. Herron, L. Zhang, J. Park, J. Wang, M. J. Kim, Z. Xie, M. Mavrikakis, Y. Xia, *Nano Letters* **2014**, 14, 3570
2. J. Park, L. Zhang, S. Choi, L. T. Roling, N. Lu, J. A. Herron, S. Xie, J. Wang, M. J. Kim, M. Mavrikakis, Y. Xia, *ACS Nano* **2015**, 9, 2635.

462615 Accelerating Nanoparticle Synthetic Development Using in Situ Small Angle X-Ray Scattering

Abstract:

The fine control of monodisperse nanoparticles (NPs) over their size, shape, and structure is essential for understanding their properties, especially catalytic properties. In the past two decades, significant developments have been achieved in the colloidal synthesis of monodisperse NPs. However, synthesizing well-defined NPs in a predictable way is still challenging. Here, we use synchrotron-based in situ small-angle X-ray scattering (SAXS) to experimentally monitor the formation of NPs during solution phase synthesis. We choose Pd NPs as an example. Through quantitative analysis of the SAXS data, the number, size, and size distribution of Pd NPs during the synthesis are determined. By real-time probing the evolution of Pd NPs, we systematically study the effects of different surfactants (e.g., oleylamine, trioctylphosphine) on the kinetics of NP nucleation and growth. Taking advantage of this in-situ SAXS technique, we are able to better understand the growth mechanism of Pd NPs and as a result optimize the synthetic parameters to precisely control the Pd NP size from 2 to 10 nm, and these well-controlled Pd NPs serve as a model system for studying their size-dependent catalysis for methane combustion reaction.

464480 Selective Oxidation of Methane to Methanol Using Pd/Au Bimetallic Nanoparticles

Abstract:

Selective Oxidation of Methane to Methanol using Pd/Au Bimetallic Nanoparticles

Cody Wrasman[1], Joshua J. Willis[1], Jong Suk Yoo[1], Jens K. Nørskov[1], Matteo Cargnello[1]

[1]*Department of Chemical Engineering and SUNCAT Center for Interface Science and Catalysis, Stanford University, Stanford, CA 94305, USA*

Methane, which makes up the largest fraction of natural gas, is an underutilized natural resource. In petroleum operations, byproduct natural gas is often flared rather than collected. The difficulty in utilizing methane fully lies in transporting gases from scattered wells to a central location for processing. Therefore, a method to convert methane to liquid products must be developed to fully utilize this resource. One promising liquid product is methanol, which is already used in the industrial chemical industry in massive quantities, can be converted into gasoline or other transportation fuels, and has potential to serve as a future transportation fuel itself. Despite significant research efforts, methods to convert methane to methanol at economically viable conditions, ($T<300$ °C and near ambient pressure) none have been identified. One of the largest challenges in developing processes to carry out this conversion is the fact that the oxidation of the first C-H bond in methane is the least energetically favorable step in the methane total combustion reaction that converts methane to carbon dioxide. Once one C-H bond is oxidized, the others are increasingly easy to oxidize, thus creating the undesired byproduct carbon dioxide. One possible way to address this is to synthesize a catalyst with active regions small enough to only activate and oxidize a single C-H bond in methane per catalytic cycle. Computational simulations predict that this may be accomplished by dispersing Pd onto Au nanoparticles.

In this work, controlled amounts of Pd are added to monodisperse Au nanoparticles of several sizes using colloidal synthesis methods. Pd is one of the most active methane oxidation catalysts while Au is relatively inert to the reaction. The aim of the syntheses was to add controlled coverages of Pd to Au NPs from atomic dispersion to complete Pd shells. These particles were deposited onto porous TiO_2 supports and tested for their catalytic activity for the gas phase methane to methanol reaction. These experiments allow for the elucidation of structure-activity relationships for Pd/Au systems in the methane to methanol reaction and provide a starting point for future catalyst design.

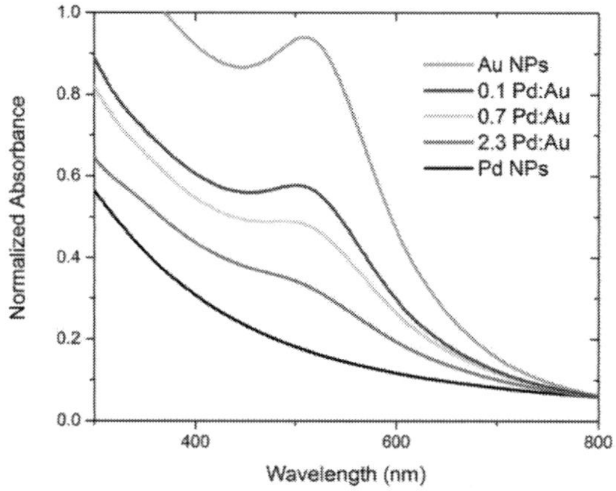

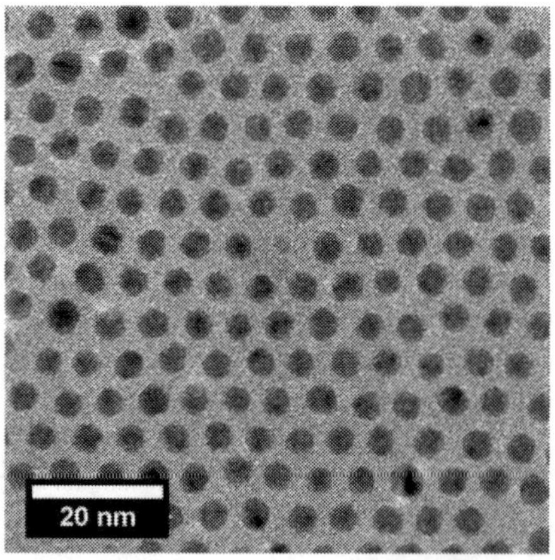

457439 Using Bound Organic Ligands to Direct Particle Size and Understand Catalytic Activity of Gold Nanoparticles

Abstract:

The environment around an enzymatic active site is precisely controlled through a combination of geometric confinement as well as precise placement of organic functional groups around the active site. In a nature-inspired approach [1], organic ligands are utilized to surround a metallic active site in order to learn more about how these ligands control access to the active site as well as how they control the activity of the active site.

To investigate the effects that different organic ligand environments have around a catalytic active site, 4-nitrophenol reduction and resazurin reduction were studied using a variety of different phosphine and thiol-containing ligands bound to the gold cluster surface. 4-nitrophenol reduction is widely used to assess the catalytic properties of metallic nanoparticles, and there is debate in the literature as to where the active site is located. [2,3] This comparative study uses these organic ligands with different geometric and electronic properties to gain further insight into the factors that control catalysis in the 4-nitrophenol reduction system.

In the synthesis of metallic clusters, the steric properties of the ligands that are bound to the metal in the precursors affect the final cluster size and accessibility of the final metal clusters. This work uses different phosphine ligands to direct the synthesis of gold clusters/nanoparticles through novel synthetic procedures. [4] These different particles exhibit different amounts of accessibility to the gold cluster surface, depending on the packing of the ligand on the cluster surface. A relationship between the particle size, ligand size, and its accessibility to small probe molecules is demonstrated for these clusters.

References:

1. Trogadas, P.; Nigra, M.M.; Coppens, M.-O.; *New J. Chem.* 2016, DOI: 10.1039/C5NJ03406J
2. Nigra, M.M.; Ha, J.-M.; Katz, A.; *Catal. Sci. Technol.* **2013**, *3*, 2976.
3. Mahmoud, M.A.; Garlyyev, B.; El-Sayed, M.A.; *J. Phys. Chem. C* **2013**, *117*, 21886.
4. Nigra, M.M.; Yeh, A.J.; Okrut, A.; DiPasquale, A.G.; Yeh, S.W.; Solovyov, A.; Katz, A.; *Dalton Trans.* **2013**, *42*, 12762.

481293 Enhancing the Ni/YSZ Catalytic Performance in the Activation of Methane: The Influence of the Electric Field and Interface Carbon

Abstract:

Enhancing the Ni/YSZ Catalytic Performance in the Activation of Methane: The Influence of the Electric Field and Interface Carbon

Fanglin Che[1], *Su Ha*[1], *and Jean-Sabin McEwen*[1]*

[1]Washington State University, Pullman, WA 99163 (United States)

**js.mcewen@wsu.edu*

Understanding the oxygen vacancy formation at the triple phase boundary (TPB) of a yttrium-stabilized zirconia supported Ni nanocluster (Ni/YSZ) is of importance since such vacancies are the active sites for coke formation [1-3] and sulfur poisoning [4]. In this way, one can make a step toward designing catalysts that decrease the deactivation of the Ni/YSZ electrode via the formation of coke in a solid oxide fuel cell (SOFC) and sulfur poisoning in a solid oxide electrolysis cell (SOEC). Our previous field-dependent microkinetic model and the corresponding experimental evidence have shown that a positive electric field can enhance the methane conversion and reduce coke formation in important ways during the Ni-based methane steam reforming (MSR) process [5-10]. Therefore, considering the field effects on the oxygen vacancy formation is necessary as such fields exist in both cells [11] and can potentially be used to alter the Ni/YSZ system so as to directly modify its electrocatalytic performance.

Our computational results [12] show that a positive electric field (less than 0.6 V/Å) can significantly increase the TPB oxygen vacancy formation energy in the Ni/YSZ+O system (**Figure 1(a)**). This suggests that the presence of such a positive electric field can reduce the oxygen vacancy concentration at the TPB region, and subsequently suppress the sulfur poisoning at the TPB region of the Ni/YSZ+O electrode in a SOEC. A negative electric field in a SOFC could lead to more active TPB vacancies by reducing the vacancy formation energies in the Ni/YSZ+O model. Both charge distribution and effective dipole moments verify the qualitative findings with regard to how the field influences the formation of an oxygen vacancy in Ni/YSZ. Overall, this investigation provides guidance for designing a Ni/YSZ electrode with an improved electrocatalytic performance via a simulated positive electric field.

In addition, there has been some debate over the past decades on the role of low concentrations of interface carbon complexes in the activation of hydrocarbons over transition metal surfaces [13]. It is also a mystery as to whether or not electric fields can enhance the conversion of methane in a fuel cell or during electro-reforming at temperatures that are lower than the normal reforming conditions [14, 15]. To provide a 'bottom-up' fundamental understanding, we present the first C-H cleavage in methane over the Ni/YSZ catalysts as a case study since such reaction is the rate-limiting step during the MSR process. Our theoretical results show that the presence of carbon or carbide-like species at the interface between the Ni cluster and its metal oxide support, as well as the application of an external positive electric field, can significantly increase the local oxidation states of the Ni nanocluster. Furthermore, the first C-H cleavage in methane is thermodynamically and kinetically favored when the oxidation state of Ni is increased (**Figure 1(b)**). As a result, the presence of a low concentration of surface carbon, interfacial carbide species, or the addition of a positive electric field will facilitate the methane activation process.

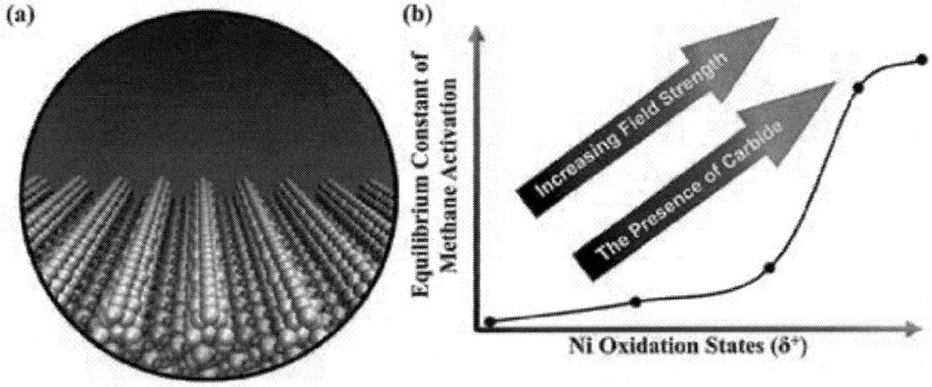

Figure 1. (a) Optimized structures used for the oxygen enriched Ni/YSZ (Ni/YSZ+O) model. Atoms with the light green, dark green, silver, and red colors represent the Zr, Y, Ni, and O species. (b) DFT results of the equilibrium constants at 873 K of the first C-H cleavage over Ni/YSZ as a function of the Ni oxidation state, which can be altered by the applied electric fields or altered by the presence of carbon species.

References

[1] Selman, J.R., *Science* **2009,** *326*, 52-53.

[2] Suzuki, T.; Hasan, Z.; Funahashi, Y.; Yamaguchi, T.; Fujishiro, Y.; Awano, M., *Science* **2009,** *325*, 852-855.

[3] Zhang, Y.; Lu, Z.; Yang, Z.; Woo, T., *J. Power Sources* **2013,** *237*, 128-131.

[4] Ebbesen, S.D.; Mogensen, M., *J. Power Sources* **2009,** *193*, 349-358.

[5] Che, F.; Ha, S.; McEwen, J.-S., *Appl. Catal. B* **2016,** *195*, 77-89.

[6] Che, F.; Gray, J.; Ha, S.; McEwen, J.-S., *J. Catal.* **2015,** *332*, 187-200.

[7] Che, F.; Zhang, R.; Hensley, A.J.; Ha, S.; McEwen, J.-S., *Phys. Chem. Chem. Phys.* **2014,** *16*, 2399-2410.

[8] Che, F.; Hensley, A.; Ha, S.; McEwen, J.-S., *Catal. Sci. Technol.* **2014,** 4020-4035.

[9] Che, F.; Gray, J.; Ha, S.; McEwen, J.-S., *In preparation* **2016**.

[10] Che, F.; Gray, J.; Ha, S.; McEwen, J.-S., *Submitted to ACS Catal.* **2016**.

[11] Stuve, E.M., *Chem. Phys. Lett.* **2012,** *519-520*, 1-17.

[12] Che, F.; Ha, S.; McEwen, J.-S., *J. Phys. Chem. C* **2016,** *120*, 14608–14620.

[13] Teschner, D.; Borsodi, J.; Wootsch, A.; Révay, Z.; Hävecker, M.; Knop-Gericke, A.; Jackson, S.D.; Schlögl, R., *Science* **2008,** *320*, 86-89.

[14] Périllat-Merceroz, C.; Gauthier, G.; Roussel, P.; Huvé, M.; Gélin, P.; Vannier, R.-N., *Chem. Mater.* **2011,** *23*, 1539-1550.

[15] Sekine, Y.; Haraguchi, M.; Tomioka, M.; Matsukata, M.; Kikuchi, E., *J. Phys. Chem. A* **2009,** *114*, 3824-3833.

465433 Metal Nanocatalysts, Their Synthesis and Size Dependent Covalent Bond Catalysis. Oxide Metal Interfaces As Active Sites for Acid-Base Catalysis

Abstract:

Colloidal chemistry is used to control the size, shape and composition of metal nanoparticles usually in the 1-10 nm range. In-situ methods are used to characterize the size, structure (electronic and atomic), bonding, composition and oxidation states under reaction conditions. These methods include sum frequency generation nonlinear optical spectroscopy (SFG), ambient pressure X ray photoelectron spectroscopy (APXPS) and high pressure scanning tunneling microscope (STM). The catalytic behavior depends on the size, oxidation state, coordination number, crystallographic orientation of metal sites and bonding and orientation of surface adsorbates.

When metal nanoparticles are placed on different mezoporous or microporous oxide supports, the catalytic turnover rates and selectivities markedly change. The charge flow between the metal and the oxide ionizes the adsorbed molecules at the oxide-metal interfaces and alters the catalytic chemistry (acid-base catalysis).

The oxidation state of metal nanoparticles becomes less metallic and assumes higher oxidation states with decreasing size. The small nanoclusters behave similar to transition metal ions that are active homogeneous catalysts. Hybrid systems composed of enzymes, homogeneous and heterogeneous catalysts are constructed to study molecularly unified catalytic schemes for the future.

458640 Titania Powders in the Selective Laser Sintering Processes

Abstract:

In selective laser sintering (SLS) a laser beam is used to partially melt particles in a layer of powders. With subsequent increments of powder layers it is possible to create three dimensional structures. This technique can be used in prototyping applications able to produce customized objects with different shapes and materials, e.g. polymer, ceramics and metals. In SLS technique the strength of the final structures increases with the energy transferred by laser. However, it also produces a volume contraction of the sintered material reducing the precision of the final object. Since the energy required for sintering depends on the particle size, sintering energies can be reduced using smaller particles, which also allow the production of more complex and precise structures due to reduced melting.

In this work, mixtures of titania powders with different particle size distributions were employed. They were obtained by mixing a coarser grained powder (particles in the range 10 - 100 microns) with a finer grained powder (particles in the submicron range). The powder composition and the powder laying procedure were studied and optimized in order to obtain specimens with of the desired mechanical properties. The equipment used is a three dimensional laser sintering equipment using a 40W CO_2 laser beam purposely built. The density and the mechanical resistance of the specimens are studied as a function of the fines content and of the amount of energy released by the laser beam on the unit surface of the lighted area. A simple theoretical modelling approach is also used to estimate the strength of the single sintered contact.

467501 The Crystal Structure and Surface Composition of Coalescing Ag-Au Nano-Alloys By Molecular Dynamics Simulations

Abstract:

Abstract

Bimetallic nanoparticles have gained significant commercial interest because of their superior or even novel electronic, chemical and plasmonic properties compared to the monometallic counterparts making them excellent candidates for many biomedical, sensory or catalytic applications. For example, gold nanoparticles that are used extensively in catalysis exhibit reduced electron transfer on Au(111) surface, hindering the adsorption of O_2. However, formation of gold-based bimetallic nanoparticles (e.g. with Ag) can enhance the affinity with O_2 compared to pure gold nanocatalysts. Furthermore, Ag nanoparticles exhibit remarkable antibacterial properties but are highly cytotoxic rendering their use as therapeutic agents challenging. Alloyed Ag-Au is an interesting example in biomedicine with potential in theranostic applications since addition of Au in Ag nanoparticles improves biocompatibility without destroying the antibacterial activity of nano-silver (Sotiriou et al., 2014).

Here, the evolution of surface composition of free-standing but coalescing Ag-Au nanoparticles is investigated for different particle sizes and temperatures by atomistic molecular dynamics (MD) simulations. The MD method is validated by the attainment of the melting point of Ag-Au core-shell nanoparticles that increases with increasing particle size and follows closely the trend of the size-dependent melting temperature of pure Au (Goudeli et al., 2016) and Ag nanoparticles (Buesser et al., 2015). Silver atoms exhibit increased mobility upon coalescence and occupy gradually the surface of the segregated particle, consistent with experiments. When Ag nanoparticles are sufficiently smaller than Au ones, a patchy Ag layer forms at the Au particle surface. Sintering of equally-sized Ag and Au nanoparticles results in the formation of segregated nano-alloys with Ag-enriched surface, consistent with the literature. The initial particle morphology affect the sintering rate but also the particle crystallinity. The X-ray diffraction patterns of Ag-Au nanoparticles are calculated during sintering, revealing that even though core-shell configurations and segregated structures exhibit the characteristic peaks of pure Ag and Au, the alloyed nanoparticles exhibit only the (111) and (200) peaks.

Keywords: crystallinity, XRD, surface composition, silver-gold, molecular dynamics.

References

Buesser, B., and Pratsinis, S.E. (2015) *J. Phys. Chem. C*, 119, 10116-10122.

Goudeli, E., and Pratsinis, S.E. (2016) *AIChE J*, 62, 589-598.

Sotiriou G A, Etterlin G D, Spyrogianni A, Krumeich F, Leroux J-C, and Pratsinis S E. (2014). *Chem. Commun.* 50: 13559-13562.

470187 Model Reduction of Population Dynamics of Particles in Bi-Component Granulation Processes

Abstract:

Wet granulation in pharmaceutical plants is a broadly used process, in which binder droplets facilitate binding of the small drug particles together and larger granules formation. The governing equation of these dynamic systems is derived by a population balance study which turns in a complex mathematical model. However, in order to control these processes in real time, development of reduced order models for these processes is necessary. This work proposes a new model reduction method for two-component coagulation systems by exploiting the method of moments and orthogonal projection approach.

The two-component agglomeration process considered in this paper has no chemical reactions. The population balance analysis for this granulation system is in form of an Integro-Differential Equation (IDE). Also, the rate of coagulation namely the coagulation kernel is dependent on size and composition. This kernel considers two steps of agglomeration; collision and binding. Either of these two steps happens with a probability, called *collision probability* and *collision success factor*, respectively. However, there exists no explicit analytical solution of the IDE for this type of coagulation kernels.

In order to simplify the results of the method of moments, some papers such as [1], exploit Taylor expansion and derive a closed finite-dimensional ordinary differential equation set. However, this approach cannot be used for composition-dependent models such as the one used in [2]. To address this issue, this work proposes a new model reduction approach using the method of moments in conjunction with Laguerre polynomials. In this way, we expand the distribution function over the set of orthogonal Laguerre polynomials which are function of moments. Also, we evaluate our new reduced model with the results obtained from a Monte Carlo simulation as a bench mark. This model will be the foundation for efficient observer and controller design for such bi-component agglomeration processes.

References:

[1] N. Hashemian and A. Armaou, "Simulation, model-reduction and state estimation of a two-component coagulation process,"Â AIChE J., 2016, 62, pp.1557-1567, DOI:10.1002/aic.15146.

[2] N. Hashemian, M. Ghanaatpishe and A. Armaou, â€œBi-Component Granulation Processes via Laguerre Polynomials,â€ Proceedings of the American Control Conference,Â accepted, MA, USA, 2016.

452250 On Optimising Particle Breakage for a Downstream Process: An Attainable Region Approach

Abstract:

The use of the Attainable region technique to milling data is presented. We considered optimising material of the desired size class for a mill by changing size class of interest and analysing the optimum test conditions. Using a feed of 1700-850μm, two size classes (850-150μm and 150-75 μm) are selected to evaluate the optimum interstitial filling to be used for the mill. Values of 1.75 and 1 are obtained respectively as optimal. Results show that optimal operating conditions are different for different objective functions. The authors show that the AR may be used to determine optimal conditions (using different objective functions) that may implemented during particle reduction processes. Included also is an extension of AR boundary were optimising using 'mixing principle' is employed.

472348 Thermodynamic Model for the Prediction of Janus Particle Morphology and Interfacial Tensions

Abstract:

Composite Janus particles exhibit many unique structural, chemical, electrical, and optical properties owing to their two distinct surfaces. Phase separation of two dissolved polymers within O/W emulsion droplets is a straightforward approach to large-scale synthesis of Janus particles. An array of Janus particle morphologies can be obtained by changing experimental conditions such as surfactant type and concentration, total polymer concentration, solvent evaporation rate, and solvent type. Of the aforementioned factors, it was found that the type and concentration of ionic surfactant have the most significant effect on particle morphology. If kinetic factors (i.e., solvent evaporation rate) are adequately controlled, then the morphology of Janus particles can be predicted based on thermodynamics alone.

The starting point of the model is the creation of a Gibbs free energy equation containing terms for each of the three interfaces: 1) Polymer 1-Water, 2) Polymer 2-Water, and 3) Polymer 1-Polymer 2. The Janus particle is modeled as two overlapping spheres, and the free energy equation is minimized using Lagrange multipliers to ensure constant Polymer 1 and Polymer 2 volumes. The end result is a system of equations that can quantitatively predict Janus particle morphology given interfacial tensions. The model accurately predicted PLGA/PCL Janus particle morphology for three case studies using PVA, SDS, and SDBS surfactants.

In reverse, the model can be applied to predict interfacial tensions based on Janus particle morphology. This is an important feature of the model, as the interfacial tension between two solid polymers cannot be experimentally measured. Results from the PLGA/PCL system were compared to literature values of calculated PLGA-PCL interfacial tensions with excellent agreement.

466619 Cellulosic Stabilized Amorphous Solid Dispersions of HMX and CL-20

Abstract:

Amorphous solid dispersions of small organic molecules offer a number of advantages in comparison to their crystalline counterparts for practical application. When considering pharmaceuticals, glassy materials exhibit enhanced bioavailability, making them beneficial both for cost and effectiveness. In the field of energetics, amorphous solid dispersions of explosive compounds are desirable because they display exciting properties such as optical transparency, enhanced reactivity, and glass castability. While possessing useful properties, stability of amorphous solid dispersions is problematic, as the form is necessarily strained, and thus of higher energy than the ordered crystalline state. Stability can be achieved by using polymeric excipients, which inhibit crystallization, and in this application, cellulosics have been found to be particularly effective. In this work, the use of cellulosic polymers to inhibit the crystallization of a mixture of octogen (HMX) and hexanitrohexaazaisowurtzitane (CL-20) is explored. The difference in stability of nitrocellulose, cellulose acetate, and cellulose acetate butyrate stabilized dispersions of HMX and CL-20 is elucidated. Of particular interest is crystallization kinetics and glass transition temperature. These amorphous solid dispersions are characterized with the following techniques: X-Ray diffraction, Raman spectroscopy, thermomechanical analysis, differential scanning calorimetry. Overall, nitrocellulose has been found to be particular effective for inhibiting crystallization, which can be tentatively attributed to its moieties which match that of HMX and CL-20.

V. Stepanov, R. B. Patel, R. Mudryy, H. Qiu, Investigation of Nitramine-Based Amorphous Energetics, Prop. Exp. Pyro. 2016, 41, 142-147. DOI: 10.1002/prep.201500118

R. Knepper, K. Browning, R. R. Wixom, A. S. Tappan, M. A. Rodriguez, M. K Alam, Microstructure Evolution during Crystallization of Vapor-Deposited Hexanitroazobenzene Films, Prop. Exp. Pyro. 2012, 37, 459 â€“467. DOI: 10.1002/prep.201200013

N. Shamin, Y. P. Koh, S. L. Simon, G. B. McKenna, The glass transition of trinitrotoluene (TNT) by flash DSC, Thermochimica Acta, 2015, 36-39. DOI: 10.1016/j.tca.2015.10.003

466105 Spray Drying of Ternary Solid Dispersions of Itraconazole: Amorphous Stability By Design of Experiment

Abstract:

SPRAY DRYING OF TERNARY SOLID DISPERSIONS OF ITRACONAZOLE: AMORPHOUS STABILITY BY DESIGN OF EXPERIMENT

Mark T. Davis[1]*, Catherine B. Kelly[1], Ahmad B. Albadarin[1], Maryam Mohammadpour[1]â€Ž
Mustafa Sajjia[2] & Gavin Walker[1]

[1]Synthesis and Solid State Pharmaceutical Centre (SSPC), Bernal Institute,
University of Limerick, Limerick, Ireland
[2]Pharmaceutical Manufacturing Technology Center (PMTC), University of Limerick, Limerick, Ireland
*Tel.: +353-(0)-61- 213096, E-mail: mark.davis@ul.ie

Abstract

Itraconazole is a BCS Class II drug with very limited aqueous solubility (<1 Âµg/ml) and poor dissolution kinetics. Many studies have been conducted in an effort to produce formulations of this drug with higher bioavailability. In this context, we have examined ternary amorphous solid dispersions (ASD) of itraconazole with excipient polymers Soluplus® and HPMCP. Evidence exists to suggest that ternary formulations can be more stable and have better release profiles than binary systems alone. Soluplus® is a structurally diverse excipient with excellent solubility and melt extrusion characteristics. HPMCP is used as a complimentary excipient, owing to the phthalate groups which help protect against gastric juices.

In this study we have taken a design of experiments (DoE) approach to optimise ternary formulations of itraconazole-Soluplus-HPMCP, producing 17 outcomes. A pre-formulation solubility test was run to ensure complete solubility of each combination. The solutions were then converted into powders by spray drying in a Büchi B-290 mini spray drier, equipped with a pneumatic nozzle, inert loop condenser and high efficiency cyclone. Yields were consistent and could be increased to as high as 83% by varying parameters.

The powders were split into three parts and stored at (i) RT and 0% RH, (ii) 30 °C and 45% RH and (iii) 40 °C and 75% RH for up to one year. PXRD measurements revealed samples to be amorphous at day 0 and importantly still amorphous after a year of storage. Low angle XRD was used to ensure that samples had not formed a meso phase, which is possible with solid state itraconazole. Differential scanning calorimetry (DSC) confirmed that the samples were amorphous and demonstrated how the glass transition temperature varied with content. Particle size distribution measurements was carried out on using laser differaction and the data was analysed by cumulative and histogram plots. A statistical analysis revealed a relationship between PSD and formulation content. This result was rationalised by hydrogen bonding in HPMCP, the only component capable of self-association.

455766 Engineering the Mechanical Properties of Amorphous Solid Dispersion Particles

Abstract:

Spray-dried amorphous solid dispersions (SDDs) is one of the most robust and mature available technologies to improve oral absorption of drug candidates with low bioavailability due to poor aqueous solubility and dissolution rate. As an active intermediate of the drug substance, SDDs are readily integrated into oral solid dosage forms, the most common being tablets. However, due to the dilution of the active ingredient in the SDD and then further dilution in the tablet formulation, pill burden can become a problem for less potent drugs. One possible solution is to engineer improved SDD particle mechanical properties through the spray drying process thereby reducing the amount of tablet fillers and improving the robustness of the tableting process. This concept of SDD particle property optimization is a multifaceted optimization problem between physical state and chemical stability of the drug, downstream processing of the resulting powder, and in vivo performance. Although this talk focuses on one aspect of downstream processing, the mechanical properties specifically related to tableting, there have been several recent reviews on this subject [1-2]. The current pharmaceutical literature is rich with information on engineering SDD particles, but there lacks the connection to how these engineering principles influence the compression-related mechanical properties.

The aim of the current study is to make that connection by engineering the mechanical properties of SDD particles through the spray-drying process. Ketoconazole spray-dried with HPMCAS-M or PVP VA-64 at varying concentrations was used as a model SDD system. Atomization and drying kinetics were adjusted to vary surface area by altering particle size distribution, particle density, and morphology. The mechanical properties were studied using compression profiles, SEM imaging, and yield pressure determination with a modified in-die compression analysis [3]. The results showed that the compression pressures required to reach a 2.0MPa tensile strength compact varied with

SDD composition and were in the range of typical fillers. Yield pressures also varied with drug loading and polymer selection which could be reasonably predicted by the pure material behaviors composing the SDD particles. The deformation and tensile strength mechanisms were further evidenced by the SEM micrographs. The results suggest that a tablet formulation using rationally engineered particles could reduce the amount of fillers needed to decrease pill burden.

[1] Patel, B.B. et al. Revealing facts behind spray dried solid dispersion technology used for solubility enhancement. *Saudi Pharm. J.* **2015**, *23 (4)*, 352-365.

[2] Singh, A. and Van den Mooter, G. Spray drying formulation of amorphous solid dispersions. *Adv. Drug Deliv. Rev.* **2015**.

[3] Katz, J. M., Roopwani, R., Buckner, I. S. A Material-Sparing Method for Assessment of Powder Deformation Characteristics Using Data Collected During a Single Compression-Decompression Cycle. *J. Pharm. Sci.* **2013**, *102 (10)*, 3687-3693.

466729 Investigation of Phase Separation Phenomena in a Supersaturated Ternary Amorphous Solid Dispersion

Abstract:

Investigation of phase separation phenomena in a

supersaturated ternary amorphous solid dispersion

Catherine B. Kelly[1*], Ahmad B. Albadarin[1] & Gavin M. Walker[1]

[1]Synthesis and Solid State Pharmaceutical Cluster (SSPC), Bernal Institute,

University of Limerick, Limerick, Ireland *Tel.: +353-(0)-61-237732, E-mail: Catherine.Kelly@ul.ie Abstract

Many new chemical entities (NCEs) are failing to progress through drug-development pipelines due to a display of insufficient bioavailability, stemming from the extremely low aqueous solubility properties of many of these materials. The overwhelming majority of these NCEs are small-molecule, crystalline solids so lend themselves to formulation in the higher solubility but thermodynamically unstable amorphous form. A common approach to the formulation of amorphous APIs is through the use of an amorphous solid dispersion. In order to stabilise the thermodynamically unstable amorphous form an excipient to which the API exhibits an affinity through molecular interactions is added during formulation.

Recent publications have suggested that even in cases in which interactions between an API and a biocompatible polymer has been detected, in order to produce an acceptable solid dosage form the amorphous API could be present at a loading above that which is thermodynamically stable. At the required higher drug loadings successful long-term stability appears to be achieved through a combination of thermodynamic and kinetic mechanisms.

Additionally, the action of the excipients in an amorphous solid dispersion is twofold. As well as the stabilisation of the amorphous form in the solid state, the excipients should also be capable of maintaining the supersaturated concentration of the API which occurs during dissolution due to the presence of the amorphous form. Due to the dual-action properties required of the polymeric excipient the use of two excipients is proposed, with this study investigating the first mechanism, namely the stabilisation of the amorphous form in the solid state.

Production of the drug-polymer temperature-composition phase diagram

Itraconazole (ITR) was used as a model poorly aqueous soluble drug and Soluplus and hydroxypropyl methylcellulose phthalate (HPMCP) were investigated as compatible excipients for the formulation of a ternary amorphous solid dispersion. The value of the Flory-Huggins interaction parameter and its dependence on temperature was determined for Itraconazole with each polymer as well as a 1:1 blend of both polymers using the dissolution end point method. This method determines the liquid-solid boundary in the API-polymer binary system by measuring the endset of melting during a heating ramp in a differential scanning calorimeter (DSC). The melting point value gained from this method was further investigated using a 16-hour annealing method which indicated that the liquid-solid curve had been underestimated for both polymers. The annealing method involved the use of hyper DSC and a heating rate of 100°C/min to search for the presence of residual crystalline API following a 16 hour hold at or below the temperature gained using the dissolution end point method.

It is known that the dissolution end point method of determining the liquid-solid curve in high viscosity systems such as these can overestimate the drug melting temperature, however the greater accuracy associated with the annealing method should be considered against the much greater period of time required to gain a phase diagram data point.

Investigation of phase separation

Following production of the three phase diagrams a number of amorphous solid dispersions with a range of drug-polymer compositions were produced through ball milling and a subsequent melt-quench. Hyper DSC was used to measure the temperature and width of the glass transition and confirm the absence of crystalline API. Itraconazole is also known to form a liquid-crystal state but the thermal events associated with this state were not apparent in the any of the thermograms of the solid dispersions. The dispersions were stored at 100°C which is within the region of the phase diagram where spontaneous spinodal decomposition should occur (Figure 1). After one week phase separation had occurred in all samples but the development of two distinct demixing mechanisms was apparent. At higher drug loadings demixing occurred by a release of the API by recrystallisation. At lower drug loadings the amorphicity of the formulations was maintained, however two distinct amorphous phases appeared, identified by two separate glass transition events. The change in demixing from recrystallisation to phase separation was observed in all of the three systems examined but shifted in drug-polymer ratio depending on the polymer(s) involved.

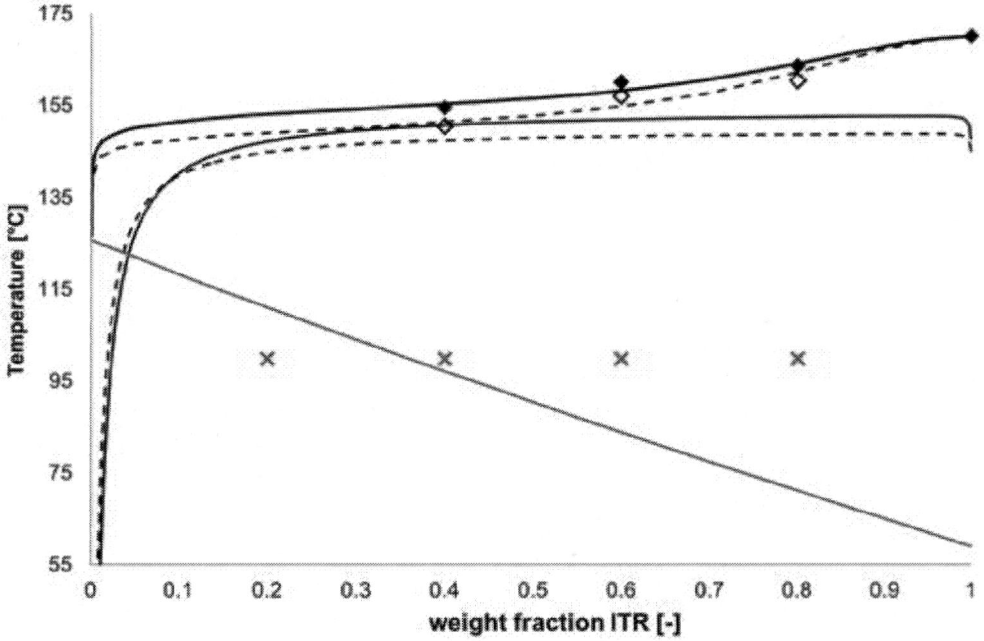

Figure 1: Phase diagram of the ITR-HPMCP-Soluplus system indicating the boundaries identified using the dissolution end point (full line) and annealing (dashed line) methods and the points at which phase separation was investigated (crosses). The upper boundary marks that between liquid and solid API while the lower is the spinodal curve.

473925 Effects of Intermediate Spray-Drying Conditions on Downstream Processing of Pharmaceutical Solid Dosage Formulations

Abstract:

As enabled formulations continue to assume an expanded role in the pharmaceutical industry, a better understanding of how spray-dried particle characteristics influence granulation and tablet compression are necessary to enable reproducible results across spray-drying scales. In this effort, drying kinetics were varied during the spray drying process to produce a spray-dried intermediate (SDI) with varying particle properties. One property, particle density, has been shown to strongly influence the compaction properties of oral solid dosage formulations containing high SDI compositions. It was observed that responses to uniaxial compaction could be successfully employed to project granulation performance at pilot and commercial scale manufacture. Tablet hardness was found to be inversely proportional to SDI particle density, which translates to significant differences in dissolution response. For bilayer tablets compressed with this SDI, extent of granulation and SDI particle density enhance interfacial adhesion. These findings enable flexibility in future spray drying campaigns by establishing how measureable SDI properties impact downstream processing and quality attributes.

467625 Chemical Layer Deposition of Porous Alumina Overcoats Increases Activity and Stability in Liquid Phase Catalytic Conversion of Biomass-Derived Chemicals

Abstract:

Biomass-derived chemicals represent an attractive sustainable alternative to petrochemicals. However, the competitiveness of these processes has been limited by the low stability of metal catalysts under the severe conditions used during biomass upgrading. In order to prevent metal leaching and sintering, we developed a chemical layer deposition (CLD) overcoating technique based on the control over aluminum alkoxide surface reactions kinetics and stoichiometry. After deposition of 30 cycles of an alumina overcoat onto alumina supported copper nanoparticles, imaging using high angle annular dark field scanning transmission electron microscopy (HAADF STEM) revealed a conformal porous overcoat. This porosity was was characterized and quantified by nitrogen physisorption. The overcoat suppressed irreversible catalyst deactivation during the liquid-phase hydrogenation of furfural. While a significant loss of activity due to mass transfer limitations and metal coverage by the overcoat was observed by O'Neill *et al.* when using atomic layer deposition (ALD),[2] most of the surface metal sites remained accessible with CLD as confirmed by chemisorption, illustrating the benefits of overcoats porosity that can be targeted by tuning synthesis parameters. Current work is focused on tuning this porosity for the synthesis of shape selective catalyst overcoats.

(1) Héroguel, F.; Rozmysłowicz, B.; Luterbacher, J. S. *Chim. Int. J. Chem.* **2015**, *69* (10), 582–591.

(2) O'Neill, B. J.; Jackson, D. H. K.; Crisci, A. J.; Farberow, C. A.; Shi, F.; Alba-Rubio, A. C.; Lu, J.; Dietrich, P. J.; Gu, X.; Marshall, C. L.; Stair, P. C.; Elam, J. W.; Miller, J. T.; Ribeiro, F. H.; Voyles, P. M.; Greeley, J.; Mavrikakis, M.; Scott, S. L.; Kuech, T. F.; Dumesic, J. A. *Angew. Chem.* **2013**, *125* (51), 14053–14057.

468951 Controlled Growth of Noble Metal Nanoparticles on Graphene Nanoplatelets Via Scalable Atmospheric Pressure Atomic Layer Deposition

Abstract:

Graphene-supported nanoparticles (NPs) are emerging as a novel class of high-performance catalysts [1,2]. For this reason, a great deal of research has been devoted to the fabrication of graphene-supported NPs of several noble and base metals for various applications such as electrocatalysis, photocatalysis, water treatment and splitting, and energy generation and storage. A number of methods have already been developed for the synthesis of metal/graphene. However, most of these methods rely on wet chemistry, in which often the metal dispersion is hard to control, and a high level of impurities arising from residual solvent and reaction byproducts. Therefore, a solvent-free and scalable method capable of depositing NPs on graphene in a controlled fashion, is highly desirable.

We demonstrate the deposition of crystalline palladium and platinum NPs on graphene nanoplatelets via atmospheric pressure atomic layer deposition (ALD). We carry this out in a reactor in which the graphene is dispersed in the gas phase, enabling the production of large amount of material [3]. The nucleation and growth of noble metal nanoparticles on the inert graphene surface was enabled by applying an ozone pretreatment step, without significantly affecting the graphene crystalline structure. Uniform nucleation on both basal planes and edges of the graphene was obtained. By analysing the evolution of the particle size distribution and spatial density, we obtained insights into the nucleation and growth of metal ALD on graphene. Such insights enabled us to tailor the particle size distribution and loading of the noble metal clusters. Finally, our approach provides a 100% solvent-free, controllable and scalable process for producing the bulk quantities of functionalized graphene. We will demonstrate the high catalytic activity of the obtained materials.

[1] Zhang, N.; Zhang, Y.; Xu, Y.-J., Recent Progress on Graphene-Based Photocatalysts: Current Status and Future Perspectives, Nanoscale 4 (2012) 5792– 5813

[2] Huang, C.; Li, C.; Shi, G., Graphene Based Catalysts, Energy Environ. Sci. 5 (2012) 8848– 8868

[3] Van Bui, H.; Grillo, F.; Helmer, R.; Goulas, A.; van Ommen, J. R., Controlled Growth of Palladium Nanoparticles on Graphene Nanoplatelets via Scalable Atmospheric Pressure Atomic Layer Deposition, J. Phys. Chem. C 120 (2016) 8832–8840

464043 Computational Studies of the Dry Reforming Reaction on Al_2O_3-Coated Transition Metal Catalysts

Abstract:

More than 50 percent of man-made greenhouse gases are composed by carbon dioxide (CO_2), which makes urgent the development of effective strategies for remediation of CO_2 emissions. A promising route for chemical conversion of CO_2 is the dry reforming reaction (DRR) between CO_2 and methane (CH_4) to produce Syngas. Syngas is a mixture of H_2 and CO used as intermediate production of synthetic natural gas and ultra-clean fuels, or as precursor for ammonia, methanol, and other valued-added chemicals. Thus, the DRR reaction opens a window for remediation processes of CO_2 with high promise of producing value-added chemicals.

So far the screening of catalysts has shown that metallic catalysts from Group VIII are active for the DRR reaction. Among these catalysts the Rh-based seem to be the most active but the Ni-based catalysts have the best compromise between performance and cost. However, all of these catalysts suffer of sintering and coke formation due to the high temperature required for the DRR reaction (~800 °C), which still hinders the implementation at commercial scale of this route for chemical conversion of CO_2. In case of Ni-based catalysts two main causes are associated to sintering and coke formation. The first cause is the inability to precisely control the synthesis parameters and the interactions between the active site and the surface acting as support. The second cause is the exposure of the active catalytic sites to the extreme temperature at which the DRR reaction takes place. While the exposure of the catalysts to the high temperatures at which the DRR is performed is necessary, the minimization of sintering and coke formation can be performed through development of structures able to control the continuous structural reconstruction of the active phase.

Sintering and coke formation can be minimized by synthesis of Ni nanoparticles (NPs) surrounded by atomically precise assemblies protecting the active site. Recent experiments have reported increased resistance to sintering and coke formation for Pd-based catalysts overcoated with an Alumina layer using atomic layer deposition (ALD) methods. It was suggested that this Alumina layer may offer better protection to the active phase thanks to stabilization of the Ni nanoparticles. Those Pd catalysts covered by the Alumina layer resulted in only 5% of carbon accumulation in comparison with bared Pd catalysts, indicating increased resistance to catalyst coking as the ALD overcoating is applied. However, there is still debate regarding the effect of overcoating thickness on catalysts activity, as thinner coatings have been also associated with better overall catalytic performance.

So far the above mentioned experiments have been performed using structures composed of monometallic Pd NPs deposited on non-modified Al_2O_3 substrates without additives to stabilize the overcoating ALD AL_2O_3 structure. However, additional improvements on these structures can potentially enhance the catalyst stability and selectivity. In this work we examined a new class of Ni catalysts coated with Alumina using ALD methods. The reasons for the enhanced catalytic efficiency were investigated through state-of-the-art DFT/AIMD analysis. The active phases of Ni nanoparticles were represented by terrace and stepped slabs in order to evaluate the earlier steps of the Alumina layer formation. It is shown that the Alumina overcoating of Ni nanoparticles is viable but also Alumina deposition is preferred on stepped regions, leaving available the catalytically active terrace regions. The effects of Alumina overcoating on the mechanistic pathways for CO_2 activation were also studied, unveiling the effect of shape changes and nanoparticles reconstruction under support-nanoparticle and nanoparticle-overcoat interactions, and the possible appearance of side reactions and coke formation.

470177 Experimentally Probing Ligand-Strain Effect Via a Novel Catalyst Platform

Abstract:

In this presentation, a novel catalyst fabrication platform was designed and investigated. Using a phase-pure transition metal carbide (TMC) nanotube as support [1], noncontiguous platinum group metal (PGM) particles were deposited from atomic level to 2-3 nanometers (nm) onto the TMC support by atomic layer deposition [2]. This platform enables experimental investigation on ligand-strain effect of resultant PGM/TMC catalysts. Characterization of metal-adsorbates bond strength, via diffuse reflectance infrared Fourier transform spectroscopy (DRIFTS) [3] and temperature-programmed oxidation (TPO), informs catalyst surface properties and model reaction mechanism [4]. Temporal analysis of products (TAP) technique was first applied to this catalyst platform revealing the existence of different active sites for Pt particle sizes (<3 nm). Additionally, the correlation between catalyst reactivity and change of deposited particle size was further investigated using water-gas-shift (WGS) reaction. The rarely investigated PGM particle size range and the ease of surface characterization by high resolution transmission electron microscopy (HRTEM), X-ray photoelectron spectroscopy (XPS) and other techniques, provides a unique PGM/TMC platform, which can be utilized closely with theoretical predictions to further our understanding on the interaction between PGM and non-PGM metals.

References

[1] Wan, C.; Regmi, Y. N.; Leonard, B. M. *Angew. Chemie Int. Ed.* **2014**, *53*(25), 6407â€"6410.

[2] McCormick, J. A.; Cloutier, B. L.; Weimer, A. W.; George, S. M. *J. Vac. Sci. Technol. A Vacuum, Surfaces, Film.* **2007**, *25*(1), 67.

[3] Ding, K.; Gulec, A.; Johnson, A. M.; Schweitzer, N. M.; Stucky, G. D.; Marks, L. D.; Stair, P. C. *Science*. **2015**, 350 (6257), 1688â€"1690.

[4] Schweitzer, N. M.; Schaidle, J. a; Ezekoye, O. K.; Pan, X.; Linic, S.; Thompson, L. T. *J. Am. Chem. Soc.* **2011**, *133*, 2378â€"2381.

464481 Cobalt/Alumina Interactions in ALD Synthesized Catalysts for Fischer-Tropsch Synthesis

Abstract:

Cobalt metal was deposited onto an alumina (Al_2O_3) support using atomic layer deposition (ALD) to create dispersed catalysts for Fischer-Tropsch Synthesis (FTS). These catalysts are highly active for the FTS reaction, exhibiting triple the cobalt mass activity of standard incipient wetness catalysts at similar reaction conditions. HRTEM imaging shows that cobalt has simultaneously deposited during ALD in both crystalline plane and nanoparticle regimes, regardless of the number of ALD cycles. However, it was found that the number of ALD cycles greatly alters the activity of these catalysts, with 4 and 8 cycle catalysts being much more active than the 1 cycle catalyst. The known structure sensitivity of FTS indicates that the crystalline cobalt regions are the active regions for FTS. The ALD process allows deposition of metallic cobalt onto Al_2O_3 with protective organic capping ligands, preventing cobalt oxidation and reconstruction in air prior to reaction. Additional ALD experiments on different phases of Al_2O_3 followed by FTS testing proves the influence of support structure on the cobalt deposition regime and consequently FTS activity. Finally, density functional theory (DFT) calculations were done to understand the metal/support properties responsible for planar vs. nanoparticle growth during ALD and explain the high activity shown by these catalysts.

467730 *Catalytic Performance of Supported Nano-Iron Catalysts Synthesized By a Gas-Expanded Liquid Deposition Technique in Fischer-Tropsch Synthesis*

Abstract:

Supported iron catalysts for low temperature Fischer Tropsch Synthesis (FTS) as prepared by impregnation of iron salt precursors on support materials have met with limited success. While the support materials provide high surface areas, iron species tend to interact strongly with traditional support materials such as SiO_2 and Al_2O_3. As such, the activity and selectivity towards long chain hydrocarbons for these catalysts were found to be inferior to those prepared by a precipitation method. However, supported iron catalysts provide a few potential advantages over precipitated catalysts, such as 1) better mechanical strength in order to withstand degradation, 2) enhanced dispersion of the iron, and 3) better control over the size and distribution of the iron species on the surface, etc. Therefore, there is a need to develop new synthesis methodologies to create improved supported iron catalysts in order to further enhance catalyst activity and selectivity towards hydrocarbons.

In this study, a new catalyst was prepared using pre-synthesized iron oxide nanoparticles that were controllably deposited onto a SiO_2 support using a gas-expanded liquid precipitation technique (GXL).Â In general, gas expanded liquids are solvent systemsâ€"typically a mixture of an organic solvent and a compressible gas, such as CO_2â€"that enable controlled deposition of pre-synthesized nanoparticles, based on their size, onto a support surface.Â By controlling the applied CO_2 pressure, and hence the solvent strength, one can systematically deposit/precipitate nanoparticles of desired sizes onto the support materials. The iron oxide nanoparticle dispersion used in this study contains oleic acid stabilized iron oxide nanoparticles dispersed in an *n*-hexane solvent. We induced the precipitation of larger sized particles from the initial nanoparticle dispersion, thereby depositing these larger particles onto the

surface of the support while enabling the removal of the smaller sized particles by removal of the residual hexane solution. The ability to exclude the smaller sized particles from those that are deposited onto the catalyst support is important since supported iron FT catalysts with nanoparticles of small size have lower activity due to their stronger interaction with the support surface. As an added benefit, the lower viscosity and higher mass transfer coefficient in the GXL can enhance the diffusivity of the iron particles into the catalyst pores during deposition, resulting in a more uniform distribution of the nanoparticles compared to traditional methods.

The objective of this work is to explore the benefits of using the GXL technique in preparing SiO_2 supported catalysts with moderate iron loadings for Fischer-Tropsch synthesis applications. The catalytic performance of GXL prepared Fe/SiO_2 catalysts with different iron loadings were evaluated in terms of CO conversion, CH_4 selectivity, C5+ selectivity and C5+ productivity. In addition to the effect of iron loading, we have also investigated the effect of reaction conditions, such as reaction temperature, on the catalytic performance of our GXL catalysts in order to understand the influence of operating conditions on the selectivity and productivity of desired hydrocarbon products.

474539 Photocatalysis for Z-Scheme Solar Water Splitting Using New Reactor Designs

Abstract:

Particle suspension reactors for solar water splitting can be an economical alternative to photovoltaic-driven electrolysis. One design resembles Nature's Z-scheme where together two photocatalyst nanoparticle reactor beds that are adjoined side-by-side and on the meters length scale, drive overall solar water splitting.[1,2] Electronic charge is mediated between the beds by a dissolved redox shuttle that undergoes oxidation or reduction at the particles and transports between the beds through a nanoporous material. While this design facilitates product separation (i.e. separation of H_2 and O_2) and therefore circumvents formation of an explosive mixture of gases, active transport of the redox shuttle over these distances has been projected to account for about half of the capital cost of the reactor.[1,2]

Our team is evaluating the feasibility of new reactor designs where the beds are stacked. This generates a true tandem light-absorbing reactor where the theoretical maximum solar-to-hydrogen conversion efficiency is over 50% larger than a side-by-side or single light-absorber design. Each bed is projected to be < 10 cm tall and therefore, this design greatly decreases the distance required for redox shuttle transport therefore reducing or even eliminating the need for forced convection.

In my presentation I will report on our team's progress on this design. We used finite-element numerical methods to model and simulate in two dimensions the transient mass transport processes, light absorption, and electrochemical kinetics in the proposed reactor. The developed model provided insights into the influence of the reactor geometry and operating conditions on the overall performance. The Beer–Bouger–Lambert law was applied to obtain the spatial light-intensity field and volumetric reaction rates were obtained by coupling solid-state photodiode expressions with Butler–Volmer kinetics on the surface of the particles. Model results suggested that a reactor operating at a ~1% solar-to-hydrogen conversion efficiency can operate for greater than half a year without complete loss of redox shuttle at any location in the reactor.

Experimentally, we investigated nanomaterials over many size scales, from single nanoparticles (~10 nm to ~1 μm in diameter) to mesoporous thin films (~10 μm thick) to laboratory-scale prototype particle-suspension reactors (on the scale of feet). On the single particle level we used bipolar electrodeposition to create Janus-type particles consisting of model carbon particles with metal and metal-oxide electrocatalysts for H_2 evolution and O_2 evolution at the poles. We also jammed and covalently bound TiO_2 nanoparticles into a single nanopore in a plastic sheet, wetted the particles with liquid electrolyte on both sides, and measured photovoltages that resulted from excitation of few particles. This was significant because it allowed, for the first time, in situ photoelectrochemical characterization of particle(s). We also synthesized, characterized, and evaluated the photo(electro)chemical performance of $BiVO_4$ and Rh-modified $SrTiO_3$ nanocrystallites as mesoporous thin films and particles in model reactors, and evaluated the transport properties of several redox shuttles. The $BiVO_4$ and Rh-modified $SrTiO_3$ were grown by sol gel methods and characterized by XRD, XPS, transmission-mode or diffuse-reflectance ultraviolet–visible electronic absorption spectroscopy, He II UPS, DLS (as particles), Raman, and photoelectrochemical performance. These nanocrystallites exhibited some of the best performance reported to date.

Collectively, our efforts represent strides toward achieving a high-level of techno-economic viability in solar water splitting reactors.

Acknowledgments: This work was supported by the U.S. Department of Energy, Office of Energy Efficiency and Renewable Energy, Fuel Cell Technologies Incubator Program under Award No. DE-EE0006963 and Lawrence Berkeley National Laboratory under Contract No. DE-AC02-05CH11231.

References:

(1) B. D. James, G. N. Baum, J. Perez and K. N. Baum, *Technoeconomic Analysis of Photoelectrochemical (PEC) Hydrogen Production*, Directed Technologies Inc., (US DOE Contract no. GS-10F-009J), Arlington, VA, **2009**.

(2) B. A. Pinaud, J. D. Benck, L. C. Seitz, A. J. Forman, Z. Chen, T. G. Deutsch, B. D. James, K. N. Baum, G. N. Baum, S. Ardo, H. Wang, E. Miller, and T. F. Jaramillo, *Energy & Environmental Science*, **2013**, *6*, 1983–2002.

460526 Effect of Heating Rate on Combustion of Nano-Composite Thermites Prepared By Arrested Reactive Milling

Abstract:

Stoichiometric nanocomposite thermites were prepared by Arrested Reactive Milling (ARM). These thermites utilized aluminum or zirconium as the fuel and were milled with various oxidizers (CuO, MoO_3, WoO_3, Fe_2O_3, and Bi_2O_3). The powders were ignited by both CO_2 laser beam and electrostatic discharge (ESD). The two ignition methods provided different heating rates with the laserâ€™s heating rate (10^6K/s) being three orders of magnitude lower than that achieved in the ESD (10^9K/s). The first set of experiments dealt with small particle ignition events in ESD; powders were spread in a monolayer across a carbon film on a brass substrate and ignited by a high voltage spark. The optical emission from these combustion events was recorded using a single photomultiplier tube (PMT) filtered at 568 nm. In different experiments, the same powders were fed in the CO_2 laser beam using an air jet feeder system. The resulting particle combustion emission was recorded using the same, 568-nm filtered PMT. In addition, a couple of PMTs filtered at 700 and 800 nm were used to obtain time-resolved color temperature traces for the burning particles. It was observed that particles ignited by the ESD had shorter burn times than those ignited by the laser. These observations are interpreted assuming that the nano-composite structure of the particles was maintained following their very rapid heating by ESD; however, the scale of mixing between metal and oxidizer was not retained for the particles heated by the CO_2 laser beam. The correlation between the scale of mixing between metal and oxidizer and burn rate was supported by scanning electron microscopy analyses of partially burned and quenched particles of different materials.

472472 Tunable Aluminum and Liquid Fluorocarbon Pyrotechnics

Abstract:

The ability to tune energetic material combustion rate through modification of the condensed phase requires enhanced energy feedback to the condensed phase by either thermal energy generation from condensed phase reactions or from an external source. Enhancement of surface emissivity and enhancement of condensed phase thermal conductivity are two particularly attractive techniques. In this effort, we investigate the enhancement of energy deposition from a pyrotechnic flame through carbon nanotube doping of the liquid phase of an aluminum/liquid fluorocarbon (LFC) pyrotechnic. We report an experimentally observed burning rate of Al/LFC compositions of 1.1 cm/s at 6.89 MPa (1000 psi) with a power law burning rate pressure dependence of 0.66. below 13.8 MPa (2000 psi). Doping of compositions with up to 3 wt.% of a variety of radiometric and thermally enhancing susceptors, including single/multiwall nanotubes and graphene nanoplatelets can result in a sixteen-fold increase in burning rate (16.6 cm/s) at 6.89 MPa with no change to pressure exponent. To explore the mechanism responsible for enhancement, we further quantify the improvements to condensed phase spectral absorptivity and thermal conductivity, estimate flame temperature using two-color high speed video pyrometry and discuss techniques with which thermal conductivity and burning rate enhancement of such formulations can be switched â€˜on commandâ€™ to tune energetic output in situ. Resulting equilibrium calculations and demonstrated burning rate enhancements suggest that in addition to its use as a pyrotechnic, such compositions may be attractive for use as on-command tunable combustion rate microscale propellants.

468660 In-Situ Imaging of Particles during Rapid Thermite Deflagrations

Abstract:

The dynamic behavior of rapidly deflagrating thermites is a highly complex process involving rapid decomposition, melting, and outgassing of intermediate and/or product gases. Few experimental techniques are capable of probing these phenomena *in situ* due to the small length and time scales associated with the reaction. Here we use a recently developed extended burn tube test, where we initiate a small pile of thermite on the closed end of a clear acrylic tube. The length of the tube is sufficient to fully contain the reaction as it proceeds and flows entrained particles down the tube. This experiment was brought to the Advanced Photon Source, and the particle formation was X-ray imaged at various positions down the tube. Several formulations, as well as formulation parameters were varied to investigate the size and morphology of the particles, as well as to look for dynamic behavior attributed to the reaction. In all cases, we see evidence of particle coalescence and condensed-phase interfacial reactions. Furthermore, we observe the formation of gas bubbles attached to the oxidizer interface, which we suggest forms as the oxidizer decomposes to release intermediate gases into the aluminum melt. The results improve our understanding of the procession of reactants to products in these systems.

This work was performed under the auspices of the U.S. Department of Energy by Lawrence Livermore National Laboratory under Contract DE-AC52-07NA27344. LLNL-ABS-691140

454758 Applying Pressure and Shear on Nitrogen-Rich Energetic Materials to Create Metamaterials

Abstract:

Applying Pressure and Shear on Nitrogen-rich Energetic Materials to Create Metamaterials

Farhad Forohar, Vasant Joshi, and Daniel Wilson

Research, Development Test and Engineering Department, Naval Surface Warfare Center Explosive Ordnance Disposal Technology Division, 4104 Evans Way Suite 102, Indian Head, MD 20640

Efforts to obtain metastable high pressure and high temperature states of nitrogen using Diamond Anvil Cell (DAC) techniques have indicated that some high density compounds may physically exist, but recovery of these materials at atmospheric pressure and temperature is still elusive. Stable poly-nitrogen compounds can be theoretically achieved by attaching them to non-nitrogen atoms. Application of combined pressure and shear to transform a nitrogen rich precursor material to a metastable condition is a new approach that should be easier than long duration-pure pressure application. This new method is being applied in an attempt to synthesize and recover novel materials from nitrogen rich precursors.Â Nitrogen rich precursors used in the present study include ammonium azide (N_4H_4), di-amino-tetra-azidocyclotriphosphazene ($P_3N_{17}H_4$), hexa-azidocyclotriphosphazene (P_3N_{21}) and 1,1â€™-azobistetrazole ($C_2H_2N_{10}$). In order to get intramolecular interaction, co-crystallizations of mixtures were also made and were subjected to pressure-shear loading. Decomposition of materials at low pressure has been achieved for some precursors.Â Experimental fixture, method, results and analysis of recovered products will be presented.

461372 Combustion Characteristics of Boron and Boron Based Composites in Different Flow Configurations

Abstract:

Combustion characteristics of boron and boron based composites in different flow configurations

Song Wang, Mirko Schoenitz, Edward L. Dreizin

New Jersey Institute of Technology

Newark, NJ 07103

Abstract

Due to its high heat of combustion, boron is widely considered as an attractive fuel. Recently, ternary B·Mg·I_2 and B·Al·I_2 composite powders were prepared as reactive materials generating biocidal combustion products. The materials contained 20 wt % of I_2 were shown to be effective in inactivating aerosolized microorganisms. The aim of this paper is to characterize the effect of flow conditions on combustion dynamics of particles of boron and composite B·Mg·I_2 and B·Al·I_2 materials. For all materials, powder particles were carried by a laminar air flow to cross a focused beam of a CO_2 laser. A cylindrical air knife was used to generate a turbulent flow pattern above the laser beam. In different experiments, particles ignited by the laser continued burning in either laminar or turbulent flows. Particle streaks were photographed at different flow conditions. For boron particle, staged combustion was observed, consistently with previous work. Streaks were brighter for the aluminum-containing composite as compared to that containing

magnesium. Oscillatory emission patterns were observed at the end of combustion for boron and aluminum containing composite, but not for the magnesium containing composite. Particle burn times and combustion temperatures were obtained for all experimental configurations. For boron, the particle burn times in the laminar flow were nearly twice as long as those in the turbulent flow. Boron combustion temperature was close to 3100 K in the laminar flow and decreased to 1500 K at the highest turbulence intensity. However, the combustion characteristics of composite materials were not altered by the flow conditions. The burn times, which were slightly lower than for boron in a laminar flow and combustion temperatures (close to 2000 and 2500 K for B·Al·I_2 and B·Mg·I_2, respectively) remained nearly the same for all flow configurations.

461239 Effect of Metal-Based Catalysts on Ignition and Combustion of Boron

Abstract:

There is considerable interest in improving boron ignition and combustion because of its high energy density and potential use as an additive in explosives and propellants. Previous work done by pre-treating amorphous boron with acetonitrile and other hydro-carbons prove that dissolution of the naturally occurring, and inhibiting hydroxide/oxide layer reduces ignition delay. However, combustion times and temperatures remain unaffected for the micron-sized agglomerates. Flame temperature measurements suggest that boron is far from its boiling point and thus the so-called full-fledged combustion is rate limited by heterogeneous processes on the surface of burning particles. In order to increase the rate of combustion reaction on the surface of a burning boron particle, this work focuses on the preparation of boron doped with catalytic additives. The additives are transition metals such as iron, nickel, and manganese, which are readily oxidized and reduced and provide kinetic pathways to assist the boron oxidation reaction. The dopants are introduced by mechanical milling, dry powder coating, and coating with organic salts to create composites with varying quantities of catalysts. The materials are characterized and combustion experiments are conducted to determine their combustion characteristics. The findings from the experiments will be discussed in detail.

454750 Application of the Peng–Robinson Equation of State to Energetic Materials RDX and TNT: Pure Components, Liquid Mixtures, and Solid Mixtures

Abstract:

Continuum simulations of energetic materials, such as cookoff models, rely on thermodynamic information to close the governing equations. Thermodynamic models of unreacted energetic materials, particularly mixtures, have been relatively limited despite the importance of these substances to national security and to a number of different industries. In this presentation of our recently published work, we show that the well-known Peng-Robinson equation of state can be applied to energetic materials, whether they are pure components, liquid mixtures, or solid mixtures. We are specifically interested in two compounds: hexahydro-1,3,5-trinitro-1,3,5-triazine (RDX) and 2,4,6-trinitrotoluene (TNT). We model RDX and TNT in both their liquid and solid phases, as well liquid and solid mixtures of the two compounds. Our work examines temperatures and pressures as high as about 500 Kelvin and 2500 bar, respectively. These are conditions relevant to cookoff before explosion occurs. The Peng-Robinson equation of state provides a good representation of experimental volumetric (e.g., density and bulk modulus), thermal (heat capacity), and phase behavior (melting temperature and solubility) data. It can be applied to other energetic materials — ranging in complexity from pure components to multiphase, multicomponent mixtures — by adapting the procedures described in this presentation.

464215 Determination of Enthalpy of Fusion of an Energetic Material, HMX, That Undergoes Simultaneous Fusion and Decomposition

Abstract:

Cyclotetramethylene tetranitrate (HMX), a commonly used energetic material, undergoes a nearly simultaneous transition from melting to decomposition, which makes direct measurement of enthalpy of fusion from calorimetric study very difficult. The enthalpy of fusion reported in the literature is very high, 69.9 kJ/mol, when compared to those of chemically similar molecules. For example enthalpy of fusion of trimethylene trinitramin (RDX) is 34.5 kJ/mol. In this work, we investigate melting properties of HMX with a differential scanning calorimetry (DSC) and theories of solution thermodynamics. A series of experiments with a DSC at the heating rates 0.25â€“20 K/min show the melting onset of HMX starts appearing at heating rate 5.5â€“6.0 K/min. Since the melting endotherm is convoluted with the decomposition exotherm, we extend the investigation to determine enthalpy of fusion of HMX from melting point depression of HMX in presence of RDX. The enthalpy of fusion value determined from the melting point depression are further compared with the corresponding value obtained from the thermodynamic modeling of HMX solubility in solvents based on NRTL-SAC activity coefficient model. The melting point depression experiments and the solubility modeling work show that enthalpy of fusion of HMX is 31.9 ± 3.9 kJ/mol and 31.5± 9.5 kJ/mol, respectively.

470757 Diffusiophoretic Effects on Coffee Ring Patterns

Abstract:

Coffee rings are a natural phenomena ubiquitous in the evaporative drying of colloidal suspensions. They are thought to be associated with only evaporation-induced capillary convection. In applications related to spray painting, coating and ink-jet printing, non-equilibrium convection-driven particle aggregation causes peripheral ring structures. The ring structure can be suppressed using higher aspect ratio particles or Marangoni convection. On the other hand, various applications require formation of strong coffee rings, while other applications want to avoid the ring structure altogether. Despite several technological advances to control the coffee ring effect, modulation from well-defined ring to completely dispersed has not been achieved successfully. Here we show both experimentally and theoretically that diffusiophoresis occurs during the evaporation of saline droplets and is efficient to concentrate or disperse the coffee rings according to the salt generated intrinsic electric fields. The coffee ring structure can be controlled by choosing various salts, whose ions have different relative diffusivities, by adjusting the particle zeta potential, or by changing the rate of evaporation. In this talk the essential physics of the process is shown, an application to particle separation is shown, and the incorporation of active colloidal motors in the coffee ring structures is explored.

458128 Microdynamics and Elasticity of Fractal Gels with Embedded Active Colloids

Abstract:

Colloidal gels are a state of soft matter with attractively bonded particles in a network; these networks display rheological properties such as linear elasticity at low-frequency and a yield stress. Controlling these properties is important in industries where gels are used, such as pharmaceuticals, agricultural formulations, and consumer products. We present a method to control the microscopic dynamics of colloidal gels through active motion. We study how the introduction of active particles affects the dynamics of fractal colloidal gels comprised of passive, Brownian particles. Platinum coated Janus particles are embedded within the gel network and activated by the addition of hydrogen peroxide through a porous hydrogel membrane. We visualize gel structure and dynamics through two channel confocal microscopy. Through image analysis, we determine the positions of particles within the gel and compute their mean squared displacement (MSD). We find that the MSD of particles within the gel increases after the embedded Janus particles are activated. We analyze the MSD enhancement as a function of hydrogen peroxide concentration and the ratio of active to passive colloids. The active to passive gel MSD ratio increases as we vary the peroxide concentration from 0% to 12% at a fixed active to passive colloid ratio of 0.05. At a fixed peroxide concentration of 5%, the gel MSD increases as we increase the amount of active particles within the gel. We explore the implications of the enhanced microscopic dynamics for the macroscopic linear elasticity of the active gels.

472809 Investigation of the Motion of Patchy Particle Swimmers Near Fluid/Fluid and Solid/Liquid Interfaces

Abstract:

Recently, active colloids have been the subject of growing interest among scientists, due to their vast variety of applications including drug delivery, particle separation and Lab on a chip particle transport. Active colloids known as self-propelled particles swim in a hydrogen peroxide solution by decomposition of H_2O_2 into oxygen and H_2O. The asymmetric distribution of the product molecules around the particle leads to the particleâ€™s autonomous motion. Many studies have focused on the motion and behavior of theses particles both in bulk and near interfaces, for example, some of these studies have examined the directionality of the motion of Janus particles and others have examined how the particles align with respect to the wall.

In this context, we have previously probed the motion of patchy particles with varying sizes near a solid/liquid interface. We found evidence that the patchy particlesâ€™ rotational dynamics and their orientation with respect to the wall impact their swimming velocity and have attributed it to a preferential rotation of the particle around its swimming axis. [1]

Here, we present our approach to visualize the cap orientation of particles partially coated with platinum with respect to the fluid/fluid and solid/liquid interfaces. Further, we will show statistically relevant data on the orientation, fluctuation, and translation of colloidal patchy particles and their self-propulsion behavior in the presence of fluid/fluid and solid/liquid interfaces. The orientation and rotational motion of patchy particles will be analyzed and compared with respect to the type of interface they are interacting with.

[1] Z. Jalilvand, A. B. Pawar, and I. Kretzschmar, Patchy particle swimmers: a controlled self- diffusiophoretic motion, In Preparation, 2016.

472994 Acitve Toroids

Abstract:

The study of multicellular sheets is of great interest because of their important role in tissue development, morphogenesis, and wound healing. In confluent layer of epithelia at sufficiently high densities, cells coordinate their behavior and can collectively respond to physical and chemical stimuli. For example, heterogeneity of tissue tension within a planar 2D monolayer has been shown to correlate with the collective cell migration pattern as well as with the local increase in cell proliferation. However, much less is known about how cells respond to out-of-plane curvatures that are on the order of 1 to 100 cell lengths. Here, we investigate the proliferation and structure of MDCK on toroidal hydrogel substrates; unlike spherical or planar substrates, toroidal substrates have regions of both positive and negative Gaussian curvature. We can control the range of curvature via tuning the size and aspect ratio of the hydrogel torus, allowing us to explore the relation between substrate curvature, cell arrangement, and cell proliferation.

468362 Simultaneous Aggregation and Height Bifurcation of Colloidal Particles Near Electrodes in Oscillatory Electric Fields

Abstract:

Micrometer-scale particles suspended in aqueous NaCl solutions aggregate laterally near the electrode upon application of a low-frequency (100 Hz) field, but the same particles suspended in NaOH solutions are instead observed to separate laterally. The underlying mechanism for the electrolyte dependence remains obscure. Recent work by Woehl et al. (PRX, 2015) revealed that, contrary to previous reports, particles suspended in NaOH solutions indeed aggregate under some conditions while simultaneously exhibiting a distinct bifurcation in average height above the electrode. Here we elaborate on this observation by demonstrating the existence of a critical frequency (25 Hz) below which particles in NaOH aggregate laterally and above which they separate. Furthermore, we establish that the height of the upper layer of particles scales as the inverse square root of the applied frequency ($h \sim \omega^{-1/2}$), with particles reaching stable heights as high as 30 µm off the electrode at sufficiently low frequencies. The results indicate that the current demarcation of electrolytes as either aggregating or separating is misleading, and that the key role of the electrolyte instead is to set the magnitude of a critical frequency at which particles transition between the two behaviors.

472296 Sticking and Slipping of Active Bacteria on Silanized Glass Surfaces

Abstract:

Bacterial biofilms foul a wide range of engineered surfaces, from pipelines to membranes to biomedical implants, and lead to deleterious costs for industry and for human health. Designing strategies to reduce bacterial fouling requires fundamental understanding of mechanisms by which active and motile bacteria attach to surfaces. Here, we correlate the near-surface motion of *Escherichia coli* bacteria with transient adhesion and deposition onto silanized glass surfaces. We image bacteria as they flow through a linear channel and attach to the surface using confocal microscopy, and track their position and orientation over time using image processing algorithms. The deposition rate of bacteria on surfaces of varying silane chemistry and chain length does not correlate with surface energy or zeta potential. Instead, we find that a metric based on the degree of near-surface flagella-driven motion is inversely correlated with the rate at which bacteria deposit on these surfaces. On very smooth surfaces these bacteria exhibit mobile adhesion, in which surface-associated cells slip along the surface in the direction of flow, but this behavior cannot be described by colloidal models that predict detachment from hydrodynamic and/or adhesive interactions. We find that *E. coli* bacteria with and without flagella exhibit mobile adhesion, indicating that this behavior is not driven by these appendages, but cells that express fimbriae do not exhibit mobile adhesion. These experiments suggest that surface-associated motion enabled by surface appendages alters bacterial adhesion to surfaces in the earliest stages of biofilm formation.

471940 A Quantitative Theory of Magnetic Janus Particle Assembly

Abstract:

Under a magnetic field, Janus particles with magnetic caps (magJP) assemble into 2D chains. These chains give the fluid they are immersed in interesting rheological properties. More specifically, the longer the chains, the greater the viscosity. The rate of the magJP assembly is dependent on the dipole strength of the particles and is studied in this project. Predicting and controlling the magJP assembly rate will enable the engineering of materials that have desirable properties.

We have carried out experiments that measure the assembly rate of the magJP chains. The particles are synthesized, confined to a 2D environment, placed in a magnetic field, and tracked as they assemble. Particle tracking software has been developed for the project, and has been used to collect information from the videos including: chain concentrations vs time, particle trajectories, and dipole-dipole interaction energy. The results are fit to the Smoluchowski coagulation equations to obtain rate constants for chain aggregation.

The talk will discuss the relevance of these rate equations. A theory is proposed for prediction of the magJP rate constants when varying: concentration, dipole strength, dipole shift, viscosity, and particle size. The theory is shown to predict the magJP rate constants and future experiments are proposed to further test the theory.

460835 Active Janus Motors in Acoustic Confinement

Abstract:

We analyze the active random motion of self-propelled Janus motors confined in an external, near-Harmonic acoustic trap. The external trap behaves as an "osmotic barrier" that confines the swimmers inside the trapping region, analogous to semipermeable membranes that confine passive Brownian particles inside a boundary. We discover that active particles exposed to a weak trap have a familiar Boltzmann-like probability distribution – analogous to that of passive, equilibrium Brownian suspensions – except with the thermal energy k_BT replaced by an intrinsic kinetic 'activity' scale of active matter. However, this equivalence breaks down for stronger traps, where the trap size becomes a macroscopic length scale that confines the run length of the microswimmers. From the swimmers' restricted motion inside the trap, we calculate the unique mechanical 'swim pressure' generated by active systems originating from the force required to confine them by boundaries. Finally, we analyze the crossover from ballistic to diffusive behavior of active matter by tracking the explosion process of a dense active crystal.

452947 Shape-Directed Micro-Rotors Powered By Ultrasound

Abstract:

Shape, symmetry and other geometric features are widely used for programming self-assembly of passive colloids. These parameters can also be used as powerful strategies for engineering motion of active colloids and controlling their interactions and collective dynamics. Unlike self-propulsion directed by anisotropic surface (*e.g.*, catalytic, magnetic and light-driven Janus particles, acoustic bimetallic nanorods), shape-directed motion of active colloids is far less explored. A recent experimental study reported that gold (Au) nanorods (length 2μm, diameter 330nm) with convex and concave tips propel linearly toward their concave end at a speed of 200μm/s in megahertz (MHz) acoustic field. Other shapes and symmetry of individual particles can lead to more complex forms of locomotion than simple linear motion. In this talk, we present shape-directed rotational motion of Au micro-gears (spinners) driven by ultrasound. Both clockwise (CW) and counterclockwise (CCW) rotation can be achieved depending on the chirality of the spinners. We investigate the role of asymmetry, size and number of fins on the rotational speed of individual spinners. Shape-directed rotation can also be harnessed in other external stimuli such as chemical fuel and electric field. Shape- and symmetry-dependent actuation offers new opportunities for designing active colloids with complex forms of locomotion for enhanced functionality and studying their collective dynamics.

469471 Influence of Electrolyte Concentration on the Aggregation Ofcolloidal Particles Near Electrodes in Oscillatory Fields

Abstract:

Micron-scale particles suspended in various aqueous electrolytes have been widely observed to aggregate near electrodes in response to oscillatory electric fields, a phenomenon believed to result from electrically induced flows around the particles. Previous work has focused on elucidating the effects of the applied field strength, frequency, and electrolyte type on the aggregation rate of particles, with less attention paid to the ionic strength. Here we demonstrate that an applied field causes micron-scale particles in aqueous NaCl to rapidly aggregate over a wide range of ionic strengths, but with significant differences in aggregation morphology. Optical microscopy observations reveal that at higher ionic strengths (1 mM) particles arrange as hexagonally closed-packed (HCP) crystals, but at lower ionic strengths (0.05 mM) the particles arrange in randomly closed-packed (RCP) structures. We interpret this behavior in terms of two complementary effects: an increased particle diffusivity at lower ionic strengths due to increased particle height over the electrode and the existence of a deep secondary minimum in the particle pair interaction potential at higher ionic strength that traps particles in close proximity to one another. The results suggest that electrically induced crystallization will readily occur only over a narrow range of ionic strengths.

458427 Covalent Surface Modification of Carbon Coated Nanomagnets Allow Stable Dispersions in Aqueous Solutions and Specific Post-Modification

Abstract:

For many applications of magnetic nanoparticles such as for medical diagnostics, biotechnology, biomedical applications and magnetic biosensors, stable dispersions of magnetic nanoparticles in aqueous solutions are crucial[1]. Ferromagnetic pure metallic nanoparticles possess a very high saturation magnetization but also show a considerable tendency to agglomerate. Thus, it is highly desirable to create a material which combines the high saturation magnetization but still forms stable dispersions in aqueous solutions.

Here, we describe how specific covalent surface coating can lead to the formation of a metal cobalt nanoparticle based ferromagnetic fluid with unprecedented dispersion stability and chemical functionality [2]. More detailed, graphene-like carbon protected metallic nanomagnets, prepared by flame spray synthesis[3], which are covalently functionalized on the particle surface *via* diazonium chemistry are shown. A block-copolymer is grown directly from the particle surface by SI-ATRP. This block-copolymer consists of a negatively charged polymer block that is responsible for the colloidal stability of the particles and a second polymer block, which allows post-modification of the particles. The length of both polymer blocks is crucial for the colloidal stability of the particles and is studied in this work. Post-modification of this material by introduction of azide-functionality followed by the well-known "click"-reaction, allows the attachment of various desired target molecules. These nanoparticles form stable dispersions in various aqueous solutions, including biological relevant buffer systems.

1 A. H. Lu, E. L. Salabas and F. Schuth, *Angew. Chem.-Int. Edit.*, 2007, **46**, 1222.

2 C. J. Hofer, V. Zlateski, P. R. Stoessel, D. Paunescu, E. M. Schneider, R. N. Grass, M. Zeltner and W. J. Stark, *Chem. Commun.*, 2015, **51**, 1826.

3 R. N. Grass, E. K. Athanassiou and W. J. Stark, *Angew. Chem.-Int. Edit.*, 2007, **46**, 4909.

462710 Scalable Manufacturing of Nanostructured Particles By Gas-Phase Deposition Techniques

Abstract:

Core-shell nanoparticles and other nanostructured particles have high potential in applications such as catalysis, energy storage and pharma. However, a hurdle in their utilisation is that typically large amounts of such materials are required. Current liquid-phase and gas-phase synthesis methods often lack the high precision required or do not lend themselves to large-scale production. Gas-phase coating can be used to provide the surface of a particle with either a thin continuous coating or a decoration of nanoclusters. Coating techniques that can be used are chemical vapour deposition (CVD) [1], atomic layer deposition (ALD) [2], and molecular layer deposition (MLD) [3].

When carried out in a so-called fluidized bed, gas-phase coating is an attractive way of producing nanostructured particles with excellent scale-up potential. In such a system, the particles are suspended in an upward gas flow. We can do this both for nanoparticles and micron-sized particles. Nanoparticles â€“ contrary to what is typically observed for larger particles â€“ are fluidized as very dilute agglomerates with distinctive fluidization characteristics. I will discuss the challenges related to coating of nanoparticles using CVD, ALD, and MLD with such a system. Moreover, I will give a number of examples of the applications of nanostructured particles produced in this way.

[1] Valdesueiro, D., Meesters, G., Kreutzer, M., and van Ommen, J.R., 'Gas-phase deposition of ultrathin aluminium oxide films on nanoparticles at ambient conditions', Materials 8: 1249-1263 (2015).

[2] Goulas, A. and van Ommen, J.R., 'Atomic layer deposition of platinum clusters on titania nanoparticles at atmospheric pressure', Journal of Materials Chemistry A 1(15): p. 4647-4650 (2013).

[3] Vasudevan, S.A., Xu, Y., Karwal, S., van Ostaay, H.G.M.E., Meesters, G.M.H., Talebi, M., SudhÃ¶lter, E.J.R.,

and van Ommen, J.R., 'Controlled release from protein particles encapsulated by molecular layer deposition', Chemical Communications 51: 12540-12543 (2015).

468768 Manipulating Ligand-Nanoparticle Interactions and Catalytic Activity through Organic-Aqueous Tunable Solvents Recovery

Abstract:

Dispersed gold nanoparticles demonstrate much higher catalytic activities compared to their supported counterparts. Current methods of nanoparticle recovery are difficult to scale up, can alter the morphology of the nanoparticles, require large amounts of energy or solvents, and/or need highly specialized syntheses. Herein, we propose a facile system for the recovery of nanoparticles from reaction mixtures using Organic-Aqueous Tunable Solvents (OATS). OATS are homogeneous mixtures of an organic solvent and water which have the inherent property to phase separate into a heterogeneous mixture under moderate CO_2 pressure. A 60 vol% acetonitrile and 40 vol% water mixture was chosen to disperse gold nanoparticles for these experiments. Poly(vinylpyrrolidone) (PVP) was selected because it is water soluble and is believed to allow access to the entire metal surface for reactive compounds. We synthesized gold nanoparticles of average size 9.4 ± 1.4 nm and performed four different thermal treatments (no thermal treatment, 40, 50, 60ĖšC). The phase separation of the OATS mixture was determined to occur between 9.6-11.3 bar of absolute CO_2 pressure. Complete recovery of gold nanoparticles was achieved for all thermal treatments based on UV-vis absorbance of the localized surface plasmon collected from each phase. Catalytic activity of the nanoparticles was benchmarked using the hydrogenation of 4-nitrophenol. Reduction of catalytic activity occurred after the nanoparticles underwent thermal treatments or were subjected to pressure separation in OATS, but still remained highly active. The decrease in catalytic activity can be attributed to additional PVP functional groups having passivated the surface of the nanoparticle and blocked active sites as the system minimized free energy indicating that PVP does not allow access to the entire nanoparticle surface. The additional surface-bound ligands also prevented precipitation and growth in OATS, but slowed dynamic surface restructuring, ultimately decreasing the apparent catalytic activity.

472077 Protein Amyloid Fibrils As Template for the Synthesis of Silica Nanofibers

Abstract:

The preparation of silica fiber-like materials has captured the attention of scientists for the last decades. Their preparation requires the use of suitable templates. In this work, we show how protein fibrils can be used as templates for the preparation of long silica fibers, with a variety of aspect ratios. Protein amyloid fibrils are self-assembled structures resulting from partial denaturation of certain proteins, and their resulting aggregation mediated by β-sheets interactions. These structures are characterized by an extremely large aspect ratio, with a diameter that changes depending on the age of the fibrils, and on the degree of association of the different strands. Amyloid fibrils have an abundance of amino acid side chains at their surface that permit their dispersion in water and offer the chance to perform a wide range of simple derivatizations. We take advantage of these favorable features of amyloid fibrils and present a very simple procedure to coat them with silica. Starting with b-lactoglobulin fibrils with typical diameters of about 20nm , and a length of several micrometers, and by following two different strategies to grow silica (either in water at acidic pH values, or in ethanol-water mixtures under Stöber conditions) we have been able to control both the thickness and the roughness of the silica layer. Silica nanofibers with a thickness ranging from a few nanometers to hundreds of nanometers have been prepared. As an example of the flexibility offered by our approach, some of the rough silica-coated nano-fibers have been further functionalized to render them hydrophobic.

Such fibers have been further used for the fabrication of lotus-like surfaces. The surfaces have been prepared with a process consisting of four simple steps. First, ten microns colloidal particles have been arranged in either a single colloidal crystal, or in a binary colloidal crystal by combining them with smaller sub-micrometer particles. Large scale substrates have been coated with colloidal crystals. The hydrophobicity of the substrate has been measured by water contact angle, and found to be superior to that of the papillar structure of the lotus leaf deprived of the wax tubular structure. Then, silica nanotubes have been prepared by coating b -Lactoglobulin amyloid fibrils with silica by means of a Stober process. These silica nanotubes have been deposited on the binary colloidal crystal surfaces through a layer-by-layer deposition procedure, using positively-charged Ludox silica particles to reverse the sign of the charge on the surface of the substrate. Finally, the substrates have been hydrophobized by binding an alkyl chain on the surface of the fibrils. The final surfaces are superhydrophobic, with a water contact angle of 165.5 degrees, which is higher than that of the original Lotus leaf. The presented approach has two main advantages. It is easily scalable, and is based on a multi-step approach, which allows one a rational design of the nanostructure and microstructure impact on the final hydrophobicity of the surface.

463672 Conformal Layer-By-Layer Coatings on Spiky Hedgehog Particles

Abstract:

In order to disperse inorganic catalytic nanoparticles in apolar media, chemical camouflage with surfactants is required which greatly hinders catalytic activity. Due to this barrier, there has been little investigation of inorganic catalysts in apolar solvents. This study seeks to create functional dispersible catalysts by utilizing surface corrugation to circumvent this requirement. Hedgehog Particles (HPs), recently developed in the Kotov lab, consist of polystyrene microparticles with nanoscale zinc oxide spikes that result in marked reduction of Van der Waals interactions and enable dispersion in both polar and nonpolar solvents. Layer-by-layer (LbL) assembly allows for the production of functional thin films on microscale colloids incorporating a diverse set of catalytic nanoparticles and polyelectrolytes. By expanding LbL films to more complex colloidal surface geometries featuring high corrugation such as HPs, one can engineer a new generation of colloids with high dispersion stability, a unique balance of Van der Waals, electrostatic, and other interactions as well as a diversity of chemical/biological functionalities. Conformal coating of nanocatalysts and other functional materials was achieved by LbL films, as demonstrated by electron and confocal microscopy. Furthermore, zeta potential measurements confirmed the deposition of alternately charged LbL layers. Dispersion stability in nonpolar solvents is maintained with these surface modifications indicating promise in forming a new class of omnidispersible materials. These hybrid HPs are being investigated as a novel catalytic system which can greatly expand on catalysis in hydrophobic solvents.

470252 Non-Covalent Functionalization of Single Wall Carbon Nanotubes with Engineered Proteins for Targeted Subcellular Delivery

Abstract:

Single wall carbon nanotubes (SWCNTs) can be non-covalently dispersed with surfactants, polymers and proteins. We have observed that protein coated SWCNTs enter numerous cell types by the millions, more than an order of magnitude more than polymer coated SWCNTs. In addition to increasing uptake, protein coatings can impart function to nanomaterials. While many contemporary studies have focused on using proteins to target SWCNTs and nanomaterials to specific cells, we utilize proteins to target SWCNTs to regions within cells. Here, we will show the utility of modified albumin proteins and engineered native cellular proteins as dispersing agents for SWCNTs. To localize SWCNTs to the nucleus we dispersed SWCNTs with the tail domain of a nucleoskeletal protein lamin B1 (LB1) engineered from the full-length *LMNB1* cDNA. The low molecular weight globular protein has a central hydrophobic core for nanotube association and stabilization and an exposed nuclear localization sequence to promote active nuclear import. SWCNTs-LB1 enter HeLa cells and localize to the nucleus of cells; we visualize localization with Raman spectroscopy and NIR fluorescence imaging of SWCNTs and interaction with DNA by fluorescence lifetime imaging microscopy.

We have used similar techniques to show cellular uptake of bovine serum albumin (BSA)-coated SWCNTs within cells. We see localization of SWCNTs-BSA within the endosomal and metabolic processing compartments. Similarly to LB1, BSA is a small globular protein, and the hydrophobic cleft of the BSA binds to SWCNTs. BSA has smaller hydrophobic regions, independent of the hydrophobic cleft, that can be loaded with drugs or fluorophores. In vitro, we show that denaturation or enzymatic processing of the BSA releases the small molecules. SWCNTs-BSA in which the BSA are pre-loaded with rhodamine drastically increases small molecule delivery in culture. We have determined spatial and concentration distribution of rhodamine within the cells, and signal is both coincident and distinct from SWCNTs as measured by NIR fluorescence and Raman. We demonstrate efficacy of this approach by delivering a fluorescent chemotherapeutic drug daunomycin that reduces proliferation in cancer cells. Together, our results demonstrates a pathway to increase the delivery of a wide variety of drugs to cells through SWCNTs coated with albumin pre-loaded with drug molecules.

The complexity of protein structures allows for multimodal modification and manipulation of cellular processes in addition to SWCNT dispersion. The modification of native cellular proteins as non-covalent dispersing agents to provide specific transport opens new possibilities to utilize both SWCNT and protein properties for multifunctional subcellular targeting applications. Specifically, nuclear targeting will allow delivery of anticancer therapies, genetic treatments, or DNA, which in turn will promote development of novel cellular therapies. Using the transport properties of albumins and intracellular processing by enzymes allows delivery of native molecules and drugs.

460880 Effect of PMMA Particle Size on Coating Behavior in Combined Gliding Arc Discharge and Spouted Bed with Solid Precursor

Abstract:

A novel particle coating method using zinc particle as a solid precursor and gliding arc discharge was used to coat poly methyl methacrylate (PMMA) particles in spouted bed. An AC voltage was applied to two diverging electrodes to generate the arc discharge. A flow rate of 1 $L \cdot min^{-1}$ of argon gas was used to elongate the arc discharge as well as to establish a spouted bed condition inside the reactor. By using a combination of gliding arc discharge and spouted bed condition, a particle coating performance was achieved. Coating process was performed in a small scale reactor with a dimension of 75 mm × 30 mm × 3 mm. Both zinc solid precursor and PMMA particle were pre-loaded inside the reactor before the coating process. The coated samples were collected after a certain time intervals and were analyzed by scanning electron microscope (SEM) to observe the coating morphology and thickness. The energy dispersive X-ray spectroscopy (EDS) was also used to calculate the coverage of the coating layer, while an inductively coupled plasma optical emission spectrometry (ICP-OES) was used to measure the amount of zinc coating and indirectly used to calculate the thickness of the coating layer.

In previous study, two coating mechanisms were proposed for this novel particle coating process, which are the coating by physical contact with the partially melted zinc particle and coating by vapor deposition. For this research, we will focus on confirming the proposed coating mechanisms by studying the effect of PMMA particle size on the coating behavior. Two different sizes of PMMA particles were used, 0.3 and 0.6 mm in diameter. The increase in the PMMA particle size decreased the terminal settling velocity of zinc over PMMA (TSV) ratio from 1.91 to 1.35. The frequency of collision between zinc and PMMA particle will be affected by the change in the TSV ratio. Two applied voltages, 1.5 and 4.5 kV, was also studied to observe the changed in the coating mechanisms.

A fine layer of zinc coating was observed by SEM after 5 min. The results showed that all conditions tested exhibited similar trend for the amount of zinc coating as a function of coating time. The amount of zinc coating increased in the early stage of the coating process and became relatively steady afterward due to the peeling off of the coating layer. The amount of zinc coating dropped by approximately 50% when the diameter of the PMMA was doubled, regardless of the applied voltage. This is due to the decrease in the surface area to volume ratio of the PMMA particle.

The coating behavior depended significantly on the applied voltage. At 1.5 kV, the amount of zinc coating was higher than that of the coated sample obtained at 4.5 kV. Moreover, the coating seemed to be thicker when 1.5 kV of applied voltage was used to generate the gliding arc discharge. This is because, the energy provided can partially melt and partially vaporized the solid precursor allowing a mixed of coating mechanisms, coating by physical contact with the partially melted zinc particle and coating by vapor deposition. Due to the significant amount of coating obtained by physical contact with the partially melted zinc particle, it resulted in thick and rough coating layer. In addition, at low applied voltage, the amount of zinc coating was observed to increase when increasing the TSV ratio as the frequency of collision between zinc and PMMA was increased when TSV ratio was increased. This result in an increase in the rate of coating by physical contact with the partially melted zinc particle. On the other hand, at 4.5 kV, the energy provided by the gliding arc discharge was high enough to totally vaporize the solid precursor allowing mainly coating by vapor deposition. Hence, at this applied voltage, the amount of coating as well as coating thickness was independent of the TSV ratio.

The coated samples had an average thickness in the range of 100 to 200 nm. In the case of zinc percentage coverage, it is indicated that both applied voltage and TSV ratio did not have significant effect on the zinc percentage coverage. This means that the major contributor for the peeling off of the coating layer was the collision between PMMA particles and attack from gliding arc discharge rather than the collision between zinc and PMMA particles.

470935 Controllable Solution Combustion Synthesis of Nanoscale α/β-Bi_2O_3 and Its Catalytic Application

Abstract:

Among the four different polymorphs of bismuth trioxide, α- and β- Bi_2O_3 show unique physical and chemical properties . In the present work, Bi_2O_3 nanoparticles of different phases (α and β) were selectively synthesized via simple, rapid and energy efficient solution combustion synthesis (SCS). Bismuth nitrate was used as the oxidizer, while tartaric acid, glycine, EDTA and sucrose were employed as fuels. By changing types of fuels, ratios of fuel/oxidizer and heating temperature, specific phases (α and β) of Bi_2O_3 were controlled and particle sizes (5-50 nm) of these materials were tailored. The formation mechanism was proposed by thermodynamic calculations. Stable α-Bi_2O_3 forms when the combustion temperature is higher than 350 °C while metastable β-Bi_2O_3 forms at temperature lower than 300 °C. Using the SCS products as catalytic supports, series of Pt/SCS-Bi_2O_3 catalysts were prepared and tested for glycerol oxidation. These results were compared to our prior studies of Pt-Bi catalysts placed on activated carbon and other supports.

472321 A Combined Theoretical and Experimental Investigation on the Degradation of Organophosphorus Chemical Warfare Agents on ZnO_{1-X}

Abstract:

Metal oxide nanoparticles (MONP) have long been known to act as reactive sorbents for the degradation of chemical warfare agents (CWA). However, contrary to nature, which is capable of producing enzymes that can detoxify the same CWA at ambient conditions, MONP are stoichiometric and require elevated temperatures in order to remove the products from the surface. Herein vacancy induced MONP are examined for the degradation of organophosphorus agents by both experimental and theoretical means. Oxygen vacancies are introduced into MONP of zinc oxide (ZnO_{1-X}) and characterized by Raman spectroscopy, x-ray photospectroscopy (XPS), UV-vis spectroscopy, transmission electron microscopy (TEM) and x-ray diffraction (XRD). The reactivity of ZnO_{1-X} is evaluated using diisopropyl fluorophosphate (DFP) with solid state NMR, indicating that the vacancy induce structures exhibit a 25% increase in reactivity compared to the parent MONP. In parallel, a detailed mechanistic study revealing the path of degradation is investigated utilizing plane-wave density functional theory (DFT). Comparative studies are performed evaluating the stability of an oxygen vacancy on either the 002 or 100 surface planes of zinc oxide, as well as in the bulk and surface, revealing the latter to be the most stable. Binding energies were calculated using phosphine as a surrogate molecule to represent the surface reaction between ZnO_{1-X} and organophosphorus CWA's. The results obtained here not only provide a mechanistic insight towards CWA degradation, but will aid in the future development of more efficient catalyst against CWA's.

JOC The Journal of Organic Chemistry

Article

pubs.acs.org/joc

Magnetic Superbasic Proton Sponges Are Readily Removed and Permit Direct Product Isolation

Elia M. Schneider, Renzo A. Raso, Corinne J. Hofer, Martin Zeltner, Robert D. Stettler, Samuel C. Hess, Robert N. Grass, and Wendelin J. Stark*

Institute for Chemical and Bioengineering, Department of Chemistry and Applied Biosciences, ETH Zürich, Wolfgang-Pauli Strasse 10, CH-8093 Zurich, Switzerland

Supporting Information

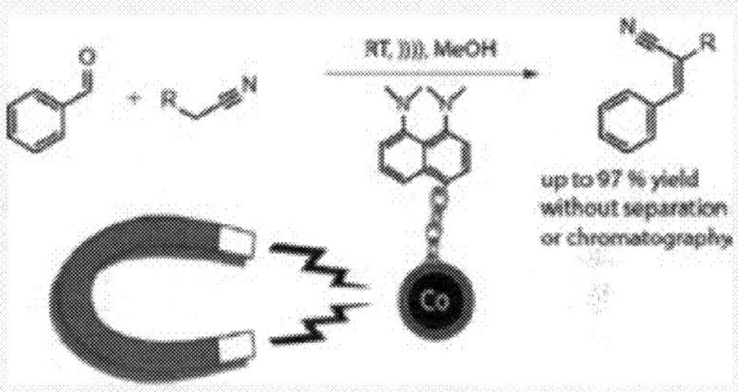

ABSTRACT: Workup in organic synthesis can be very time-consuming, particularly when using reagents with both a solubility similar to that of the desired products and a tendency not to crystallize. In this respect, reactions involving organic bases would strongly benefit from a tremendously simplified separation process. Therefore, we synthesized a derivative of the superbasic proton sponge 1,8-bis(dimethylamino)naphthalene (DMAN) and covalently linked it to the strongest currently available nanomagnets based on carbon-coated cobalt metal nanoparticles. The immobilized magnetic superbase reagent was tested in Knoevenagel- and Claisen–Schmidt-type condensations and showed conversions of up to 99%. High yields of up to 97% isolated product could be obtained by simple recrystallization without using column chromatography. Recycling the catalyst was simple and fast with an insignificant decrease in catalytic activity.

INTRODUCTION

Organosuperbases, such as the classic proton sponge 1,8-bis(dimethylamino)naphthalene (DMAN),[4] Verkade's base (proazaphosphatrane), and 1,8-bis(tetramethylguanidino)-naphthalene (TMGN) have become very important reagents in organic chemistry in past years.[7] Their exceptional basicity is associated with high levels of kinetic activity in proton-exchange reactions. These properties are often manifested by low nucleophilicity, making this kind of compound interesting for a wide scope of reactions,[8−11] and even heterogeneous cases are known.[1,2] However, the workup of such reactions on the laboratory level can be difficult, time-consuming, and expensive. Similarly, in industry, a broad variety of condensation reactions are applied. At present, free bases, such as sodium hydroxide, potassium hydroxide, and various organic bases, are the most commonly used, and they often lead to corrosion, significant waste production, and difficult workup. Thus, a chemically stable magnetic base with the attributes necessary for rapid quantitative separation would be very useful for industrial applications, resulting in significant solvent savings, a reduced amount of time spent, a decreased need for expensive equipment, and reusability of the separation reagent. Moreover, a ready-to-separate reagent is interesting for its potential use in the synthesis of high-quality products, where impurity carryover defines product performance. Magnetic nanoparticles have fascinated scientists for several decades and have been used in a plethora of applications,[14] such as for drug delivery,[15] in cancer treatments,[16] and as contrast agents for magnetic resonance imaging.[17] In chemical synthesis, magnetic nanoparticles have recently gained attention in the field of catalysis[18] because they combine a large surface area with the potential for a simple separation process. However, many magnetic reagents and surface linkers for reagent attachment are unstable in acidic-, basic-, or organic-solvent-containing reaction media. The exceptional chemical stability of carbon-coated metal nanoparticles[19] is based on a crystalline, graphene-like carbon surface that effectively prevents core oxidation. The all-carbon surface further allows covalent functionalization of the particle surface by using commercially available aryl diazonium salts.[20] At present, a number of promising recyclable, stable, metal-based magnetic catalysts have been proposed.[21−31] Most recently, alkene hydrogenation palladium catalysts[32,33] (so-called "catch-and-release" systems[34]) have been developed. In this work, we present an organic superbase, coupled to magnetic nanoparticles, with stability amenable for use under such challenging reaction conditions. We also show that such an easy-to-separate reagent is useful in a number of condensation reactions and simplifies workup and product isolation.

RESULTS AND DISCUSSION

Preparation and Characterization of the Magnetic Base. Magnetic, nanosized DMAN was prepared from chemically modified magnetic nanoparticles, a linker, and a derivative of DMAN. For covalent attachment of the base to the magnetic particle (1), the bifunctional linker hexamethylene diisocyanate (4) was first reacted with the primary amine of 4′-

Received: August 18, 2014

benzylamine-derivatized, graphene-coated metal nanoparticles (C/Co@amine) (3) to yield C/Co@hexamethyleneisocyanate (5). This intermediate contains a stable *N*-alkylurea bond and an active isocyanate moiety. Subsequently, C/Co@hexamethyleneisocyanate (5) was reacted in situ with 4-amino-1,8-bis(dimethylamino)naphthalene (DMAN-NH_2, 6), which was prepared according to the literature.[35] This afforded a magnetic, covalently bound DMAN (C/Co@DMAN (7)) in proper yield with a base capacity of 0.11 ± 0.01 mmol g^{-1}. For an illustration of the synthesis, see Scheme 1. The chemical identity of the herein synthesized reagent C/Co@DMAN (7) was proven by diffuse reflectance infrared Fourier transform spectroscopy (DRIFTS) and elemental microanalysis (C, H, N; Figure S1 and Table S1). The fairly strong urea-stretching vibration at 1690 cm^{-1} clearly confirms urea formation in the case of C/Co@hexamethyleneisocyanate (5). The successful attachment of DMAN to the isocyanate linker of C/Co@hexamethyleneisocyanate (5) was investigated with IR spectroscopy and elemental microanalysis (Table S1). A substantial increase in nitrogen content (ΔN = 0.82%) as well as an increase in carbon content (ΔC = 3.43%) is in line with successful functionalization. Vibrating sample magnetometry (VSM) measurements (at room temperature) showed an overall magnetization of 139.6 emu g^{-1}, which is in good agreement with the expected values, namely, a little lower than the saturation magnetization of nonfunctionalized C/Co (lower content of organics or carbon; 158 emu g^{-1}).[36] It is notable that the saturation magnetization of this material is still much higher than even the strongest conventional magnetite−silica nanoparticles (30−50 emu g^{-1}).[37]

The high saturation magnetization is of relevance because it permits rapid and easy separation.[38] To confirm the chemical stability and robustness of C/Co@DMAN (7), we recorded transmission electron microscopy (TEM) images of nanocarrier system 7 before and after reaction runs. Particles remained spherical and uniformly sized, in the diameter range of 20 to 60 nm (Figure S2). The reactivity of the herein proposed magnetic base C/Co@DMAN (7) was compared to reactions using similar catalysts from the literature. Another reference material, polystyrene-based resin-supported DMAN (S2), was synthesized (Scheme S2), fully characterized (Figure S3 and Table S2), and used for comparison.

Reliable recycling and separation of a reagent requires a robust confirmation of covalent attachment. Therefore, we used several test reactions (Scheme S1) to prove the covalent nature of DMAN's attachment to the nanomagnets. As a first control experiment, C/Co@hexamethyleneisocyanate (5), which bears one free, active isocyanate moiety, was reacted with pure (not aminated) DMAN. The lack of a free amine group in DMAN does not allow covalent attachment. Using the same synthesis and purification/washing procedure as for the herein proposed reagents, product analysis (Table S1) revealed the absence of both nitrogen incorporation and physisorption effects. As a separate control experiment, C/Co@hexamethylene-isocyanate (5) was first quenched with water, leading to a primary amine (i.e., the linker was deliberately rendered nonfunctional). Nonfunctional 5 was then exposed to functionalized, amine-bearing DMAN derivative 6; product analysis after applying the same purification/washing procedure as in the preparation of the functional magnetic base again confirmed the absence of reactivity (Table S1). These two experiments show that both the linker and the amination of the DMAN are necessary for the covalent binding of a DMAN moiety to the nanomagnet.

Scheme 1. Synthesis of Immobilized Magnetic Superbase 7[a]

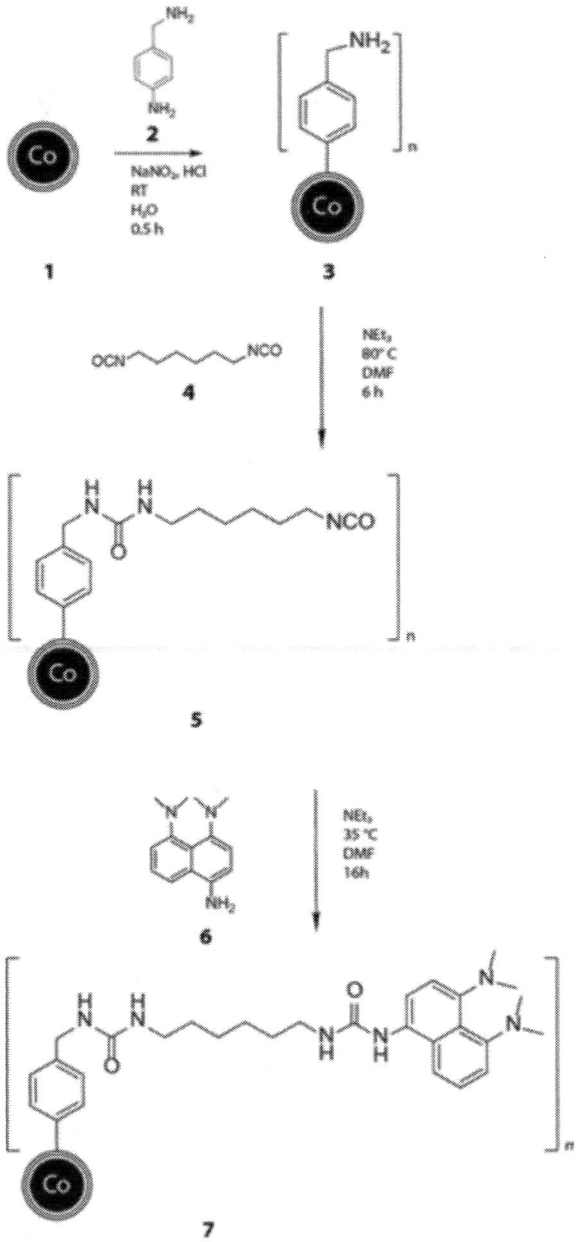

[a]C/Co (1) was coupled to 4-aminobenzylamine (2) via diazonium chemistry to yield C/Co@amine (3), which was further reacted with hexamethylendiisocyanate (4), resulting in C/Co@hexamethyleneisocyanate (5), bearing an active isocyanate moiety. C/Co@hexamethyleneisocyanate (5) was coupled with 4-amino-1,8-bis(dimethylamino)naphthalene (6) to yield C/Co@DMAN (7).

The experiments further show the absence of base physisorption, a potential experimental error in such reagent-anchoring studies.

Knoevenagel and Claisen−Schmidt Condensation. The literature states that DMAN indeed has catalytic activity in Knoevenagel condensation.[39] After identity confirmation, the catalytic activity of C/Co@DMAN (7) for the Knoevenagel

Table 1. Knoevenagel Condensation of Benzaldehyde and Malononitrile under Different Reactions Conditions

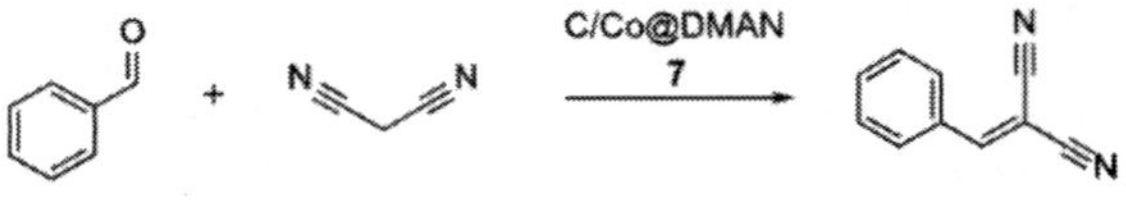

8 9 10

entry	catalyst	catalytic amount	solvent	time (h)	conversion (%)[a]	isolated yield (%)[b]	purity (%)
1	7	2 mmol %	toluene	7.5	63	61	99
2	7	2 mmol %	H_2O	7.5	99 (2)[c]	97	94
3	7	2 mmol %	MeOH	7	93 (85)	89	95
4	7	2 mmol %	MeOH[d]	4.5	98 (95)	91	96
5	7	0.2 mmol %	MeOH[d]	6	92 (90)	91	92
6	DMAN[e]	2 mmol %	MeOH	6	99	63[f]	96
7	**S2**	2 mmol %	toluene	6	99	97	98
8			MeOH	5.5	30	g	
9	C/Co	100 mg	MeOH	7.5	90	77	86
10	$CoCl_2$	5 mg	MeOH	6	33	g	

[a]Conversion was determined by HPLC with the reference product as the standard (benzaldehyde/malononitrile = 1:1). [b]Isolated yield after recrystallization. [c]Numbers in parentheses represent the percent conversion without washing the particles once with 3 mL of toluene. [d]Reaction carried out using an ultrasound bath. [e]Experiment was carried out using 2 mmol % of the free base. [f]Isolated yield after column chromatography (DCM/MeOH = 20:1). [g]No direct product isolation possible without column chromatography.

Table 2. Knoevenagel Condensation with Various Substrates Catalyzed by C/Co@DMAN (7)

entry[a]	basic substrate	aldehyde / ketone	product	temperature (°C)	conversion (%)[b]	isolated yield (%)
1	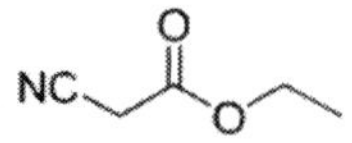	8	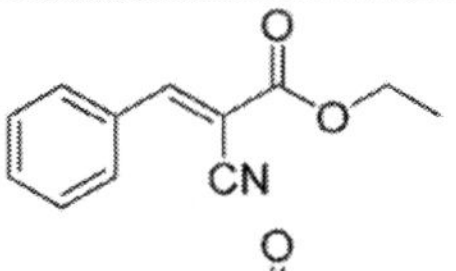	40	96[c]	95 [99][d]
2	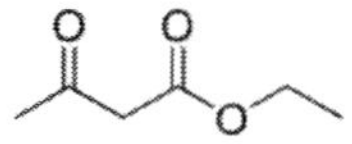	"	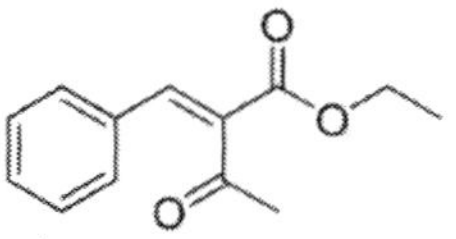	40	7	-[e]
3	9	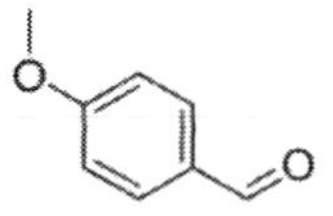	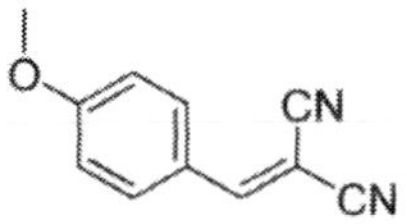	40	99	96 [91][d]
4	"	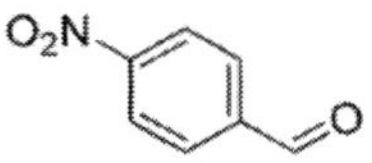	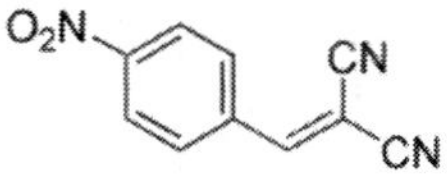	40	83	73[f] [96][d]
5	"	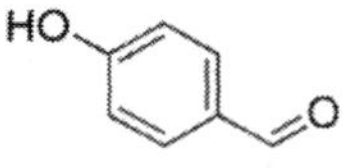	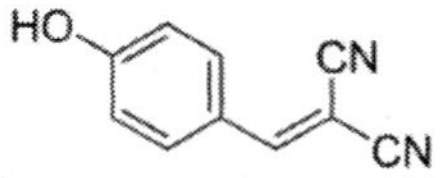	40	8	-[e]
6	"		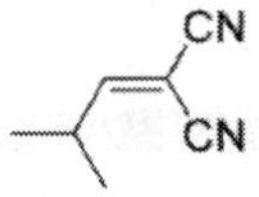	RT	81	-[e]
7	"	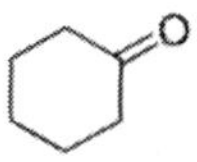	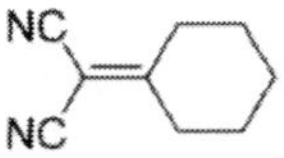	40	37	-[e]

[a]Reaction conditions: catalyst loading, 2 mmol %; basic substrate, 0.4 mmol; aldehyde or ketone, 0.4 mmol; MeOH, 5 mL; 16 h. [b]GC conversion. [c]After three consecutive cycles, conversion was still 95%. [d]Purity detected by NMR is given in brackets [ ]. [e]No direct product isolation was possible without column chromatography. [f]Isolated yield after recrystallization (DCM/hexane).

reaction of benzaldehyde (**8**) with malononitrile (**9**) was evaluated in different solvents at room temperature (Table 1, entries 1−3). If toluene was used as a solvent, then only 61% yield could be obtained. However, in water, quantitative conversion was achieved after 7.5 h. Still, the catalyst had to be washed once with toluene to release the product from the carbene surface. Therefore, methanol was tested as a solvent with promising results (Table 1, entries 3−5). The use of ultrasound during catalysis is known to ensure good reagent dispersion and consistently leads to excellent yields at shorter contact times. Indeed, when the reaction was conducted in an ultrasonication bath, a higher yield (91%) was obtained in a shorter time span (Table 1, entry 4). If the catalyst loading was decreased 10-fold, then similar yields were detected but a longer reaction time was necessary (Table 1, entry 5). Alternatively, running the same reaction using free DMAN where product workup and purification were not considered, quantitative conversion was obtained after 6 h (Table 1, entry 6). However, column chromatography had to be used to separate the catalyst residues from the product, which resulted in a significant drop in isolated yield. Magnetic base C/Co@DMAN (7) was separated within seconds, and recrystallization of the crude product (leading to the pure product) was achieved within minutes. Using polystyrene-immobilized proton sponge **S2** gave good yields as well, but filtration of the polymer was tedious and not very solvent-efficient. It is remarkable that even in the absence of any catalysts for the chosen reaction a low conversion (30%) was detected after 5.5 h (Table 1, entry 8). More surprisingly, the chosen reaction showed enhanced conversion (90%) in the presence of nonmodified (i.e., naked carbon surface) nanoparticles (Table 1, entry 9). To study further this unexpected finding, we included additional experiments. One probable hypothesis is based on leached cobalt ions functioning as a source of activity (hypothesis 1). Also plausible is that the carbon surface itself acts as an active catalyst (hypothesis 2). To exclude the first hypothesis, the reaction was carried out with a deliberate addition of cobalt(II) chloride ($CoCl_2$; Table 1, entry 10), the species that is supposedly leaching from the particles.[40] The addition of cobalt salts did not significantly affect the conversion (Table 1, entry 10; conversion = 33%); therefore, we do not assign significant activity of ionic cobalt in this kind of reaction. With regard to the second hypothesis, however, it is known that carbon-based surfaces (e.g., graphene, carbon nanotubes (CNT), carbon nanorods (CNR), etc.) indeed have been reported to support catalytic reactions (so-called carbocatalysis).[41,42] Thus, our finding is in line with similar carbon-surface activity and is subject to further extended studies by our group.

In the next part of the investigation, we tested different substrates (Table 2, entries 1−7). Ethyl cyanoacetate (Table 2, entry 1) reacted smoothly with benzaldehyde, and the product (ethyl α-cyanocinnamate) was detected in high yield and purity (96 and 99%, respectively) after 16 h at 40 °C; even after three consecutive cycles, the conversion was similarly high (95%). The less acidic and thus more demanding substrate ethyl acetoacetate (Table 2, entry 2) was not very active, and only low yields could be obtained. Also, different aldehydes and ketones were reacted with malononitrile. 4-Methoxy benzaldehyde performed very well (Table 2, entry 2), whereas the electron-poor nitro substrate was slightly less active (Table 2, entry 3). Surprisingly, 4-methoxy benzaldehyde reacted quickly, whereas 4-hydroxy benzaldehyde showed a slow conversion rate (Table 2, entry 6). An explanation may be the formation of hydrogen bonds with the solvent; one can imagine a shell of solvent molecules shielding the substrate from the catalyst. Nonaromatic substrates such as isobutyraldehyde and cyclohexanone were tested as well (Table 2, entries 6 and 7), and isobutyraldehyde performed better because of steric reasons (i.e., hindrance of nucleophilic attack on ketones).

Kinetic plots (Figures 1 and 2) confirmed the different reaction rates of C/Co@DMAN (7) when different solvents and mixing conditions (stirring vs sonication) were used. As expected, the reaction rate increased when the superbase was mixed more thoroughly with the reactants (i.e., a higher degree of mixing). To put the performance of C/Co@DMAN (7) in an appropriate context and to provide a fair comparison, we measured reaction kinetics for free DMAN (no mass-transfer limitation expected) and polystyrene-bound superbase **S2** (under optimal conditions in toluene) (Figure 2). C/Co@DMAN (7) initially catalyzes the chosen reaction at least as well as free DMAN, which leads to comparable or even slightly higher efficiency after 1 h of reaction time. The conversions converge with time, leading to high final conversion for all catalysts. The immobilized magnetic base C/Co@DMAN (7) can therefore be considered tp be directly competitive with existing or alternative catalytic systems. The minor changes in activity (i.e., rapid first part) may be a result of the altered electronic structure (note that DMAN was derivatized at position 4 to anchor the nanoparticle).

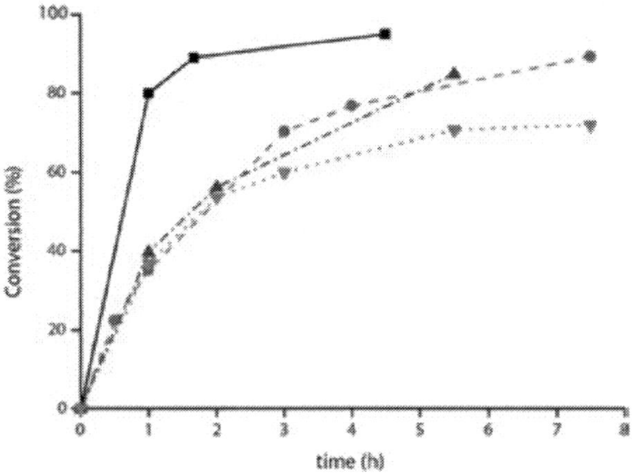

Figure 1. Kinetic plots for the Knoevenagel condensation of benzaldehyde (1.5 equiv) with malononitrile (1 equiv). All reactions used 2.5 mmol % **7** unless otherwise specified. Solvents and conditions for each reaction include ■, methanol and ultrasonication; ●, toluene and stirring; ▲, methanol and stirring; and ▼, C/Co instead of **7**, methanol and stirring.

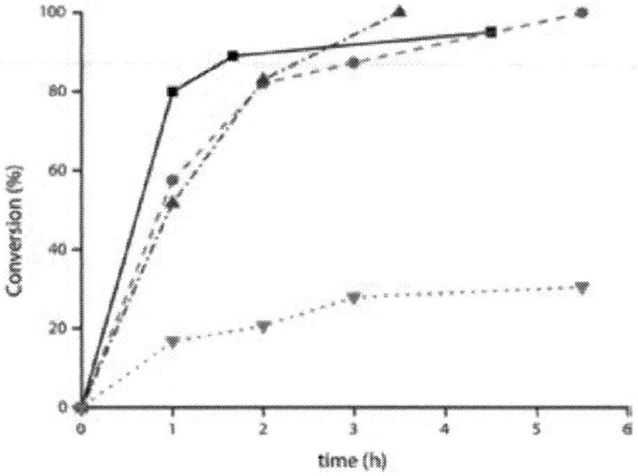

Figure 2. Kinetic plots for reactions with different catalysts (2.5 mmol %) under optimal conditions for each: ■, **7** in methanol, using ultrasonication (taken from Figure 1); ●, free DMAN in methanol; ▲, polymer-bound DMAN (**S2**) in toluene; ▼, without catalyst in methanol.

To determine the relevance of and to illustrate a broad applicability of magnetic organic bases such as C/Co@DMAN (7), another type of condensation reaction (Claisen−Schmidt) between acetophenone and benzaldehyde was tested (Table 3).

Table 3. Claisen−Schmidt Condensation of Acetophenone and Benzaldehyde

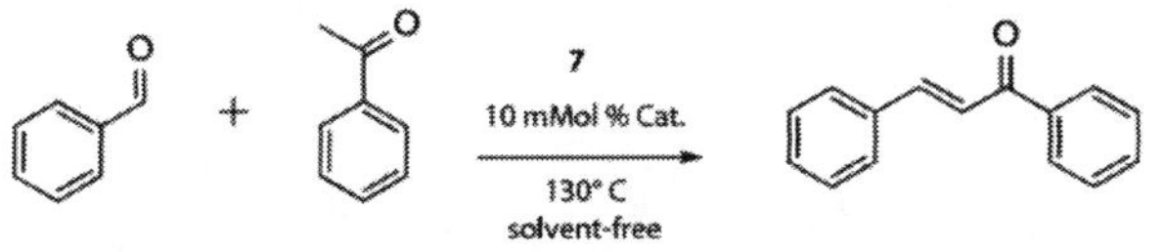

11 **12** **13**

entry	catalyst	run	conversion (%)[a]	isolated yield (%)[b]	time (h)	Co (ppm)[c]
1	7	1	66	62 [97.6]	16	19
2	7	2	78	d	20	16
3	7	3	71	d	18	5
4	7	4	68	d	18	44
5	7	5	76	75 [98.6]	20	64
6	DMAN	1	<1	e	20	
7	Co/C	1	58	e	21	42
8		1	<1	e	21	
9	**S2**	1	<1	e	20	

[a]Conversion was determined by HPLC using the reference product as the standard. [b]Isolated yield after solvent evaporation; purity as detected by NMR is given in brackets []. [c]ICP-OES detection limit for Co = 0.2 ppb. [d]No product isolation has been carried out. [e]No direct product isolation was possible without column chromatography.

This well-known reaction yields *trans*-chalcone, an important precursor for flavonoids that are widely used in medicine as anti-inflammatory,[43] antiviral,[44] and anticancer agents.[45] Immobilized magnetic base C/Co@DMAN (7) catalyzed the reaction with a yield of up to 75% and high levels of purity. Additionally, the same catalyst was reused in five iterative runs with similar yields. The harsh temperature (130 °C) leads to the conclusion that C/Co@DMAN (7) indeed remains stable using covalent-linker systems (Table 3, entries 1−5). Surprisingly, free DMAN does not catalyze the above condensation reaction (Table 3, entry 6; yield = 1%). This behavior is explainable considering the pK_a values. Because acetophenone has a higher pK_a value (pK_a = 25)[46] than does DMAN (pK_a = 12.1),[2] it can be concluded that the basicity of DMAN is not the key factor governing this catalysis. Furthermore, this result is in line with earlier work in the literature. Corma et al. experimentally investigated a similar case and reported that heterogeneous base catalysts work considerably better than free bases.[47]

Comparison to Known Solid-Supported Catalytic Base Reagents. In the recent literature, a number of excellent examples describe immobilized base catalysts.[1,2,6,13,47,48] Magnetic supports have been realized using modified magnetite and cobalt ferrite with good yields (Table 4, entries 6−7). In comparison to metal-based C/Co@DMAN (7), the oxidic supports exhibit a much lower saturation magnetization (133 emu/g for C/Co@DMAN (7), ca. 70 emu/g for magnetite[49]). This is a direct result of the inherently lower level of magnetization of most oxides. This value is important because it correlates directly with separation efficiency/speed and has recently enabled magnetic metal nanoparticles to purify water on a ton per hour scale.[50] In a typical laboratory setting, the improved magnetization facilitates workup, as demonstrated

Table 4. Comparison to Different Solid Support Base Reagents

entry	catalyst/base	saturation magnetization (emu g^{-1})	reaction	solvent	desired product	loading (mmol %)	time (h)	conversion (%)	conversion after three cycles (%)
1	7	139	Knoevenagel condensation	benzaldehyde	α-ethyl-2-cyanocinnamate	2	16	96	95
2	PS-MCM-41-DMAN[1]		Knoevenagel condensation	EtOH	α-ethyl-2-cyanocinnamate	0.5	7	99	85
3	DMAN/SiO_2-0.5[2]		Knoevenagel condensation	EtOH	α-ethyl-2-cyanocinnamate	1	6	90	85
4	TMGN/SiO_2-0.5[2]		Knoevenagel condensation	EtOH	α-ethyl-2-cyanocinnamate	1	6	100	97
5	7	139	Knoevenagel condensation	MeOH	benzylidene malononitrile	2	4.5	98	96
6	$CoFe_2O_4$@Si−N1-propylethane-1,2-diamine[3]	30[5]	Knoevenagel condensation	benzene	benzylidene malononitrile	2.5	0.33	100	100
7	Fe_2O_3@HAP@Silmidazoliumsalt[6]	55[12]	Knoevenagel condensation	H_2O	benzylidene malononitrile		1	98	97
8	7	139	Claisen−Schmidt condensation	acetophenone	*trans*-chalcone	10	20	78	71
9	DMAN/SiO_2-0.5[2]		Claisen−Schmidt condensation	acetophenone	*trans*-chalcone	1	24	95	
10	TMGN/SiO_2-0.5[2]		Claisen−Schmidt condensation	acetophenone	*trans*-chalcone	1	14	98	
11	Fe_3O_4-DABCO[13]	<60	Morita−Baylis−Hillman	MeOH	2-(hydroxy(4-nitrophenyl)methyl)acrylonitrile	20	5	88	88

using commercially available silica−iron oxide particles and the herein described metal reagent (Figure 3). The high level of magnetization of the carbon-coated cobalt allows clean, quantitative separation of the organic base within seconds. Established nonmagnetic immobilized catalyst supports are frequently based on zeolites and silica beads. The active site/moiety is predominantly attached by a siloxane-type linker.[51,52] These catalytic systems usually show excellent performance and reusability (Table 4, entries 2−4, 9−10). In the case of harsh reaction conditions (e.g., strong base), the weakest of these otherwise well-performing materials are the siloxane-type moieties. It is well known that extremely acidic or basic environments (i.e., pH <4 or >9) hydrolyze silane linkers. In contrast, the herein presented dialkyl urea moiety exhibits a high stability over a broad range of pH.

Figure 3. Recovery of organic base bound to magnetic nanoparticles using a low-cost external magnet. Model system in H_2O; same particle concentration as used in all experiments (0.7 mg mL^{-1}). A, metal-based reagents derived from Co/C; B, silica-coated Fe_3O_4. Note the different times above and below.

To confirm this statement experimentally, we directly compared siloxane linkers to the herein used C/Co@DMAN (7) in a number of representative acidic and basic environments. After 24 h at room temperature and pH 4, more than 20% of the siloxane linkers were cleaved, but less than 5% of the urea-type linkers were cleaved (Figure 4). At pH 9.5, the dialkyl urea linkers afforded lower stability (about 10% cleaved), but it is notable that under the same conditions >75% of the siloxane linkers were cleaved. Furthermore, the stability of these linkers was tested at elevated temperatures in different solvents (toluene and water). The dialkyl urea linkers performance under these conditions was superior to that of their silane-based pendants (Figure 4). The intrinsic stability of these particles in acidic or basic media has already been investigated[19] and has also been proven to be superior to that of silica-coated magnetite.

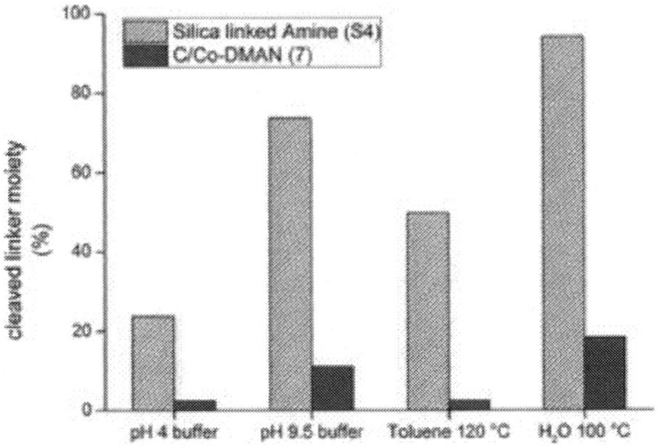

Figure 4. Immobilization linker stability tests under harsh reaction conditions overnight. The amount of cleaved linker was determined by analyzing the nitrogen content of the materials before and after exposure using quantitative elemental microanalysis. pH 4 buffer, 0.1 M potassium hydrogen phthalate with HCl; pH 9.5 buffer, 0.1 M carbonate-bicarbonate.

■ CONCLUSIONS

We have described the preparation of a covalently bound, chemically stable, magnetic nanoparticle-immobilized organic superbase and its extensive characterization using quantitative elemental microanalysis, infrared spectroscopy, vibrating sample magnetometry, and transmission electron microscopy. A number of control reactions were conducted to prove that the reagent was indeed chemically linked to the nanoparticles, and these confirmed the importance of linker type and reliability under relevant, harsh reaction conditions. We further demonstrated the experimental advantages of magnetic bases in a number of Knoevenagel condensations, reporting quantitative conversion and rapid product isolation in minutes. The demanding Claisen−Schmidt condensation afforded conversions of around 70% at 130 °C. The base catalyst was separated from the reaction mixtures in less than 1 min and could be recycled through at least five iterative runs. A quantitative comparison between siloxane linkers used in many immobilization studies and the herein described carbon shell-based chemistry revealed the important role of linker stability when designing reliable reagents for rough, everyday laboratory use.

■ EXPERIMENTAL SECTION

General Experimental Details. Carbon-coated cobalt nanoparticles[20] were suspended by the use of an ultrasonic bath and subjected to various reactions. After a reaction or a pretreatment step, nanoparticles were recovered from the reaction mixture with the aid of a conventional magnet (neodymium-based magnet, side length = 12 mm). Silica-coated magnetite nanoparticles (350 nm) were purchased from a chemical supplier. Commercially available reagents were used as received. All air-sensitive reactions were carried out under an argon atmosphere. The nanoparticles were analyzed by FT-IR spectroscopy (5% in KBr) and elemental microanalysis. HPLC measurements were performed (particle size, 5 μm; column size, 2.1 × 150 mm^2) to determine the conversion of the catalytic reactions. GC−MS measurements were performed (capillary, 30 m × 250 × 0.5 μm^2) with split injection (250 °C; ratio = 50:1; injection volume, 1 μL) and a temperature program (80 °C for 2 min; increase at 20 °C min^{-1} until 250 °C). The cobalt concentration present in solution was measured by inductively coupled plasma−atomic emission spectroscopy (ICP-OES).

Synthesis of C/Co@amine (3). Carbon-coated cobalt nanoparticles (**1**) were purchased from a chemical supplier. The particles were functionalized according to reported procedures (i.e., 3 g of **1** was suspended in 20 mL of H_2O, and 4-aminobenzylamine (**2**) (0.7 mL, 5.3 mmol) were added). Then sodium nitrite (0.8 g, 11.5 mmol) was added to the slurry. The vessel was put in the sonication bath, and 2 mL of concentrated HCl was slowly added dropwise. Upon completion of the reaction (30 min), the nanoparticles were recovered from the reaction mixture with a magnet and washed with toluene (2 × 10 mL), EtOH (1 × 10 mL), and acetone (2 × 10 mL) and dried for 24 h at 50 °C in vacuo. FT-IR: 1603, 1503, 1015, 831 cm^{-1}.

Synthesis of C/Co@hexamethyleneisocyanate (5). C/Co@amine (**3**, 2 g) was degassed three times in a Schlenk flask. Fifty milliliters of dry DMF, hexamethylendiisocyanate (HDMI) (6 mL, 37 mmol, 97 equiv, 98%), and triethylamine (NEt_3, 0.01 mL, 99%) were added under an argon atmosphere. Then the solution was dispersed in a sonication bath for 5 min. The dispersion was heated to 70 °C and stirred for 6 h. Upon completion of the reaction, the nanobeads were recovered from the reaction mixture with a magnet and washed with anhydrous DMF (2×). FT-IR: 2927, 2857, 1689, 1553, 1018 cm^{-1}.

Synthesis of C/Co@DMAN (7). Freshly synthesized C/Co@hexamethyleneisocyanate (**5**, 2 g) was used directly in a 250 mL Schlenk flask under an argon atmosphere. Forty milliliters of dry DMF and triethylamine (NEt_3, 0.01 mL, 99%) were added. The solution was dispersed in a sonication bath for 5 min. 4-Amino-1,8-bis-(dimethylamino)naphthalene (DMAN-NH_2 (**6**), 1 g, 4.3 mmol) was degassed three times in a Schlenk flask, dissolved in 10 mL of anhydrous DMF, and added dropwise to the C/Co-HMDI solution. The dispersion was heated to 35 °C and stirred overnight (16 h). Upon completion of the reaction, the nanoparticles were recovered from the reaction mixture with a magnet, washed with DMF (2 × 10 mL), EtOH (1 × 10 mL), acetone (2 × 10 mL), and dried for 24 h at 50 °C in vacuo. FT-IR: 2931, 2857, 2783, 1676, 1532, 1249 cm^{-1}.

Synthesis of Polystyrene-Supported DMAN (S2). Chloromethyl polystyrene (1 g, 2% DVB, 100–200 mesh, 0.9–1.5 mmol/g,) was used in a 50 mL Schlenk flask. Twenty milliliters of anhydrous DMF and triethylamine (NEt_3, 0.01 mL, 99%) were added under an argon atmosphere. DMAN-NH_2 (**6**) (0.6 g, 2.58 mmol) was degassed three times in a Schlenk flask, dissolved in 5 mL of anhydrous DMF, and added dropwise to the polystyrene slurry. The dispersion was heated to 40 °C and stirred overnight (20 h). Upon completion of the reaction, the functionalized polymer was filtered and intensively washed with acetone, 0.1 M NaOH, H_2O, EtOH, and acetone and dried for 24 h at 50 °C in vacuo. FT-IR: 3030, 2927, 2775, 1944, 1874, 1726, 1602 cm^{-1}.

Synthesis of Silica-Amine (S4). Silica gel 230–400 mesh (1 g) was placed in a 50 mL Schlenk flask and degassed three times. Twenty milliliters of anhydrous toluene was added, and the slurry was heated to 60 °C. *N*1-(2-Aminoethyl)-*N*2-(3-(trimethoxysilyl)propyl)ethane-1,2-diamine (0.3 mL, 11.6 mmol) was added dropwise to the solution, which then was stirred for 24 h. The resulting solid was filtered with 50 mL of toluene, 100 mL of DCM, and 20 mL of Et_2O and dried for 24 h at 50 °C in vacuo. FT-IR: 3289, 2940, 2828, 1863, 1474, 1109 cm^{-1}.

General Procedure for Knoevenagel Condensation Reactions. A mixture of solvent (5 mL) and C/Co@DMAN (**7**, 0.1 g, 0.01 mmol) was sonicated in an ultrasonication bath for 5 min in a Schlenk flask and then stirred at room temperature. Benzaldehyde (0.5 mmol) and malononitrile (0.5 mmol) were added, and the progress of the reaction was monitored by HPLC (MS to ensure product identity, UV at 280 nm for quantification) or GC-FID using a commercially available product as a reference.

Benzylidenemalononitrile (10). Isolated product was obtained by recrystallization in hot EtOH and hexane, followed by filtration. ^1H NMR ($CDCl_3$, 200 MHz, 25 °C): δ 7.86–7.82 (m, 2H), 7.71 (s, 1H), 7.57–7.43 (m, 3H).

α-Cyanocinnamic Acid Ethyl Ester. Pure isolated product was obtained by evaporation at 40 °C. ^1H NMR ($CDCl_3$, 200 MHz, 25 °C): δ 8.19 (s, 1H), 7.9 (m, 2H), 7.47–7.43 (m, 3H), 4.37–4.27 (q, 2H), 1.37–1.30 (t, 3H).

2-(4-Methoxybenzylidene)malononitrile. Pure isolated product was obtained by evaporation at 40 °C. ^1H NMR ($CDCl_3$, 200 MHz, 25 °C): δ 7.86–7.75 (m, 2H), 7.58 (s, 1H), 6.96–6.92 (m, 2H), 3.84 (s, 3H).

2-(4-Nitrobenzylidene)malononitrile. Isolated product was obtained by evaporation of the solvent, followed by recrystallization in DCM/hexane. ^1H NMR ($CDCl_3$, 200 MHz, 25 °C): δ 8.73 (s, 1H), 8.45–8.42 (m, 2H), 6.96–6.92 (m, 2H).

General Procedure for Claisen–Schmidt Condensations. A solvent-free mixture of acetophenone (2 mL) and C/Co@DMAN (**7**, 0.1 g, 0.01 mmol) was sonicated in an ultrasonication bath for 5 min and then stirred at 130 °C under a nitrogen atmosphere. Benzaldehyde (0.1 mmol) was added, and the progress of the reaction was monitored by HPLC (MS to ensure product identity, UV at 280 nm for quantification) using a commercially available product as a reference.

1,3-Diphenyl-2-propen-1-one (13). Isolated product was obtained by evaporation in high vacuum at 65 °C. ^1H NMR ($CDCl_3$, 200 MHz, 25 °C): δ 7.93–7.97 (m, 2H), 7.18–7.71 (m, 10H).

General Washing/Drying Procedure for Catalyst Recovery. Used catalyst was washed by sonication (5 min) with the reaction solvent of the catalytic reaction and acetone (2×) and dried for 4 h at 50 °C in vacuo.

General Procedure for Linker Stability Tests. A sample (30 mg) was placed in a 10 mL round-bottomed flask, and solvent (10 mL) was added. The solution was dispersed via ultrasonication bath and then shaken overnight. The solid sample was washed by either sonication (magnetic samples) or filtration (silica samples) with EtOH, H_2O, and acetone and dried for 4 h at 50 °C in vacuo.

■ ASSOCIATED CONTENT

Supporting Information

Spectroscopic data, elemental analysis data, IR spectra, TEM and SEM micrographs of the magnetic and polymeric compounds, scheme depicting the test reactions carried out to verify covalent linkage, and linker stability tests with a duration of 1 week. This material is available free of charge via the Internet at http://pubs.acs.org.

■ AUTHOR INFORMATION

Corresponding Author

*E-mail: wstark@ethz.ch. Web: www.fml.ethz.ch. Fax: +41 44 633 10 83. Tel: +41 44 632 09 80.

Notes

The authors declare the following competing financial interest(s): W.J.S. and R.N.G. declare financial interests, because they are shareholders of TurboBeads LLC, a company active in magnetic nanoparticles.

■ ACKNOWLEDGMENTS

We thank Andreas Dutly and Nikita Kobert for GC measurements, ICB/ETH Zurich, the EU-ITN network Mag(net)icFun (PITN-GA-2012-290248), and the Swiss National Science Foundation (no. 200021-150179) for financial support.

■ REFERENCES

(1) Corma, A.; Iborra, S.; Rodriguez, I.; Sanchez, F. *J. Catal.* **2002**, *211*, 208.

(2) Gianotti, E.; Diaz, U.; Velty, A.; Corma, A. *Eur. J. Inorg. Chem.* **2012**, 5175.

(3) Phan, N. T. S.; Jones, C. W. *J. Mol. Catal. A: Chem.* **2006**, *253*, 123.

(4) Alder, R. W.; Bryce, M. R.; Goode, N. C.; Miller, N.; Owen, J. *J. Chem. Soc., Perkin Trans. 1* **1981**, 2840.

(5) Rondinone, A. J.; Samia, A. C. S.; Zhang, Z. J. *J. Phys. Chem. B* **1999**, *103*, 6876.

(6) Zhang, Y.; Zhao, Y.; Xia, C. *J. Mol. Catal. A: Chem.* **2009**, *306*, 107.

(7) Ishikawa, T. *Superbases for Organic Synthesis*; John Wiley & Sons: Chichester, U.K., 2009; p 1.

(8) Gwaltney, S. L.; Sakata, S. T.; Shea, K. J. *J. Org. Chem.* **1996**, *61*, 7438.

(9) Ireland, R. E.; Liu, L. B.; Roper, T. D.; Gleason, J. L. *Tetrahedron* **1997**, *53*, 13257.

(10) Nielsen, M.; Jacobsen, C. B.; Jørgensen, K. A. *Angew. Chem., Int. Ed.* **2011**, *50*, 3211.
(11) Tur, F.; Saá, J. M. *Org. Lett.* **2007**, *9*, 5079.
(12) Mori, K.; Kanai, S.; Hara, T.; Mizugaki, T.; Ebitani, K.; Jitsukawa, K.; Kaneda, K. *Chem. Mater.* **2007**, *19*, 1249.
(13) Luo, S.; Zheng, X.; Xu, H.; Mi, X.; Zhang, L.; Cheng, J.-P. *Adv. Synth. Catal.* **2007**, *349*, 2431.
(14) Gao, J. H.; Gu, H. W.; Xu, B. *Acc. Chem. Res.* **2009**, *42*, 1097.
(15) Xu, Z. P.; Zeng, Q. H.; Lu, G. Q.; Yu, A. B. *Chem. Eng. Sci.* **2006**, *61*, 1027.
(16) Pankhurst, Q. A.; Connolly, J.; Jones, S. K.; Dobson, J. *J. Phys. D: Appl. Phys.* **2003**, *36*, R167.
(17) Huh, Y.-M.; Jun, Y.-w.; Song, H.-T.; Kim, S.; Choi, J.-s.; Lee, J.-H.; Yoon, S.; Kim, K.-S.; Shin, J.-S.; Suh, J.-S.; Cheon, J. *J. Am. Chem. Soc.* **2005**, *127*, 12387.
(18) Schätz, A.; Reiser, O.; Stark, W. J. *Chem.—Eur. J.* **2010**, *16*, 8950.
(19) Schaetz, A.; Zeltner, M.; Michl, T. D.; Rossier, M.; Fuhrer, R.; Stark, W. J. *Chem.—Eur. J.* **2011**, *17*, 10565.
(20) Grass, R. N.; Athanassiou, E. K.; Stark, W. J. *Angew. Chem., Int. Ed.* **2007**, *46*, 4909.
(21) Baruwati, B.; Guin, D.; Manorama, S. V. *Org. Lett.* **2007**, *9*, 5377.
(22) Gleeson, O.; Tekoriute, R.; Gun'ko, Y. K.; Connon, S. J. *Chem.—Eur. J.* **2009**, *15*, 5669.
(23) Keller, M.; Collière, V.; Reiser, O.; Caminade, A.-M.; Majoral, J.-P.; Ouali, A. *Angew. Chem., Int. Ed.* **2013**, *52*, 3626.
(24) Polshettiwar, V.; Baruwati, B.; Varma, R. S. *Chem. Commun.* **2009**, 1837.
(25) Riente, P.; Yadav, J.; Pericas, M. A. *Org. Lett.* **2012**, *14*, 3668.
(26) Schätz, A.; Long, T. R.; Grass, R. N.; Stark, W. J.; Hanson, P. R.; Reiser, O. *Adv. Funct. Mater.* **2010**, *20*, 4323.
(27) Stevens, P. D.; Fan, J. D.; Gardimalla, H. M. R.; Yen, M.; Gao, Y. *Org. Lett.* **2005**, *7*, 2085.
(28) Stevens, P. D.; Li, G. F.; Fan, J. D.; Yen, M.; Gao, Y. *Chem. Commun.* **2005**, 4435.
(29) Zeltner, M.; Schaetz, A.; Hefti, M. L.; Stark, W. J. *J. Mater. Chem.* **2011**, *21*, 2991.
(30) Cantillo, D.; Moghaddam, M. M.; Kappe, C. O. *J. Org. Chem.* **2013**, *78*, 4530.
(31) Zheng, Y.; Stevens, P. D.; Gao, Y. *J. Org. Chem.* **2005**, *71*, 537.
(32) Kainz, Q. M.; Linhardt, R.; Grass, R. N.; Vile, G.; Perez-Ramirez, J.; Stark, W. J.; Reiser, O. *Adv. Funct. Mater.* **2014**, *24*, 2020.
(33) Linhardt, R.; Kainz, Q. M.; Grass, R. N.; Stark, W. J.; Reiser, O. *RSC Adv.* **2014**, *4*, 8541.
(34) Wittmann, S.; Schätz, A.; Grass, R. N.; Stark, W. J.; Reiser, O. *Angew. Chem., Int. Ed.* **2010**, *49*, 1867.
(35) Ozeryanskii, V. A.; Pozharskii, A. F. *Russ. Chem. Bull.* **1997**, *46*, 1437.
(36) Zeltner, M.; Grass, R. N.; Schaetz, A.; Bubenhofer, S. B.; Luechinger, N. A.; Stark, W. J. *J. Mater. Chem.* **2012**, *22*, 12064.
(37) Lu, A. H.; Salabas, E. L.; Schuth, F. *Angew. Chem., Int. Ed.* **2007**, *46*, 1222.
(38) Schumacher, C. M.; Herrmann, I. K.; Bubenhofer, S. B.; Gschwind, S.; Hirt, A.-M.; Beck-Schimmer, B.; Guenther, D.; Stark, W. J. *Adv. Funct. Mater.* **2013**, *23*, 4888.
(39) Rodriguez, I.; Sastre, G.; Corma, A.; Iborra, S. *J. Catal.* **1999**, *183*, 14.
(40) Schumacher, C. M.; Grass, R. N.; Rossier, M.; Athanassiou, E. K.; Stark, W. J. *Langmuir* **2012**, *28*, 4565.
(41) Machado, B. F.; Serp, P. *Catal. Sci. Technol.* **2012**, *2*, 54.
(42) Schaetz, A.; Zeltner, M.; Stark, W. J. *ACS Catal.* **2012**, *2*, 1267.
(43) Gomes, A.; Couto, D.; Alves, A.; Dias, I.; Freitas, M.; Porto, G.; Duarte, J. A.; Fernandes, E. *BioFactors* **2012**, *38*, 378.
(44) Cushnie, T. P. T.; Lamb, A. J. *Int. J. Antimicrob. Agents* **2005**, *26*, 343.
(45) Woo, H. D.; Kim, J. *PLoS One* **2013**, *8*, e75604.
(46) Novak, M.; Loudon, G. M. *J. Org. Chem.* **1977**, *42*, 2494.
(47) Climent, M. J.; Corma, A.; Iborra, S.; Primo, J. *J. Catal.* **1995**, *151*, 60.
(48) Karaoğlu, E.; Baykal, A.; Şenel, M.; Sözeri, H.; Toprak, M. S. *Mater. Res. Bull.* **2012**, *47*, 2480.
(49) Goya, G. F.; Berquó, T. S.; Fonseca, F. C.; Morales, M. P. *J. Appl. Phys.* **2003**, *94*, 3520.
(50) Rossier, M.; Schreier, M.; Krebs, U.; Aeschlimann, B.; Fuhrer, R.; Zeltner, M.; Grass, R. N.; Günther, D.; Stark, W. J. *Sep. Purif. Technol.* **2012**, *96*, 68.
(51) Corma, A.; Garcia, H. *Adv. Synth. Catal.* **2006**, *348*, 1391.
(52) Han, Y.-J.; Stucky, G. D.; Butler, A. *J. Am. Chem. Soc.* **1999**, *121*, 9897.

461122 Selective Hydrogenolysis of Furfuryl Alcohol to 2-Methylfuran over Reduced Co-Fe-Al Mixed Metal Oxides

Abstract:

Taylor Sulmonetti, Bo Hu, Pradeep Agrawal, Christopher W. Jones

Selective Hydrogenolysis of Furfuryl Alcohol to 2-Methylfuran Over Reduced Co-Fe-Al Mixed Metal Oxides

Furfural and furfuryl alcohol have shown promise as major biomass platform chemicals, since they can be converted into a variety of value added products. Through side chain hydrogenolysis, furfuryl alcohol can be converted into 2-methylfuran, a potential fuel additive. Few monometallic catalysts show selective conversion of furfuryl alcohol towards 2-methylfuran; therefore, bifunctional and bimetallic catalysts have been widely pursued for this purpose in recent years. In this investigation, low cost, earth-abundent metal/metal oxide composites were utilized for the creation of 2-methylfuran through vapor phase conversion of furfuryl alcohol. An array of Co-Fe-Al catalysts were synthesized through the calcination of layered double hydroxides, and subsequently reduced in H2 to produce materials that contained both metallic and oxide domains. Vapor phase flow reactions were conducted after reduction to evaluate the reactivity of the reduced mixed metal oxides with varying Co/Fe ratios. The introduction of Fe into the mixed metal oxide matrix facilitated further reduction of the Co species to produce more catalytically active sites. Additionally, with the incorporation of Fe, the selectivity of 2-methylfuran reached about 85%, which is comparable or superior to prior reports using other catalytic compositions. Further characterization utilizing an array of spectroscopic experiments, including XAS and XPS, indicated the emergence of both metallic Co and Fe phases upon reduction. Lastly, the morphology was investigated through STEM to support the hypothesis that the catalysts were composed of well-dispersed mixed metal oxides containing metallic domains.

471191 Supported, Homogeneously Alloyed Bimetallic Nanoparticles By Electrostatic Adsorption

Abstract:

Bimetallic catalysts are often important for chemical reactions due to their enhanced performance over monometallic catalysts [1]. Incorporating a secondary metal into a system can benefit the catalytic system through ensemble, bifunctional, and electronic effects. The most prevalent method for synthesizing supported catalysts is dry impregnation (DI), also known as incipient wetness, in which a solution containing a specific amount of metal precursor is added to just fill the pore volume of the support [2]. While the procedure is simple and the metal loading is precisely known, limitations of DI preparations often include larger particle sizes and less intimate metal-support interactions. Moreover, bimetal catalysts produced through DI often exhibit a poor degree of alloying between the metals.

Our goal was to demonstrate a simple, effective method to synthesize highly dispersed, well alloyed bimetallic nanoparticles using strong electrostatic adsorption (SEA). SEA is achieved by charging the support by controlling the pH of the solution. A metal precursor that is oppositely charged can then be adsorbed onto the surface of the charged support. Using a simultaneous adsorption of either two cationic or anionic precursors allows excellent bimetallic interactions. The combinations of multiple bimetallic systems (Cu, Co, Ni, Pd, and Pt) prepared by SEA were compared to equivalent DI catalysts. Particle sizes and the degree of alloying were determined through electron micrographs, elemental mapping of nanoparticles, and powder XRD.

Results from XRD show the bimetallic catalysts produced by SEA were highly dispersed (<2 nm), while the DI catalysts exhibited large XRD peaks (>5nm). Initial elemental maps and STEM images indicate the SEA catalysts had a higher degree of alloying compared to the DI catalyst, and particle agglomeration was avoided for the SEA catalyst. A complete comparison of the bimetal systems prepared by DI and SEA will be discussed.

References

[1] Sankar, M., et al. (2012). "Designing bimetallic catalysts for a green and sustainable future." Chemical Society Reviews **41**(24): 8099-8139.

[2] Cho, H.-R. and J. R. Regalbuto (2015). "The rational synthesis of Pt-Pd bimetallic catalysts by electrostatic adsorption." Catalysis today **246**: 143-153.

460867 Chemoselective Transfer Hydrogenation of Nitroarenes Catalyzed By Highly Dispersed, Supported Nickel Nanoparticles

Abstract:

Aromatic amines are useful intermediates in the preparation of dyes, pharmaceuticals, and agricultural chemicals, which can be obtained by the reduction of aromatic nitro compounds. A variety of procedures involving metal catalysts and other reducing agents are available for this purpose; however, the selective reduction of a nitro group in the presence of other reducible functional groups in a molecule is a challenging task. In addition, reduction of aromatic nitro compounds often stops at an intermediate stage, yielding hydroxylamines, hydrazines, azoarenes, or azoxyarenes as byproducts. In this study, a highly active nickel (Ni) nanoparticles catalyst supported on porous silica gel particles was prepared by atomic layer deposition. Chemoselective reduction of nitroarenes was studied using the prepared Ni/SiO_2 as the catalyst and hydrazine hydrate as a hydrogen donor. Different kinds of nitroarenes were converted to the corresponding anilines with high yields. The high activity of the catalysts could be a result of the highly dispersed Ni nanoparticles.

Controllable Synthesis of Highly Dispersed Cu/SiO_2 Catalysts in Supercritical CO_2 and their Application in the Hydrogenation of Dimethyl Oxalate to Ethylene Glycol

*Qin-Qin Xu, Guo-Yue Qiao and Jian-Zhong Yin**

State Key Laboratory of Fine Chemicals, School of Chemical Machinery, Dalian University of Technology, Dalian 116024, China

* Corresponding Author: Jian-Zhong Yin, jzyin@dlut.edu.cn ; Tel. and Fax. +86-411-84986274

Abstract: Highly dispersed Cu nanostructures were synthesized within mesoporous silica in supercritical CO_2 using inorganic Cu precursors. Co-solvent was found to have significant influence on the metal loading and the nanostructure morphologies of the nanocomposites. Although the inorganic precursor has large solubility in ethanol, 1,3 propylene glycol, n-butyl alcohol and methanol, only big nanoparticles outside the nanochannels were obtained when the first three co-solvents were used, while zero metal loading was obtained when methanol was used as co-solvent. However, when ethylene glycol or the mixture of ethylene glycol and water were used as co-solvents, large amounts of short nanotubes were found inside the nanochannels with large metal loading. It indicates that the role of the co-solvent may have many aspects, such as enhancement of the precursor dissolution, change of the interaction between the precursor and the substrates and so on.

Key Words: supercritical carbon dioxide; catalyst; mesoporous silica; co-solvent; ethylene glycol

1. Introduction

Ethylene glycol (EG) is widely used as antifreeze and precursors in the production of polyesters. One of the most promising approach to EG is coal-based production technology via dimethyl oxalate (DMO) hydrogenation. Preparation of highly active catalysts plays a key role in the synthesis of EG through the coal chemical industry. Silica supported copper catalyst was found to be the most efficient one in this reaction[1-3]. There are many methods to prepare these supported metal nanocomposites, such as impregnation[4], sol-gel processing[5], chemical vapor impregnation[6] and so on. There are some disadvantages for the traditional methods to prepare these copper-based catalysts, for example, the active components are usually poorly dispersed for the impregnation method and the metal loading is quite low for the ion exchange method. Ammonia-evaporation method is characterized by highly dispersed active components with high metal loading, however, ammonia is very toxic. Herein, it is very important to develop some new and effective methods to prepare these catalysts.

Supercritical fluid deposition (SCFD) is an effective and environmentally benign technique to synthesize supported nanocomposites. This method involves the dissolution of metal precursor in supercritical fluids (SCFs), the diffusion and impregnation of SCF solution into the pores of the substrates and adsorption of metal precursor onto the walls of the channels, followed by thermal or chemical reduction in or after depressurization process. The low surface tension of SCF will favor the impregnation of precursors into the nano-scale channels of the substrates. Furthermore, the ordered pores of the substrate won't be collapsed because no organic solvents are left after depressurization, and the post-treatment process is quite simply compared to other preparing methods.

Supercritical carbon dioxide ($scCO_2$) is the most widely used SCF due to its abundance, low cost, non-flammability, non-toxicity and in particular the mild critical pressure and temperature ($T_C = 31.1°C$, $P_C = 7.38$ MPa). SCFD method was first proposed by Watkins based on chemical vapor deposition (CVD) technique in 1995[7]. After that, it has attracted growing attentions by many researchers. They used this method to deposit thin metal films (Cu, Au, Pt, Pd, Co, Ni etc.) onto a wide range of substrates such as silicon wafer, metal, ceramic, polymer and so on[6,8-12]or incorporate metallic nanoparticles (Pt, Pd, Au etc.) on different inorganic substrates such as carbon aerogel, carbon black, Al_2O_3, silica aerogel[13-16], graphene[17,18] etc.

However, organometallic compounds were usually used as precursors because they have relatively large solubilities in $scCO_2$. In order to overcome the limitation of using the expensive and toxic organometallic compounds, inorganic salt was used as precursor with the addition of appropriate co-solvent by some groups. Liu and coworkers used metal nitrates dissolved in $scCO_2$/ethanol system to deposit metal and metal oxides on carbon nanotubes[16,19,20] . Hu et al. prepared 4A-zeolite supported silver nanoparticles with the assistance of $scCO_2$ and hydrogen reduction using $AgNO_3$ as precursor[21]. Yin and his co-workers have made much effort towards the cheap SCFD process based on inorganic precursors and appropriate co-solvents in recent years [22-26].

In this study, $Cu(NO_3)_2$ was successfully deposited into mesoporous silica SBA-15 using $scCO_2$ as the solvent and different co-solvents. Single co-solvent including methanol, ethanol, ethylene glycol, 1,3 propylene glycol and n-butyl alcohol was investigated to find out the most proper one. After that, considering the competition adsorption agent, a small amount of water or dilute acid was added into the most proper co-solvent and formed the dual co-solvent to prepare the different samples. TEM, N_2 adsorption-desorption isotherm and ICP analysis were performed to characterize the metal loading and the microstructure of these samples.

2. Experimental

2.1 Materials

Mesoporous SBA-15 was prepared according to the published literatures[27]. The synthesized SBA-15 has a BET surface area of 807 m^2/g, total pore volume of 0.87 m^3/g and average pore diameter of 4.3 nm. $Cu(NO_3)_2 \cdot 3H_2O$ was purchased from TianJin DaMao chemicals Co. Ltd.. Carbon dioxide (>99%) was obtained from DaLian GuangMing gas Co. Ltd.. Ethanol, methanol, ethylene glycol, 1,3 propylene glycol and n-butyl alcohol of analytical grade were supplied by TianJin FuYu fine chemicals Co. Ltd..

2.2 Deposition process and characterizations

The schematic of the experiment and the details of the deposition process were described elsewhere [23,24,28]. Briefly, a certain amount of precursor and co-solvent were placed at the bottom of the reactor. The substrate was held in a stainless steel basket which was fixed in the upper part of the reactor. Then the reactor was connected to the lines, preheated to the experimental temperature and charged into CO_2 using a piston pump until the pressure of the system got up to the required pressure. Keeping the pressure and temperature for a required time, and then the reactor was depressurized slowly. After that, $Cu(NO_3)_2$@SBA-15 was obtained and subjected to calcinations (CuO@SBA-15) and H_2 reduction (Cu@SBA-15). The products were characterized by transmission electron microscopy (TEM), N_2 adsorption-desorption isotherms and Inductively coupled plasma (ICP).

3. Results and Discussion

3.1 Synthesis of CuO@SBA-15 using different co-solvents

SCFD method involves the dissolution of precursor in $scCO_2$, the diffusion and impregnation of the supercritical solution into the pores of the substrates and adsorption of the precursor onto the walls of the channels, followed by depressurization. Therefore, the dissolution of precursors in $scCO_2$ is essential to the successful preparation in this method. However, inorganic salts are not soluble in $scCO_2$. As we know that, $scCO_2$ has a strong ability to dissolve organic solutes and these organic solutes may act as co-solvent to enhance the ability of $scCO_2$ to dissolve the inorganic salts.

In this study, five kinds of organic solvents were tried to use as co-solvents respectively. Five samples were prepared using 200 mg SBA-15, 200 mg $Cu(NO_3)_2 \cdot 3H_2O$, 1 mL co-solvent at deposition temperature of 50°C, pressure of 18 MPa and deposition time of 3h. The metal loading were 14.4%, 8.3%, 0%, 16.0%, and 7.6% for the samples prepared using ethanol, ethylene glycol, methanol, 1,3 propylene glycol and n-butyl alcohol as co-solvents respectively. However, it was interesting that there were only big nanoparticles aggregated outside of the nanochannels for the three samples when ethanol, 1,3 propylene glycol and n-butyl alcohol were used as co-solvent (Figure 1a, 1c and 1d), while there were some nanoparticles and short nanorods inside the nanochannels with ethylene glycol as co-solvent (Figure 1b).

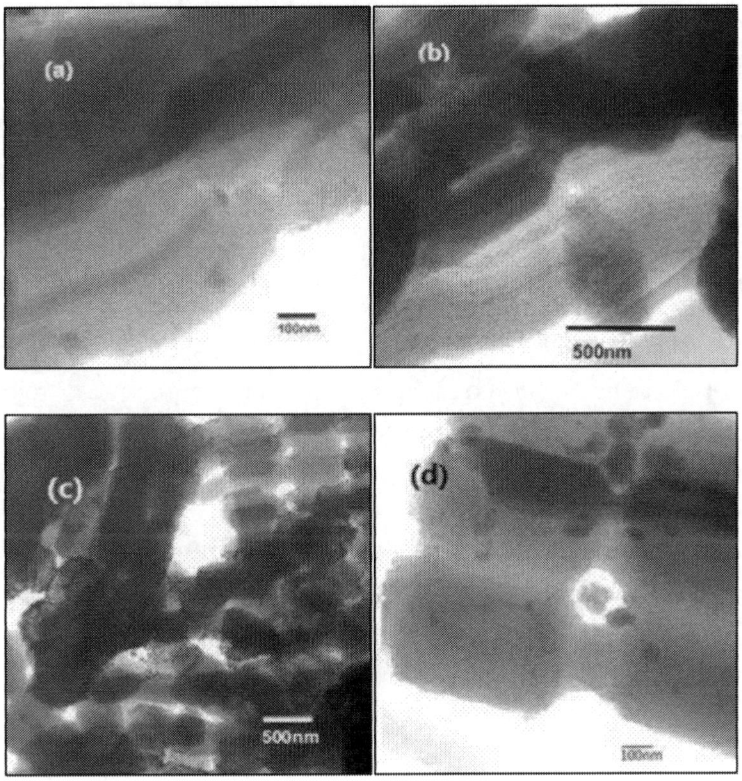

Figure 1 TEM images of samples prepared by different co-solvents (a) ethanol; (b) ethylene glycol; (c) 1,3 propylene glycol; (d) n-butyl alcohol

As stated above, the dissolution of inorganic salts in $scCO_2$ is essential to the deposition process and all the five co-solvents could play the role of the solubility enhancement. However, nanostructures could be confined within the nanochannels only when the ethylene glycol was used as co-solvent. It seems that the deposition process was not only dependent on the solubility of the precursor in the system, but also greatly influenced by the interaction between the co-solvent and the substrates. This phenomenon was also observed when silver nanostructures supported nanocomposites were prepared in $scCO_2$ and co-solvent system[25].

3.2 Synthesis of CuO@SBA-15 using dual co-solvent

Yin and his co-workers[24] found that compared to the case of only ethanol as the co-solvent, the employment of dual co-solvent of ethylene glycol and ethanol led to the significant reduction of deposition pressure and time when Ag@SBA-15 was prepared in $scCO_2$. In their subsequent work[25], they found that well dispersed and continuous nanowires were anchored into the nanochannels of SBA-15 in only 5 minutes at 8.5 MPa with a metal loading of 5.8 wt%. Furthermore, the metal loading increased with the increase of deposition time while the morphology was not changed. In summary, co-solvent played an important role in the adsorption of the precursor on the substrates. Besides the solubility enhancement, the dielectric constant, surface tension, viscosity of the solution and the interaction among the precursor, co-solvent and substrate were different when different co-solvents were used, which would affect the adsorption and dispersion of the precursor on the substrate.

Water is usually thought as a nonideal co-solvent, because it has a small solubility in $scCO_2$ due to its strong polarity. However, the solubility of $Cu(NO_3)_2$ in water is quite large. Therefore, the mixture of ethylene glycol and water was applied as co-solvent. Water was used to enhance the dissolution of precursor in co-solvent and ethylene glycol was added here to enhance the dissolution of the co-solvent in $scCO_2$. Furthermore, Yin et al.[29] enlightened by the traditional impregnation method and considered the addition of a small amount of dilute acid as the competition adsorption agent. Herein, we investigated the feasibility of preparing Cu/SBA-15 using the dual co-solvent of ethylene glycol and water as well as ethylene glycol and dilute acid.

When the sample was prepared at the temperature of 50°C, pressure of 18MPa, deposition time of 3h using the mixture of 1mL ethylene glycol and 0.5mL water as co-solvent, more uniform and dispersed nanoparticles were found inside the nanochannels of SBA-15, and the metal loading was increased to 12.7% (8.3% for only ethylene glycol co-solvent). It seems that addition of water will favor the deposition of precursors onto the substrates. When 1 mL ethylene glycol and 0.5 mL dilute HNO_3 was used as co-solvent, the metal loading was 13.8%. As can be seen in the TEM images (Figure 2), the nanochannels of the substrates were more complete filling after addition of water or dilute acid as co-solvent. What's more, the nanostructures were more uniform dispersed using the dual co-solvents.

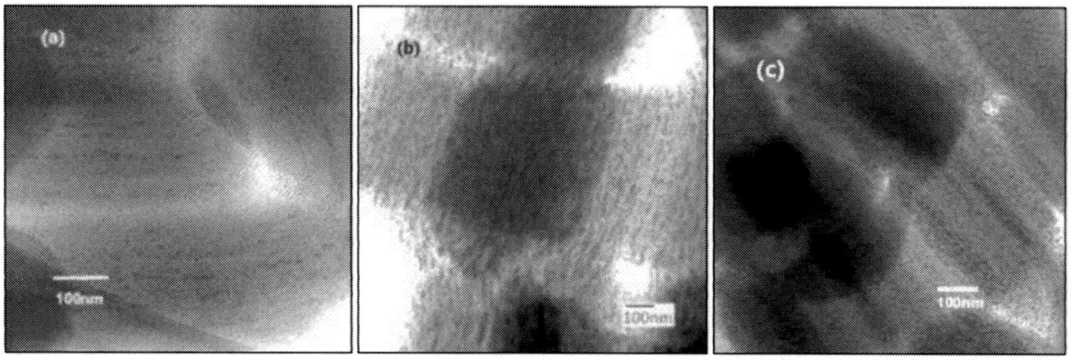

Figure 2 TEM images of samples prepared by different co-solvents: (a) ethylene glycol; (b) ethylene glycol and water; (c) ethylene glycol and dilute acid

The N_2 adsorption isotherms and the pore size distribution of SBA-15 and CuO@SBA-15 with different co-solvents shows typical steps in the 0.5-0.8 relative pressure(p/p_0) range which is characteristic of a highly ordered mesoporous porous lattice. The BET surface area, pore volume and the most probable pore size derived from the BJH adsorption branch were shown in Table 1.

Table 1 The N_2 adsorption-desorption analysis results of the samples

Sample	Co-solvent	BET surface area (cm^2/g)	Pore volume (cm^3/g)	Average pore size (nm)
SBA-15	-	807	0.87	4.3
CuO@ SBA-15	ethylene glycol	473	0.61	5.2
CuO@ SBA-15	ethylene glycol +HNO_3	519	0.62	4.8
CuO@ SBA-15	ethylene glycol +water	547	0.64	4.7

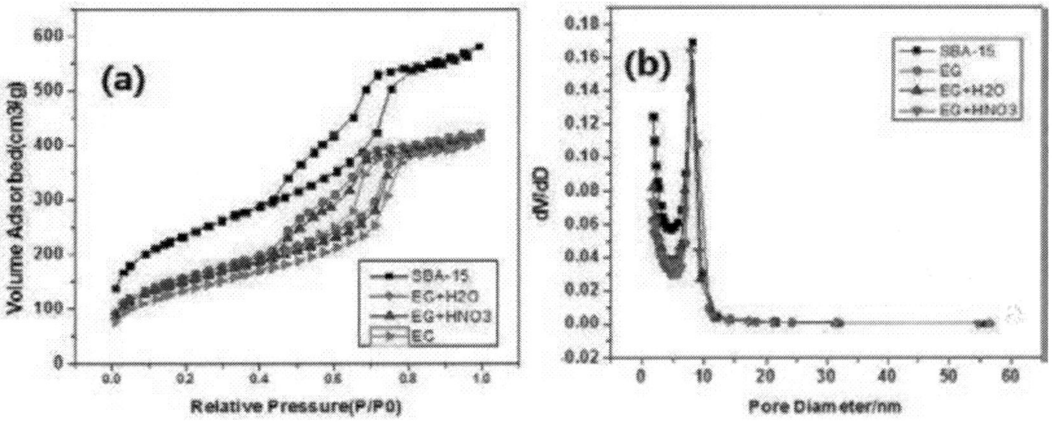

Figure 3 The N_2 adsorption isotherm(a) and pore size distribution curve(b) of the nanocomposites prepared using different co-solvents

3.3 The hydrogenation of dimethyl oxalate to ethylene glycol

This part of the work is just underway and the results will be reported later on.

4. Conclusions

In summary, SCFD is an effective method to deposit inorganic precursor into the nano-scale channels of mesoporous silica using appropriate co-solvent. It was found that co-solvent had a significant influence on the metal loading and the morphology of the nanostructure. Ethylene glycol or the mixture of ethylene glycol and water, the mixture of ethylene glycol and dilute acid are proved to be effective co-solvent to deposit $Cu(NO_3)_2$ into the nanochannels of SBA-15, while ethanol, methanol, 1,3 propylene glycol and n-butyl alcohol were not beneficial to the deposition. The underlying reason for different co-solvents, the operating parameters influencing the deposition results and the hydrogenation of dimethyl dxalate to ethylene glycol are underway.

Acknowledgments

The work was financially supported by the National Natural Science Foundation of China (21376045, 21506027), Chinese Postdoctoral Science Foundation (2015M571307) and the Fundamental Research Funds for the Central Universities.

References

1. Gong J, Yue H, Zhao Y, et al. Synthesis of ethanol via syngas on Cu/SiO2 catalysts with balanced Cu0-Cu+ sites. *Journal of the American Chemical Society.* Aug 29 2012;134(34):13922-13925.
2. Ma X, Chi H, Yue H, et al. Hydrogenation of dimethyl oxalate to ethylene glycol over mesoporous Cu-MCM-41 catalysts. *AIChE Journal.* 2013;59(7):2530-2539.
3. Yue H, Zhao Y, Zhao L, et al. Hydrogenation of dimethyl oxalate to ethylene glycol on a Cu/SiO2/cordierite monolithic catalyst: Enhanced internal mass transfer and stability. *AIChE Journal.* 2012;58(9):2798-2809.
4. Huang MH, Choudrey A, Yang PD. Ag nanowire formation within mesoporous silica. *Chem. Commun.* 2000(12):1063-1064.
5. Bekyarova E, Kaneko K. Structure and physical properties of tailor-made Ce,Zr-doped carbon aerogels. *Adv. Mater.* Nov 2000;12(21):1625-1628.
6. Miller JM, Dunn B, Tran TD, Pekala RW. Deposition of ruthenium nanoparticles on carbon aerogels for high energy density supercapacitor electrodes. *J. Electrochem. Soc.* Dec 1997;144(12):L309-L311.
7. Watkins JJ, McCarthy TJ. Polymer/Metal Nanocomposite Synthesis in Supercritical CO2. *Chem. Mater.* Nov 1995;7(11):1991-1994.
8. Blackburn JM, Long DP, Cabanas A, Watkins JJ. Deposition of conformal copper and nickel films from supercritical carbon dioxide. *Science.* Oct 2001;294(5540):141-145.
9. Cabanas A, Long DP, Watkins JJ. Deposition of gold films and nanostructures from supercritical carbon dioxide. *Chem. Mater.* May 2004;16(10):2028-2033.
10. Morere J, Tenorio MJ, Torralvo MJ, Pando C, Renuncio JAR, Cabanas A. Deposition of Pd into mesoporous silica SBA-15 using supercritical carbon dioxide. *J. Supercrit. Fluids.* Mar 2011;56(2):213-222.
11. Karanikas CF, Watkins JJ. Kinetics of the ruthenium thin film deposition from supercritical carbon dioxide by the hydrogen reduction of Ru(tmhd)(2)cod. *Microelectron. Eng.* Apr 2010;87(4):566-572.
12. Haruki M, Li SK, Qian G, Watkins JJ. Reactive deposition of cobalt using bis(2,2,6,6-tetramethyl-3,5-heptanedionato) cobalt(II) from supercritical carbon dioxide. *Journal of Supercritical Fluids.* Jan 2016;107:189-195.
13. Saquing CD, Cheng TT, Aindow M, Erkey C. Preparation of platinum/carbon aerogel nanocomposites using a supercritical deposition method. *J. Phys. Chem. B.* Jun 2004;108(23):7716-7722.

14. Zhang Y, Kang DF, Saquing C, Aindow M, Erkey C. Supported platinum nanoparticles by supercritical deposition. *Ind. Eng. Chem. Res.* May 2005;44(11):4161-4164.

15. Erkey C. Preparation of metallic supported nanoparticles and films using supercritical fluid deposition. *J. Supercrit. Fluids.* Jan 2009;47(3):517-522.

16. Sun ZY, Liu ZM, Han BX, Miao SD, Miao ZJ, An GM. Decoration carbon nanotubes with Pd and Ru nanocrystals via an inorganic reaction route in supercritical carbon dioxide-methanol solution. *J. Colloid Interface Sci.* Dec 2006;304(2):323-328.

17. Meng Y, Su FH, Chen YZ. Synthesis of nano-Cu/graphene oxide composites by supercritical CO2-assisted deposition as a novel material for reducing friction and wear. *Chemical Engineering Journal.* Dec 2015;281:11-19.

18. Zhao J, Yu H, Liu ZS, Ji M, Zhang LQ, Sun GW. Supercritical Deposition Route of Preparing Pt/Graphene Composites and Their Catalytic Performance toward Methanol Electrooxidation. *Journal of Physical Chemistry C.* Jan 2014;118(2):1182-1190.

19. Liu ZM, Han BX. Synthesis of Carbon-Nanotube Composites Using Supercritical Fluids and Their Potential Applications. *Adv. Mater.* Feb 2009;21(7):825-829.

20. Sun ZY, Liu ZM, Han BX, An GM. Supercritical carbon dioxide-assisted deposition of tin oxide on carbon nanotubes. *Mater. Lett.* Sep 2007;61(23-24):4565-4568.

21. Hu XS, Bai J, Hong HL, Li CP. Supercritical carbon dioxide anchored highly dispersed silver nanoparticles on 4A-zeolite and selective oxidation of styrene performance. *Crystengcomm.* 2016;18(14):2469-2476.

22. Yin J-Z, Xu Q-Q, Wang A-Q. Controlled Growth of Copper Nanoparticles and Nanorods in the Channels of Sba-15 by Supercritical Fluid Deposition. *Chemical Engineering Communications.* 2010;197(4):627-632.

23. Xu Q-q, Wang Y-q, Wang A-q, Yin J-z. Systematical study of depositing nanoparticles and nanowires in mesoporous silica using supercritical carbon dioxide and co-solvents: morphology control, thermodynamics and kinetics of adsorption. *Nanotechnology.* 2012;23(28):285602.

24. Xu Q-Q, Zhang C-J, Zhang X-Z, Yin J-Z, Liu Y. Controlled synthesis of Ag nanowires and nanoparticles in mesoporous silica using supercritical carbon dioxide and co-solvent. *The Journal of Supercritical Fluids.* 2012;62:184-189.

25. Xu QQ, Ma YL, Gang X, Yin JZ, Wang AQ, Gao JJ. Comprehensive study of the role of ethylene glycol when preparing Ag@SBA-15 in supercritical CO2. *Journal of Supercritical Fluids.* Aug 2014;92:100-106.

26. Xu QQ, Xu G, Yin JZ, Wang AQ, Ma YL, Gao JJ. Preparation of Superhighly Dispersed Co3O4@SBA-15 with Different Morphologies in Supercritical CO2 with the Assistance of Dilute Acids. *Industrial & Engineering Chemistry Research.* Jun 2014;53(25):10366-10371.

27. Zhao D. Triblock Copolymer Syntheses of Mesoporous Silica with Periodic 50 to 300 Angstrom Pores. *Science.* 1998;279(5350):548-552.

28. Xu Q-Q, Ma Y-L, Gang X, Yin J-Z, Wang A-Q, Gao J-J. Comprehensive study of the role of ethylene glycol when preparing Ag@SBA-15 in supercritical CO2. *The Journal of Supercritical Fluids.* 2014;92:100-106.

29. Xu Q-Q, Xu G, Yin J-Z, Wang A-Q, Ma Y-L, Gao J-J. Preparation of Superhighly Dispersed Co3O4@SBA-15 with Different Morphologies in Supercritical CO2with the Assistance of Dilute Acids. *Industrial & Engineering Chemistry Research.* 2014;53(25):10366-10371.

456379 Crystallization of Calcium Sulfate Dihydrate during Phosphoric Acid Production in the Presence of Magnesium Ions

Abstract:

Crystallization of Calcium Sulfate Dihydrate during Phosphoric Acid Production in the Presence of Magnesium Ions

You Peng

Zhilong Zhu

Kamal Samrane

Richard D. Braatz

Allan S. Myerson

The production of phosphoric acid from phosphate mineral rock involves the addition of phosphate rock to a concentrated sulfuric acid solution. The induced reactive crystallization process produces a side product of calcium sulfate hydrates which become the filter media in the subsequent acid separation process. For most industrial processes, the dihydrate form of calcium sulfate crystals (gypsum) precipitates and its shape and size distribution are key factors in determining the downstream filtration efficiency [1]. Particularly, the metal ion impurities coming from raw phosphate rock plays an important role as shape modifiers. The presence of impurities in the acid mixture has an impact both thermodynamically and kinetically, although most of the available literature focuses on their role as growth inhibitors and has neglected their potential impact on solution speciation [2].

Past studies on gypsum crystallization in phosphoric acid solutions usually involve the study of crystal growth and nucleation kinetics [3, 4]. However, past works did not use the correct definition of supersaturation when fitting kinetic parameters. The high concentrations in this multicomponent electrolyte system implies that supersaturation which be written in terms of the solubility product ratio, as governing by nonideal thermodynamics, which requires the com- putation of activity coefficients as well as free ion concentrations. For this purpose, the mixed solvent electrolyte (MSE) model is utilized to capture the solution speciation in order to properly quantify supersaturation at any given condition. The MSE model is a first-principles model that determines solid-liquid equilibrium by calculating excess Gibbs energy from additive pairwise interactions [5]. When impurities are present, additional binary interactions need to be included in the databank, which is carried out by regression analysis using solubility measurements.

Continuous reactive crystallization experiments are carried out with and without additives using a mixed-suspension, mixed-product removal (MSMPR) crystallizer. Crystal size dis- tribution and supersaturation are measured once the process reaches steady state. Different conditions are imposed to acquire both the temperature and supersaturation dependency of the crystallization kinetics [6]. A two-dimensional growth model is developed in order to capture the needle-like crystal morphology and the varying crystal aspect ratio with temperature, which is made possible by performing multi-scale image segmentation [7] and edge detection using the Canny method [8]. Growth inhibition models such as developed by Kubota and Mullin [9] are used for numerical quantification of step advancement retardation in the presence of impurities.

Experimental and numerical results are obtained for the base system and in the presence of Mg^{2+} ions. The results indicate that, while retarding crystal growth, Mg^{2+} ions tend to increase the crystal aspect ratio and lead to elongated needle-like crystals. The experimental results are consistent with experimental observation reported in the literature [10]. This study goes beyond past studies by providing a full two-dimensional population balance model for a highly concentrated ionic system that includes crystallization kinetics and thermodynamically correct driving force accounting non-ideality as well as the effects of an impurity.

References

[1] Pierre Becker. *Phosphates and Phosphoric Acid: Raw Materials, Technology, and Eco- nomics of the Wet Process*. Marcel Dekker, New York, 2nd edition, 1989.

[2] Gilles Févotte, Nesrine Gherras, and Jacques Moutte. Batch cooling solution crystalliza- tion of ammonium oxalate in the presence of impurities: study of solubility, supersatura- tion, and steady-state inhibition. *Crystal Growth & Design*, 13(7):2737–2748, 2013.

[3] E.T. White and S. Mukhopadhyay. Crystallization of gypsum from phosphoric acid so- lutions. In Allan S Myerson and Ken Toyokura, editors, *Crystallization as a Separations Process*, chapter 23, pages 292–315. American Chemical Society, Washington, DC, 1990.

[4] M El Moussaouiti, R Boistelle, A Bouhaouss, and JP Klein. Crystallization of calcium sul- phate hemihydrate in concentrated phosphoric acid solutions. *Chem. Eng. J.*, 68(2):123–130, 1997.

[5] P Wang, A Anderko, and RD Young. A speciation-based model for mixed-solvent elec- trolyte systems. *Fluid Phase Equilib.*, 203(1):141–176, 2002.

[6] You Peng, Zhilong Zhu, Richard D Braatz, and Allan S Myerson. Gypsum crystalliza- tion during phosphoric acid production: Modeling and experiments using the mixed- solvent-electrolyte thermodynamic model. *Industrial & Engineering Chemistry Research*, 54(32):7914–7924, 2015.

[7] J Calderon De Anda, XZ Wang, and KJ Roberts. Multi-scale segmentation image anal- ysis for the in-process monitoring of particle shape with batch crystallisers. *Chemical Engineering Science*, 60(4):1053–1065, 2005.

[8] John Canny. A computational approach to edge detection. *Pattern Analysis and Machine Intelligence, IEEE Transactions on*, (6):679–698, 1986.

[9] Noriaki Kubota and JW Mullin. A kinetic model for crystal growth from aqueous solution in the presence of impurity. *Journal of Crystal Growth*, 152(3):203–208, 1995.

[10] M.M. Rashad, M.H.H. Mahmoud, I.A. Ibrahim, and E.A. Abdel-Aal. Crystallization of calcium sulfate dihydrate under simulated condition of phosphoric acid production in the presence of aluminum and magnesium ions. *J. Cryst. Growth*, 267:372–379, 2004.

464898 Modifiers of Calcium Oxalate Monohydrate Crystallization: Tailoring Modifier–Crystal Interactions for Rational Drug Design

Abstract:

Modifiers of calcium oxalate monohydrate crystallization: Tailoring modifier–crystal interactions for rational drug design

Bryan G. Alamani, Jihae Chung, and Jeffrey D. Rimer

Chemical and Biomolecular Engineering, University of Houston, Houston, TX, 77204 USA

Calcium oxalate monohydrate (COM) is a major constituent of kidneys stones. Current therapeutics block solute (Ca^{2+} and $C_2O_4^{2-}$) attachment to crystals through either complexation in solution or binding to specific sites on crystal surfaces.[1] Despite advancements in drug design, the incidence rate of stones is on the rise, thus emphasizing the need for alternative therapies. When designing new modifiers as drugs, it is important to consider the underlying fundamental physical interactions of modifiers (ions or molecules) with COM crystal surfaces. These interactions may be elucidated through a combination of experimental and theoretical studies. In this talk, we will describe systematic studies of bulk crystallization and in situ characterization of surface growth in the presence and absence of modifiers.[2] We have explored the effect of metal ions (both monovalent and divalent), small organics, and native proteins that are commonly observed in human urine and kidney stone matrix. The majority of modifiers act as growth inhibitors with varying efficacy.[3] Interestingly, we have identified modifiers that function as COM growth promoters.[4]

Understanding the interactions between growth modifiers and crystal interfaces requires a combination of experimental techniques capable of probing a wide range of length and time scales.[5] Analysis of macroscopic properties (bulk crystal growth) provides information on modifier changes to crystal size and habit. Kinetic studies were employed to assess the effect of modifiers on the rate of COM growth through ion-selective electrochemistry (ISE). To gain molecular insights, it is important to study the interface that captures surface phenomena leading to changes in crystal morphology. For this task, we use *in situ* atomic force microscopy (AFM) to observe time-resolved dynamics of surface growth in the presence of modifiers.[6] The culmination of these studies provide insight on how we can potentially engineer modifiers with tailored properties to take advantage of specific modifier-crystal interactions to suppress growth, or even lead to crystal dissolution. To this end, we will present our discovery of modifiers that are viable candidates to replace current therapeutics of pathological COM crystallization.

References1. Wesson, J. A.; Ward, M. D., *Elements* **2007,** *3*(6), 415-421.

2. Farmanesh, S.; Chung, J.; Chandra, D.; Sosa, R. D.; Karande, P.; Rimer, J. D., *Journal of Crystal Growth* **2013,** *373*, 13-19.

3. Farmanesh, S.; Alamani, B. G.; Rimer, J. D., *Chemical Communications* **2015,** *51*(73), 13964-13967.

4. Farmanesh, S.; Ramamoorthy, S.; Chung, J.; Asplin, J. R.; Karande, P.; Rimer, J. D., *Journal of the American Chemical Society* **2014,** *136*(1), 367-376.

5. Rimer, J. D.; An, Z.; Zhu, Z.; Lee, M. H.; Goldfarb, D. S.; Wesson, J. A.; Ward, M. D., *Science* **2010,** *330*(6002), 337-341.

6. De Yoreo, J. J.; Chung, S.; Friddle, R. W., *Advanced Functional Materials* **2013,** *23* (20), 2525-2538.

464204 Combined Cooling and Anti-Solvent Crystallization of L-Asparagine Monohydrate

Abstract:

Crystallization is a key unit operation especially in fine chemical and pharmaceutical industries to produce high value solids with desired purity and size distribution [1]. The cooling and anti-solvent crystallization are common operational modes used in industry and the optimization in such processes is traditionally sought with respect to optimal cooling profile, anti-solvent feed-rate profile, and seeding characteristics. Essentially, the control is implemented by manipulating the supersaturation trajectory during the course of crystallization due to its relationship with the fundamental crystallization phenomena such as nucleation, growth, etc. Cooling crystallization is advantageous when the solubility of the compound strongly depends on temperature. Anti-solvent crystallization is particularly advantageous when the solute to be crystallized has a solubility that is a weak function of temperature and/or the solute's thermal stability is low. Sometimes the solubility of a compound is significantly affected by both temperature and addition of an anti-solvent. Therefore, combining cooling and anti-solvent crystallization can be advantageous as it offers two manipulative variables that can be varied to exert greater control on the supersaturation profile and thereby on the outcome of the crystallization process [2]-[3].

The objective of this research is to perform a systematic study on how the cooling and anti-solvent addition should be combined using three open-loop conventional cooling policies and three feeding policies that are inspired by these well-known cooling policies [4]. Specifically, we perform a series of combined cooling/anti-solvent crystallization of L-asparagine monohydrate (LAM), an important amino acid, from its aqueous solution using isopropanol as anti-solvent. First, solubility of LAM in water-isopropanol mixture is determined gravimetrically as a function of temperature (25 to 45 °C) and weight fraction of isopropanol (0 to 80%). Next, combined cooling/anti-solvent crystallization experiments are conducted where a saturated

aqueous solution of LAM at 45 °C is cooled using a given cooling policy from 45 to 25 °C in 2 h while isopropanol is added simultaneously with a given feeding policy such that its weight fraction changes from 0 to 40 wt% by the end of the experiment. All combined cooling and anti-solvent crystallization experiments recorded improvement in crystal size distribution characteristics and significant enhancement in the yield compared to only cooling crystallization.

Experiments with various combinations of cooling and feeding policies generate a wealth of data for understanding the dynamics of the process and estimating kinetic parameters for developing a predictive model. The concentration of the solute is measured at regular sampling intervals by high performance liquid chromatography and the final crystal size distribution is measured by laser diffraction. We develop a dynamic model based on population balance equation coupled with mass balance equation; and nucleation and growth kinetics are simultaneously estimated from experimental data using a nonlinear parameter estimation technique. The fully validated model is then used to determine the optimal temperature and anti-solvent feeding profiles. A comparison is then made among the optimal policy and conventional policies to understand the convenient and optimal path of combining the two modes of operation.

References:

[1] J. W. Mullin, Crystallization. Woburn, MA: Butterworth-Heinemann, 2001.

[2] Z.K. Nagy, M. Fujiwara, and R.D. Braatz, J. Process Control 18, 856–864, 2008.

[3] C. Lindenberg, M. Krattli, J. Cornel, and M. Mazzotti, Cryst. Growth Des. 9, 1124–1136, 2009.

[4] H. Hojjati, M. Sheikhzadeh, and S. Rohani, Ind. Eng. Chem. Res. 46, 1232-1240, 2007.

462129 A Semi-Empirical CLD-to-Psd Model for Crystallization Process Monitoring

Abstract:

Controlling the particle size distribution of the final API is often one of the main objectives of a pharmaceutical crystallization process since API particle properties impact the subsequent formulation process and the ability to produce a drug product with the desired quality attributes. Thus, API crystallization process monitoring and control have become more and more important as part of the increasing QbD efforts in the pharmaceutical industry.

Among the possibilities to characterize particle populations in situ and in real-time, the Focused Beam Reflectance Measurement (FBRM) has become broadly accepted mainly due to its ease of process implementation as well as due to its performance robustness. However, one major limitation of the FBRM technique is that its measurement signal, the Chord Length Distribution (CLD), cannot be easily transferred into the particle size distribution (PSD) despite the application of several different resolution approaches over the last 15 years. Thus, the CLD data is often used qualitatively rather than quantitatively and generally provides limited value for advanced crystallization process modeling purposes.

In this work a semi-empirical method is introduced that translates CLD to PSD data based on a training set of several particle sieve cuts at increasing particle concentrations. Unlike matrix-based methods, this approach is not limited to compact particle morphologies but holds as long as changes to the particles shape distribution do not exceed certain ranges. This optimization-based method consists of two steps. First, the â€œspectral decompositionâ€ of the

experimental CLD is generated which yields an idealized particle size distribution. In the subsequent optimization step, fuzzy-trained sets are used to correlate the idealized PSD with the actual experimentally measured PSD. This work highlights the application of this model to several relevant compounds and discusses the possibility of using in situ FBRM data for crystallization process modeling.

468457 Optimizing a Distillative Crystallization and Evaluating Form Risk

Abstract:

Distillative crystallizations represent a class of particle formation with unique challenges in establishing control. In contrast to antisolvent addition rate-controlled or temperature ramp rate-controlled crystallizations, the evolution of supersaturation with time is both scale and equipment dependent for distillative crystallizations. Thus, controlling the nucleation and growth processes that contribute to particle size and form is more challenging. This presentation provides a case study of a distillative crystallization which exhibited poor and unreliable filtration rates upon scale up, leading to a process bottleneck. The variable nature of this unit operation also resulted in the appearance of several polymorphs. Further process understanding was needed to establish a robust process with respect to filtration rate and polymorph control to a desired form. The crystallization was characterized through monitoring trends in nucleation and growth via FBRM, evaluating particle size via photomicroscopy, and determining relative supersaturation via HPLC. Water content, temperature, and age time were identified as critical parameters in maximizing the particle size of the crystals. Competitive slurry experiments and measurement of relative solubilities were used to develop the thermodynamic phase map and establish optimized conditions to enable generation of the desired form.

466961 Application of Ehrfeld® Valve Assisted Micromixer for Generation of Micron Sized Particles of Poorly Water Soluble Drug Compounds

Abstract:

Production of micron and submicron size particles for hydrophobic compounds finds a great application in the pharmaceutical industry due to their enhanced surface properties which aid in overcoming dissolution and bioavailability issues. The key to obtaining fine particles is inducing homogeneous precipitation in supersaturated systems. However, batch systems prove to be inefficient for homogenous nucleation due to non-uniform mixing conditions and continuous reduction in driving force as the process progresses over time. Favorable conditions for homogenous nucleation can be easily established in micromixers, due to presence of microchannels they are able to achieve superior mixing conditions with small mixing times (τ_{mix}). Over the past few years, micromixers have played an instrumental role in the transition of processes from batch to continuous systems in several sectors, yet very few studies focus on their applications for carrying out liquid antisolvent precipitative reactions of pharmaceutical compounds.

The current work investigates the application of Ehrfeld® valve assisted micromixer (VAMM) technology for generation of micronsized drug compounds which are poorly water soluble. The VAMM has emerged as a promising technology due to its unique construction which splits the fluid stream into multiple microchannels triggering impingement at multiple sites. In addition the unidirectional restrictor valve excludes reverse flow of fluid stream (therefore having an advantage over conventional T-mixers) and the passage of the thin fluid lamellae through the body of the micromixer ensures a well-mixed and highly supersaturated solution. First the mixing efficiency of the VAMM was characterized by developing an empirical correlation for mixing time (τ_{mix}) [1], followed by experimentation using a set of competitive parallel reactions (whose kinetics have been well studied) and establishing conversion of key reactant as a function of varying jet velocity. These values were then used, to estimate Damköhler number (Da) and τ_{mix}

which enables a direct comparison of mixing efficiencies between VAMM and other well established micromixer systems [2].

Model Biopharmaceutical classification system (BCS) II compounds were identified and highly supersaturated solutions of same were subjected to anti-solvent crystallization process via VAMM in the presence of suitable stabilizer system [3]. Similar crystallizations were also carried out in batch process, where the results clearly indicated the superior performance of VAMM in generation of fine particles when compared with the batch . In both the cases particles as small as d_{90} < 5μm were achieved, therefore establishing the Ehrfeld® Valve Assisted Micromixer as a feasible platform for precipitation of fine pharmaceutical compounds.

References:

1. Wolf A., etl al., Precipitation in a Micromixer –From Laboratory to Industrial Scale. Chemical Engineering & Technology (2015), 2017-2024

2. Johnson K., and Prudhomme R., Chemical processing and micromixing in confined impinging jets, AIChE J, 2003, 2264-2282

4. Flash nanoprecipitation of organic actives and block copolymers using a confined impinging jets mixer, Australian Journal of Chemistry (2003), 1021-1024.

IPDO communication number: IPDO IPM-00468

455229 Linking Particle Size and Shape Distributions to Filtration Performance Using Measurements and Models

Abstract:

Chemical products sold as solids typically undergo at least one crystallization step in their respective production process[1]. Crystallization processes are primarily used for purification, but also as a means to establish the downstream processing properties of the generated crystals, for instance their filterability, flowability and ultimately – if relevant, such as for pharmaceuticals – their tabletability. These properties are thus of vital importance when moving from a suspension towards a dry product.

It is widely accepted that the particle size and shape distribution (PSSD) originating from a crystallization process is chiefly responsible for a product's downstream processing properties[2]. Considering the processing step immediately following crystallization, filtration, the PSSD affects the efficiency and effectiveness of the filtration step, i.e., how fast and how complete the separation of solids from the liquid is achieved. For example, there is ample experimental evidence that small particles and wide particle size distributions lead to longer filtration times (lower filtration efficiency)[3], while it is generally observed that non-isotropic particle shapes (needles, platelets) lead to higher mother liquor retention (lower filtration effectiveness)[4].

Current state of the art approaches describing the relationship between particle size and shape and filtration performance make strong simplifications with respect to the particles' size and shape, i.e., the crystals are often assumed to be monodispersed, all of the same shape or both. In this work, we investigate whether (and when) more accurate information about the particles, such as the full PSSD, is beneficial to predict filtration performance. To this end, several different particle size and shape distributions of β-L-glutamic acid crystals were prepared by different crystallization, milling and sieving operations. The particles were generally needle-like, but with varying aspect ratios and sizes. On these materials two types of

measurements were carried out: first, we report measurements of complete, quantitative PSSDs obtained with a custom-built stereoscopic measurement device[5]. Second, we report measurements of the mother liquor flow rate through cakes of these materials at constant pressure drops up to 5 bar. By correlating the full PSSDs as well as simplified particle characterizations (such as 1D particle size distributions and their moments/quantiles), with filtration performance, we aim at providing insight which level of particle characterization allows for a (reasonably) quantitative link.

References:

[1] Variankaval, N., Cote, A.S., Doherty, M.F., "From form to function: Crystallization of active pharmaceutical ingredients", AIChE J., 2008, 54, 1682-1688.

[2] Wakeman, R., Tarleton, S., "Solid Liquid Separation: Principles of Industrial Filtration", Elsevier, Amsterdam, Netherlands, 2005.

[3] Wakeman, R., "The influence of particle properties on filtration", Sep. Purif. Technol., 2007, 58, 234–241.

[4] Garside, J. (ed.), "Separation Technology: The next ten years", Institution of Chemical Engineers, Rugby, UK, 1994, p. 81.

[5] Schorsch, S., Ochsenbein, D.R., Vetter, T., Morari, M., Mazzotti, M., "High accuracy online measurement of multidimensional particle size distributions during crystallization", Chem. Eng. Sci., 2014, 105, 155-168.

454631 The Influence of Squeeze Films in Particle-Impeller Breakage Events

Abstract:

Existing theories surrounding secondary nucleation mechanisms in stirred batch crystallisations predominantly include varying forms of breakage kernels for particle-impeller collisions. The foundations of such theories lie with the assumption that high impeller tip-speeds are linked to the fragmentation of crystals as they come into contact with the blade surface. What will be outlined here is a mechanism by which crystals near the impeller may not even contact its surface due to a squeeze film boundary layer at the impeller blade's surface. Through the use of an impingement-jet particle impactor, direct imaging and tracking of particle-plane collisions was captured outlining the behaviour of crystals as they approach a target surface with initial velocities of up to 10m/s. Furthermore, A theoretical description of the squeeze film is detailed alongside its implications for future secondary nucleation kinetics dealing with breakage or attrition events near reactor surfaces.

A GPU-BASED MONTE CARLO TECHNIQUE FOR THE SIMULATION OF SIMULTANEOUS NUCLEATION, COAGULATION AND GROWTH BASED ON WEIGHTED SIMULATION PARTICLES

Gregor Kotalczyk, Ivan Skenderovic and Frank Einar Kruis, Faculty of Engineering, University Duisburg-Essen, Duisburg, Germany

Abstract

A hybrid Monte Carlo method is investigated, which combines discrete coagulation and nucleation events with the continuous growth or evaporation of particles. An efficient parallel implementation of this hybrid method on the GPU is discussed, taking into account the coupling of the growth mechanisms to a continuous monomer concentration in the gaseous (or liquid) phase. Exemplary simulations of the gas-phase synthesis of particles are presented, which lead to multimodal initial particle size distributions which evolve into a self-preserving form for longer simulation times. This self-preserving form is described by the competing mechanisms of coagulation and evaporation.

Introduction

The synthesis and production of particles is usually modelled with the population balance equation (PBE) [1]. The Monte Carlo (MC) simulation is one method among others to solve this equation numerically. It has the advantage that it can be easily adapted to particle populations described by several particle properties. It can render single particle events and track thus the history of single particles. The usage of weighted particles in the scope of the MC-simulation allows a higher accuracy [2] and the simple incorporation of nucleation processes into the simulation in the scope of a hybrid approach [3,4] constituting thus an attractive alternative to the usually used constant-number schemes [5].
The MC-simulations require in general large computational times which arise due to the demanding coagulation process. MC simulations are therefore not well suited for the coupling to CFD- or compartmental models. Several techniques have been proposed to overcome this problem. One of those recent approaches made use of a GPU and a fast approximation of the mean coagulation rate [6]. Speed-ups of a factor of 200 were reported by the mere use of the GPU for the coagulation process [7].

We present in the following an extension of the constant-number algorithm for the simulation of coagulation [6]. The operator splitting technique [8] is used to simulate in a hybrid approach the growth of particles and nucleation of new ones during one MC time step. The implementation of the nucleation algorithm is based on a parallel merging algorithm, which keeps the number of the simulation-particles constant. The growth of single simulation particles is simulated by the parallel solution of the corresponding differential equations describing the growth rates. This makes the simulation of condensation and evaporation processes possible. We present in this context the fast parallel summation technique in order to account for the mass-balance (i.e. the coupling to the gaseous phase). This coupling influences in turn the nucleation and condensation (or evaporation) rates of the simulated particles, so that Ostwald-ripening (typical for crystallization processes) can be simulated by the algorithm as well.

We present the application of this algorithm to the simulation of particles in a hot wall reactor and discuss the dependency of the particle properties on the initial saturation of the system.

Population Balance Equation

The PBE describes the temporal change of the particle size distribution (PSD) $n(v)$ for particles with the volume v. The simultaneous process of coagulation, nucleation and growth are described by the following equation:

$$\frac{\mathrm{d}n(v,t)}{\mathrm{d}t} = \frac{1}{2}\int_0^v \beta(v', v-v')n(v',t)n(v-v',t)\mathrm{d}v' - n(v,t)\int_0^\infty \beta(v,v')n(v',t)\mathrm{d}v \\ -\nabla_v(G(v,n_G)n(v,t)) + N_R(n_G)\cdot\delta(v - v_{\mathrm{nuc}}(n_G)) \tag{1}$$

The first two terms on the r.h.s. in equation (1) describe the coagulation process for a given coagulation kernel $\beta(v,v')$, which is the collision frequency between two particles with the volumes v and v'. The third term on the r.h.s. describes the growth (if $G(v,n_G) > 0$) or evaporation (if $G(v,n_G) < 0$) of a particle with the size v with the corresponding rate $G(v,n_G)$. The last term describes the nucleation of new particles with the volume $v_{\mathrm{nuc}}(n_G)$ which are introduced with a nucleation rate of $N_R(n_G)$ into the simulated system.

The growth and nucleation rates $G(v,n_G)$ and $N_R(n_G)$ (as well as the volume of the nucleated particles, $v_{\mathrm{nuc}}(n_G)$) depend on the concentration n_G of the monomers in the continuous gaseous (or liquid, if precipitation processes are described) phase. A general mass-balance equation has to be added in order to account for the conservation of the total mass (neglecting optional terms, which describe an additional in or out flow of n_G into the simulated system):

$$\frac{\mathrm{d}n_G}{\mathrm{d}t} = -N_R(n_G)\frac{v_{\mathrm{nuc}}}{v_M} - \frac{1}{v_M}\int_0^\infty n(v')G(v',n_G)dv' \tag{2}$$

The atomic (resp. molecular) volume of the material is denoted by v_M.

Parallel, Hybrid Monte Carlo Algorithm

Instead of treating all simulated processes as competitive Monte Carlo events (see. [5] for example), the coagulation is combined in a hybrid manner with the growth and nucleation processes based on the operator splitting technique. This technique decouples the coagulation and the nucleation and growth for small intervals of time. These intervals are chosen in this work to be equal to the event-driven MC time steps t_{MC}. The following simulation algorithm describes this notion:

- Initiate the time $t = 0$ and the nucleation volume buffer $V_B = 0$.
- Repeat:
  1. Perform a parallel MC-event-driven step (one coagulation event) and estimate the time step value t_{MC}. This procedure is described in [6] in detail.
  2. Simulate the growth (or evaporation) of all particles in parallel for the interval of time t_{MC} by the repetition of small steps $\Delta t \le t_{MC}$:

     Initiate the time $t_{\mathrm{Growth}} = 0$

     Repeat:

     a. For each MC-particle i (which has a statistical weight W_i and a volume v_i) the following equation is solved:

$$\frac{\mathrm{d}v_i}{\mathrm{d}t} = G(v_i, n_\mathrm{G}) \Rightarrow v_i^{(new)} = v_i + \Delta t \cdot G(v_i, n_\mathrm{G}) \quad (3)$$

b. Remove particles r from the simulation, if $v_r^{(new)} < v_\mathrm{M}$.

c. Take the coupling to the continuous phase into account (Equation (2)):

$$n_\mathrm{G}^{(new)} = n_\mathrm{G} - \Delta t \cdot N_\mathrm{R}(n_\mathrm{G}) \frac{v_\mathrm{nuc}(n_\mathrm{G})}{v_\mathrm{M}} - \frac{\Delta t}{v_\mathrm{M}} \sum_{\substack{\text{not evaporated} \\ \text{particles } i}} W_i \cdot G(v_i, n_\mathrm{G}) + \sum_{\substack{\text{evaporated} \\ \text{particles } r}} W_r \frac{v_r}{v_\mathrm{M}} \quad (4)$$

d. Add the volume-concentration of newly nucleated particles to the buffer V_B:

$$V_\mathrm{B} = V_\mathrm{B} + \Delta t \cdot N_\mathrm{R}(n_\mathrm{G}) v_\mathrm{nuc}(n_\mathrm{G}) \quad (5)$$

e. If V_B is greater than a specific threshold value V_T, then

   i. include a new simulation particle into the simulation with the volume v_nuc and the statistical weight $W_\mathrm{nuc} = V_\mathrm{B} / v_\mathrm{nuc}$.

   ii. Reset the buffer which stores the nucleated volume: $V_\mathrm{B} = 0$.

f. Increase the time $t_\mathrm{Growth} = t_\mathrm{Growth} + \Delta t$

Until the time t_MC is reached: $t_\mathrm{Growth} = t_\mathrm{MC}$

3. Increase the time $t = t + t_\mathrm{MC}$

Until the simulation end time t_end is reached: $t \geq t_\mathrm{end}$

The parallel implementation of step 2a and 2c of the algorithm is depicted in figure 1, first the updates of the volumes of each simulation particle are calculated, and then a parallel adding technique is used in order to estimate the new value of the concentration n_G. Only 14 parallel computational steps are necessary for the summation of 10 000 simulation particles.

Figure 1 and the described algorithm depict a Euler-implementation of the solution of the differential equation, in the here presented work a 4-5 Runge-Kutta-method has been used. This method comprises of the calculation of 7 consecutive v values, the algorithm in figure 2 is thus repeated 7 times for one Runge-Kutta time step Δt.

Simulated Test Cases

Small particle sizes are encountered in early stages of the gas-phase synthesis. The coagulation kernel describing the free-molecular regime can be used in order to describe the corresponding coagulation process:

$$\beta(v, v') = \left(\frac{3}{4\pi}\right)^{\frac{1}{6}} \sqrt{\frac{6k_B T}{\rho_p}} \cdot \sqrt{\frac{1}{v} + \frac{1}{v'}} \cdot \left(v^{\frac{1}{3}} + v'^{\frac{1}{3}}\right)^2 \quad (6)$$

Table 1 summarizes the parameter values and constants used for the simulation, k_B is the Boltzmann constant.

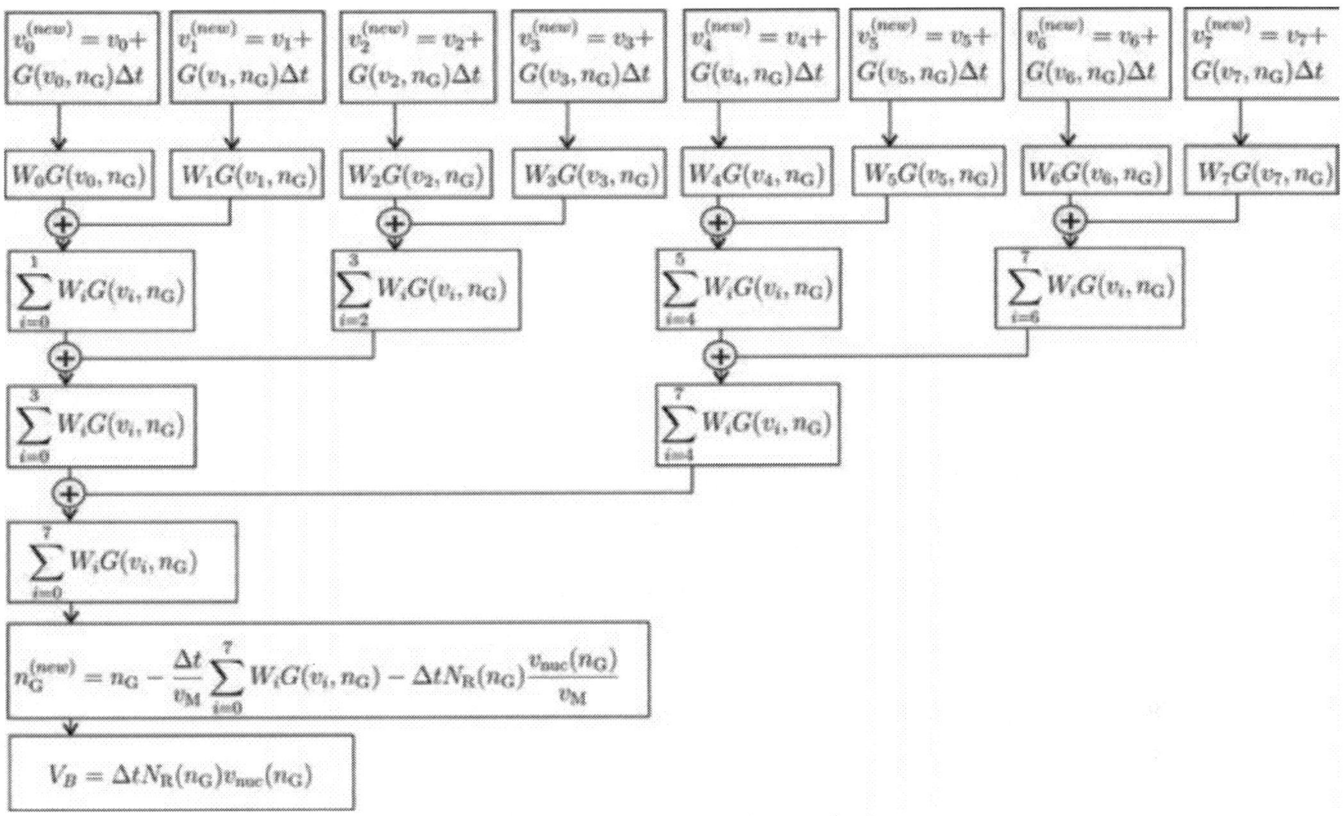

Figure 1. The parallel algorithm calculates the new volumes $v_i^{(new)}$ of the simulation particles i in the first row. The evaluated values $\Delta t \cdot G(v_i, n_G) \cdot W_i$ (resp. v_i / v_M in the case of evaporation) are reused in the consecutive computational steps for the parallel calculation of $n_G^{(new)}$.

The growth rate in the free-molecular regime is described by:

$$G(v, n_G) = \frac{v_M \cdot \pi \cdot \left(\frac{6v}{\pi}\right)^{\frac{2}{3}}}{\sqrt{2\pi \cdot m_1 \cdot k_B \cdot T}} \cdot \left(k_B \cdot T \cdot n_G - p_s \cdot \exp\left\{ \frac{4 \cdot \gamma \cdot v_M}{k_B \cdot T \cdot \left(\frac{6v}{\pi}\right)^{\frac{1}{3}}} \right\} \right) \tag{7}$$

The nucleation term is described by the classical kinetic theory:

$$N_R(n_G) = n_G \cdot \sqrt{\frac{2\gamma}{\pi \cdot m_1} \frac{p_s}{k_B \cdot T}} \cdot \exp\left(-\frac{16 \cdot \pi \cdot \gamma^3 \cdot v_m^2}{3 \cdot k_B^3 \cdot T^3 \cdot \ln(S)^2} \right) \quad \text{, with } S = \frac{n_G k_B T}{p_s} \tag{8}$$

The volume of the nucleated particles is given by the Kelvin diameter d^*, with

$$v_{nuc}(n_G) = \frac{\pi (d^*)^3}{6} = \frac{64 \cdot \pi^4 \cdot v_M^3}{6 \cdot k_B^3 \cdot T^3 \cdot (\ln(S))^3} \quad \text{, with } S = \frac{n_G k_B T}{p_s} \tag{9}$$

The initial conditions for the simulated system consist of a lognormal PSD, with the geometric mean value d_g=2 nm and the geometric standard deviation σ_g=1.2, rendering a total number-concentration of 10^{16} m^{-3}. The change of the PSD depending on the initial saturations of S_0=1.5, 2.5 and 4.5 are discussed in the results section.

Table 1. Constants and parameters used for the simulation.

Name	Value
Temperature	$T = 1380$ K
Density (Fe)	$\rho_p = 7.874$ g cm^{-3}
Atomic volume (Fe)	$v_M = 1.178 \cdot 10^{-29}$ m^3
Atomic weight (Fe)	$m_1 = 9.27 \cdot 10^{-26}$ kg
Surface tension (Fe , 1380 K)	$\gamma = 0.29$ J m^{-2}
Vapor pressure (Fe , 1380 K)	$p_s = 0.805$ Pa

Simulation Results

The nucleation rates depend strongly on the saturation of the system, as shown in figure 2a and 2b, a slight drop of the saturation from 2.5 to 1.5 leads to a drop of $N_R(n_G)$ which comprises nearly 20 orders of magnitude. This leads in turn to dramatic changes of the number-concentration, as can be seen in figure 2c.

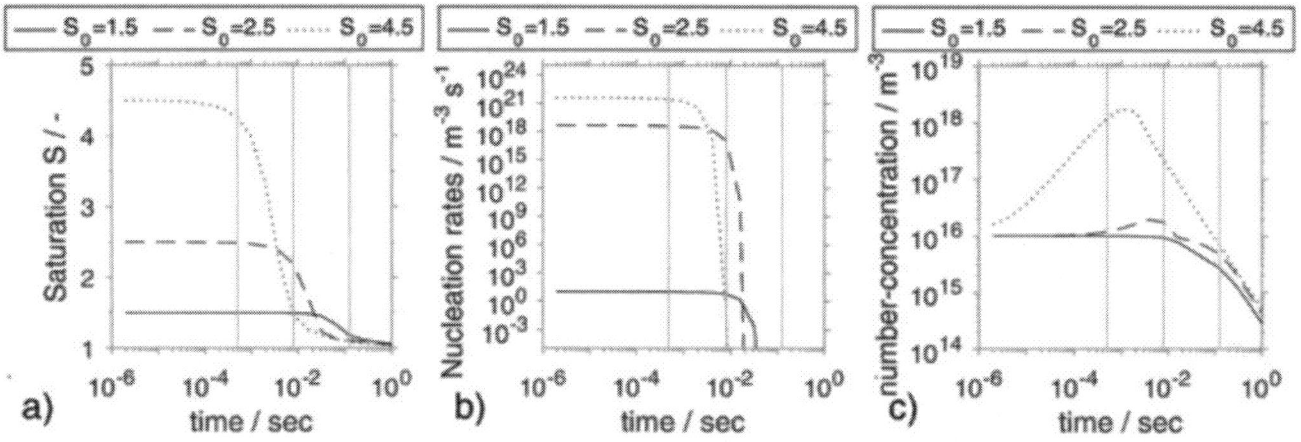

Figure 2. The saturation S (a), nucleation rates $N_R(n_G)$ (b) and the number-concentration of all simulation particles (c) are shown in dependency of the initial saturation S_0.

Exemplary PSDs are shown in figure 3 for the time points marked by the horizontal lines in figure 2. A growth of the initial PSD is accompanied with the nucleation of novel particles in both presented cases leading to a multimodal PSD. The higher nucleation rates for S_0=4.5 (figure 3b) lead to a nucleation peak which describes particle concentrations, which are 3 orders of magnitude higher than the particles rendered by the initial PSD concentration. It can be seen, that the usage of weighted particles allows to render even such extreme PSDs, which are only reproducible if a very large number of equally weighted simulation particles is used.

The moments of the PSDs are shown figure 4 it can be seen, that for longer simulation times the geometric standard deviation σ_g reaches a constant value (large noise levels are found for this value due to the rendition of evaporation by discrete particles, the left side of the PSD in figure 5(b)), while the mean geometric diameter d_g increases linearly, this marks the characteristics of a self-preserving particle size distribution (SPD), whose form is the same for all simulated initial saturations S_0 and is discussed for the case S_0=1.5 in more detail.

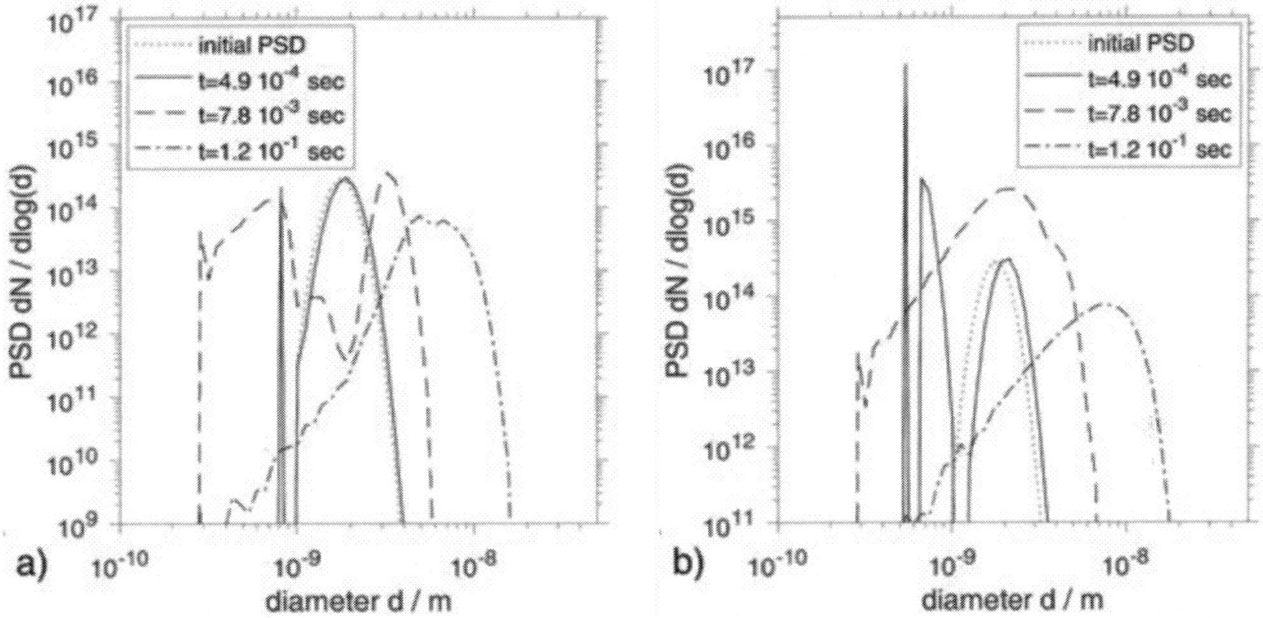

Figure 3. The PSDs for the initial saturation S_0=2.5 (a) and S_0=4.5 (b) are shown.

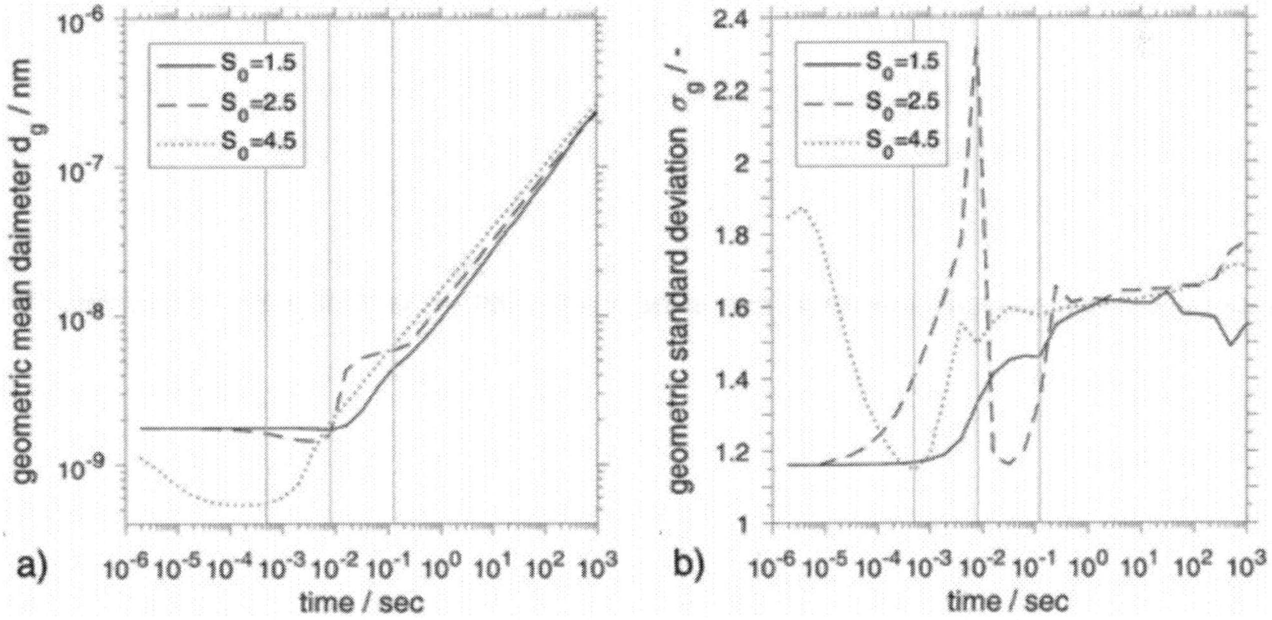

Figure 4. The geometric mean diameter (a) and standard deviation (b) are shown.

Low initial saturation values S_0 lead to negligible nucleation rates, so that the development of the given initial PSD is described by the coagulation and growth only. Exemplary simulation results for this scenario are shown in figure 5 for S_0=1.5. It can be seen, that the PSDs attain a self-preserving distribution (SPD), which is slightly different from the SPD which is attained for the coagulation only. It can also be seen, that the major part of the SPD is smaller than the corresponding Kelvin diameter and should therefore evaporate in the absence of the coagulation process. The simulation particles have therefore to be simulated as long as they are bigger than the monomer (atomic) size of the simulated material – instead of removing them already from the simulation when they reach sizes which are smaller than the Kelvin diameter.

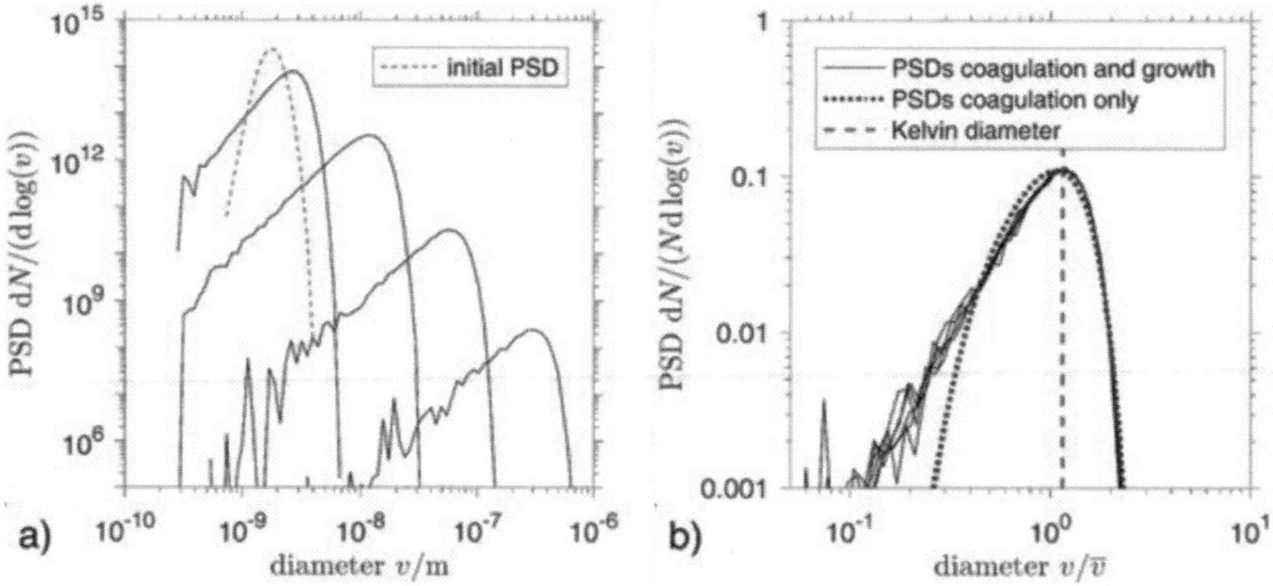

Figure 5. The PSDs for the simulation of combined coagulation and growth are shown in (a) for an initial saturation of S_0=1.5. The PSDs are normed to the mean diameter $\bar{v}$ and the particle number-concentration N in (b) in order to highlight the self-preserving character of the PSDs.

The computational times are summarized in table 2, the results show the required simulation times for 100 simulations which are executed in parallel, each simulation consists of 10^3 simulation particles. This means, that 10^5 coupled differential equations have been solved in total (combined with the event driven Monte Carlo code). This finding underscores the computational efficiency of the presented algorithm. The simulations were performed on a NVIDIA GTX 980 graphics processing unit.

Table 2. Computational times required for the simulation of t= 1000 seconds.

	S_0=1.5	S_0=2.5	S_0=4.5
Computing time	45 sec	74 sec	145 sec

Conclusions

A hybrid GPU-based simulation technique for the simultaneous simulation of coagulation, nucleation and growth (resp. evaporation) with a coupling to the continuous phase is presented. The combination of a parallel, event-driven Monte Carlo simulation with a Runge-Kutta based parallel solution of differential equations is discussed for this purpose. The resulting Algorithm can be used to describe the gas-phase synthesis of particles. The usage of weighted simulation particles allows to render multimodal PSDs, for which each mode represents different particle concentration which vary several orders of magnitude and are typically encountered at the initial stages of the simulation. The detailed rendering of the evaporation process of the particles allows the description of a self-preserving size distribution which results from the competing process of evaporation and coagulation and is encountered at the later stages of the simulation. Although the explicit calculations of several thousands of coupled differential equations poses a severe computational task, we show, that an efficient GPU implementation of this problem is able to produce relevant results which describe the simulated process completely within just a few minutes. The algorithm is therefore applicable in the framework of compartmental models or flow-sheet simulations.

Acknowledgement

The financial support of DFG (Deutsche Forschungsgemeinschaft) within the priority program SPP 1679 "Dynamic flowsheet simulation of interconnected solids processes" is gratefully acknowledged.

References

1. Ramkrishna, D. (2000), "Population balances: Theory and applications to particulate systems in engineering", New York,Acad. Press
2. Zhao, H., Kruis, F. E. and C. Zheng (2009), "Reducing statistical noise and extending the size spectrum by applying weighted simulation particles in Monte Carlo simulation of coagulation", *Aerosol Science and Technology*, 43(8), pp. 781–793.
3. Menz, W. J., Akroyd, J. and M. Kraft (2014), "Stochastic solution of population balance equations for reactor networks", *Journal of Computational Physics*, 256, pp. 615–629.
4. Hao, X., Zhao, H., Xu, Z. and C. Zheng (2013), "Population balance-Monte Carlo simulation for gas-to-particle synthesis of nanoparticles", *Aerosol Science and Technology*, 47(10), pp. 1125–1133.
5. Khalili, S., Lin, Y., Armaou, A. and T. Matsoukas (2010), "Constant number Monte Carlo simulation of population balances with multiple growth mechanisms", *AIChE Journal*, 56(12), pp. 3137–3145.
6. Wei, J. and F. E. Kruis (2013), "A GPU-based parallelized Monte-Carlo method f or particle coagulation using an acceptance–rejection strategy", *Chemical Engineering Science*, 104, pp. 451–459.
7. Wei, J. (2014), "Comparison of computational efficiency of inverse and acceptance–rejection scheme by Monte Carlo methods for particle coagulation on CPU and GPU", *Powder Technology*, 268, pp. 420–423.
8. Celnik, M., Patterson, R., Kraft, M. and W. Wagner (2007), "Coupling a stochastic soot population balance to gas-phase chemistry using operator splitting", *Combustion and Flame*, 148(3), pp. 158–176.

460855 Development and Validation of Fundamental Closures for the Population Balance Using DEM

Abstract:

Flows of cohesive particles are subject to the formation and break-up of agglomerates. One approach to the continuum modeling of such flows is the population balance equation coupled with kinetic theory balances. The population balance includes sink and source terms representing agglomeration and breakage. These sink and source terms depend on the success factors of agglomeration and breakage, which quantify the fraction of collisions that result in agglomeration and breakage. Though known to be important, previous works have largely ignored the effect of particle properties and particle impact velocities on the success factors. The focus of this work is to develop a fundamental link between the success factors and the granular temperature found in kinetic theory by relating the distribution of normal particle impact velocities to the granular temperature. This distribution of impact velocities is then used to determine an effective coefficient of restitution for use in kinetic-theory closures. The resulting theoretical expressions derived for both the impact velocity distribution and the effective restitution coefficient are validated via discrete element method (DEM) simulations and shown to be critical for accurate predictions. DEM simulations are then utilized to determine the critical velocities of aggregation and breakage for the given particle-particle cohesion model. Finally, the full set of closed balances (population, mass, momentum, and granular energy) are validated in the spatially uniform homogeneous cooling and simple shear flow systems using DEM.

463920 Modeling and Simulation of Hydrothermal Synthesis of Nanoparticles in a Continuous Flow Reactor

Abstract:

Continuous hydrothermal synthesis of nanoparticles is gaining attention because of its simple and efficient operation, use of inexpensive solvent, short process time and inherent scalability. Typically, supercritical water is used as the reaction medium while an aqueous metal salt solution is used as the precursor. Streams of supercritical water and aqueous metal salt solution are continuously supplied to the reactor and the product is also continuously withdrawn. Upon mixing of reactants, product monomers form as a result of hydrolysis and condensation reactions, subsequently growing to nanoparticles via nucleation, diffusional growth, coagulation, coalescence, and Ostwald ripening. Now, physical, thermodynamic and transport properties of water under supercritical conditions are significantly different from those under ambient conditions. This results in enhanced rates of hydrothermal reactions and particle formation and growth. Besides the implementation of appropriate process conditions (temperature, pressure, pH, flow rates and concentrations of the reactants etc.), mixing of reactants also has a strong influence on the particle size distribution (PSD) and morphology of the product.

Various attempts, by both experimental and numerical investigations, have been made by several research groups to understand the mixing of reactants in different flow reactors. Particle size distribution (PSD) is the key product quality parameter but very little focus has been made to study the effect of mixing on the product PSD. The effect of mixing on temperature and velocity profiles inside the reactor has been studied using computational fluid dynamics (CFD) simulations. Similarly, population balance models (PBMs) have also been developed to predict the PSD. A few attempts have also been made to couple the CFD with PBM to predict the effect of mixing on PSD, but testing and validation of these coupled models with real experimental data has not been demonstrated so far. A few research groups have attempted to only correlate numerical results of flow and temperature fields inside the reactor with experimentally measured PSD. However, prediction of the PSD and its validation with experimental data is limited.

In the present work, we have developed a coupled computational fluid dynamics (CFD) and population balance model (PBM) to study the effect of reactor configuration and process parameters on the product PSD. The CFD model consists of mass, momentum, species and energy balance equations. The PBM is one-dimensional in nature and takes into account particle nucleation, diffusional growth and growth by particle coagulation and coalescence. The coupled CFD-PBM model was implemented in a commercial CFD software (FLUENT 14.5) and was tested with published experimental data for three different reactor configurations with a T-mixer, used for synthesis of ceria nanoparticles. The predicted thermal profiles are reasonably close to the published data obtained using neutron radiography. Similarly, the predicted velocity profiles are in qualitative agreement with the measured fluid density profiles. Experimental data were also published for the effects of different reactant flow rates and temperatures at a constant pressure on the PSD and the predictions of the coupled CFD-PBM for product PSD are in good numerical agreement with the data.

The validated model was subsequently used to study the effect of important process parameters. When the flow rates of supercritical water and aqueous metal salt solution are comparable, it was observed that vortex formation and circulation occur around the T-junction. This is primarily because of the buoyancy force which occurs due to the density difference between the streams. The mean size of the particles increases as residence time of particles increases due to recirculation. In other cases, where the difference between the flow rates is significant, the supercritical water slides over the metal salt solution and the mean particle size decreases. Since the proposed model can predict the effect of mixing on product PSD, it can be employed beneficially for reactor scale-up and for designing novel reactors including features targeted towards process intensification.

463976 Methodology and Pitfalls When Calibrating a PBM: The Case of Twin-Screw Wet Granulation

Abstract:

Traditional pharmaceutical processes comprise of a series of batch-wise operations. Nowadays, a shift is being made from these batch processes to continuous manufacturing, to cope with the inefficiencies and high cost involved in process development. Twin-screw wet granulation is an up-and-coming continuous granulation process that is being assessed for its performance in the solid dosage manufacturing. However, since these continuous processes are fairly new in the pharmaceutical industry, detailed process knowledge and understanding is still lacking. Application of mechanistic models can help in bridging this gap by assessing the experimental data and uncovering the underlying mechanisms. In this work, a Population Balance Model (PBM) is developed for predicting the granule size distribution inside the granulator starting from the pre-blend, up until the wet granules at the end of the process. In PBM, physical processes like aggregation and breakage of granules are represented by kernels which are often empirical in nature with fitting parameters. Model equations are solved using the Cell Averaging Technique (CAT) which is able to deal with aggregation and breakage. Different aggregation and breakage mechanisms can be implemented using different kernels. However, models are only useful when calibrated/validated using experimental data. Calibration of a PBM presents unique challenges in various intermediate steps.

First, the model grid needs to be defined (i.e. number and location of size classes). Different measurement techniques (laser diffraction, QICPIC, sieve analysis) use different grids. As each measurement technique has its own peculiarities, it is difficult to compare measurements performed with different techniques. Second, an objective function needs to be defined. The current practice to calibrate PBM is similar to how this was done for time series and using deterministic models. Sum of Squared Errors (SSE) and Root Mean Squared Error (RMSE) are thus typical gold standards. Similarly, the information of a whole particle size distribution can be condensed into some characteristic numbers, such as mode, mean, span, and the Sauter diameter. Proper evaluation is needed to confirm whether this is

indeed the way to go to obtain a good modelling practise for PBM. A lot of freedom is available when dealing with particle size distribution, but the question is how to deal with this kind of freedom and how it is best coupled to the modelling objective. Third, a technique to find the minimal value of the objective function has to be selected. In this study, predictions of a selected PBM formulation are generated using a global parameter space exploration by means of a large set of Monte Carlo simulations. Changing the aggregation and breakage kernel parameters yields different size distributions. Different objective functions are evaluated to determine the objective function whose optimum yields the optimal agreement between the simulated and measured particles bearing in mind the objective. This optimal agreement, within a predefined error range, is called the calibrated model for that process setting. This study is repeated for different process settings of the twin-screw granulator. By comparing the calibrated results for different process settings, the aggregation and breakage mechanisms can be identified, and the most dominant regimes can be found. Guidance on the different choices to be made during the calibration process will be provided.

468709 A Population Balance Model for Calculation of Total Evaporation Rate during Fluid Bed Granulation or Coating

Abstract:

It is known that the granule moisture is a critical property of particles during fluid bed granulation (FBG) or fluid bed coating (FBC) processes. It is an important attribute for the quality of coating and has a strong impact on the granule growth. The granule moisture is usually measured by taking samples and gravimetrically evaluating the moisture (e.g. in terms of LOD = Loss on Drying). Relatively newer NIR method can measure the same quantity continuously [1].

Mathematical models of different complexity have been developed and used for simulation of FBG processes. The models use different expressions for calculation of the evaporation rate of water present in the sprayed binder [1 -3]. The models assume presence of a continuous film on the granule surface and the evaporation rate is then calculated by standard expressions for the mass and heat transfer coefficients. Most procedures employ either the Antoine equation or the experimentally measured sorption equilibria for calculation of the equilibrium moisture content at the particle surface [3]. Based on our experience with employing the above methods for different FBG and FBC systems it is challenging to find a meaningful correlation between the parameters of the evaporation rate expressions and process conditions [1, 2].

A total mass and heat balances (TMHB) model employing the above expressions was developed and presented before [1, 2]. In this contribution a different approach for calculation of total evaporation rate is proposed and tested. The approach assumes that the total evaporation rate is a summation of evaporation rates of individual droplets of different sizes deposited on the particle surface. This assumption is justified by the fact that many FBG and particularly FBC processes are run at dry conditions and the particle LOD is typically between 1% -6%. So, it is unlikely to have a continuous wet layer of the binder solution on the particle surface. A new population balance (NPB) model (based on volume approach) calculating the total evaporation rate as summation of evaporation rates of shrinking and drying droplets was developed by using Direct Quadrature Method of Moments (DQMOM). The method is simple and fast (by adding only 4-6 ordinary differential equations for the PSD moments tracking) and calculates the total evaporation rate as a change of the first moment (volume) of the PSD of all droplets delivered into the system. During the presentation we will share simulation results and comparison of models of different complexity (1D or 2D) and also comparison with experimental results. The TMHB model and the DQMOM model (NPB) are then combined with another population balance (PB) model calculating the granule growth [3]. The PB model employs the granule moisture calculated by the TMHB+ NPB combination for calculation of the agglomeration kernel. In summary, the combined model predicts granule moisture, temperature and PSD from the process conditions [3].

References:

[1] Rajniak, P. et al.: Development of a mechanistic model for fluid bed granulation. AIChE Annual Meeting, 2015, Salt Lake City.

Â [1] Ricart, B. and P. Rajniak: Experimental and theoretical study of simultaneous agglomeration and drying in fluidized bed granulator. AIChE Annual Meeting, 2013, San Francisco.

[3] Peglow, M. et al.: A generic population balance model for simultaneous agglomeration and drying in fluidized bed. Chem. Eng. Sci. 62 (2007) 513 â€“ 532.

470306 Three-Compartmental Population Balance Model and Experimental Validation of a Continuous Twin-Screw Wet Granulation Process

Abstract:

A novel modelling methodology for a continuous wet granulation process is presented, combining population balance model (PBM) and multi-compartment concept to predict the critical quality attributes of the granule product, such as granule size distribution and mean granule size, along with the barrel length. By dividing the granulator screw barrel into three compartments along barrel length: wetting compartment (WC), mixing compartment (MC) and steady growth compartment (SC) and assuming different granulation mechanisms on each compartment, the proposed model could consider spatial heterogeneity of the twin screw granulation system and improve prediction accuracy in estimating particle size distribution. It was assumed that each compartment is homogeneous and can be described by a one-dimensional PBM. DOE with 36 experiments were carried out for model development and model validation. 16 combinations of aggregation and breakage mechanisms are compared in predicting the experimental particle size distribution and it is found that aggregation is heavily dependent upon particle length/area and all the four breakage mechanism do not have influence on the model prediction capability. The proposed three-compartmental PBM is compared with the one-compartmental PBM in predicting granule critical quality attributes (CQAs) at different experimental operating conditions and is proved to have advantage over one-compartmental PBM. Average residence time of granules on each compartment has been investigated pertaining to their influence on granules properties. It was shown that the longer residence time generates the larger granules on WC and SC, while generating small average particle and large fraction of trivial particle on MC. This novel three-compartmental population balance model has advantages of both considering the system heterogeneity and affordable computational cost, which can facilitate process optimization and model-based control for the TSWG process.

469431 Scaling of Heat Transfer and Temperature Distribution in Granular Flows in Rotating Drums

Abstract:

Rotating drums are one of the most common devices used for industrial processes that involve thermal treatment of granular materials. To improve efficiency, several factors, including the physical, mechanical, and thermal properties of the powder & calciner, as well as, operating conditions such as feed rate and speed of rotation, need to be considered. A powerful approach towards achieving an efficient process is to develop predictive models for the particlesâ€™ temperature distribution (space & time), particularly for the purposes of scale-up from laboratory scale experiments. We used discrete element method (DEM) simulations to study the scaling of particlesâ€™ temperature distribution in rotating drums at low temperatures, where conduction through particle-particle and particle-wall contact area is the dominant heat transfer mechanism. A wide range of several particle properties and operating conditions were considered in the simulations. Based on the results, we show that the most important timescales relevant to the heat transfer process are the time to heat up a single particle to the drum wall temperature and the time of contact between a particle and the drum wall. Then, using these timescales, we developed a model to predict the particlesâ€™ temperature distribution at any time during the heating process. All of the simulations can be collapsed into one relationship. This model can be used to predict the required time to heat up all particles to the desired temperature. The results from this model also compare very well with experimental results conducted at low temperatures.

460038 Parametric Analysis of Fluidized Bed Drying of Pharmaceutical Materials

Abstract:

Parametric Analysis of Fluidized Bed Drying of Pharmaceutical Materials

Hao Chen, Xue Liu, Cody Bishop, Benjamin J. Glasser

Department of Chemical and Biochemical Engineering, Rutgers University, Piscataway, NJ, 08854

Fluidized beds are extensively used in the pharmaceutical and other chemical industries either as a batch or continuous process for drying moist powders and granular solids because of good mixing of solids and intensive heat and mass transfer between the solid and hot gas phases in the system. It has been reported that an average of 12% of all energy consumed is used on drying, and the cost of drying could reach up to 60%-70% of total cost of investments [1]. Therefore, optimal operation of the drying process is sought to meet the requirements for cost-effective manufacturing.

In this work, parametric analysis of four fluidized bed dryer operating parameters â€" initial moisture content, material loading, heating air temperature and air flux â€" and their effect on drying were studied using two factorial experimental designs. Two different dryer scales, the Glatt GPCG-1 and MiniGlatt, were used to dry Dibasic Calcium Phosphate powders. According to a Pareto analysis, initial moisture content and air flux are the most significant variables. For the larger GPCG-1, the four variables used in the factorial tests were found to be independent. For the smaller MiniGlatt, the interaction of the initial moisture content and the air flow rate was also significant. From the parametric analysis, it was observed that drying is dominated by the constant drying stage, and zero-order kinetics, for the mild drying conditions (appropriate for many pharmaceutical products) we considered in this work. A comparison of the two dryers and an evaluation of energy efficiency were also carried out. It was found that drying should be

carried out with relatively low temperatures and high air fluxes for both the large and small scale dryers in order to maximize the energy efficiency.

In the second part of this work, the effect of particle size on drying was investigated. It is commonly believed that during the constant drying stage, where convective drying dominates, small particles dry faster than larger particles because the former have larger surface-area-to-volume ratio. However, it was observed for the DCPA that large (150-212 micron) and small (53 â€" 106 micron) particles have the same drying rate during the constant-rate drying stage. Similar phenomena were also found for the MCC (microcrystalline cellulose) PH101 and the wet granules mainly consisting of lactose and MCC. This indicates that surface area may not be a key variable controlling the drying rate during the constant-rate stage. A comparison of fluid bed drying and oven drying is also carried out in this work.

Our work serves to characterize performance of fluidized bed dryer systems and provide physical insight into the fundamentals of drying of porous materials.

[1] M. S. H. Sarker, M. N. Ibrahim, N. Abdul Aziz, and M. S. Punan, "Energy and exergy analysis of industrial fluidized bed drying of paddy," *Energy,* vol. 84, pp. 131-138, 5/1/ 2015.

459043 Estimating Thermal Diffusivity of Bulk Solids

Abstract:

Conductive heat transfer in packed beds is of industrial interest for numerous processes. Packed beds are widely used in catalytic reactors, gas-cooled nuclear reactors, drying, mixing, adsorption processes, pebble bed heaters, high performance cryogenic insulations, absorbers, evaporators and thermal energy storage devices. The large surface area to total volume ratio in packed beds lends itself to excellent heat & mass transfer along with good mixing of the continuous phase. To operate packed bed based processes, one must have a fundamental understanding of packing structure, porosity distribution, fluid flow, pressure drop and heat transfer mechanisms.

For engineering purposes, it is desirable to represent the composite heterogeneous media by an equivalent (or effective) continuous and homogenous media. This approach is justifiable as long as the macroscopic scale of the physical process is significantly larger than the microscopic scale describing the structure of the composite media. Similar approach is taken to describe the thermal conductivity of a packed bed with stagnant fluid. The overall thermal conductivity of a packed bed can be lumped into a single representative value called effective thermal diffusivity.

In steady state methods, the packed bed of well defined simple geometry (cylindrical, cube or parallelepiped) is allowed to achieve steady state in heat flux and temperature gradient. By knowing the precise boundary conditions, temperature gradient and heat flux, and applying Fourier or Laplace equations, the value of thermal conductivity of packed bed can be estimated. The transient method extends the above analysis by imposing a sudden change in boundary condition, either as a step or sinusoidal input, and observing the dynamic thermal response of the packed bed. The thermal conductivity of the packed bed can be determined by fitting the experimental data to the predicted unsteady state thermal response.

In this work, we have measured the effective thermal diffusivity of packed beds (with stagnant fluids) using both steady state and transient methods, and compared the results with available correlations in the literature. Further, the effect of bed voidage has been investigated.

463581 Effect of Particle Size and Shape on Drying and Stress Development in Colloidal Coatings

Abstract:

Coatings prepared from colloidal particle suspensions have a wide range of applications including barrier coatings, antireflective coatings and printed electronic devices. For coatings prepared from colloidal suspensions, major changes in microstructure and properties occur during drying. As-deposited, the liquid suspension is stress-free, but during drying stress develops and may cause defects, such as cracking, curling or delamination. The study of coatings stress development and microstructure change during drying provides a basis for developing formulations and processing conditions to control stress and defects and provide for high performance.

Mineral pigments like calcium carbonate, clay and talc are extensively used in the paper industry. Latex like styrene butadiene, styrene acrylic and polyvinyl acetate are used to provide good cohesion of the porous structure formed by rigid pigments. Different pigments size, shape and blending ratio can significantly impact dispersion properties, as well as the appearance and physical properties of the coated products.

Past research has demonstrated that coatings made from suspensions of large particles develop lower stresses and are less likely to crack compared with those made from suspensions of fine particles. In this work, the effect of particle size is further explored by studying coatings with similar average particle size but different particle size distribution. In addition, particulate coatings with clays of different shape, relative size and content ratio to other pigments in the coating formulation are also investigated. This research addresses the effects of particle size and shape on coatings stress development, microstructure formation and mechanical properties.

In this research, we studied aqueous coating systems containing ground calcium carbonate particles, latex particles and clays. Coating drying behavior, including weight loss and stress development, was monitored under controlled conditions (temperature and humidity). Drying stress was determined by substrate deflection method. SEM and cryo-SEM techniques were used to relate the coating stress development and cracking behavior to microstructure changes. The cracking behavior was studied by measuring the critical cracking thickness and mechanical properties (i.e., hardness, modulus and fracture toughness) using nanoindenter. Critical cracking thickness was found to increase with clay content in certain systems. The objective of this study is to understand this phenomenon and the effects of particle sizes and shape on coating stress development and mechanical properties, which are responsible for coating performance.

467349 Drying of Spent Coffee Grounds

Abstract:

The objective of this work is to delve into the intricacies of drying spent coffee grounds for purposes such as the facilitation of transport and yield increase for biodiesel production, using a tray dryer as a preliminary study. The research first considers and develops a consistent approach to account for periodic variance in the utilized testing apparatus. This allows for the subsequent analysis of pertinent drying parameters for spent coffee grounds: the convective heat and mass transfer coefficients, the effective diffusivity, and the possible presence of constant rate drying periods to facilitate large-scale drying operations. In order to do this, inlet and outlet airstream qualities were periodically measured in a tray dryer operation concerning spent coffee grounds for over three hours per test. The environmental variance was measured by analyzing two minutes of second-by-second data. Ultimately, temperature and humidity was predictably periodic, and air flow speed proved to be mostly erratic. The results for the convective transfer coefficients seemed reasonable but no studies were not found in pertinent literature to compare against, while the diffusivity was mostly consistent with prior studies. The consistent appearance of two constant rate drying periods, with similar qualities across tests, directly contradicts with prior claims that no such period exists and therefore warrants further investigation.

472489 Assessing the Effectiveness of Dry Product Recycling

Abstract:

Product build up on the dryer surface (i.e. rind formation) may significantly decrease the overall heat transfer and increase the drying time. In industrial practice, dry product recycling (i.e. leaving a dry product heel in the dryer at the start of the operation) is utilized in contact dryers to reduce the adhesion and build up of the wet mass on the dryer. Mixing the wet mass with the dry product decreases the overall liquid content immediately prior to the drying and increases the flowability of the wet mass. Increased flowability means better turnover of the powder bed, and better contact with the heating surface. In this study, we investigate the validity of the thermal conductivity values of particulate materials presented at the AIChE Annual meeting over the last couple of years and the extent of enhancement in the drying time with dry product recycle in a laboratory scale contact drying. The results will be presented for powders with varying physical properties such as particle size and porosity.

462698 Drying of Almonds: Single Particle

Abstract:

Abstract

Drying is important in many food processing applications, and particularly so in the dry fruits industry. We have experimentally as well as computationally investigated drying of almonds. Drying of almonds exhibit many characteristic key features: non-spherical shape, swelling/ shrinkage as a function of moisture content, uneven drying because of peculiar shape and proximity of other almonds and so on. Mettler Toledo Halogen Moisture Analyzer unit was used to quantify drying kinetics of a single almond particle. In this widely used equipment, internal flow patterns and therefore heat and mass transfer depend on natural circulation.

In this work, we have used detailed three dimensional computational fluid dynamics (CFD) model to simulate flow, heat and mass transfer in Mettler Toledo unit while drying of almond particles. Systematic experiments with a single re-wetted almond were conducted in Mettler Toledo Moisture Analyzer Excellence Plus HX204 unit at three set temperatures of 55, 65 and 75 °C. Acquired transient moisture data was used to estimate key parameters of interest (drying kinetics and effective diffusivity).

A comprehensive CFD model was developed and used to simulate natural convection within the moisture analyzer. Radiation heat transfer was modelled using the discrete ordinate mode (ANSYS Inc). The computational model was validated by comparing simulated and measured temperature near almond nut. The temperature near almond surface was found to be lower (by about 5 C) than the temperature set point of the moisture analyzer. Actual measured temperature near the almond surface was used while obtaining drying kinetics. Simulated flow results were used to estimate local heat transfer coefficients for almond surfaces. Based on estimated effective diffusivity and analogy between thermal and concentration boundary layers, distribution of mass transfer coefficient around almond surface was estimated. The results provide quantitative information about non-uniform heat and mass transfer and therefore non-uniform drying of a single almond particle.

The presented approach, models and results will be useful for enhancing utility of the results obtained with moisture analyzer units and improve ability to extend the results obtained with such units to industrial systems. The results and models presented here will also provide useful basis for further work on multiple almond particles and on tray dryers.

Acknowledgements

Authors would like to Thank CSIR for funding this research via its Indus MAGIC (CSC123) project.

References

ANSYS® ANSYS Fluent, Release 14.5

460625 A Generalized Rule for a Grid Independent Solution for TFM Simulations of Geldart B Bubbling Fluidized Beds

Abstract:

Abstract

Bubbling fluidized beds (BFBs) of Geldart B particles are commonly encountered in combustion and gasification applications. Recently, they have been used as fuel reactor in the Chemical-Looping Combustion (CLC) process, a leading alternative to reduce the economic cost of CO_2capture. Computational Fluid Dynamics (CFD) modeling can aid in design and scale-up of fuel reactor used in the CLC process. The Eulerianâ€"Eulerian CFD approach, commonly known as kinetic theory-based two fluid model (TFM), is extensively used for small and large-scale simulation of gas-solid fluidized beds due its relatively low computational cost. However, application of TFM to industrial-scale BFBs is limited due to requirement of fine spatiotemporal discretization to solve the mesoscale structure. The characteristic properties such as bubble formation, bubble size and shape of bubbling fluidized bed are largely dependentÂ on grid resolution,Â and poor resolution leads to inaccurate prediction of hydrodynamic behavior and inaccurate prediction of reactor performance. We have reviewed several published reports showing grid independent solutions for Geldart B BFBs, and it is apparent that grid independence depends on particle size.

In this study, a large number of two dimensional CFD simulations were performed for two pilot-scale BFBs at two diameters: 30.8 cm and 58.0 cm. We have evaluated the effect of particle diameter on mesh requirements for grid independence. All simulations were performed using MFIX, an open source code developed by NETL, USA. To identify a grid independent solution for the TFM simulation of BFBs, carefully computed bed expansions from CFD were compared with available experimental result (Geldart, D. *Powder Technology* **1968**, *1*, 355) and empirical predictions. Grid independent solutions are achieved with a mesh size directly proportional to particle size, irrespective of the scale of the fluidized beds. Our results show that, for the range of Geldart B systems studies, grid independent solutions are achieved with mesh size approximately equal to 19 particle diameters. Results are compared to numerous published results, and are in good agreement. Up until now, common practice is to use a rule of mesh size equal to 10 particle diameters, a rule derived for riser flow simulations. With a somewhat relaxed constraint for mesh size, our conclusion will allow for simulation of industrial scale reactors. This will serve as guideline for choosing the appropriate grid size and for avoiding the time consuming grid independent study that requires for all numerical simulations.

461156 Numerical Simulations of Fluidized Beds in Fine-Grid Geometries Using Source Smoothing with Cell Clustering Method for DEM-CFD Coupling

Abstract:

Discrete Element Method (DEM) proves to be often useful and accurate for modeling flow of particles in fluidized bed [1,2]. In typical industrial cases the number of particles being large imposes computational cost constraint on the choice of the 2-way coupling method between fluid (Eulerian) and particle (DEM) phases. This constraint singles out the coupling approach which requires particle size being smaller than cell size in order to maintain the smoothness in particle solid fraction distribution which is used to calculate the momentum sources in Eulerian phase. Unfortunately this requirement for cell size with respect to particle size is not easy to realize for typical fluidized bed geometries which often include high curvature boundaries and fluidized bed internals, resulting in inaccuracy or instability of the solution.

In this work we introduce the source smoothing approach for 2-way DEM-CFD coupling which allows to run efficient, accurate and stable 2-way coupled simulation without requirement of cell size being larger than particle size. Within this approach the second coarser grid is generated from original fine grid by clustering neighboring cells. The fluid solution is solved on the original fine grid. The particle volume fraction is calculated on the coarser grid of clustered cells and mapped onto the fine grid to be used by fluid solver. Compared to the two-grid formulation for fluid-particle systems proposed by Deb and Tafti [3] our approach has an advantage of being applicable for complex geometries that resolve fine details. This is because of adopting the novel algorithm for clustering cells which minimizes the difference in volume and shape between cells of resulting coarse grid. In addition, our approach is implemented within commercial software STAR-CCM+ version 11.06 by CD-adapco and allows running DEM coupled CFD models in parallel on many cores, as well as running coarse grain particle model [1] at the same time.

In this work we first validate the source smoothing method for 2-way coupling by reproducing the fluidized

bed results obtained in [1]. We then show that modifying geometry to include new small scale details (by including the tube bundle or other internals into the model) does not render the model unstable or inaccurate even though the coarse grain particles are larger than cells throughout the volume of geometry.

[1] O. Baran, A. Newale, D. Greening, DEM simulations of industrial size fluidized bed using coarse grain model with particle size distribution, AIChE 2014
[2] R. H. Aglave, O. Baran, M. Tandon, A. Karnik and . Lo, Numerical Simulation of Dense Gas-Solid Fluidized Beds: Comparison between Eulerian Multiphase and Discrete Element Methods, AIChE 2015
[3] S. Deb, D. Tafti, A novel two-grid formulation for fluid–particle systems using the discrete element method, Powder Technology 246 (2013) 601–616

462159 A Dynamic Drag Model Using Sub-Grid Scalar Variance of Solid Volume Fraction for Gas-Solid Suspensions

Abstract:

Filtered drag models for coarse-grid, two-fluid model simulations of gas-solid flows in industrial-scale applications have been under investigation for many years [1-5]. These models are designed to capture the effects of unresolved sub-filter-scale flow on the resolved flow.

In the present study, we have performed highly resolved Computational Fluid Dynamics (CFD) – Discrete Element Method (DEM) simulations of gas-fluidization of mono-disperse particles with three different diameters (75, 150, 300 μm) in periodic domains at various solid volume fractions. Simulations of this kind are being done by several research groups to learn more about meso-scale structures in gas-particle flows; for example, see [6,7].

We first mapped the Lagrangian results onto the Eulerian field and then filtered them by volume averaging in order to evaluate the sub-filter contribution of Eulerian drag force. We found that the sub-filter contribution of drag force can be captured via a model relating the filtered drag coefficient to the filtered particle volume fraction, the sub-filter scalar variance of solid volume fraction, and the particle Froude number. The sub-filter scalar variance is a measure of the degree of local inhomogeneity of the solid volume fraction within the filter. The Froude number is based on particle diameter, terminal settling velocity, and gravitational acceleration.

As the sub-filter scalar variance of solid volume fraction cannot be obtained from the resolved field, one must develop a closure model or an additional transport equation for it. In this study, we have formulated an algebraic closure for this scalar variance in terms of the filter size and filtered solid volume fraction. Towards this end, we have analyzed the CFD-DEM simulation results and extracted the functional dependence of the sub-filter scalar variance of the solid volume fraction on the filtered volume fraction and

filter size to within an unspecified multiplicative constant. It is then proposed that this constant be determined dynamically in coarse simulations by using a scale similarity assumption [8], and a test filter following the approach proposed by Germano et al. [9].

We assessed the accuracy of the model by computing correlation coefficients between model predictions and exact values calculated from mapped results. The correlation coefficients are around 0.7 even for large filter sizes, indicating that the sub-filter contribution is well captured by the model.

As a further study, we plan to implement the proposed model into a two-fluid model in order to assess a posteriori performance of the model.

References

[1] Igci, Y. and Sundaresan, S., 2011. Constitutive models for filtered two-fluid models of fluidized gas–particle flows. *Industrial & Engineering Chemistry Research*, *50*(23), pp.13190-13201.

[2] Schneiderbauer, S., Puttinger, S. and Pirker, S., 2013. Comparative analysis of subgrid drag modifications for dense gas particle flows in bubbling fluidized beds. *AIChE Journal*, *59*(11), pp.4077-4099.

[3] Parmentier, J.F., Simonin, O. and Delsart, O., 2012. A functional subgrid drift velocity model for filtered drag prediction in dense fluidized bed. *AIChE Journal*, *58*(4), pp.1084-1098.

[4] Milioli, C.C., Milioli, F.E., Holloway, W., Agrawal, K. and Sundaresan, S., 2013. Filtered two fluid models of fluidized gas particle flows: New constitutive relations. *AIChE Journal*, *59*(9), pp.3265-3275.

[5] Ozel, A., Fede, P. and Simonin, O., 2013. Development of filtered Euler–Euler two-phase model for circulating fluidised bed: high resolution simulation, formulation and a priori analyses. *International Journal of Multiphase Flow*, *55*, pp.43-63.

[6] Capecelatro, J. and Desjardins, O., 2013. An Euler–Lagrange strategy for simulating particle-laden flows. *Journal of Computational Physics*, *238*, pp.1-31.

[7] Capecelatro, J., Desjardins, O. and Fox, R.O., 2014. Numerical study of collisional particle dynamics in cluster-induced turbulence. *Journal of Fluid Mechanics*, *747*, p.R2.

[8] Bardina, J., Ferziger, J.H. and Reynolds, W.C., 1983. Improved turbulence models based on large eddy simulation of homogeneous, incompressible, turbulent flows. Stanford University.

[9] Germano, M., Piomelli, U., Moin, P. and Cabot, W.H., 1991. A dynamic subgrid scale eddy viscosity model. *Physics of Fluids A: Fluid Dynamics (1989-1993)*, *3*(7), pp.1760-1765.

462178 Staging Effect By a Baffle in a Deep Fluidized Bed

Abstract:

A commercial deep fluidized bed such as a fluid coking reactor is dominated by the large scale internal recirculation patterns that are much faster than the external solids circulation rate. When hot solids are introduced to the reactor, the solids are well mixed throughout and maintain a rather uniform temperature profile in the fluidized bed reactor. As a result, there is a need to separate the reactor into two stages/zones operating at different temperatures to increase its flexibility, e.g., running the top stage at lower temperature for higher liquid yield. A CFD simulation was applied to demonstrate the feasibility to separate the fluidized bed reactor into two distinct mixing zones. It indicated that without an annular baffle present, the solids were distributed uniformly in the reactor and no staging occurred; with an annular baffle, the solids injected below the baffle remained confined within the zone below the baffle. A cold flow testing in a 3ft ID unit was also used to demonstrate this behavior. Solids tracer testing was conducted to test the solids distribution with and without the annular baffle. The cold flow testing results were consistent with the CFD simulation. This paper describes how cold flow testing and advanced modeling are used to develop technology solutions for commercial application.

462748 Investigation of Continuum Two-Fluid Models in Simulating Fluidised Beds in Openfoam

Abstract:

There is no single continuum based method that can simulate a range of rapid granular flow types in a single simulation. In this work we present modifications to an existing two-fluid model in OpenFOAM that allow for both compressible and incompressible flows to be simulated using a single model. A non-equilibrium multiphase compressible flow can be considered as having both a convective mass flux and volume flux. Volume flux may evolve independently from the mass flux. This gives rise to volume diffusion driven forces within constitutive equations. These corrections are here used to extend classical continuum two fluid models. Our approach also investigates Korteweg diffuse interphase constitutive model type to improve the two-fluid models. A gas-solid fluidised bed configuration is simulated and results are compared with experiments, with and without the modifications. The fluidised bed configuration is of a pseudo 2D bed with a uniform flow distributor at the base; various inlet velocities are simulated. The modification is applied to each phase which is then simulated independently to observe the effect of volume diffusion on each phase. Effects of the modifications are observed at 4 times the minimum fluidisation velocity. A decrease in the time averaged pressure drop for both modifications with the Korteweg diffuse interface model is shown showing the greatest decrease. To further test the capabilities of the modifications, a simulation of a circulating fluidised bed with six cyclones is conducted. It is a scaled model of a 660MWe super-critical pressure circulating fluidised bed. It consists of a 5.8m high riser section and has six cyclones arranged in axis â€"symmetric configuration which separates the particles at the top of the riser and are fed back to the riser near the base. The bottom of the riser has a distributor for the gas inlet which is represented by a uniform gas velocity inlet and a measured pressure drop. Simulations using the new models are compared with previously published data.

467704 Extension of Fluidized Catalytic Cracking Regenerator Modeling to Improve Emissions Performance

Abstract:

Fluidized Catalytic Cracking (FCC) units have been used by refiners for over 70 years to upgrade low valued feed stocks into much higher valued LPG olefins and motor fuels. The FCC regenerator processes the spent catalyst returning from the reactor, cleaning coke off the catalyst via combustion, while providing process heat.

Computational Fluid Dynamics (CFD) models of FCC regenerators have been introduced in recent years, with primary focus on erosion and afterburn reduction. This presentation considers recent regenerator modeling work, both published and unpublished, and shows how available information can be used to extrapolate to refinery operation decision making beyond the direct CFD model scope.

Computational modeling was performed using Barracuda Virtual Reactor® software employing the Computational Particle Fluid Dynamics (CPFD®) numerical approach. Both generalized and localized regenerator afterburn are discussed. The connection between localized afterburn and maldistribution of the spent catalyst entering the regenerator is shown as impacted by overall catalyst fluidization patterns. An example of how the CFD modeling enabled informed decision-making for potential modification of regenerator internals, with extrapolation to impact of maldistribution and mixing on emissions, is presented. Finally the case of a US gulf coast refiner is used to highlight how modeling enabled a reduction in excess O2 and combustion promotor use while positively impacting CO and NOx flue gas emissions following a 2016 refinery shut down.

Comparison of the Filtered Two Fluid Model and Dense Discrete Phase Model for Large-Scale Fluidized Bed Reactor Simulations

Schalk Cloete[1], Jan Hendrik Cloete[2] & Shahriar Amini[1,2]
[1]SINTEF, Trondheim, Norway
[2]Norwegian University of Science and Technology, Trondheim, Norway

Abstract

This paper compares the performance of two filtered two fluid model (fTFM) formulations and two dense discrete phase model (DDPM) formulations in two different geometries. Comparisons to lab-scale experimental data with a fine Geldart A powder showed the ability of both the fTFM and DDPM approaches to deliver reasonable predictions of fluidized bed behavior on relatively coarse grids. Large-scale simulations then clearly showed the advantage of recently developed two-marker fTFMs where the drag correction depends on both the volume fraction and the slip velocity. It was also shown that both DDPM formulations become inaccurate on very coarse grids, suggesting clear limits to this approach when simulating large-scale fluidized beds.

Introduction

Fluidized bed reactor technology is broadly deployed in industry, but the complex gas-solid flow dynamics characterizing these units continue to pose many design and operating challenges. For this reason, multiphase flow modelling of fluidized bed reactors has been a popular research topic to gain a better understanding of these highly versatile process units. The Eulerian two fluid model (TFM) closed by the kinetic theory of granular flows (KTGF) [1, 2] has been developed to a high level of maturity since its introduction three decades ago. This approach has been proven to accurately predict the flow dynamics in the fluidized bed [3], but requires very fine spatial and temporal resolution to achieve numerical accuracy [4, 5]. For this reason, the TFM is generally restricted to laboratory scale systems and/or 2D simulations. Several approaches have been proposed to overcome this limitation, two of which will be investigated in this paper: the filtered two fluid model (fTFM) and the dense discrete phase model (DDPM).

The fTFM approach [6-8] utilizes multiscale modelling to capture the effects of dynamic particle structures (clusters and bubbles) on the transport phenomena inside the bed. In this way, the fluidized bed can be correctly modelled using computational cells that are larger than the particle clusters themselves, thereby reducing computational costs by several orders of magnitude. This approach can theoretically simulate fluidized bed reactors of any size, but requires a broad range of new closures. Substantial progress has been made in this field over the last decade, but significant uncertainty remains about the accuracy and generality of currently available closures.

The DDPM approach [9] tracks particles through the domain in the form of computational parcels. It is similar in nature to another popular approach, the multiphase particle-in-cell (MP-PIC) method [10]. Each parcel might consist of thousands or millions of individual particles, thereby enabling the simulation of large-scale beds where the particle count can easily exceed 10^{12}. Aside from the more natural inclusion of particle size distributions, the major advantage of this approach over the TFM is that it does not suffer from any numerical diffusion of the volume fraction field. As such, it can achieve grid independent results on substantially coarser meshes [11, 12]. However, particle clusters still have to be reasonably resolved to achieve accurate solutions, suggesting clear limits to the maximum achievable reactor size.

This paper will compare the performance of the fTFM and the DDPM in a 3D laboratory scale fluidized bed using a fine particle size where model performance can be experimentally validated. Subsequently, the reactor size is scaled up to increase the volume by a factor of 125 and allow the grid dependence behavior of the models to be evaluated. Based on the results from this investigation, different avenues for application of these large-scale fluidized bed reactor modelling approaches will be discussed.

Simulations

Filtered two fluid model

Two fTFM formulations will be evaluated in this work. The first formulation has already been validated in the reactor geometry to be tested in this work [13]. This formulation is attractively simple, but requires specialized wall functions to function properly. The need for these wall functions affects model generality and complicates model implementation, especially in more complex domains. More advanced filtered models, which do not require any wall functions, have recently been presented. Whereas the original fTFM formulation was derived only as a function of the filter size and the filtered volume fraction, recently proposed models have added additional independent variables (generally called "markers") to the filtered corrections. This work will evaluate the formulation of Sarkar et al. [8] where the filtered slip velocity is added as a marker for the drag correction and the filtered strain rate as a marker for the solids stresses. For details on the model setup, the reader is referred to the aforementioned works.

Dense discrete phase model

In addition, two different DDPM formulations will be evaluated. The first formulation is the default implementation in ANSYS FLUENT that was used in the validation studies mentioned earlier [11, 12]. This formulation is simple and computationally stable, but lacks a substantial amount of modelling for particle collisions and sustained frictional particle contacts. More recently, this formulation was improved by including the full stress tensor (including frictional stresses) modelled according to the KTGF [14]. This more complete DDPM formulation will also be included in the evaluation. More details on the model setup can be found in the aforementioned works.

Geometry and boundary conditions

The simulated geometry is based on the work of Zhu et al. [15, 16] and is described in more detail in the aforementioned fTFM validation study [13]. The bed region of the geometry is 0.267 m in diameter and 2.5 m in height with an expanding freeboard on top. No freeboard was included in the fTFM simulations where the bed typically did not expand beyond 2.5 m. The bed is filled with 65 μm particles to a height of 0.9 m.

The geometry was meshed with three mesh sizes using a Cartesian cut-cell method with a targeted cell size of 1 cm, 1.4 cm and 2 cm respectively. Only the case with a fluidization velocity of 0.9 m/s will be considered in this study. Simulations completed in these geometries were compared to experimental results in a validation study. Subsequently, the three meshed geometries were scaled up by a factor of five (increasing the reactor volume by a factor of 125) in order to assess large-scale performance of the respective models.

Results and discussion

Validation study

The most easily recognized problem with using the TFM on coarse grids (especially for fine particle sizes as used in this study) is a large over-expansion of the bed. For this reason, the pressure gradient predictions of different models are compared to experiments in Figure 1.

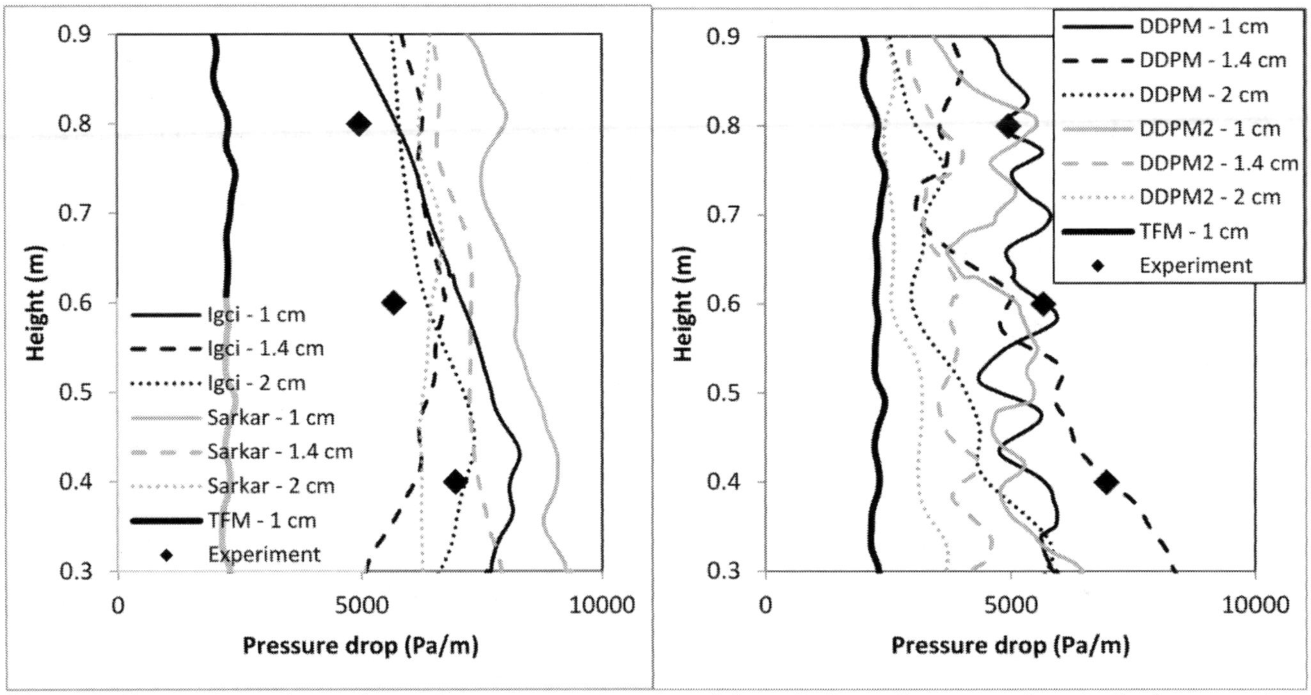

Figure 1: Comparison of experimental measurements and model predictions of pressure drop in the lower regions of the bed. "Igci" refers to the original fTFM formulation validated earlier [13], "Sarkar" is the latest fTFM formulation [8], "DDPM" is the standard DDPM formulation in ANSYS FLUENT, "DDPM2" refers to the DDPM with complete implementation of particle stresses [14], and "TFM" is the conventional TFM closed only by the KTGF. The numbers refer to the three different grid sizes considered in this study.

The first and most obvious impression from Figure 1 is that all the models perform significantly better than the standard TFM. Both DDPM formulations show some grid dependence behavior, especially the formulation with the implementation of the full stress tensor. However, the finest cases of both DDPM formulations predict similar results and may result in even more accurate solutions upon further refinement of the grid.

The fTFM formulations generally over-predict the pressure gradient in the bed (implying an under-prediction of the bed expansion). This is especially the case for the Sarkar model that increasingly over-predicts the pressure gradient as the grid is refined. It therefore appears that predictions with this model will become increasingly inaccurate as the grid is refined. This may be due to the rather weak filter size dependence of this model, implying that the strength of the filtering does not reduce sufficiently as the grid is refined and more flow details are directly resolved. It can indeed be shown that when the grid size in the coarse grid simulations equals the grid size at which the Sarkar model was derived, a substantial drag correction is still predicted. It is therefore important that the filtered models should respect the limit that the corrections should become small when the grid size in the coarse grid simulations approach the grid size used in the fTFM derivation. As was the case before, the original fTFM formulation (Igci) performs well on all grid sizes [13].

A further comparison can be made to radial time averaged solids volume fraction profiles (Figure 2). It is clear that all models captured the radial solids segregation inside the bed reasonably well. Aside from this, the conclusions that can be drawn from Figure 2 are similar to the prior discussion of Figure 1.

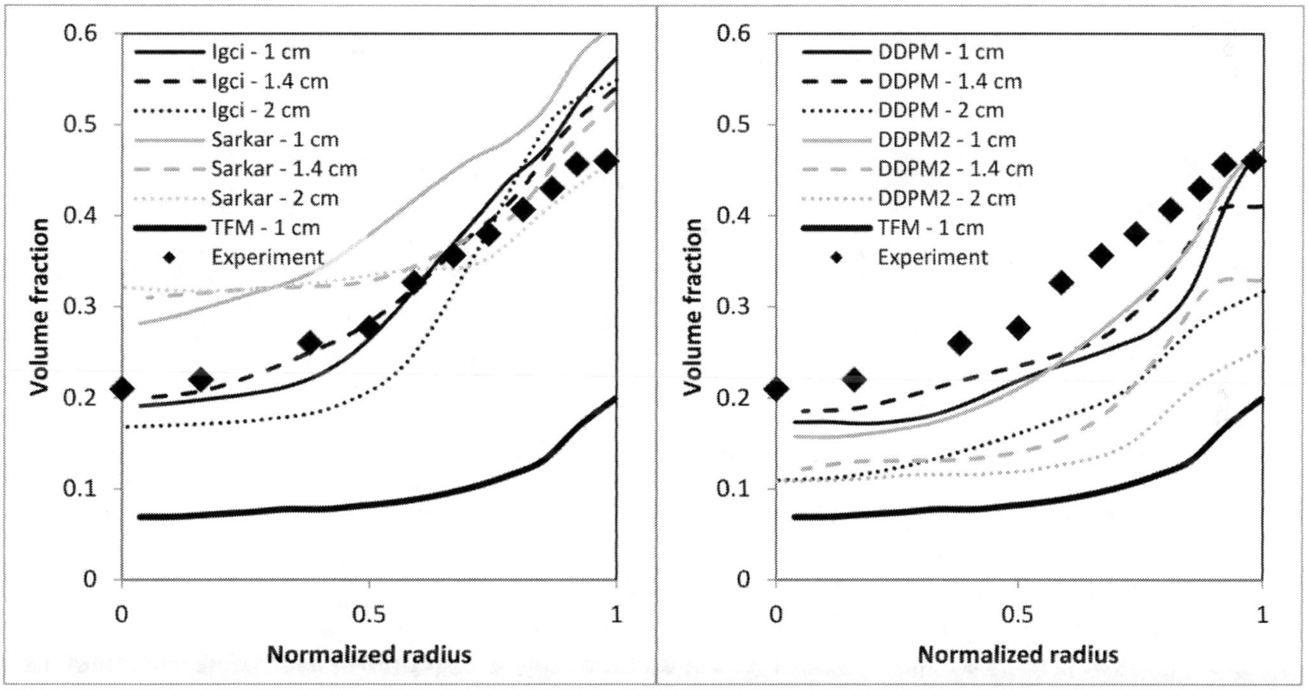

Figure 2: Comparison of experimental measurements and model predictions of radial solids volume fraction at a height of 0.6 m in the bed. See the caption of Figure 1 for an explanation of the labelling in the graphs.

To gain further insight into model behavior, instantaneous volume fraction profiles of all cases are shown in Figure 3. It is again very clear that the standard TFM (rightmost image pair) fails to capture the cluster formation correctly, thereby greatly under-predicting the solids loading in the bed. Both DDPM formulations (third and fourth image pairs) do a much better job at resolving the clusters on the 1 cm mesh. There is no clear difference between the two DDPM formulations except that the densest regions are at maximum packing for the standard DDPM, but limited below maximum packing by the frictional pressure model implemented in the improved DDPM (DDPM2).

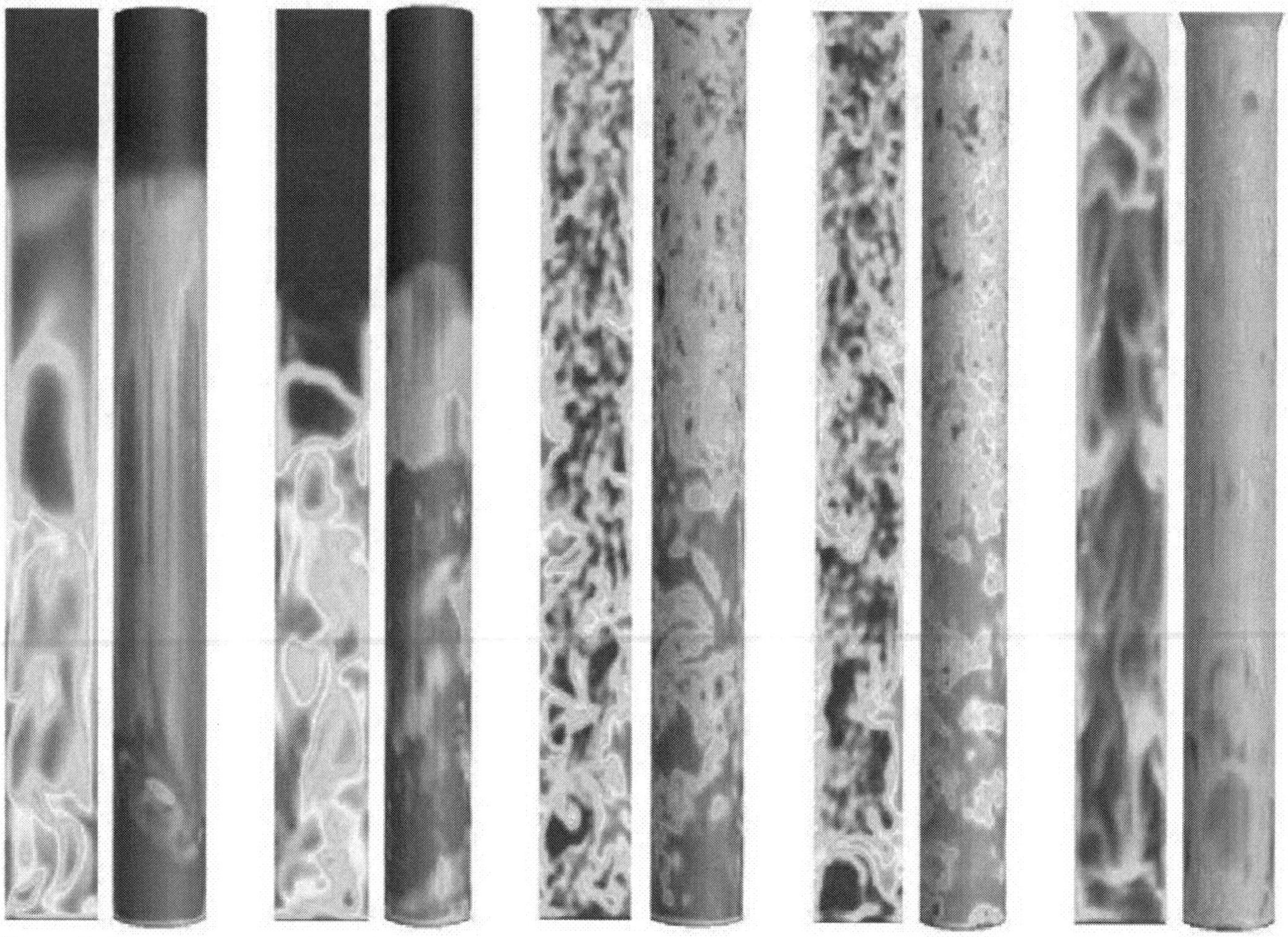

Figure 3: Instantaneous volume fraction contours on the 1 cm mesh using a blue-green-red color scale over the range of 0-0.63. The left and right images of each image pair display the contours on a central plane and on the walls of the geometry respectively. Image pairs from left to right area arranged as follows: Igci, Sarkar, DDPM, DDPM2 and TFM (see the caption of Figure 1 for more details on these names).

A lower degree of flow detail is resolved by the two fTFM formulations (first and second image pairs in Figure 3), especially in the Igci case. Despite this lower degree of cluster resolution, the filtered drag model greatly reduces the drag force experienced by the solids phase, thereby achieving a reasonable prediction of the bed expansion and the associated pressure drop.

Large-scale simulations

The simulations discussed above were repeated in a geometry scaled by a factor of five in all three Cartesian directions. No experimental data is available for this case, but the experimental pressure drop measurements in Figure 1 can still be used as an indicator of a reasonable bed voidage. Using this approximation, Figure 1 is repeated for the new large-scale geometry in Figure 4 below.

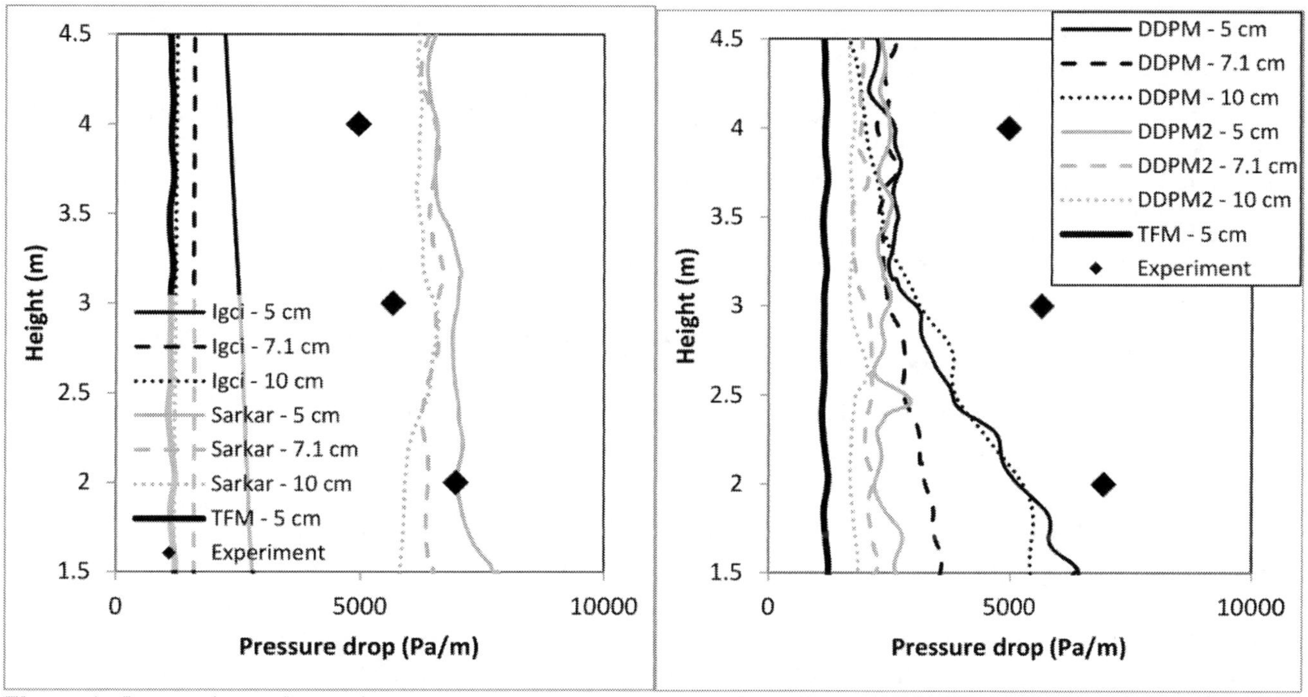

Figure 4: Comparison of experimental measurements and model predictions of pressure drop in the lower regions of the large-scale bed. Experimental measurements are from small-scale experiments and should only serve as a rough indicator. See the caption of Figure 1 for an explanation of the labelling in the graphs.

It is immediately clear from the left hand side of Figure 4 that the Igci model completely fails to predict the bed expansion, especially for coarse grids, while results from the recent Sarkar model appear much more reasonable (albeit still with significant grid dependence effects).

Three main reasons can be identified for the failure of the Igci model. Firstly, the wall functions required by this model were not derived as a function of the filter size and therefore did not become stronger in this simulation where the cell size was increased by a factor of five. As shown in the earlier validation study of this model [13], these wall effects have a large positive effect on model accuracy. Secondly, the maximum drag reduction that can be achieved by the Igci model is about a factor of 5, whereas the Sarkar model can reduce the drag by a factor of 33. This is very important because the majority of cells in the large-scale simulations were at the maximum possible drag correction. Thirdly (and related to the previous point), the Sarkar model is a two-marker model where the drag correction at a specified filter size is a function of both the filtered volume fraction and the filtered slip velocity. This is the main reason for the much stronger maximum drag correction: the maximum drag correction in the Igci model is an average over all slip velocities, whereas the Sarkar model can further increase the drag correction in case of high slip velocities.

Both DDPM formulations clearly under-predict the pressure drop, especially the more advanced DDPM2 formulation. The DDPM formulation with very little modelling of solids stresses (particle collisions) actually produced improved pressure gradient predictions in the bottom regions of the bed in some cases. This is the result of some large densified regions being formed on the gas distributor with gas slipping by in the form of a central channel (as could be expected from fluidizing a Geldart C powder). Even though this effect improves the pressure drop prediction in the lower bed regions, it is probably an inaccurate representation of the bed dynamics caused by the missing modelling of particle collisions and prolonged frictional contacts. Interestingly, this densified zone phenomenon could be sustained on the 5 cm and 10 cm grids, but did not form on the intermediate 7.1 cm grid. This raises further questions about the validity of this modelled phenomenon.

The DDPM2 formulation behaves more as expected. Even with the increased cluster resolution capabilities of the DDPM relative to the TFM, the cell sizes employed in this large-scale simulation are simply far too large to reasonably represent cluster formation. The model therefore greatly over-predicts gas-particle momentum transfer, thus under-predicting the axial pressure gradient. A substantial improvement over the TFM is still clearly visible in Figure 4 though.

Conclusions

This work has evaluated two fTFM and two DDPM formulations. In principle, these models are capable of achieving much improved accuracy relative to the conventional TFM approach for large, 3D fluidized bed simulations.

Comparisons to experimental results from a lab-scale 3D fluidized bed filled with fine Geldart A particles showed reasonable performance from all four models. Three main conclusions could be drawn: 1) all models greatly improve performance relative to the TFM, 2) the most recent fTFM formulation appears to have too weak filter size dependency and therefore shows significant grid dependence effects, and 3) both DDPM formulations can give good predictions, but require mesh sizes that are already quite expensive when simulating a lab scale fluidized bed.

Subsequently, the simulation geometry was scaled up by a factor of five (increasing the reactor volume by a factor of 125). Only the most recent fTFM formulation could achieve reasonable results in this case. The superior performance of this model is due to the drag correction that is a function of both the filtered volume fraction and the filtered slip velocity, whereas the original fTFM drag correction only depends on the filtered volume fraction. The DDPM formulations could not capture sufficient flow detail to return reasonable predictions of the axial pressure drop in the bed, suggesting a clear limit of using this approach in large-scale 3D simulations.

From these results, it can be concluded that the fTFM is a better candidate than the DDPM for general large-scale simulation tasks. This includes large-scale fluidized beds using fine powders such as this study, inherently transient processes requiring hours or days of flow time to be simulated, and technology implementation studies requiring many simulations to investigate process performance. At its current level of development, however, the fTFM introduces significant levels of uncertainty. The DDPM, on the other hand, should be significantly more accurate, but computational cost limits its large-scale application to troubleshooting and basic design of fluidized beds using larger particle sizes.

Acknowledgement

The authors gratefully acknowledge financial support from the Research Council of Norway under the FRINATEK programme (Acronym: CSRproject number: 221902) and the European Commission under the NMBP programme (Acronym: NanoSim, project number: 604656).

References

[1] J.T. Jenkins, S.B. Savage, Theory for the rapid flow of identical, smooth, nearly elastic, spherical particles, Journal of Fluid Mechanics, 130 (1983) 187-202.

[2] D. Gidaspow, R. Bezburuah, J. Ding, Hydrodynamics of Circulating Fluidized Beds, Kinetic Theory Approach, in: 7th Engineering Foundation Conference on Fluidization 1992, pp. 75-82.
[3] S. Cloete, A. Zaabout, S.T. Johansen, M. van Sint Annaland, F. Gallucci, S. Amini, The generality of the standard 2D TFM approach in predicting bubbling fluidized bed hydrodynamics, Powder Technology, 235 (2013) 735-746.
[4] S. Cloete, S.T. Johansen, S. Amini, Grid independence behaviour of fluidized bed reactor simulations using the Two Fluid Model: Effect of particle size, Powder Technology, 269 (2015) 153-165.
[5] S. Cloete, S.T. Johansen, S. Amini, Grid independence behaviour of fluidized bed reactor simulations using the Two Fluid Model: Detailed parametric study, Powder Technology, 289 (2016) 65-70.
[6] Y. Igci, A.T. Andrews, S. Sundaresan, S. Pannala, T. O'Brien, Filtered two-fluid models for fluidized gas-particle suspensions, AIChE Journal, 54 (2008) 1431-1448.
[7] Y. Igci, S. Pannala, S. Benyahia, S. Sundaresan, Validation Studies on Filtered Model Equations for Gas-Particle Flows in Risers, Industrial & Engineering Chemistry Research, 51 (2011) 2094-2103.
[8] A. Sarkar, F.E. Milioli, S. Ozarkar, T. Li, X. Sun, S. Sundaresan, Filtered sub-grid constitutive models for fluidized gas-particle flows constructed from 3-D simulations, Chem. Eng. Sci., 152 (2016) 443-456.
[9] B. Popoff, M. Braun, A Lagrangian Approach to Dense Particulate Flows, in: 6th International Conference on Multiphase Flow, Leipzig, Germany, 2007.
[10] D.M. Snider, An Incompressible Three-Dimensional Multiphase Particle-in-Cell Model for Dense Particle Flows, Journal of Computational Physics, 170 (2001) 523-549.
[11] S. Cloete, S.T. Johansen, M. Braun, B. Popoff, S. Amini, Evaluation of a Lagrangian Discrete Phase Modeling Approach for Resolving Cluster Formation in CFB Risers, in: 7th International Conference on Multiphase Flow, Tampa, FL USA, 2010.
[12] S. Cloete, S.T. Johansen, M. Braun, B. Popoff, S. Amini, Evaluation of a Lagrangian Discrete Phase Modelling Approach for Application to Industrial Scale Bubbling Fluidized Beds, in: 10th International Conference on Circulating Fluidized Bed and Fluidized Bed Technology, Sunriver, Oregon, USA, 2011.
[13] S. Cloete, S.T. Johansen, S. Amini, Evaluation of a filtered model for the simulation of large scale bubbling and turbulent fluidized beds, Powder Technology, 235 (2013) 91-102.
[14] S. Cloete, S. Amini, The dense discrete phase model for simulation of bubbling fluidized beds: Validation and verification, in: 9th International Conference on Multiphase Flow, Firenze, Italy, 2016.
[15] H. Zhu, J. Zhu, G. Li, F. Li, Detailed measurements of flow structure inside a dense gas–solids fluidized bed, Powder Technology, 180 (2008) 339-349.
[16] H. Zhu, Turbulent Fluidized Bed vs. High Density Riser - Regimes and Flow Characterizations, in, The University of Western Ontario, 2006.

450998 Lattice Boltzmann Based Discrete Simulations for Gas-Solid Two-Phase Flows

Abstract:

Gas-solid two-phase flows are widely encountered in many industrial processes, and computational fluid dynamics (CFD) simulations of gas-solid two-phase flows are useful for both practical and academic interests. In this work, we present an alternate Euler-Lagrange approach based on the lattice Boltzmann method (LBM) to describe the gas flow and transport behaviors, which is completely different from the finite-difference, finite-volume and finite-element numerical technique are used to solve gas flows based on the Navier-Stokes equations or volume-averaged Navier-Stokes equations (Zhang et al., 2014). For exploring the flow, transport and reaction details at the gas-solid interface, and to explore the constitutive laws for the simulations at the scales above, a lattice Boltzmann based direct numerical simulation (DNS) of the gas flow around each particle and its interaction with the particle surfaces is proposed by Wang et al. (2010). It does not have to establish boundary links at all, thus avoids a stair-step representation of the particle surface in Ladd's (1994) method and produces a smooth transition across the stepwise interface with sub-grid resolution at the moving boundaries (Zhou et al., 2011). Taking advantage of the inherent parallelism of LBM and the attractive Flops/Price ratio of graphics processing units (GPUs), we implemented GPU parallel computing of this algorithm and conducted large-scale DNS of gas–solid suspension, with 1,166,400 solid particles for a two-dimensional system and 129,024 particles for a three-dimensional system on a Mole-8.5 system. Large-scale DNS reveals that the standard drag correlation expressed in terms of the solid volume fraction and the Reynolds number is not adequate, the magnitude and form of the heterogeneity should be considered in the drag correlation explicitly (Zhou et al., 2014), and meso-scale structures singificantly influence the statistical properties of particles in gas-solid flows. Due to DNS extremely limited by the computational cost and practically not feasible for industrial applications, we also proposed a lattice Boltzmann based discrete particle simulation (DPS) for gas-solid two-phase flows (Wang et al., 2013). The partial saturation concept has been extended to model the objects much smaller than the cell spacing (i.e. porous media), and both the linear and nonlinear drag effects of the solid phase (media) have been considered in the lattice Boltzmann based DPS for the first time (Wang et al., 2013). It is noteworthy that the similar computational strategy has been implemented in lattice Boltzmann based DNS (Wang et al., 2010) of gas-solid two-phase flows where the modified LBE was used with particles size much larger than the cell spacing, and in lattice Boltzmann based DPS (Wang et al., 2013) of gas-solid two-phase flows where the modified LBE was used model the objects much smaller than the cell spacing. Finally, two different Euler-Lagrange numerical methods in hydrodynamic modeling and simulation of gas-solid two-phase flows at different levels have been developed and implemented in the framework of LBM.

References

Ladd A.J.C. Numerical simulations of particulate suspensions via a discretized Boltzmann equation. Part 1. Theoretical foundation. J. Fluid Mech.271:285-309 (1994).

Wang L.,Zhou G.,Wang X.,Xiong Q.,Ge W. Direct numerical simulation of particle-fluid systems by combining time-driven hard-sphere model and lattice Boltzmann method. Particuology 8, 379-382(2010).

Wang L., Zhang B., Wang X., Ge W., Li J. Lattice Boltzmann based discrete simulation for gas-solid fluidization. Chem. Eng. Sci. 101:228 -239 (2013).

Xiong Q., Li B., Zhou G., Fang X., Xu J., Wang J., He X., Wang X., Wang L., Ge W., Li J. Large-scale DNS of gas–solid flows on Mole-8.5. Chem. Eng. Sci. 71: 422–430(2012).

Zhou G., Wang L., Wang X., Ge W. Galilean-invariant algorithm coupling immersed moving boundary conditions and Lees-Edwards boundary conditions, Phys. Rev. E 84, 066701(2011)

Zhou G., Xiong Q., Wang L., Wang X., Ren X., Ge W. Structure-dependent drag in gas–solid flows studied with direct numerical simulation. Chem. Eng. Sci. 116: 9–22(2014).

Zhang J., Wang L., Ouyang J. Lattice Boltzmann model for the volume-averaged Navier-Stokes equations. Europhys. Lett. 107(2):20001(2014).

472155 Gold-Layersome Nanoshells: Preparation and Characterization for Photothermal Therapy

Abstract:

Gold nanoshells have attracted great attention in medicine due to their surface plasmon resonance (SPR) bands, nontoxic properties and ease of surface functionalization. A unique property of gold nanoshells is the ability to tune the optical resonances within the near-infrared (NIR) region. The high absorption efficiency of gold nanoshells and their ability to efficiently convert absorbed light into heat has enabled them to be an effective candidate for photothermal therapy for cancer treatment. In this work we present a new approach to synthesize gold nanoshells with variable shell morphologies using soft hollow templates; anionic liposomes coated with the cationic polyelectrolyte poly-l-lysine (PLL) (ca. 100 nm hydrodynamic diameter). PLL was used to enrich the template surface by adsorbing negatively-charged gold species in solution.

Two different approaches to create gold nanoshells were compared. In both approaches the nanoshells were characterized by transmission electron microscopy, field-emission scanning electron microscopy and UV-vis/NIR spectroscopy. The first method was in situ reduction of chloroauric acid in presence of layersomes, which has been reported for gold nanoshell formation on liposomes. This approach led to a packed structure of gold nanoparticles on the layersome surface with little control over nanoparticle or shell morphology. To gain this control and achieve a more continuous gold shell, a second approach based on a seeded growth method was used. Small gold nanoparticles (1-3 nm diameter) were synthesized and anchored onto the surfaces of layersome via electrostatic interactions to nucleate the growth of the gold overlayer. The resulting gold nanoshells exhibited controllable shell thickness by adjusting the gold ions concentration. Upon changing the shell thickness, the absorption wavelength corresponding to these plasmonic layersomes could be tuned to the biologically safe near-infrared wavelength. Changing chloroauric acid concentration also led to gold nanoshells with rough structures (branched, star or flake-like) that may have been due to the shape directing effect of PLL. It is hypothesized that the key role of PLL as a structure directing agent is originated from the electrostatic interaction between positive amine groups in PLL and the negatively charged AuNPs. The prepared gold nanoshells have the potential to be used for photothermal therapy and photothermal mediated drug delivery.

471453 Chiral Nanoparticles for Enantioslective Adsorption and Separations

Abstract:

The surfaces of chemically synthesized Au nanoparticles have been modified with *D*- or *L*-cysteine to render them chiral and enantioselective for adsorption of chiral molecules. Their enantioselective interaction with chiral compounds and pharmaceuticals has been probed by optical rotation measurements during exposure to mixtures with varying enantiomeric excess of chiral compounds such as *R*- and *S*-propylene oxide, *R*- and *S*-2-butanol, or *R*- and *S*-propranolol. The ability of optical rotation to detect enantiospecific adsorption arises from the fact that the specific rotation of polarized light by *R*- and *S*-propylene oxide is enhanced by interaction with Au nanoparticles. This effect is related to previous observations of enhanced circular dichroism by Au nanoparticles modified by chiral adsorbates. More importantly, chiral Au nanoparticles modified with either *D*- or *L*-cysteine selectively adsorb one enantiomer of the probe molecules such as propylene oxide from a racemic solution, thus leaving an enantiomeric excess in the solution phase. Au nanoparticles modified with *L*-cysteine (*D*-cysteine) selectively adsorb the *R*-propylene oxide (*S*-propylene oxide).

A simple adsorption model and accompanying experimental protocol have been developed to enable optical rotation measurements to be analyzed for quantitative determination of the ratios of the enantiospecific adsorption equilibrium constants of chiral species on the surfaces of chiral nanoparticles, $K_L^S / K_D^S = K_D^R / K_L^R$. This analysis is robust in the sense that it obviates the need to measure the absolute surface area of the absorbent nanoparticles, a quantity that is somewhat difficult to obtain accurately. This analysis has been applied to optical rotation data obtained from solutions of *R*- and *S*-propylene oxide, *R*- and *S*-2-butanol, or *R*- and *S*-propranolol, in varying concentration ratios, with *D*- and *L*-cysteine coated Au nanoparticles, in varying concentration ratios.

Recently, tetrahexahedral (THH, 24-sided) Au nanoparticles modified with *D*- or *L*-cysteine (Cys) have been used as enantioselective separators of the chiral pharmaceutical propranolol in solution phase. Polarimetry has been used to measure the rotation of linearly polarized light by solutions containing mixtures of propranalol and Cys/THH-Au NPs with varying enantiomeric excesses of each. Polarimetry yields clear evidence of enantiospecific adsorption of propranalol onto the Cys/THH-Au NPs. This extends prior work using propylene oxide as a test chiral probe, by using the crystalline THH Au NPs with well-defined, crystal facets to separate a real pharmaceutical. This work suggests that chiral nanoparticles, coupled with a density separation method such as centrifugation, could be used for enantiomeric purification of real pharmaceuticals. The model developed earlier has also been used to extract the enantiospecific equilibrium constants for *R*- and *S*-propranalol adsorption onto the *D*- and *L*-Cys/THH-Au NPs.

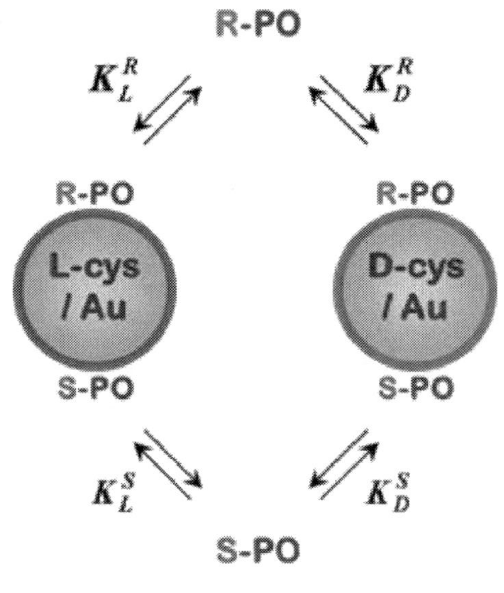

The figure illustrates the reversible equilibrium adsorption of *R*- and *S*-propylene oxide (PO) on Au nanoparticles modified with *L*- or *D*-cysteine. The equilibrium constants are enantiospecific, $K_L^S = K_D^R$, $K_D^S = K_L^R$.

460345 Triggered Ligand Clustering on Lipid Nanoparticles Enables Selective Targeting and Killing of Untargetable Cancer Cells: The Case for 'sticky Patches'

Abstract:

Objective: HER2-targeted nanoparticles encapsulating chemotherapeutics have shown promise for reducing the severity of side effects associated with traditional chemotherapeutics. However, the ability of HER2-targeted nanoparticles for effective targeting stops to hold on cancer cells expressing less than 200,000 HER2 receptors per cell (<1+ HER2 by immunohistochemistry). For cases of cancer with HER2 expression below this threshold, there is a lack of targeted treatment options because tumors are still considered untargetable. To address this issue, we have developed pH-responsive lipid nanocarriers (liposomes) that present a high local multivalency (â€˜very sticky patchesâ€™) on the surface of the liposome containing HER2-targeting lipopeptides. Sticky patches are formed by inducing preferential partitioning of lipopeptides into lipid raft-like domains that are triggered to form on the liposome membrane at acidic pH values (6.5<pH<6.7) matching the tumor interstitial pH.

Methods: Using HER2-targeting sticky liposomes loaded with doxorubicin, on a variety of cells expressing a wide range of HER2 (>1M copies per cell to 54,000 copies per cell), we evaluate the: 1) targeting selectivity, 2) killing efficacy, 3) endocytic pathway, 4) intracellular localization, 5) mechanism of action, and 6) potential toxicity to human cardiomyocytes and normal human mammary epithelial cells. In all studies, we compare to liposomes uniformly functionalized with HER2-targeting ligands via (or not) a PEG tether, and to non-targeted liposomes.

Results: Our studies show that sticky liposomes selectively and effectively kill otherwise untargetable breast cancer cells with as few as 54,000 HER2 receptors per cell utilizing both clathrin- and caveolar-mediated endocytosis, and are nontoxic to normal breast cells and cardiomyocytes which minimally express HER2. Sticky liposomes also demonstrate K_D values which are independent of HER2 expression, and long enough residence times on the cell surface ($t_{1/2}$ of dissociation ranged from 82 to 140 minutes) to allow for internalization of vesicles ($t_{1/2}$ of internalization ranged from 27 to 29 minutes).

Conclusions: These studies demonstrate the potential of sticky liposomes to selectively target and effectively kill cancer cells considered otherwise untargetable.

Acknowledgements: NJCCR, American Cancer Society RSG-12-044-01, NSF DMR-1207022, NSF CBET-1510015, GAANN.

470220 Superparamagnetic Nanoparticles for Triggered Drug Release from Alginate Hydrogels

Abstract:

While magnetic nanomaterials have already been used in clinics for contrast enhancement in magnetic resonance imaging (MRI) [1], there has been no clinical approval for a drug delivery system containing such nanoparticles, yet. Magnetic nanoparticles currently investigated for their possible application in biomedicine are predominantly different crystalline polymorphs of iron oxides. For small enough crystal sizes, iron oxides exhibit the so-called "superparamagnetic" behavior, a feature combining high magnetization with very low coercive forces. Such superparamagnetic nanoparticles show great potential in therapeutic applications due to their ability to transform the energy of an alternating magnetic field to thermal energy in what is often called magnetic fluid hyperthermia [2]. These particles dissipate thermal energy by magnetic relaxation through the Brownian and Neel mechanisms. Therefore, aqueous suspensions at relatively higher nanoparticle concentrations (g/L), the so-called "ferrofluids", also increase their temperature in the presence of an alternating magnetic field [3]. In this work, a composite multi-scale structure consisting of the biopolymer alginate, functional nanoparticles and a model drug is fabricated and analyzed. We examine the potential of flame-made SiO_2-coated Fe_2O_3 nanoparticles [4] as the stimuli-responsive material in the multi-scale composite structure. We perform detailed physicochemical and magnetic characterization on the hybrid alginate hydrogel beads and evaluate their potential in magnetic fluid hyperthermia and enhanced biomolecule release in the presence of an external AMF. The possibility to externally stimulate drug release will open up new possibilities in intelligent, on-demand drug administration [5].

References

[1] A. Singh, S. K. Sahoo, *Drug Discov. Today*, **19**, 474-481 (2014).

[2] A. Jordan, R. Scholz, P. Wust, H. Fahling, R. Felix, *J. Magn. Magn. Mater.*, **201**, 413-419 (1999).

[3] G. A. Sotiriou, M. A. Visbal-Onufrak, A. Teleki, E. J. Juan, A. M. Hirt, S. E. Pratsinis, C. Rinaldi, *Chem. Mater.*, **25**, 4603-4612 (2013).

[4] A. Teleki, M. Suter, P. R. Kidambi, O. Ergeneman, F. Krumeich, B. J. Nelson, S. E. Pratsinis, *Chem. Mater.*, **21**, 2094-2100 (2009).

[5] A. Teleki, F. L. Haufe, A. M. Hirt, S. E. Pratsinis, G. A. Sotiriou, *RSC Adv.* **6**, 21503-21510 (2016).

472317 Dual Drug-Loaded Janus Particles for Co-Delivery of Hydrophobic and Hydrophilic Compounds

Abstract:

The desire for drug delivery systems with tunable properties and multiple functionalities has spawned a new generation of particulate carriers, including Janus particles. Biocompatible Janus particles with internal compartmentalization are attractive as drug delivery vehicles because they offer a platform for co-delivery of two drugs with decoupled release kinetics.

Curcumin (CUR) and quercetin (QCT) were chosen as the model hydrophobic compounds for this study, while acetaminophen (APAP) and naproxen (NPX) were chosen as the model hydrophilic-hydrophobic drug pair. Curcumin and quercetin are bioflavonoids with anti-inflammatory, anti-proliferative, and immunosuppressive properties used to treat a host of diseases, including multi-drug resistant cancer and arthritis. Acetaminophen and naproxen are often used in combination due to their additive effects in pain management and treatment of rheumatoid arthritis.

The goal of this work is to obtain dual-loaded Janus particles with a different drug in each compartment, and to determine their encapsulation efficiency and drug content. Hydrophobic drugs are encapsulated by the single O/W emulsion technique. Hydrophilic compounds require special modifications for co-encapsulation with hydrophobic drugs due to their poor oil solubility and tendency to escape to the outer aqueous phase during the emulsification and solvent evaporation steps. Three different strategies for incorporating hydrophilic drugs were employed: 1) O/W emulsion with partially-water miscible solvent, 2) O/W emulsion with co-solvent (i.e., acetone, methanol, ethanol), or 3) W/O/W double emulsion. The encapsulation efficiencies and drug loading percentages are measured using UV/Vis spectroscopy and compared for the different synthesis methods.

461094 Preparation of Colloidal Microcapsules for Controlled Drug Delivery Systems

Abstract:

Nanoparticles (NPs) are one of the most used and studied materials in nanotechnology. Among the other usages, nanoparticles can also be used as a drug delivery agent in biomedical applications. They can be exploited as a sensor in the diagnosis and a delivery agent in the treatment of diseases. NPs can be synthesized by using a variety of constituents, primarily metallic, polymer, composite and lipid.. Solid lipid nanoparticles (SLNs), prepared using various physiological lipids, emulsifiers and water, have been used since the 1990s as colloidal nanocarrier systems, either in combination with or as an alternative to liposomes, emulsions and polymeric nanoparticles [1]. Possessing adjustable release kinetic, determined by the method of synthesis used, its level of biocompatibility, its stability and the degree to which active materials may be encapsulated within its structure, SLNs are ideal for use within the body and may be administered via different routes (parenteral, topical, ocular, etc.). There are several methods that could be utilized to synthesize SLNs, including high pressure homogenization, high shear homogenization, etc. However, if the incorporation of selected compounds is intended, particularly drugs, genes, proteins and other biomolecules, methods such as high pressure or high shear homogenization which create harsh environments for the material, need to be avoided. By using microemulsion method, the nanoparticle formulations of interest can be prepared in vials rapidly, reproducibly and cost effectively in a two-step process involving only mild operating temperatures. Moreover, the uniform nanoparticles produced become more biocompatible due to the elimination of organic solvent use during its preparation. Furthermore, the solubilization properties of hot microemulsions enable SLNs to be loaded with various drugs and is particularly useful for drugs with poor water solubility.

As well as NPs are being used in several areas including medical and pharmaceutical individually, more versatile systems can be established via self-assembly of NPs. Colloidal microcapsules are one of the most interested part of these self-assembled structures whose shell side are composed of nanoparticles [2]. Layer-by-layer method (LbL) is one of the most used methods in the production of self-assembled structures. These self-assembled structures, independent from their single form, possess attractive properties such as high stability, enhanced functionality and increased loading capacity. Additionally, using pharmaceutically active ingredients such as drug crystals or drug loaded nanoparticles as the core material of colloidal microcapsules provides controlled release of the encapsulated material with the help of shell layers [3]. In this study, novel self-assembled colloidal microcapsules by using Ibuprofen crystals or $CaCO_3$ nanoparticles as core material and SLNs as shell material were prepared by using LbL method. The performance of establishing the layers was monitored by measuring the change of zeta-potential values. Also the self-assembled structures were visualized with SEM. Controlled drug release from the microcapsules was monitored by using UV-visible spectroscopy. Due to several advantageous properties such as being composed of physiologically related lipids, possessing properties like small and adjustable structure, high surface area, enhanced stability, stimuli-responsivity and increased encapsulation efficiency, these unique systems will be an ideal alternative to the conventional nanocarriers, which are especially used in pharmaceutical and biomedical applications as controlled drug delivery agents.

[1] Mehnert, W., Mader, K., 2001, Advanced Drug Delivery Reviews, 47, 165-196.

[2] Bae, J. et al. 2013, Current Organic Chemistry, 17, 3-13.

[3] Sukhorukov, G. B. et al. 2004, Langmuir, 20, 3398-3406.

461183 Stimuli Responsive Lipid Nanotubes for Drug and Protein Delivery

Abstract:

Lipid nanotubes (LNT) are one of the most attention drawing systems due to their superiorities provided by their asymmetric geometry such as high interior volume, high drug loading capacity, easy surface modifications and high circulation persistence as well as their high biocompatibility. Thus, they have high potential to be used in many different applications ranging from electronics to bionanotechnology. One of the most important applications of LNTs is their usage as drug delivery agents. Due to their exceptionally high interior volume compared with most of the other carrier systems they are expected to effectively entrap both small drugs and biomacromolecules.

Recently, we designed a molecule called as AQUA (AQ-NH-$(CH_2)_{10}COOH$, where AQ is anthraquinone group) that can form self-assembled lipid nanotubes. The molecular structure has been designed very carefully so that each group in the structure has key properties for both nanotube formation and its functions. AQUA LNTs have dual responsive character provided by COOH and AQ groups and its morphology can be controlled reversibly by external stimuli (pH and redox reaction). In the design step of AQUA molecules, we especially took into consideration to use them as delivery agents for some selected pharmaceuticals such as Doxorubicin (DOX), Hemoglobin (Hb) and BSA.

Total Hb entrapment capacity is determined by UV-vis spectroscopy from the absorbance value of the samples which were treated with Drabkinâ€™s reagent at 420 nm. The entrapment capacity is significantly affected by the solution pH and the loaded Hb amounts are constant and at its maximum, in the pH range of 5.0-7.0. AQUA nanotubes can hold Hb approximately two times that of their own weight. (i.e loading ratio is ~2) However, loading ratio decreases to 1.1 as the pH is increased to 9.0 due to the reduced electrostatic attraction. However, even at the high pH values AQUA nanotubes can carry significant amount of Hb. The capability to make Ï€-Ï€ stacking interactions of the AQ group in the AQUA increase the loading capacity and helps to obtain high loading ratios in a wide pH range.

Hb release profile to PBS buffer at pH 7.0 changes with loading pH and release temperature however, even at the maximum measured release conditions, AQUA nanotubes keep more than 80 % of the initially loaded Hb after three days and from this time the Hb release almost vanish. This such a low Hb release at the physiological conditions along with the high loading capacity, is very important and advantageous for the artificial blood application potential of AQUA nanotubes.

In the BSA loading and release studies Biuret method is used and the BSA amount is determined from absorbance value at 540 nm by UV-vis spectroscopy. Maximum BSA loading is achieved at pH 3.0 and pH 6.0 with the loading ratio of 0.8. When the BSA loading ratios of AQUA nanotubes are compared with the systems in the literature, it is seen that AQUA nanotubes are loaded with 10-50 times more BSA than most of them systems in the literature. At room temperature, the highest release is observed at pH 7.0 and totally 59 % ±0.5 of the loaded BSA is released after 48 hours which is determined as 18 %±0.5 for pH 5.0 and 20 %±0.5 for pH 6.0. When the temperature is increased to 37 °C at pH 7.0, the total released amount in the 48±0.5 hours increases to 68 %±0.5 due to the increase of the water solubility of BSA with temperature [1]. Unlikely to Hb, high release rates can be reached for BSA. The in vitro release profiles from AQua nanotubes shows opposite behaviors for Hb and BSA and both of those profiles are advantageous for different purposes. The very low release percentages for the Hb even when it is at its maximum, is a desired result for the artificial blood applications while the very high and pH and temperature dependent release profiles for BSA are pleasing for the controlled release applications.

The entrapment and release studies carried out with two different model proteins showed that the special functional groups of the AQUA molecule are quite effective to obtain strong attraction interactions with the guest molecules and it is possible to control these interactions on demand with the help of the environmental triggers due to the stimuli responsive character of our nanotubes. After these promising results the entrapment and release studies of the small drugs was carried out by using doxorubicin (DOX) as the model drug.

DOX is an anthraquinone based antibiotic which is commonly used for cancer treatment. However, it has various important side effects such as cardiotoxicity, supressing blood cell production, low therapeutic index, etc. So, developing effective carrier systems for DOX is a widely interested topic [2]. One of the most common cancer types for which DOX is used as the therapeutic agent is the colorectal cancer (CRC). Although parenteral ways are generally preferred for the treatment of the colorectal cancer, oral or intracolonic administrations can also be used.

AQUA and DOX have some similarities in their molecular structure and they both contain anthraquinone groups which we believe affects the affinity of DOX to AQUA LNTs via Ï€-Ï€ stacking interactions. In addition, the presence of carboxylic acid group in the molecular structure of AQUA helps us to encapsulate significant amount of DOX due to the hydrogen bonding and electrostatic interactions. There are some studies in the literature in which DOX is added to molecule structure/carrier via a covalent bond. These kinds of approaches will require more complex synthesis procedures which will prevent their use in large scale applications. Here we present a drug delivery system formed by relatively simple, but smartly tailored molecules.

The pH sensitive character of the AQUA LNTs is advantageous for increasing their efficiency on drug entrapment and release as well as providing selective release of the DOX to the tumor cells. DOX loading capacity is effected by the parameters such as drug concentration, pH, drug to carrier mass ratio. The maximum loading capacity is reached at pH 9 when the DOX/LNT ratio is 1/1 (g/g). In this condition AQUA LNTs can entrap almost as much DOX as their own mass. This loading capacity is significantly higher than the ones that could be reached in the literature [3]. Also it has been found that the release of DOX from the AQUA LNTs changes depending on the temperature and pH. The AQUA LNTs releases significantly more DOX at pH 5.5 which is the pH of the many cancer tissues than physiological pH (i.e. 7.0) This shows that AQUA LNTs selectively releases the DOX to the cancer tissue which is very advantageous for their usage as the carrier system in the cancer treatment.

The abovementioned results show that the rationally designed new stimuli responsive nanotube system is versatile as a highly functional carrier system for various different guest molecules changing from small drugs to biomacromolecules.

[1] M. A. Masuelli, Advances in Physical Chemistry, 2013, **2013**, 360239

[2] B.D. Weinberg, H. Ai, E. Blanco, J.M. Anderson, J. Gao, J Biomed Mater Res., 2007, **81**, 161.

[3] S. Ilbasmis-Tamer, H. Unsal, F. Tugcu-Demiroz, G.D. Kalaycioglu, I.T. Degim, N. Aydogan, Colloids Surf B Biointerfaces, 2016, **143,** 406.

472995 Prediction of Flowability of Methocel^{TM} Direct Compression Grades Using a Limiting Flow Rate Calculation

Abstract:

To enhance the productivity in the manufacturing of solid oral dosage forms it is important to ensure that powders flow reliably from hoppers and into tablet dies at sufficient rate for high throughput and low variation to minimize product variability. This paper evaluates the flowability of powders by fundamentally describing discharge rates based on Johanson's limiting flow rate approach [1]. The findings for hopper discharge rate are then related to tablet properties.

The evaluation of flowability of fine powders is often addressed by standard flow testers. The flow functions from shear testers, discharge rate through cones, orifice plates or a compaction ratio are used as descriptors of the flowability of powders. However, these testers can fail to address the discharge characteristics from a hopper at commercial scale to predict the actual discharge rate of a fine powder, because they do not consider the effects of interstitial air pressure.

This paper discusses the discharge rate of fine powders from a hopper using permeability, compressibility, and shear testing in combination with the calculation of the limiting discharge rate using Johanson's approach [1]. The predictive concept for powder discharge rate using the limiting flow rate calculation is correlated with experimental discharge rates from a model hopper. The use of permeability and bulk density was found to be the most important powder descriptors to predict discharge rates of fine powders. Based on the calculation of the limiting flow rate, predictive correlations were developed based on permeability and bulk density.

The impact of flowability of a material on tableting was evaluated experimentally with a Piccola bilayer press. It was found that the standard deviations of tablet weight, compaction force and radial tensile strength can be described as a function of flowability. With increasing flowability, variation in all three metrics was significantly reduced. Thus, the characterization of powder flowability can be applied to understand how variation in the tableting process might be reduced.

[1] Johanson, J. R., "Two-Phase Flow Effects in Solids Processing and Handling," *Chemical Engineering*, pp. 77–86 (Jan. 1, 1979).

452038 Choking Modifications to a Conveyor Feed Chute

Abstract:

For this work, a chute choke was designed to provide a constant bulk solids feed rate to a bucket elevator under conditions of sporadic impulse feed to the chute. The choke was designed using modifications to the Beverloo correlation obtained via an adjustable chute simulator built for experimentation. The simulator was constructed in order to investigate the effects of orifice size, chute angles, and tilt angle on the flow rate of the bulk solid. This talk will present the results of these experiments, along with the proposed interpretation of the effects of the experimental parameters on the Beverloo correlation.

452565 Minimum Pickup Velocity

Abstract:

Minimum pickup velocity – the transition from nano-scale to micro-scale particles

Aditya Anantharaman[a], J. Ruud van Ommen[b], Jia Wei Chew[a,c]

[a]School of Chemical and Biomedical Engineering, Nanyang Technological University, Singapore

[b] Chemical Engineering, Delft University of Technology, The Netherlands

[c]Singapore Membrane Technology Center, Nanyang Environment and Water Research Institute, Nanyang Technological University, Singapore

Pneumatic conveying is an important operation in a wide range of chemical industries from pharmaceutical to oil and gas. A critical parameter for ensuring the efficient operation of pneumatic conveying systems is the minimum pickup velocity, U_{pu}, defined as the minimum superficial gas velocity required to initiate rolling or suspend a particle initially at rest. Despite the growing popularity of nanoparticles, most of the past studies [1-3] on U_{pu} have focused on micron sized particles (0.5-3000 µm); with only a recent work [4] studying U_{pu} of nanoparticles in the particle diameter (d_p) range of 13-21 nm. This work is aimed at bridging the knowledge gap between the nano- and micro-scale particles by investigating the U_{pu} of particle diameters traversing both scales.

Experiments were conducted to determine the U_{pu} of nine different particle diameters (d_p) of alumina

(Al_2O_3) particles from 5 nm to 110 µm using the weight loss method and categorized according to the established pickup zones [3] (which is analogous to the Geldart Groups [5]). The following are the key findings. Firstly, U_{pu} varied non-monotonically (namely, increased, then decreased) with increasing particle diameter (d_p) in the range investigated, which spans the nano-scale and micro-scale. Secondly, while the larger micro-scale particles agreed well with the established zone, the smaller nano-scale particles agreed with a zone that they did not belong in. Thirdly, the intermediate particle diameter (d_p) of 200 nm did not correspond to any zone, and represented the transition between the nano- and micro-scales. Collectively, the results reveal the missing link on the pneumatic conveying behavior of particles between d_p = 21 nm [4] and d_p = 0.53 µm [2] in terms of U_{pu} trends.

References:

1. Cabrejos, F.J. and G.E. Klinzing, ***Incipient motion of solid particles in horizontal pneumatic conveying. .*** Powder Technology, 1992. **72**: p. 51-61.

2. Hayden, K.S., K. Park, and J.S. Curtis, ***Effect of particle characteristics on particle pickup velocity.*** Powder Technology, 2003. **131**(1): p. 7-14.

3. Kalman, H., et al., ***Pickup (critical) velocity of particles.*** Powder Technology, 2005. **160**(2): p. 103-113.

4. Anantharaman, A., J.R. van Ommen, and J.W. Chew, ***Minimum pickup velocity (U pu) of nanoparticles in gas–solid pneumatic conveying.*** Journal of Nanoparticle Research, 2015. **17**(12).

5. Geldart, D., ***Types of Gas Fluidization.*** Powder Technology, 1973. **7**(5): p. 285-292.

471388 Caking of Bulk Materials - a Systematic Approach to Understanding Caking Mitigation

Abstract:

Caking of bulk solids is a serious problem that leads to significant lost production, silo damage, and health and safety concerns. It is, perhaps, one of the most misunderstood problems in handling of bulk solids. Caking is the gain in bulk strength of a powder or a granular material with storage time. Caking is induced by a change in the surface condition of adjacent contacting particles to cause crystal growth, cold creep, or sintering effects. The net effect is cementation of the adjacent particles in a bulk material via solid bridge formation mechanisms. In some cases, caking is caused by the dissolution of soluble components in the bulk and subsequent concentration, pooling and recrystallization of solid material between connected particles. In other cases, caking is the result of glass transition temperature effects that convert contacting particle surfaces to a soft amorphous structure at a prescribed temperature. Stress effects near the soft surface cause creep and allow the surface of particle A to fuse to the surface of particle B. The subsequent change in local temperature then induces a crystallization event that forms solid bonds between particles and results in caking. Glass transition temperature can be a function of the moisture content of the surfaces or be completely independent of moisture effects. It is obvious that many caking mechanisms require the presence and movement of moisture. Thus, understanding moisture migration can help in understanding caking events in equipment. In some cases the material itself generates or stores the moisture needed to cause the caking reaction. Some reactions generate moisture as a product of the reaction. Some materials contain crystalline bound moisture that is released as the temperature is changed. These types of materials can provide both the source or sink for the moisture in caking reactions. There are a few conditions that all caking reactions have in common. They all have a series of time constants that describe the reactions that induce caking. It is important to note that there may be multiple reactions occurring to cause caking, and these reactions may have independent time constants. Measurement of time constants is one means of characterizing caking behavior. Determining the time constants can help in the design of process methods to mitigate caking behavior, or help in identifying the root cause of caking events. Each caking mode is associated with a characteristic time constant that depends on some external caking stimulus (temperature cycle, moisture migration cycle, etc.). The magnitude or severity of the caking event depends on the relationship between the caking cause reaction rate and the time the caking stimulus is applied. For caking events to become serious, some threshold value (such as local moisture content or local temperature) must be exceeded. This paper examines the various caking causes and identifies the time constant for caking, the mode of strength increase, and the potential mitigation tools needed to reduce these effects. Several systems comprising the key mechanisms were studied and data for each of these systems reported.

462652 Fluidized-Bed Measurements of Carefully-Characterized, Mildly-Cohesive (Group A) Particles

Abstract:

Fluidization is sensitive to mild cohesion levels, as evidenced by the different flow behaviors of Geldart Group A and Group B particles. Additionally, cohesion is extremely sensitive to surface roughness. Here, mildly cohesive (Group A) spherical particles were carefully characterized and used in high-fidelity experiments of defluidization, bed expansion, and bubble characterization. The experiments were conducted such that van der Waals forces were the only source of cohesion, and thus the results presented can be used for validating DEM and/or continuum models. Particle characterization includes surface roughness and adhesion (pull-off) force, which can be used for modeling the van der Waals force. Additionally, the particle friction coefficient, coefficient of restitution, shape (sphericity), and diameter were measured via a suite of simple experiments. This level of particle characterization, which is critical for the validation of cohesive models (DEM and continuum), has not been reported in previous fluidization studies. Accordingly, the characterization experiments employed here are described in sufficient detail for straightforward adaption elsewhere.

471536 Predicting Feeder Performance Based on Material Flow Properties

Abstract:

Purpose: Gravimetric powder feeders have been recognized as an important unit operation for many powder-based processes, including food, detergents, pharmaceuticals, minerals, and many others. Accurate and consistent delivery of materials by well-designed feeders ensures overall process stability. Importantly, feeder performance is strongly dependent on powder flow properties. The purpose of this study is to develop a methodology that identifies a predictive correlation between material flow properties and feeder performance.

Method: The proposed methodology includes techniques to characterize material flow properties, methods to quantify feeder performance of a loss-in-weight feeder, and predictive multivariate analysis. Materials with varying flow properties were firstly characterized. The flow properties of each material was represented by 30 flow indices. Two approaches to correlate feeding performance and material flow properties were examined in the study: principal component analysis followed by similarity scoring (PCA-SS), and partial least squares regression (PLSR).

Results: Experimental results showed that selection of the optimal feeder screw to achieve optimal feeder performance is heavily dependent on material flow properties. Both approaches were validated by testing an additional material. The predicted feeder results were generally in good agreement with the experimental results. In addition, a strong linear relation was observed between the initial feed factor of each material and its scores for the first principal component.

Conclusion: The work presented here has shown an efficient approach to correlate material properties with process performance using multivariate analysis. This approach is especially powerful in the early phase of process and product development, when the amount of a material is limited.

469349 Experimentally Validated Computational Models to Predict Powder Flow at Different Humidity Conditions

Abstract:

The flow properties of powder are governed by its intrinsic materials properties (cohesion, surface energies, elastic moduli, coefficient of friction and yield stress) and particle properties (size distribution, morphology, porosity and roughness). Environmental variables that affect the powder flow properties include relative humidity, temperature, and electrostatic charging. This large number of variables and their mutual interactions, along with unknown material properties, make it difficult to predict the flow properties for a new entity (API) or excipient or their mixture without very detailed experimental investigation (based on Design of Experiments-DOE) requiring significant amount of time and resources.

Properties such as flow, compaction, disintegration, dissolution, hardness and chemical stability are all influenced by moisture. The effect of moisture has been introduced in a Discrete Elementary Method based model by incorporating Granular Bond numbers. The experimental appproach involves the flow discharge study of two different hopper angles at 20%, 40% and 60% RH of different grades of Lactose and Avicel. The DEM models for those hopper models, are validated by predicting the accurate granular bond numbers for different RH conditions and comparing the flow properties against the two hopper angles. After systemica compareison of the hopper flow experiments of different individual powders, further study will involve their binary mixtures , to determine the appropriate DEM parameters (bond number, friction coefficients etc).

The preliminary results suggest improved flow properties with the decrease of relative humidity.

477397 Modeling Blending of Granular Materials: a Multi-Scale Model Combining FEM Flow Field Information with Advection-Diffusion Correlations

Abstract:

A multi-scale model is presented for predicting the magnitude and rate of powder blending in blenders. The model combines diffusion coefficient correlations from small-scale discrete element method (DEM) simulations with advective flow field information from blender finite element method simulations. The multi-scale model predictions closely match results from large-scale DEM-only simulations for a rotating drum, but take only a hours to compute as opposed to taking days of computation time for the DEM-only simulations. The multi-scale model scales much more efficiently to industrial-scale blenders than DEM-only simulations.

458228 Nanoparticle Phagocytosis: Characterization and Quantification By Magnetic Cytometry

Abstract:

Abstract

Nanoparticle characterization is of importance and in high demand in medical biology research and clinical application. The uptake by cultured CHO (Chinese hamster ovary) cells of 50 and 100nm iron oxide particles (MNPs) coated with starch, aminated starch and PEG (2K, 5k and 20k Daltons) was studied to reveal the chemistry of nanoparticle phagocytosis. A particle tracking velocimeter, Hyperflux^{TM} was utilized to directly measure the magnetophoretic mobility, size and other morphological parameters of magnetic nanoparticles and labeled cells. The magnetophoretic mobility is a key parameter to describe the cell motion in a defined magnetic field and is used as a quantitative indicator of the number of particles ingested per cell. By quantitatively characterizing and determining the cell uptake kinetics as a function of particle size and surface chemistry, we have been able to reveal the dependencies of phagocytosis on cell type, incubation time, particle composition, particle size and particle toxicity. We found that surface aminated nanoparticles, which are highly positively charged, are more effectively taken up by CHO cells than starch-coated particles or uncoated particles. The PEG content in coating, though bio-friendly, will prevent the penetration of MNPs into CHO cells. In addition, the Hyperflux Velocimeter analysis provides a better understanding of cell labeling and as a tool to optimize setting of MNPs type and incubation condition.

Key Words: magnetic particles, nanoparticles, magnetophoretic mobility, phagocytosis, CHO cells, particle tracking velocimetry

464416 Non-Contact AFM Measurement of the Hamaker Constants of Solids: Calibrating Cantilever Geometries

Abstract:

The van der Waals (vdW) attractive force plays a significant role in several microscale and nanoscale phenomena, such as nanoparticle self-assembly, colloidal stability, and interfacial adhesion. The Hamaker constant, *A*, captures the effects of material composition on the vdW force, which is often needed as input for predicting the attractive interactions among particles and between particles and surfaces. Thus, developing accurate experimental methods to determine *A* reliably is important.

Experimental attempts to determine the strength of the vdW interaction using an atomic force microscope (AFM) are typically hindered by issues inherent to the cantilever tip-surface contact regime, such as surface roughness and deformation, and contact separation distance. Thus, we previously developed a new method for determining *A* of solids from the approach-to-contact regime of an AFM experiment.[1,2] In contrast to other non-contact methods that rely solely upon a quasi-static model of the tip deflection, the new method accounts for the inertial effects of the cantilever tip's motion. Specifically, an apparent Hamaker constant is determined, the value of which depends on the cantilever approach speed and AFM sampling resolution. As the approach speed goes to zero and the sampling resolution approaches the ultra-fine limit, the apparent Hamaker constant becomes nearly identical to the system's "true" Hamaker constant. The method was tested experimentally using silica and polystyrene test substrates, and was demonstrated to yield estimates of *A* for these materials that were in very good agreement with previously published Lifshitz calculations.

The accuracy of the non-contact method for measuring interaction forces relies heavily on the accuracy of the models for the geometry of the interacting surfaces that are used in the data interpretation. For the initial validation experiments of the new method, an AFM cantilever tip with a pyramidal body and a hemispherical cap was used. Although this geometry can be confirmed and the dimensions estimated via scanning electron microscopy (SEM), even high-resolution SEM analysis of the tip cannot provide sufficient detail to allow precise enough determination of the tip's geometric parameters. In addition, the numerical methods required to evaluate the Hamaker constants with high accuracy for geometries like the pyramid with spherical cap can quickly become prohibitively complex.

Therefore, we propose an adaptation of our current method in which the geometric complexity of the cantilever tip is eliminated from the determination of the Hamaker constant by describing the tip as an 'effective' perfect sphere, thereby capturing all the geometric effects in the single dimension of this sphere. This new approach requires measurements of the interaction between the cantilever and one or more ideal surfaces. These ideal surfaces have vdW interactions with the tip that are very well established, and they are extremely smooth, so that there are minimal effects of the surface topography on the execution of the method. After the forces against these well-known surfaces are measured, the interactions are modeled by assuming a simple 'effective' shape for the cantilever tip. When the dimensions of this effective shape are captured properly, the Hamaker constants for the tip-surface interactions will be correctly calculated in all cases where this cantilever is applied. In this way, the dimensions of the effective tip shape are "calibrated" based on the attractive forces of the well-characterized ideal surfaces. If a sphere is chosen as the "effective" shape for the tip, the effective "radius of curvature", R_{eff}, of the tip is therefore calibrated. Thus, this procedure utilizes a calibration surface of known *A* to generate a R_{eff} of a given tip which can then be used to determine the unknown *A* of another material of interest. We demonstrate the practicality and accuracy of this updated method by comparing the results with both the original pyramid model and Lifshitz approximations (when available) for flat substrates composed of silica, polystyrene, highly ordered pyrolytic graphite (HOPG), sapphire (α-Al_3O_2), and acrylonitrile butadiene styrene (ABS).

References

(1) Fronczak, S. G.; Dong, J.; Franses, E. I.; Beaudoin, S. P.; Corti, D. S. A New Method for Determining the Hamaker Constant of Solids Using an Atomic Force Microscope. 1. Theoretical Development. *Langmuir* **2016**, Submitted.

(2) Fronczak, S. G.; Browne, C.; Beaudoin, S. P.; Corti, D. S. A New Method for Determining the Hamaker Constant of Solids Using an Atomic Force Microscope. 2. Experimental Considerations. *Langmuir* **2016**, Submitted.

464476 Study of the Physicochemical Properties, the Aging Phenomena, and the Production Method of an Extinguishing Powder for Sodium Fire

Abstract:

The French Atomic and alternatives Energy Commission (CEA) developed an effective powder capable of extinguishing sodium fire. It is a powder based on a mixture of lithium carbonate and low-hydrated sodium carbonate in a near eutectic proportion, with a melting temperature around 500°C, associated with graphite. However, ever since the dismantling of several old sodium installations facilities, CEA has at its disposal an important stock of unused powder batches. The idea of reutilizing these powders initiates the question about their efficiency to extinguish a sodium fire after long term storage. This study proposed the physicochemical analyses of these powders in order to identify their compositions and characteristics for different batches. The results highlight the presence of lithium sodium carbonate ($LiNaCO_3$) and trôna ($Na_2CO_3.NaHCO_3.2H_2O$). The aging experiments were developed to study the role of moisture and ambient carbon dioxide during the storage. The results showed that $LiNaCO_3$ is sensitive to the presence of moisture, able to transform it into lithium carbonate and sodium carbonate monohydrate, meanwhile the latter reacts with water and carbon dioxide to form trôna. Finally, a study of the formation of $LiNaCO_3$ allowed the understanding of the fabrication method of the powder.

455945 Elucidating the Physical Stability of Griseofulvin Nanosuspensions Via Combined Use of Surface Charge, Stabilizer Adsorption, and Wettability Measurements

Abstract:

Almost 40% of the new chemical entities in drug pipeline of pharmaceutical companies have low aqueous solubility, which leads to poor bioavailability. Reduction of drug particle size down to nano-scale via wet stirred media milling (WSMM) has been an effective way for increasing the dissolution rate of poorly water-soluble drugs and thus their bioavailability. A major challenge for formulators has been to stabilize milled suspensions as drug nanoparticles are prone to aggregation and size-growth. Polymers and surfactants either alone or in combination have been used to ensure physical stability of the nanosuspensions. Selection of proper stabilizers and their optimum concentration plays a major role in formulating a stable drug nanosuspension. It is generally agreed that inadequate concentration of stabilizers cannot suppress drug nanoparticle aggregation, while their excess may promote Ostwald ripening. In this study, multiple characterization techniques were used to elucidate the impact of presence of an anionic surfactant and polymer concentration–molecular weight (MW) on the stability of drug suspensions. Particles of griseofulvin (GF), which is a poorly water-soluble drug, were wet media-milled with various polymer–surfactant formulations. Two grades of hydroxypropyl cellulose (HPC) with different MWs (i.e. ~40 and 140 kDa) were used as stabilizers from 0–7.5% (w/w) with/without 0.05% (w/w) sodium dodecyl sulfate (SDS). As a novelty of current work, rheometry with a power-law rheological model was used along with laser diffraction (LD) and scanning electron microscopy (SEM) data to assess the aggregation state of the milled drug suspensions. Moreover, complementary characterization methods such as stabilizer adsorption, electrophoresis–zeta potential, and wettability tests were used to explain the physical stability of the suspensions. We found that for a drug nanosuspension stabilized sterically or electrosterically, zeta potential measurements could be misleading, while polymer adsorption and wettability enhancements could better explain the observed physical stability. This study demonstrates the need for multiple characterization

techniques to elucidate the complex and elusive nature of the physical stability of drug nanosuspensions.

469163 The Innovative Use of Micronized Hydroxypropyl Methylcellulose As a Pore Former in Dry Powder Coatings of Ethylcellulose Barrier Membrane Coatings on Multi-Particulates

Abstract:

The Innovative Use of Micronized Hydroxypropyl Methylcellulose as a Pore Former in Dry Powder Coatings of Ethylcellulose Barrier Membrane Coatings on Multi-Particulates

Shawn Engels‡, Nick Grasman †, Jason Folkenroth † and Brian Jensen‡

‡Freund-Vector Corporation, Marion, IA 52302, USA

†The Dow Chemical Company, Midland, MI 48674 USA

Purpose: Ethylcellulose is commonly used throughout the pharmaceutical industry for barrier membrane coatings for sustained drug release. In many formulations that utilize ethylcellulose coatings, soluble pore formers are used to modify and speed up the release of API from the coated material. Hydroxypropyl Methylcellulose (HPMC) is a common pore former in these formulations that is easily blended into traditional solution preparations of ethylcellulose and applied at precise ratios to produce predictable, repeatable drug release. The recent development of a novel ethylcellulose grade for use in dry powder coatings of multiparticulates has offered the pharmaceutical industry a vastly improved method for applying barrier membrane coatings in a safe, fast and efficient process. This study investigated utilizing micronized HPMC as a dry pore former in dry powder coatings of ethylcellulose.

Methods: Acetominophen (APAP) was dry powder coated onto sugar spheres (#20 – 25; Colorcon Inc., USA) using polyvinyl pyrrolidone (K2932, ISP Inc., USA) as binder on a Granurex GXR-35 rotor process (Freund-Vector Corp., USA). A drug to binder ratio of 97:3 w/w was maintained. Two grades of METHOCEL™ (HPMC), E3 and K3 were micronized and blended at ratios of 10%, 15% and 20% with ETHOCEL™ HP(The Dow Chemical Company, USA). These blends were dry powder coated onto the APAP beads with dibutyl sebacate (DBS) as a plasticizer using the same rotor process as above. Another batch was coated with 20% weight gain of ETHOCEL HP alone as a control. Once the APAP sugar spheres were coated to 20% weight gain of ETHOCEL/METHOCEL blends, samples were cured at a 50°C product temperature for 30 and 60 minutes in the GXR-35. Dissolution and SEM were evaluated for curing influence on drug release. USP Apparatus I (baskets) was utilized for in-vitro dissolution studies and drug release data was compared using similarity factor (f_2) analysis.

Results: Initial experiments demonstrate that the addition of the micronized METHOCEL was successful in modifying the release of the APAP from the sugar beads. Initial results showed that the addition METHOCEL E3 resulted in a faster drug release than the addition of the METHOCEL K3. Regardless of the amount of HPMC addition to the ETHOCEL HP coating, the process efficiency, time and yields were unaffected from the baseline process with no addition. No agglomeration was present in any of the trials. Samples have been placed on stability, and 1 month, 3 month and 6 month data will be reported when available.

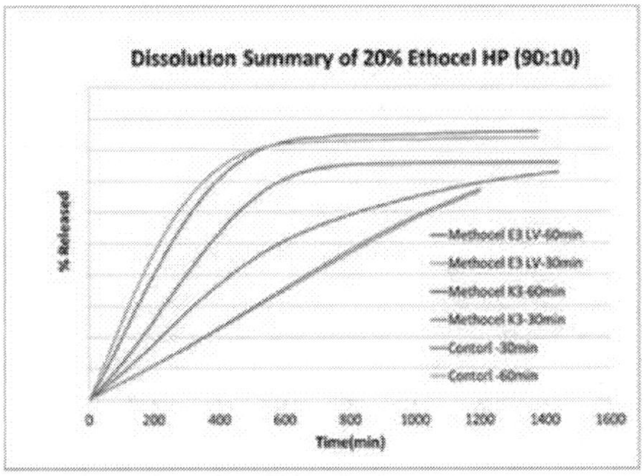

A

B

Figure 1. (A) Influence of the addition of various HPMC grades on APAP drug release and (B) View of Freund-Vector GXR-35 rotor coater with powder feeder.

Conclusion: Using the dry powder layering process on a rotor coater allows for fast, efficient and solvent-free coatings for ethylcellulose. The ability to add micronized HPMC grades to the dry powder ethylcellulose gives formulators freedom to modify the release rates from the coatings with the same flexibility that is available with the solution based coatings that are currently widely in use throughout the industry. Since the addition of the dry HPMC does not affect the coating process, the formulation and drug release can be customized precisely without modifying the overall coating process parameters.

471542 Advanced Characterisation of Pharmaceutical Powders Using Multi-Component Vapour Adsorption

Abstract:

This paper reviews recent developments in the use of DVS for the characterisation of pharmaceutical powders. One of the recent major advances is the development of instruments which can provide multicomponent solute vapour mixtures.

470391 CFD-DEM: Modeling the Small to Understand the Large

Abstract:

Gas-solid flows in industrial devices present a challenge for prediction, scale-up and optimization with numerical simulations. Direct numerical simulation (DNS), which is essentially free of closures, is limited by the computational cost associated with solving for the motion of every particle and the fluid flow around the particles. At present, DNS can reach $O(10^4)$ particles, many orders of magnitude below most industrial problems. Coarse-grain methods, such as filtered two-fluid models (TFMs) and multi-phase particle in cell (MP-PIC) method. Coarse-grained methods are able to solve larger problems by using continuum grid cells or discrete parcels each containing up to ~$O(10^5)$ particles. However, the closures, assumptions and simplifications required by coarse-grain methods may lead to large sources of uncertainty in model predictions. Coupled computational fluid dynamics-discrete element method (CFD-DEM) offers a compromise between computational overhead and sources of uncertainty. Though DEM has seen a recent explosion among academic researchers, as evidenced by the number of relevant papers doubling roughly every four years (a rate of about half of Mooreâ€™s law), a large potential exists to extend the application of DEM to industries. The challenge still lies in the relatively high computational demand for full industrial systems, which commonly contains up to $O(10^{14})$ particles, exceeding the state-of-art of DEM simulations of $O(10^7)$. To guide the future enhancement of DEM solvers toward better utilization in industry, a gauge on current and future industrial perspective on DEM was conducted by a survey of CFD champions working at 34 PSRI member companies. Interesting trends and conclusions drawn from the 18 responses will be presented, in terms of user demographics, expected expenditure on personnel, acceptable time commitment, urgency in improvement of code functionality, among others. Additionally, a few examples of how CFD-DEM can be used in a manner to provide insightful, value-added information will also be presented.

470928 Design and Troubleshooting FCC Operation Using CFD Techniques

Abstract:

Technip Stone & Webster Process Technology uses computational fluid dynamics (CFD) as an FCC design and troubleshooting tool. CFD provides qualitative as well as quantitative information which is used to visualize and understand gas/catalyst flow behavior. Technip is actively using CFD to identify unit mal-performance and evaluate hardware and operational changes required to optimize FCC units. Recently, Technip incorporated proprietary coke combustion kinetics into CFD model to troubleshoot and evaluate hardware changes on FCC regenerator performance.

This paper will highlight key areas of FCC technology where Technip has successfully used CFD modeling techniques in addition to other analytical tools for FCC technology design improvements, and for troubleshooting existing units. Examples of how Technip uses CFD for research, design, troubleshooting and evaluating FCC technology upgrades will be presented.

468899 Hybrid Lagrangian-Eulerian Simulation of Biomass Gasification in a Bubbling Fluidized Bed Reactor

Abstract:

Biomass is a promising fuel resource through gasification to produce synthesis gas. There are several types of gasifier reactors under study. Fluidized Bed Gasifiers are the one that has the potential to scale up to a production capacity more than 15 tons per day. Modeling and simulation of fluidized bed reactors is essential to scale up and optimize the operating condition. Dense Discrete Phase Model (DDPM) is a new computational fluid dynamic (CFD) approach to simulate the hydrodynamic of industrial scale fluidized bed reactors considering particle size distribution in a reasonable computational time. It is a hybrid Lagrangian-Eulerian approach tracking the particles using the Lagrangian approach while interactions between particles are estimated by kinetic theory of granular flows (KTGF).

In this work, the gas and the bed material are modeled as continuum phases whereas biomass particles are simulated as a discrete phase in Lagrangian domain using Dense Discrete Phase Model. Different drag models were used to accurately estimate the mixing of biomass and bed material. The model predictions were compared with experimental results reported for a pseudo two dimensional fluidized bed. The accuracy of motion of the biomass particles in a bubbling condition was evaluated by comparing mixing index and probability of their presence at different points with experimental results. Moreover, the effect of particle size distribution of biomass particles on gasification was studied. Finally, the results showed that the hybrid model had a good prediction of outlet gas composition.

470926 Modeling an Industrial-Scale Fluidized Bed Wurster Coating Process Using CFD-DEM

Abstract:

Industrial fluid-bed coating operations involve the flow of several hundred million beads in a large-scale Wurster-coating device. Experimental measurements available from such large scale operations are limitedâ€"very few beads per sample (10-100) can be analyzed for coating thickness uniformity. In such scenarios, computational fluid dynamics (CFD) based modeling models become important for developing a better understanding of the process and identifying the key parameters that influence coating-thickness uniformity. With recent advances in computational hardware and software, including graphics card (GPU) based processing, simulations of larger industrial-scale systems can be performed.

Previous results have shown that the CFD-DEM (discrete element method) approach is better-suited to obtain particle trajectories compared to the CFD-TFM (two-fluid model) approach. With GPU-enabled codes, ~20 million beads can be realistically simulated. Therefore, exploiting the azimuthal symmetry of the geometry to simulate a smaller 3-D sector (pie-slice) of the full unit, the flow results are shown to be reasonable with this idealization. Single-phase CFD simulations were performed to determine the air flow distribution through the distributor plate, which is subsequently used as the inlet boundary conditions for the CFD-DEM models. The bead-size dispersity is also incorporated in the model.

The influence of controllable process parameters, such as air-flow rates, and material properties can be virtually investigated using these CFD-DEM simulations. The residence time distributions of the beads in the active coating region (inside the Wurster column) and the non-coating regions are compared for the different conditions explored. These residence time distributions, especially inside the Wurster column, provide qualitative and quantitative measures of the bead-coating uniformity at present. In future, these residence time measurements can be extrapolated to predict long-term coating uniformity using compartment-based and/or population-balance models.

458127 Triple Phase Simulation of Bubbling Fluidized Beds of Geldart "a" Particles with a Bubble-Based Drag Model

Abstract:

Triple Phase Simulation of Bubbling Fluidized Beds of Geldart "A" Particles with a Bubble-Based Drag Model

Teng Wang, Caixia Chen*

Key Laboratory of Coal Gasification and Energy Chemical Engineering of Ministry of Education, East China University of Science and Technology, Shanghai 200237, China

Abstract

The Eulerian-Granular two-fluid model (TFM) has been employed in the simulations of the gas-solid hydrodynamics of fluidized beds, its application for Group "A" particles was barely successful unless very fine grid sizes and very small time steps were used in the computations. To make the TFM model work for coarse-grid simulations, Modifications of the interphase drag force are required. In this presentation, we report a triple-phase numerical model for the simulations of gas-solid bubbling fluidized beds with Geldart "A" particles. In the model, the gas-solids suspension was modeled as a primary gas phase presented in the emulsion, a secondary gas phase presented in the bubble, and a third phase consisting of solids. A bubble-based drag model was derived taking into account of the heterogeneous flow structures of the emulsion phase. The KTGF model was used to describe the solid phase stresses. Simulations were conducted for two different scales bubbling fluidized beds of Group "A" particles. Mesh independences were confirmed up to a radial grid size of 200 times particle diameter, and the sensitivity of the transitional results on the time-steps was investigated. The predicted bed expansion and the voidage distribution, axial and radial solids volume fraction profiles, and the solids velocity vectors have been validated against literature data. The bubbly motion of the gas and the gross circulation of solids were simulated. The heterogeneous properties of emulsion phase were captured using mesh size of the order of 50 particle diameters for a 0.19 m diameter bed. For a 0.296 m diameter bed, a mesh size of 200 $d_p$was capable of capturing the time-averaged hydrodynamics without losing accuracy.

Key words: triple-phase simulation; Group A particle; gas-solid bubbling bed; bubble-based drag model; KTGF closure.

Highlight

- A triple-phase model is developed for gas-solid fluidized bed of Group "A" particles.
- A bubble-based drag model is derived and validated with benchmark simulations.
- The bubbly motion of gas and the gross circulation of solids are simulated.
- Grids independence is confirmed up to a radial size of 200 times particle diameter.

*Correspondence Author. Tel: +86 021 64252057.

Email address: cxchen@ecust.edu.cn

465043 Speeding up of Gas-Particle Flow Simulations Using Non-Iterative Time Advancement (NITA) Solver

Abstract:

Kinetic theory based Two-Fluid model (TFM) for gas-particle flow simulations is routinely used from initial conceptual designs to improving the efficiency of the particle handling equipment. There has always been a demand for simulations to be faster and scalable, especially with recent advances in high performance and cloud computing. In this study we demonstrate the considerable speed-up of gas-particle flow simulations using the NITA solver and scalability of TFM over a large number of compute cores. The National Energy Technology Laboratory (NETL) small-scale fluidized bed, bubbling fluidized bed and circulating fluidized bed challenge problems were used to test the speed-up of the simulations using NITA solver. It was observed that the NITA solver both qualitatively and quantitatively captured *distinct* gas-particle flow patterns as predicted by the classical iterative time advancement solver while speeding up the simulations by as much as a factor of six. The scalability of kinetic theory based TFM was tested using NETL circulating fluidized bed problem and the model achieved excellent scalability over a large number of compute cores.

459733 Simulations of Chemical Vapor Deposition of Silicon in a Fluidized Bed Reactor with Detailed Reaction Kinetics

Abstract:

Title: Simulations of Chemical Vapor Deposition of Silicon in a Fluidized Bed Reactor with Detailed Reaction Kinetics

Authors: Caixia Chen[1], Youwen Fan

Organization: Key Laboratory of Coal Gasification and Energy Chemical Engineering of Ministry of Education, East China University of Science and Technology, Shanghai 200237, China

Abstract:

Fluidized bed reactor for silane deposition is an important technology for solar-grade silicon production. A problem that hinders a continuous operation of the FB-CVD reactor is the clogging of reactor internals because of undesired fine dusts production mainly due to homogeneous decomposition. In this presentation, we report a comprehensive numerical model for silicon deposition in fluidized bed reactors with detailed reaction kinetics. A reaction mechanism based on detailed gas phase reaction and a set of semi-empirical surface reactions was validated against published experimental data obtained in a temperature range similar to a FB-CVD reactor using a boundary layer reaction module of CHEMKIN. The results showed Ho's gas and surface mechanisms were suitable for FB-CVD.

The TFM-KTGF two-fluid model coded in Ansys Fluent platform was employed in the simulation of the gas-solid fluid dynamics at a fluidized bed silicon CVD-reactor. A 0.5 m ID cylindrical silicon fluidized bed was simulated with computational grids of 200 times particle size. The Matlab image processing toolbox was used for the post-processing of the simulation data and the bubble size distributions were obtained. The

classical Gidaspow drag force model and a literature reported sub grid scale model (SGS) were compared, and the deviation of the bubble size predicted by using the SGS model is 12.6% comparing with Mori-Wen's equation, which is smaller than the deviation of 21.4% obtained from Gidaspow model.

The reaction model has been coupled with the TFM-KTGF simulation, the preliminary results showed that the predicted silicon deposition rate was in consistence with Hsu's experimental data. Further work will be performed to investigate the effects of operating conditions such as the temperature, pressure, silane/hydrogen ratio on the deposition process. The effects of bubble size distributions on fine dusts formation will be discussed.

Key words: FB-CVD, bubble size distribution, detail reaction mechanism, TFM-KTGF simulation

---

[1] To whom correspondence should be addressed: Tel:86-021-64252057 E-mail address: cxchen@ecust.edu.cn

468735 Reducing Cut Size of Uniflow Cyclones with Optimized Swirl Vane Design

Abstract:

Uniflow cyclones are constructed as cylindrical separators with gas and particles passing through them in only one direction. Advantageous compared to reverse flow cyclones is their compact design and their easy implementation into piping systems [1]. These characteristics minimize manufacturing and installation costs. Approved calculation models for uniflow cyclones, which are valid for a wide range of applications, are missing up to now. MCI intends to develop design fundamentals for this cyclone type on the basis of systematic experimental and numerical studies. Results from these investigations provide useful insights into the operation of such devices and means to identify design criteria. The influence of the swirl vane inlet (SVI) to the separation performance is mainly associated with following effects. The direct relation between small vane angles leading in higher separation efficiencies due to higher swirl intensities [2,3]. A second effect is the flow velocity within the vane channels itself which is accounted by geometry of the SVI. Vane optimizations in this work show high separation efficiencies linked with significant lower pressure drop based on reduced flow velocities within the swirl vane channels [4]. This reduction of energy demand due to reduced flow velocities leads in optimized vane angles to force cyclones cut size to a lower level. All investigations have been carried out at a test stand with a cyclone diameter of 0.1m. The results of the performance tests showed good compliance with additional optical investigations within the separation chamber carried out with high speed camera equipment. Moreover, a critical sedimentation path within the separation chamber could be determined, which is correlating to relevant vane design parameters.

[1] Hoffmann, A.C., Stein, L.E., Gas Cyclones and Swirl Tubes - Principles, Design and Operation. Berlin, Heidelberg: Springer, 1. au. ed., 2002.
[2] Kraxner, M., Empirische Ermittlung von Auslegungskriterien für Gleichstromzyklone in Multizyklonblöcken, phd thesis, TU Munich, 2013.

[3] Kraxner, M., Muschelknautz, U., Wechner, S., Ackermann, S., Greif, V., Bolda, J., Influence of the Inlet Vane Geometry On the Uniflow Cyclones Performance (639f). AIChE 2012 Annual Meeting, 2012.
[4] Kraxner, M., Skarke, B., Kofler, T., Pillei, M., Pressure Drop in Uniflow Cyclones: Investigation on an Empirical Calculation Model. CFB11, 2014.

DEVELOPMENT OF FRAMEWORK FOR UNDESIRED AGGLOMERATION IN AN AGITATED FILTER DRYING PROCESS

Hong Lee Lim, Monash University, Clayton, Victoria, Australia
Karen Hapgood, Monash University, Clayton, Victoria, Australia
Brian Haig, SunPharmaceuticals, Port Fairy, Victoria, Australia
**karen.hapgood@monash.edu*

1.0 Introduction

In the manufacturing of active pharmaceutical ingredients (API), the final products are often isolated as solids through crystallization after a series of chemical transformation steps. The solid phase is then separated by filtration and further purified via a series of cake washing step. The drying of the API is often performed following the final filtration step to remove the solvents from the solid phase to meet the residual solvent specification. In the pharmaceutical industry, the agitated filter dryer (AFD) is often used for the filtration, washing and drying of API to avoid the need to transfer the filter cake between filtration and dryer units. The AFD also provides a well contained environment which minimizes the risks of personnel exposure to toxic API while also prevents product contamination. However, the AFD is widely known to produce undesired agglomerates during the agitated drying of wet powders. Formation of hard, unbreakable lumps as well as product degradation due to residual moisture content are two of the most important quality issues caused by this phenomenon. The entrapped moisture in the core of the agglomerates may also lead to substantial increase in drying time due to mass transfer limitation.

In recent years, there is an increased awareness in understanding the inter-dependent relationship between the structure, properties, processing and performance of the API to achieve the desired drug performance [1]. While the performance of the API is determined by its physical properties and chemical structure, the desired properties is essentially governed by how well the processing step is executed. As a result, this has spurred great interests from various researchers to investigate the impact of the filter dryer operation on the undesired agglomeration of powder. A number of investigation works has since been carried out which leads to the proposal of various agglomeration prevention strategies [2-6]. However, it is understood that the suggested methods can be specific depending on the powder-solvent combination. In addition, much less is known on the impact of some phenomena which are commonly neglected such as condensation of vapour in the equipment and detachment of wall deposits.

A lack of generalization of the suggested mitigation method and poor understanding of the different agglomeration mechanisms that may arise for different powders types has led to the need to develop a comprehensive framework which incorporates all the root causes and mechanisms of undesired agglomerations. The agglomeration framework proposed in this work was developed based on bibliographic research comprising of both experimental and theoretical results from literature (ranging from 2003 to 2016), providing the current state of understanding on the agglomeration issues faced during an agitated drying process.

2.0 Agglomeration Framework

The agitated filter drying process is a multi-step operation (e.g. cake wash, filtration, blowdown, agitated drying etc.) which comprised of a wide range of parameter that has to be carefully controlled. Agglomeration for certain powder types can be avoided if these process parameters are managed and maintained consistently at the correct level, without being offset by any disturbances. Appropriate control of temperature, agitation speed, blowdown period and fill volume are some of the most important parameters that have been studied extensively by various researchers.

2.1 Temperature

Figure 1 depicts the potential relationship between temperature control and agglomeration of powder. Many studies have reached a similar conclusion where agglomeration associated with high drying temperature is caused by the increased dissolution-recrystallization (solid bridging) effect and also the melting of crystals in some case. However, it is known that drying high temperature will also lead to drips induced nucleation and growth of agglomerates due to the condensation of excessive vapour on the cold surfaces (e.g. sight glass and spray bars) in the dryer as the drying rate was enhanced [2]. In addition, overheating due to temperature overshoot at the initial heating stage and friction caused by excessive agitation at low clearance may also contribute to the solid bridging and the formation of lumps even when the drying is conducted at a temperature low enough to avoid dissolution and melting. The initial heating of dryer can be done by ramping the temperature at a predefined rate to reach the setpoint value. By carefully controlling the preheating stage, both the temperature overshoot and generation of excessive vapour at the start of drying (where moisture content of powder bed is at its highest) can be avoided. Condensation can also be avoided by applying a "bleed" where stream of dry gas (e.g. nitrogen or air) is flown across the vessel headspace to remove the saturated solvent vapour from the dryer.

On the other hand, drying at low temperature especially for heat sensitive materials are also susceptible to agglomeration and is often neglected in most cases. Low drying temperature suggests that a prolonged period of agitated drying is required to reach the desired moisture content specification. This in turn led to excessive agitation of powder bed at critical moisture content region (where the tendency of wet powder to agglomerate is at its highest), leading to severe agglomeration. When the drying rate is limited by the heat sensitivity of the material, a change in wash solvent with lower boiling can be considered to speed up the drying process.

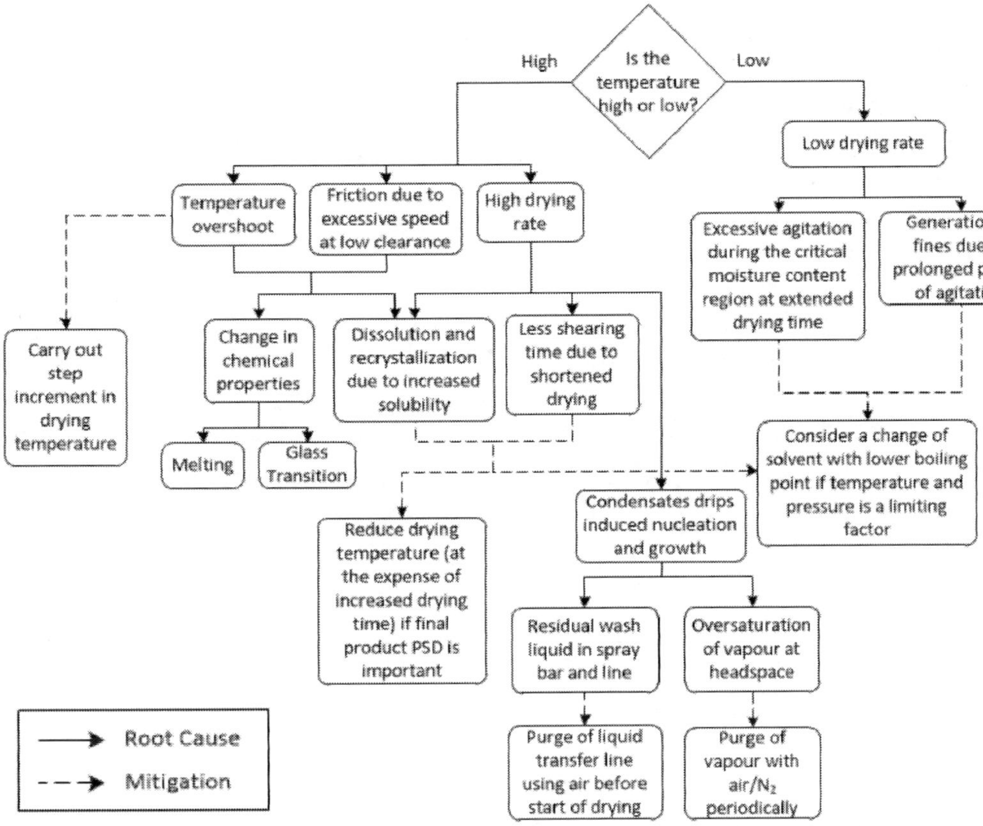

Figure 1: Framework for temperature related agglomeration.

2.2 Agitation Speed

An appropriate choice of agitation speed is also essential in preventing the formation of agglomerates as depicted in Figure 2. High agitation speed often leads to severe compression as well as generation of fines (for fragile crystals) especially when the bed moisture content is high. Excessive generation of fines may contribute to the hardening or cementing of agglomerates where fine particles filled the gaps between the larger primary particles and is further consolidated under high agitation speed. In contrast, low agitation speed tends to produce insufficient shear force for agglomerates breakage purposes while also result in poor mixing. Therefore, consideration must be given to the physical properties of the powder such as compressibility, shape and hardness when selecting the correct impeller speed. It is also worth noting that the control of agitation may not be as simple as selecting a target speed which runs constantly throughout the drying process. Specific drying protocol or regime may be required, depending on the type of powder used. This includes periodic axial raising or lowering of agitator, increase or reduction of agitation speed as well as intermittent stirring to minimize the disruption of bed without hindering its heat transfer capability. It is also important to avoid agitation when the moisture content of the powder bed is high (often at the start of the drying process). The use of intermittent stirring will also prevent the generation of excessive fines to maintain the balance between agglomeration and attrition.

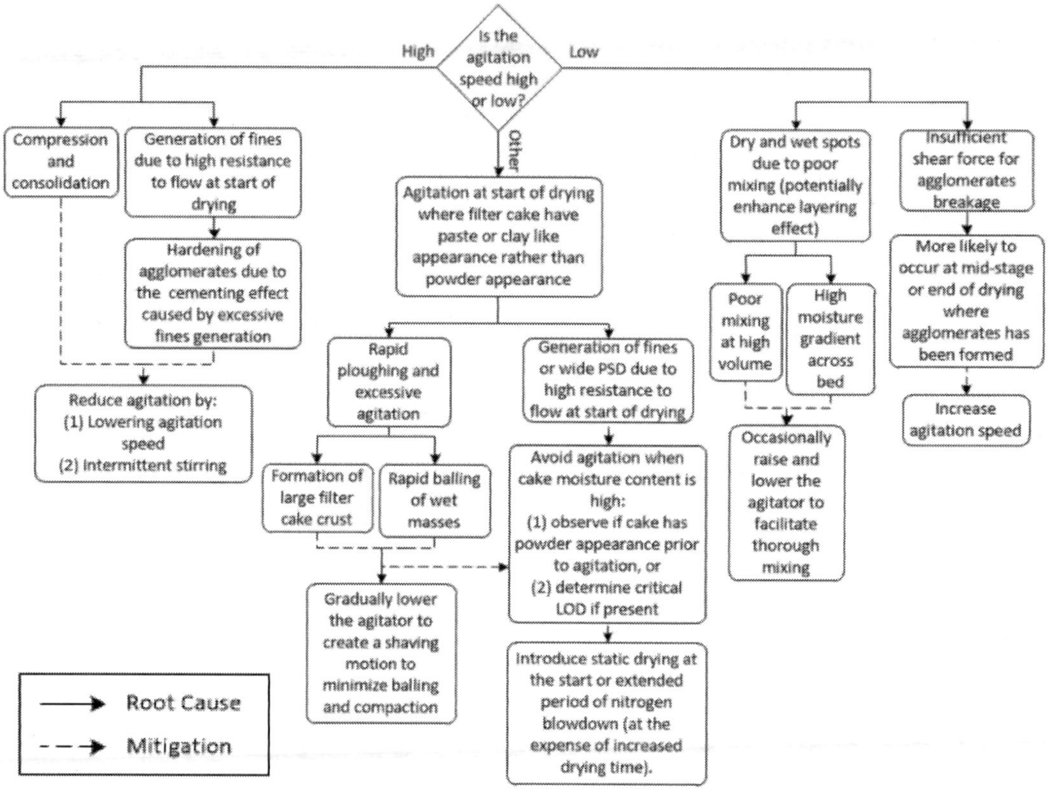

Figure 2: Framework for agitation speed related agglomeration.

2.3 Blowdown Period

The extended blowdown step often serves as an effective way to further reduce the filter cake moisture content prior to the agitated drying step. However, the selection of blowdown pressure can be highly dependent on the properties of the filter cake (see Figure 3). Highly compressible cakes subjected to excessive blowdown period or pressure may increase the cake's resistance to fluid flow (reduction in cake voidage), leading ineffective moisture removal. This in turn produces filter cake with higher initial LOD which is one of the main root causes of agglomeration in an agitated drying process. Excessive cake pasting to removal the filter cake cracking as well as high blowdown pressure may also lead to cake compaction, resulting in the formation of lumps as well. Crystals that are not susceptible to breakage and compression may be subjected to a more severe blowdown condition. If high moisture content persists, an investigation has to be carried out on the crystal structure where its high liquid retention capability may be linked to microscopic properties such as excessive hydrophilic functional group, high surface area and pore network distribution.

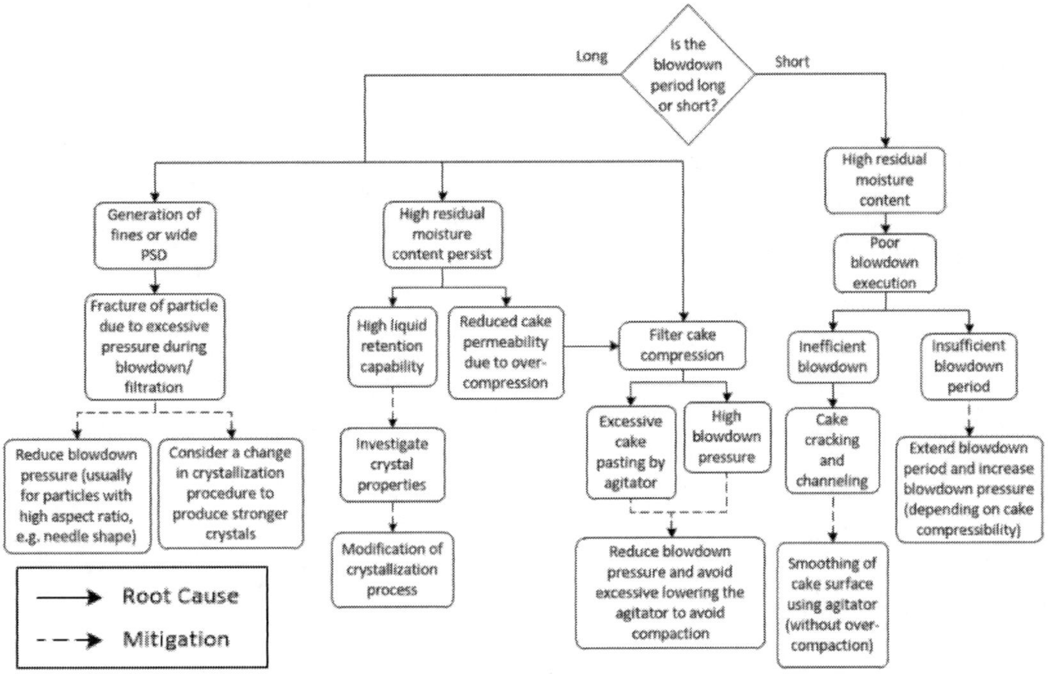

Figure 3: Framework for blowdown related agglomeration.

2.4 Fill Volume

Another aspect of the agitated drying operation which is often overlooked by powder manufacturers is the fill volume of the dryer (see Figure 4). The amount of powder if often dictated the upstream processes such as extraction and other chemical transformation process where certain batches may have a lower yield due to the variability of the raw materials. Agitation of powder bed with low fill volume may produce higher amount of lumps due to the greater collision frequency as well as its higher energy input per unit mass. A lack of hydrostatic pressure to restrict the movement of agglomerates (especially for spherical agglomerates) will also cause the agglomerates to escape the shear impact as it rolls over the impeller. On the other hand, high fill volume may reduce the contact between impeller and particles, giving rise to insufficient power input to the material as well as irregular movement of mass. The "Brazilian nut effect" will also cause the agglomerates to avoid impact of agitation as the agitator lies primarily at the bottom of the bed. It is therefore important to raise and lower the agitator periodically for optimal mixing and drying while ensuring that the agglomerates are exposed to impeller shearing for breakage. In addition, the agitator may also be positioned at various height for a set period of time to ensure powders at different sections of the dryer are subjected to sufficient agitation.

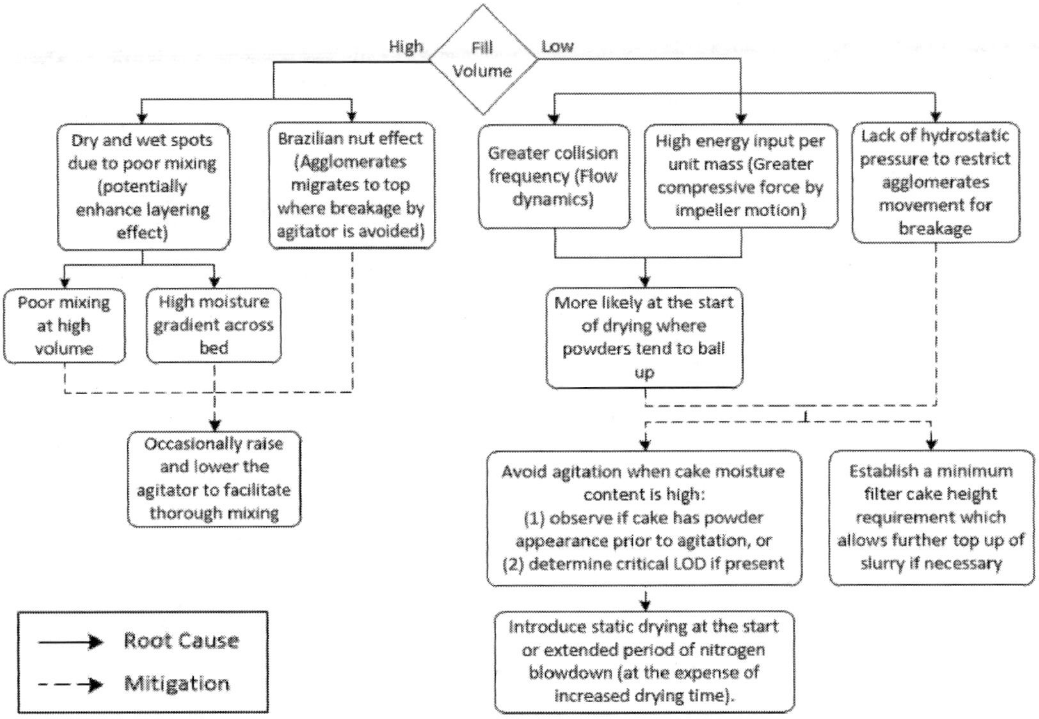

Figure 4: Framework for fill volume related agglomeration.

2.5 Filter Cake Washing and Solvent Composition

The elimination of agglomeration using different solvent combinations has gained significant attention in recent years (see Figure 5). Solvents with high particle solubility profile will promote solid bridging between particles due to the recrystallization of dissolved solids. In addition, residual solvents with high surface tension such as water may encourage the formation of liquid bridging due to their extensive hydrogen bonding capability. It was reported that a secondary solvent type can be added to reduce both the overall surface tension and solubility which was proven to be effective in minimizing the formation of agglomerates for certain powder-solvent combination. The use of multi-component solvent may also promote the removal of multiple impurities during the cake washing step. However, the choice of solvent combination has to be carefully selected based on their compatibilities in boiling point. Wash liquids comprised of solvents with varying boiling point will lead to enrichment of the less volatile solvent. Solvent properties may be reverted to its undesirable state as the concentration of the more volatile solvent decreases during drying, especially if the less volatile solvent has a higher solubility or surface tension. To prevent solvent enrichment, an additional wash can be carried out to isolate as single solvent after the main cake washing procedure has been completed.

While the contribution of dissolution for crystals with low solubility may be negligible towards solid bridge formation, it is speculated that the number of wash cycles and amount of wash solvent used may change the physical properties of the particles that will subsequently affect its downstream processing behaviour. Crystals that undergo several wash cycles may exhibit a more irregular faces with eroded edges suggesting that washing may not only remove the target impurities, but also dissolves the produced crystals. The potential impact of eroded crystal surface on agglomeration is not fully understood but the presence irregular surface due to erosion may increase the specific surface area of the crystals, allowing the crystals to retain a greater amount of liquid on their surfaces while also increases its particle interlocking tendency (or reduced flowability).

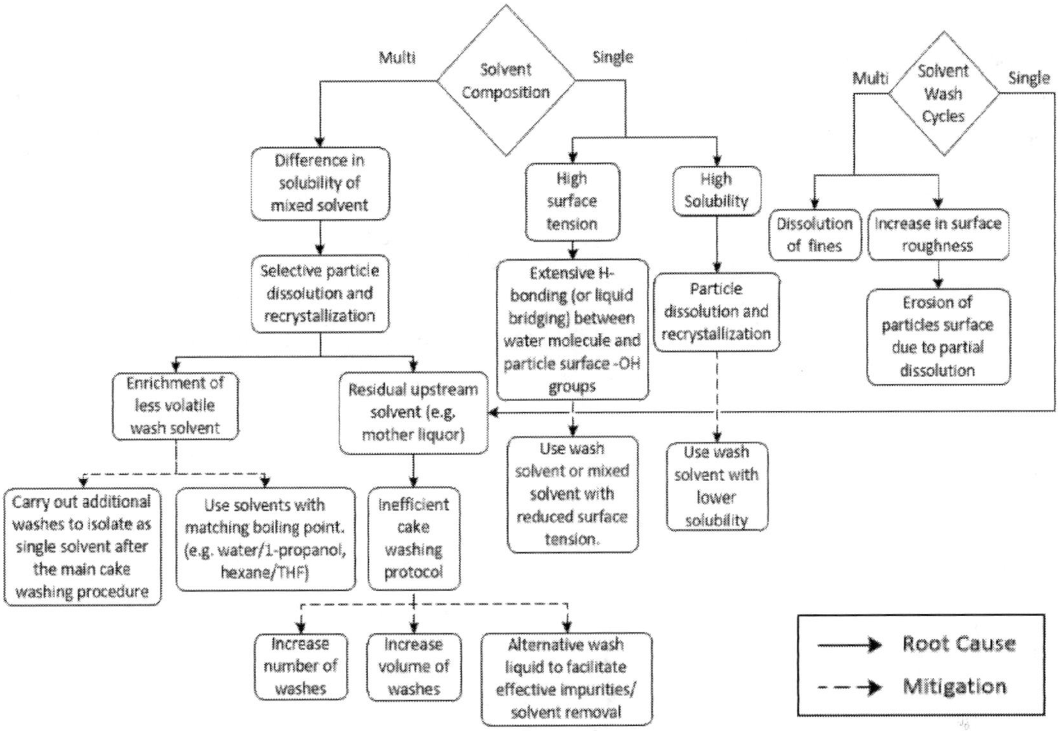

Figure 5: Framework for filter cake washing related agglomeration.

3.0 Future Direction

The framework has outlined the potential agglomeration root causes that may arise for different powder types. While this may serve as a useful tool to overcome any existing agglomeration issues faced by the manufacturers, a guide to identify the early symptoms of the various agglomerations during the initial product development stage is still lacking. Further work is required to incorporate different characterization methods or small scale testings into the framework which will serve as a comprehensive guide to identify the potential agglomeration root causes during the early development stage. Early detection of potential agglomeration prior to up-scaling allows the agglomeration issues to be addressed in the lab scale where optimization of the process condition can be carried out to ensure a smooth and successful transition to large scale production. This also prevents the over-reliance on operators to manually adjust the agitator to break any visible lumps formed during the process which can cause significant delay in production cycle.

4.0 Conclusions

This work reports the preliminary stage of ongoing studies in developing a compiled mechanism framework that will serve as a guideline in troubleshooting the agglomeration during an agitated filter drying process. While this work may not cover entirely the potential causes of powder agglomeration in the pharmaceutical industry, it is understood that this framework will provide a better understanding in controlling the occurrence of agglomeration contributed by some of the common root causes that has been identified to date. This work is the subject of further ongoing research by the authors and will be further revised as more data become publicly available.

Acknowledgement

The authors would like to acknowledge GlaxoSmithKline and Monash University for providing financial support and resources for this study. Note that since September 2015, the Port Fairy facility where this study was conducted is now owned and operated by SunPharmaceuticals.

References

1. Sun, Calvin C. (2009), "Materials science tetrahedron—A useful tool for pharmaceutical research and development," in *Journal of Pharmaceutical sciences*, 98(5), pp. 1671-1687.
2. Lim, H.L., Karen P. Hapgood, and Brian Haig (2016), "Understanding and preventing agglomeration in a filter drying process," in *Powder Technology*, 300, pp. 146-156.
3. am Ende, D., Birch, Melissa Birch, Steven J. Brenek, and Mark T. Maloney (2013), "Development and application of laboratory tools to predict particle properties upon scale-up in agitated filter-dryers," in *Organic Process Research & Development*, 17(10), pp. 1345-1358.
4. Birch, M. and Ivan Marziano (2013), "Understanding and Avoidance of Agglomeration During Drying Processes: A Case Study," in *Organic Process Research & Development*, 17(10), pp. 1359-1366.
5. Kim, Soojin, Bruce Lotz, Mark Lindrud, Kevin Girard, Terence Moore, Karthi Nagarajan, Mario Alvarez et al. (2005), "Control of the particle properties of a drug substance by crystallization engineering and the effect on drug product formulation," in *Organic process research & development*, 9(6), pp. 894-901.
6. Kougoulos, E., C.E. Chadwick, and M.D. Ticehurst (2011). "Impact of agitated drying on the powder properties of an active pharmaceutical ingredient," in *Powder technology*, 210(3), pp. 308-314.

472581 Modeling of the Pressure Drop Across Polydisperse Packed Beds in Depth Filtration

Abstract:

Depth filtration is the process of separation that involves flow of solid-liquid (or liquid-liquid) mixture through a fixed (packed) bed of porous (or particulate) materials. The flow through such a device is commonly modeled using the so-called Kozeny-Carman equation in order to predict the pressure drop during the filtration process. In this work we assess the predictive power of this approach for systems operated at low Reynolds numbers and comparably high pressure drops (on the order of several bar). We find substantial agreement between model and experiment only for systems that result in well-ordered particle packings (i.e., those that have a tight distribution of void sizes). Dramatic disagreement is observed for particle beds that exhibit wide void size distributions. The cake structure is primarily influenced by the size ratio of the particles that compose the cake; specifically, particles with a size ratio where Rsmall/Rlarge is larger than 0.5 do not typically form an ordered pore structure. We propose a modified approach, based on a bimodal void distribution, by introducing two factors: the fraction of expanded voids (kappa) and the ratio of void sizes (beta). Discrete Element Method (DEM) simulations of the packing of poly-disperse spheres are used to analyze the bed structure for different size ratios of binary mixtures. Based on the simulation results, void size distributions of the simulated beds can be extracted by means of a radical Delaunay tessellation. The void structure is quantified in terms of probability density functions of pore and constriction sizes. By fitting the simulated void size distributions to a bimodal (two normal) distribution, the factors kappa and beta can be calculated based on different mean void sizes and probability density. The predicted flow dynamics from the modified equation with factors extracted from the simulation results are found to be much more similar to the experimental flow rates than those calculated using the unmodified Kozeny-Carman equation.

465585 Measurement of the Axial Dispersion of Cohesive Powders in Rotary Kilns

Abstract:

With applications in a wide range of solids manufacturing processes including blending, drying, and disposing of waste, the rotary kiln has been established as an essential device in chemical and metallurgical industries. In the preparation of catalysts in particular, a better scientific understanding of rotary kilns will improve continuous calcination processes for increased throughput with improved product quality. While widely used in large-scale catalyst manufacturing, fundamental understanding of rotary calcination is absent from scientific literature. Thus, the goal of this research is to improve fundamental understanding of rotary calcination processes to aid in scale-up from laboratory and pilot plant scales to the manufacturing scale. For successful calcination to occur, the residence time of the particles must exceed the time required for heating and subsequent treatment. The optimal residence time therefore depends on both of these competing time scales, which are functions of feed material properties, kiln geometry and kiln operating conditions. For uniform treatment of the feed, the particles must also exhibit low axial dispersion. In this work, the residence time distribution and axial dispersion coefficient for two cohesive fluid catalytic cracking powders were measured in pilot plant kilns using a tracer study developed by Danckwerts. Results were successfully matched to the Taylor fit of the axial dispersion model and the Sullivan prediction for mean residence time. It was found that an increase in feed rate, kiln incline and rotary speed decreased mean residence time and overall dispersion. The axial dispersion coefficient was found to vary with kiln conditions. Such results have not been previously reported for cohesive powders.

464496 Residence Time Distribution of Fine to Coarse Particles in Rotary Kilns

Abstract:

Rotary kilns are gas-solid reactors commonly used in industry to achieve a wide array of material processing operations. Rotary kilns are used for applications such as reduction of oxide ore, pyrolysis of hazardous waste, calcining of petroleum coke, conversion of uranium fluoride into uranium dioxide for the manufacture of nuclear fuel, and so on. When operated at atmospheric pressure, they consist of a cylindrical shell that can be inclined, into which the solid charge is fed continuously at one end and discharged at the other. They can be equipped with lifting flights or lifters, and/or an exit dam at the kiln outlet end. They usually require very little labor to operate.

Though the operational cost of these units is usually high, their design is often conservative due to the lack of fundamental physical understanding of both solid flow and heat transfer. The objective of this presentation is to provide a new model to predict the mean residence time, axial dispersion coefficient and hold-up of fine to coarse solid particles within the rotary kiln based on a dimensional analysis. Flow of material through a rotary kiln depends on many factors: length and diameter of the kiln, design and number of lifters distributed around the circumference, rotational speed and slope of the kiln, exit dam height at the kiln outlet end, flow rate and physical properties of the material.

Experiments were carried out on two pilot-scale rotary kilns at room temperature, whether equipped or not with different kinds of lifters or fitted with a dam at the outlet end. These experiments aimed at determining the effects of most of the factors listed above on the Residence Time Distribution of solid particles. The tracer impulse-response technique was used to establish residence time distribution curves. Four granular solids having different properties were used: fine and medium sand, broken rice and beech chips. The other operational parameters were also varied, maintaining the cascading (tumbling) motion.

We have already presented* the results on the smaller kiln with two types of particles. In the present study, we will show that the correlation established with these earlier results had a good predictive ability over the larger kiln with coarser particles. New results with fine particles will also be presented. These new experiments will be used for the consolidation and validation of the previous model.

* Alex Stéphane Bongo Njeng, Stéphane Vitu, Marc Clausse, Jean-Louis Dirion and Marie Debacq « Modeling of Mean Residence Time of Solid Particles in Rotary Kilns » 14AIChE annual meeting, Atlanta, November 2014 [paper 374307]

466810 Evaluation of Small-Scale Powder Flow Characterization Tests

Abstract:

In pharmaceutical development, powder flow characteristics are notoriously difficult to predict. Although tests exist that aim to determine the flowability of powders, many have not been definitively correlated with a powder's performance in downstream processes. Part of the challenge is that powder flow must occur over a broad range of consolidating stresses, including low stress flow regimes such as the filling of a die cavity in a tablet press and high stress flow regimes such as the deformation associated with tablet compression. The objective of this work was to characterize several placebo formulations utilizing a variety of standardized and non-standardized experiments and contrasting those results with downstream tablet compression performance. Angle of repose, bulk and tap density, Hausner ratio, Carr's compressibility index, a visual powder flow assessment, a mass flow rate test, flodex, and powder shear testing were evaluated on several different blends of placebo powders to determine a predictive test for downstream process performance. The placebo powders consisted of a standard base unit formula, mixed with various levels of four different "pseudo-Active Pharmaceutical Ingredients (API's)" (i.e. excipients that mimic poorly flowing Active Pharmaceutical Ingredients). All formulations were evaluated by the individual flow tests and then compressed on a rotary tablet press. Formulations were deemed failures if the relative standard deviation of tablet weight was > 2%, material stuck to punches (low lubrication), or exhibited ejection forces high enough to trigger automatic shutdown of tablet press. The 2% RSD limit is based on internally developed best practices for transferring tablet compression processes from development into manufacturing. The results of the tablet compression process were then compared to the results of the flowability tests to detect any correlations. The Hausner ratio and powder shear test results discriminated marginally between successful and failed batches and were highly correlated with each other.

456386 Assessment and Predictive Modeling of Pharmaceutical Powder Flow Behavior in Small-Scale Hoppers

Abstract:

The propensity of powder to bridge in hoppers and bins is important for a wide variety of pharmaceutical processes, yet it remains difficult to predict large-scale performance from small-scale powder measurements. Despite the availability of various powder flow testing techniques, the prediction of powder bridging has remained elusive. To address this issue, powder flow performance was characterized for several pharmaceutical powders out of small-scale conical hoppers under various conditions. Specifically, the impacts of powder properties, hopper design, hopper fill level and vibration prior to discharge were characterized. The hopper flow regimes were categorized as mass flow, funnel flow, rat-holing or bridging. A Projection to Latent Spaces (PLS, a.k.a partial least squares regression) model was developed to predict flow behavior from easily measured powder properties such as the Hausner ratio and measured hopper dimensions. The model correctly predicted flow initiation for new materials for ~90% of tests. The variables of Hausner ratio, vibration prior to discharge, and hopper orifice diameter were seen to be most predictive of flow. Surprisingly, powder shear test data such as flow function and unconfined yield strength along with other flow properties were not predictive of hopper performance for the conditions studied. A useful metric, *tan(Ï•)/D*, was developed utilizing the hopper semi-apex angle (*Ï•*) and orifice diameter (*D*), and was considered as a first-pass scale-up model.

461681 Application of PAT-Based Feedback Control for the Crystallization of Pharmaceuticals in Porous Media

Abstract:

The development of new drug delivery methods to tailor the desired drug release have been of major interest such as substrates, polymer films and microspheres. The drug release can be affected by various attributes of the polymer properties such as molecular weight, size and size distribution, porosity, morphology and make-up. However, the purpose of this work is to evaluate the benefits of controlling the bulk crystallization in the presence of alginate porous particles. The main objective is to increase the drug loading (DL) in the alginate beads and minimizing the heterogeneous nucleation occurring in the bulk system. Two types of controlled cooling profiles and the model-free Direct Nucleation Control (DNC) strategy were evaluated. A mean DL from 49 to 69 % was achieved through the various scenarios. The application of DNC increased or maintained the mean DL achieved compared to the cooling only scenarios, while a decrease of about 75 % was observed in the batch-to-batch variability. SEM images showed that crystals of different crystal size distribution (CSD) were obtained for the various cooling scenarios, which can be related to the different operating trajectories in the phase diagram achieved during the feedback control approaches. The possibility of manipulating the CSD of particles within the polymer beads can be used to tailor dissolution profile and bioavailability. The work demonstrates the benefits of using feedback control approaches to increase DL, decrease batch-to-batch variability and control dissolution behavior of drug-loaded polymer microspheres.

470315 Optimization of the Bottom-up Precipitation of Nano-Particles of an Ultra-Fast Crystallizer through the Qualitative Characterization of API-Polymer Interactions

Abstract:

Formulations of poorly soluble drugs, seeking to improve bio-availability by API size reduction into the nano-range, typically rely on top-down comminution techniques, e.g. media-milling. While effective, media-milling is costly, time and energy-intensive as well as presenting some isolation challenges. Alternative methods, aiming to access the nano-range through control of the precipitation process, usually rely on very high-intensity mixing in specialty devices like impinging jets and multi-inlet vortex mixers.

For a special class of compounds, crystal nucleation and growth are so fast that even very rapid solvent-antisolvent mixing is insufficient to limit the final crystal size to the nano range.

This work explores the use of polymeric additives to quench the crystal growth of a very fast crystallizer in an attempt at bottom-up nano-crystal precipitation. Lack of first principles understanding of the exact process, through which polymers suppress crystal growth typically leads to high-throughput screening methods for the identification of the optimal additive. However, the large number of potential polymeric candidates, coupled with the parameters controlling the complex high-shear precipitation process lead to a design space too large to explore through brute-force techniques.

Instead, chemometric analysis of the API-polymer interactions was correlated to the performance during precipitation, decoupling polymer optimization from the precipitation process. Identification of specific functional group interactions between the polymer and API also sheds some light on the nature of the crystal growth suppression effect. The combination of high-shear solvent-antisolvent mixing in a rotor-stator wet-mill and polymer additive effect enabled the scalable bottom-up production of nano-particles.

469039 Core-Shell Nanocomposites for Fast Release of Hydrophobic Active Pharmaceutical Ingredients: Impact of Carrier Size

Abstract:

The bioavailability of a large percentage (estimated to be between 40% and 70%) of newly discovered active pharmaceutical ingredients (APIs) is limited by their poor water solubility–hydrophobicity and slow dissolution rate. Over the last decade, major research efforts have been focused on the development of API nanocrystals in suspension form and subsequently converting it into dry core-shell nanocomposite microparticles (NCMPs) to overcome the aforementioned challenges. Preparing NCMPs via fluidized bed (FB) coating of the API nanosuspensions onto inert excipient particles has been used in the manufacturing of nanoformulation based solid dosages. Recent work shows that if the suspensions are properly stabilized, core-shell NCMPs having a water-soluble core can enhance the dissolution rate of poorly water soluble APIs signficantly[1,2]. However, the impact of the carrier particle size has not been investigated. In this study, formation of core-shell nanocomposites of Fenofibrate and Itraconazole, model poorly water soluble APIs, via FB coating of their well-stabilized high drug loaded nanosuspensions is investigated. Specifically, the extent of dissolution enhancement, when fine carrier particles (sub-50 mm) as opposed to the traditional large carrier particles (>300 μm) are used, is examined. This allows testing the hypothesis that greatly increased carrier surface area and corresponding thinner shell at the same API loading for finer carriers can significantly increase the dissolution rate. Fine sub-50 μm lactose (GranuLac® 200) carrier particles were made fluidizable via dry coating with nano-silica, enabling decreased cohesion, fluidization and nanosuspension coating. For both APIs, 30% loaded suspensions were wet-stirred media milled along with a polymer and a surfactant as stabilizers. The stabilizer amounts were screened to achieve stable suspensions with small particle size. The suspensions were FB coated onto hydrophilic nano-silica (M5P) dry coated sub-50 micron lactose (GranuLac® 200) carrier particles or larger carrier particles of median size >300 μm (PrismaLac®40). The resulting finer composite powders (sub-100 mm) based on GranuLac®200 were freely flowing, had high bulk density, and exhibited much faster, immediate release of the poorly water-soluble drugs, in particular for Itraconazole. This is attributed to a much higher specific surface area of the carrier leading to corresponding thinner coating layer for fine carriers as opposed to those for large carrier particles.

Keywords: Active Pharmaceutical Ingredients/drugs, Poorly Water Soluble, Wet-Stirred Media Milling, Fluidized Bed Coating, Fine and large Carrier Particles, Fast Release/Dissolving Nanocomposite Powders

References:

1. C. Knieke, **M. Azad**, D. To, E. Bilgili, R. Dave, Sub-100 micron Fast Dissolving Nanocomposite Drug Powders, **Powder Technol.** 271 (2015) 49–60.
2. A. Bhakay, **M. Azad**, E. Bilgili, R. Dave, Redispersible Fast Dissolving Nanocomposite Microparticles of Poorly Water-Soluble Drugs, **Int. J. Pharm. Sci.** 461(2014) 367–379.

469735 *'designer'* Hydrogel Particles for Controlled Crystallization and Release of Hydrophobic Active Pharmaceutical Ingredients

Abstract:

Nearly 40% or more of drugs currently in the pipeline of the pharmaceutical companies are poorly water soluble, which introduces a significant hurdle for formulating such major class of drugs with an effective dosage. Utilization of innovative biocompatible materials in formulations of hydrophobic drugs to enhance their solubility-limited oral bioavailability and absorption is a rapidly growing area in pharmaceutical materials research. In this work, we develop a novel bottom-up approach for producing and formulating *nanocrystals* of poorly water-soluble active pharmaceutical ingredients (APIs) using core-shell composite hydrogel beads. Organic phase nanoemulsion droplets stabilized by polyvinyl alcohol (PVA) and containing a model hydrophobic API (fenofibrate) are embedded in the alginate hydrogel matrix and subsequently act as crystallization reactors. Controlled evaporation of this composite material produces core-shell structured alginate-PVA hydrogels with drug nanocrystals (500 - 650 nm) embedded within the porous confinement environment of the core. Adjustable loading of API nanocrystals up to 83% by weight is achieved with improved dissolution behavior. We also develop a quantitative model and experimentally validate that the drug release patterns of the fenofibrate nanocrystals can be modulated by controlling the thickness of the PVA shell and drug loading. In summary, our composite hydrogel enabled encapsulation system allows for controlled crystallization, adjustable drug loading, and customizable dissolution profiles ranging from fast to delayed release by utilizing the unique microstructures of coreâ€“shell hydrogels, thus provides a versatile oral delivery system.

Dissolution of Carbamazepine Crystallised directly onto Excipients

Vivek Verma, Clare M Crowley, Peter Davern, Sarah Hudson & B. Kieran Hodnett
Synthesis and Solid State Pharmaceutical Centre, Department of Chemical and Environmental Sciences, Materials and Surface Science Institute, University of Limerick, Ireland
Email ID: vivek.verma@ul.ie

For BCS Class II drugs, such as carbamazepine (CBMZ), *in vivo* bioavailability is significantly affected by dissolution rate [1]. Conventional approaches to improving the latter have relied on mechanical crystal size reduction techniques, such as attrition milling, but this can lead to degradation of active pharmaceutical ingredients (APIs), non-uniform crystal size distributions and the incorporation of impurities [2]. To circumvent these issues various 'bottom-up' approaches (*e.g.*, spray drying, antisolvent precipitation, sonoprecipitation, *etc.*) to crystal size reduction have been investigated [3].

A potential alternative approach to reducing crystal size is the heterogeneous nucleation of APIs in the presence of dispersed excipient particles. Chadwick *et al* [4] showed an increase in the nucleation rate of acetaminophen (AAP) in the presence of α-Lactose Monohydrate (α-LMH) and β-D-Mannitol (β-D-Man). Increasing nucleation rates should concomitantly reduce the crystal size distribution of the APIs and may, therefore, obviate the need to mill API batches. In this study CBMZ was selected as the model API and the influence of the process conditions (*i.e.,*time and supersaturation) and dispersed, insoluble excipients on its crystallisation from metastable CBMZ-methanol solutions and its subsequent dissolution was examined. Unlike the AAP system studied by Chadwick et al, the presence of excipients in a range of supersaturation and crystallisation time had no impact on the final crystal size distribution (CSD) of CBMZ.

PXRD and Raman spectroscopy of the isolated solids fractions confirmed the formation of CBMZ FIII from metastable CBMZ-methanol solutions (supersaturations (S) of 1.22, 1.34 and 1.55) in the absence and presence of excipients (α/β-Lactose (α/β-Lac), β-D-Mannitol (β-D-Man), microcrystalline cellulose (MCC) and carboxymethyl cellulose (CMC)). The crystal size distribution (CSD) of the FIII crystals, assessed by image analysis of the SEM micrographs, showed similar CSD in the presence and absence of dispersed excipient (Table 1 and Figure 1). However, the dissolution rate of CBMZ FIII was enhanced 2-fold upon crystallisation in the presence of dispersed excipients as compared to CBMZ FIII recrystallised in the absence of excipients under the same conditions (Figure 2). The dissolution rate of CBMZ FIII crystallised in the presence of dispersed excipients is higher than the physical mix of CBMZ FIII and excipients, mostly evident with α/β-Lac and β-D-Man (Table 2) reflecting the superiority of the process. SEM-Raman indicates the presence of low concentration of CBMZ on the surface of excipient facilitating faster dissolution rate at initial times. Figure 2 also presents the comparitive dissolution of CBMZ FIII crystallised in the presence of dispersed β-D-Man and the physicsal mix of CBMZ FIII and β-D-Man indicating the improved dissolution rate throughout the dissolution profile.

Table 1: Mean crystal size distribution (CSD) (µm) of CBMZ FIII crystals formed in the presence and absence of excipients

S	Excipient				No Excipient
	α/β-Lac	β-D-Man	MCC	CMC	
1.22	16±11	13±14	17±11	13±7	22±10
1.34	23±14	13±7	14±7	32±22	23±10
1.55	17±9	18±10	19±10	20±10	27±12

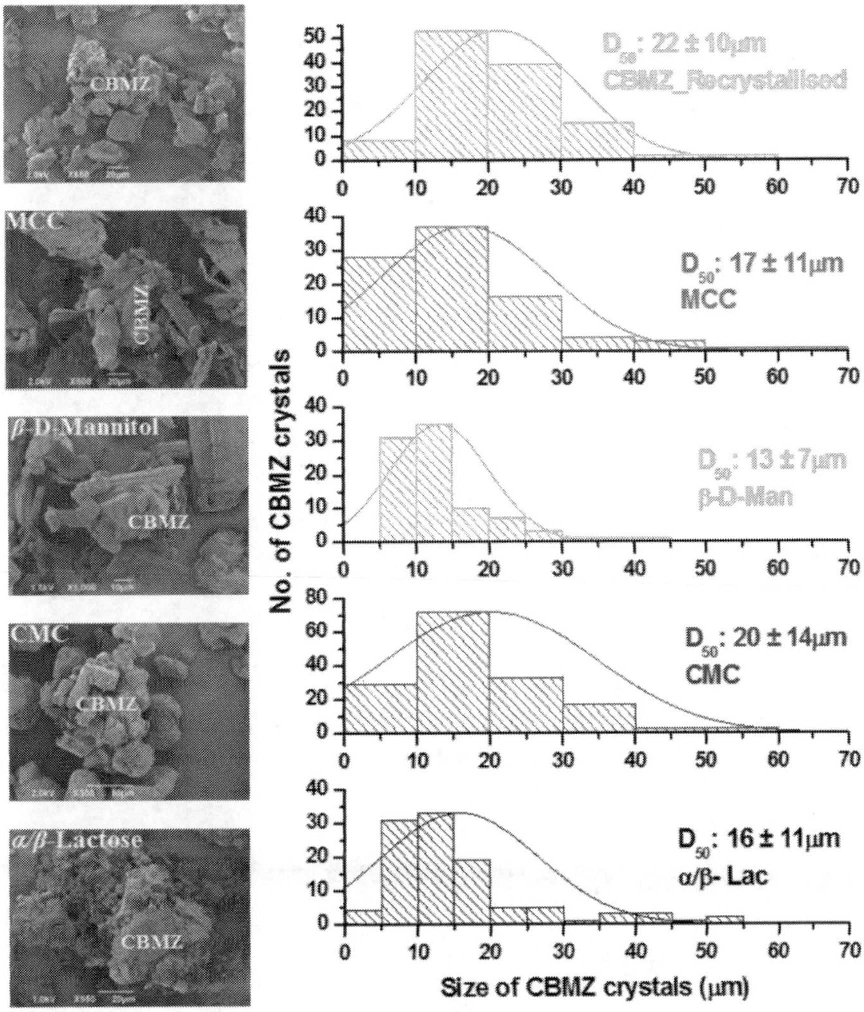

Figure 1: SEM micrographs and CSDs of CBMZ FIII crystallised in the absence and presence of excipients from metastable CBMZ-MeOH solution (t=3 hr aged samples; S = 1.22)

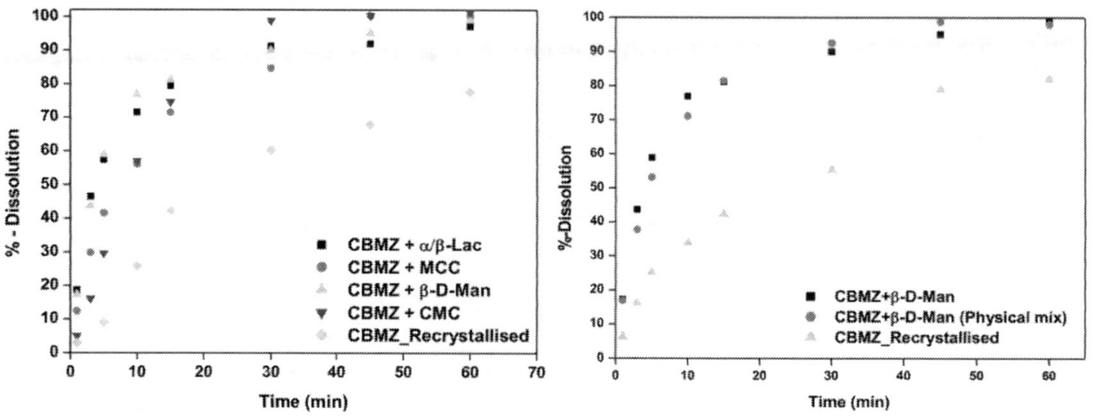

Figure 2: (Left) Comparison of the dissolution rates of CBMZ FIII crystallised in the absence and presence of excipients from metastable CBMZ-MeOH. (Right) Comparision of the dissolution rates of CBMZ crystallised in presence of β-D-Man and the physical mix of CBMZ FIII and β-D-Man; (S=1.22, t=3hr)) (dissolution media: PBS, pH = 7.4; sink conditions ($[CBMZ]_{max}$= 40 mg/L))

Table 2: Dissolution rate of CBMZ FIII over 5 min in the presence and absence of excipients

Excipients	**S**	**Dissolution rate over 5 min**
None	Recry	3.47
None	Recry	3.47
None	Recry	3.47
None	Recry	3.47
CMC	1.22	5.75
CMC	Phymix	6.01
CMC	1.55	6.12
CMC	1.34	6.45
MCC	1.34	8.36
MCC	1.22	8.85
α/β-Lac	Phymix	9.84
MCC	Phymix	10.79
β-D-Man	Phymix	11.31
β-D-Man	1.34	11.97
MCC	1.55	12.2
α/β-Lac	1.34	12.3
β-D-Man	1.22	12.65
α/β-Lac	1.22	12.72
α/β-Lac	1.55	13.62
β-D-Man	1.55	15.74

Acknowledgements

This work was funded by Science Foundation Ireland under Grant 12/RC/2275. This work was also conducted under the framework of the Irish Government's Programme for Research in Third Level Institutions Cycle 5, with the assistance of the European Regional Development fund.

References

1. Lindenberg M., S. Kopp and J.B. Dressman, European Journal of Pharmaceutics and Biopharmaceutics, 2004, **58**(2): p. 265-278.
2. Horn D. and J. Rieger, Angewandte Chemie International Edition, 2001, **40**(23): p. 4330-4361.
3. Poornachary S. L. et al., Crystal Growth and Design, 2016, **16**(2):p. 749-758.
4. Chadwick, K.; Chen, J.; Myerson, A. S.; Trout, B. L., Toward the Rational Design of Crystalline Surfaces for Heteroepitaxy: Role of Molecular Functionality. *Crystal Growth & Design* 2012, **12** (3), 1159-1166.

*** Formatting Issue - Best Available Paper/Graphic ***

467271 Modeling Spray Dried Dispersion: From Droplet to Particle

Abstract:

Modeling Spray Dried Dispersion: From Droplet to Particle

Jaime L. Curtis-Fisk[1], Priti Jain[2], Jodi Mecca[1], Bart Rijksen, [3] Shrikant Khot[1], William W. Porter III[1]
[1]The Dow Chemical Company, Midland, MI USA

[2]Dow Chemical International Private Limited, Mumbai, India

[3] Dow Benelux BV, Terneuzen, Netherlands

Purpose:

Poorly soluble APIs are often formulated as a solid amorphous dispersion in combination with an excipient polymer to enhance their solubility. A common route to producing the dispersion is through spray drying. This process involves spraying a solution of polymer, active pharmaceutical ingredient (API) and solvent into a heated drying gas where the liquid spray droplets dry into solid particles of the amorphous dispersion. Hydroxypropyl methylcellulose acetate succinate (HPMCAS), such as AFFINISOL™ HPMCAS, is one polymer shown to significantly enhance API solubility when in an amorphous dispersions. The purpose of this work is to develop a droplet drying model connecting the fundamental properties of HPMCAS polymer solutions to the spray drying particle formation kinetics and the resulting particles' morphology and compositional uniformity. With this modeling capability, it may be possible to predict and gain insight into the observed properties of amorphous spray dried dispersions.

Approach and Methods:

A particle formation model was developed considering the balance between solvent evaporation and solute diffusion to predict particle morphology and composition. The model, as illustrated in Figure 1, considers the diffusion of API and excipient towards the core of the droplet and surface recession of the solvent as the two main counteracting mechanisms in the drying of droplets. Methods for predicting particle formation are known, but do not fully address the complexities of producing spray dried dispersions from a solution containing API and polymer.[i] This model framework builds upon these references and addresses the non-ideality of the liquid phase due to the use of polymeric excipients, including the interaction between solutes and the change in evaporation rate due to non-ideal vapour liquid equilibrium. This added complexity will provide a more accurate route to predicting the impact of formulation and spray drying conditions on particle morphology.

To provide inputs to the model, fundamental properties including viscosity and diffusivity of HPMCAS-API solutions were characterized in a wide range of solvents and solvent blends. Additionally methods to characterize HPMCAS-API interactions were used to understand potential effects on mass transport within the solution droplet.

*** Formatting Issue - Best Available Paper/Graphic ***

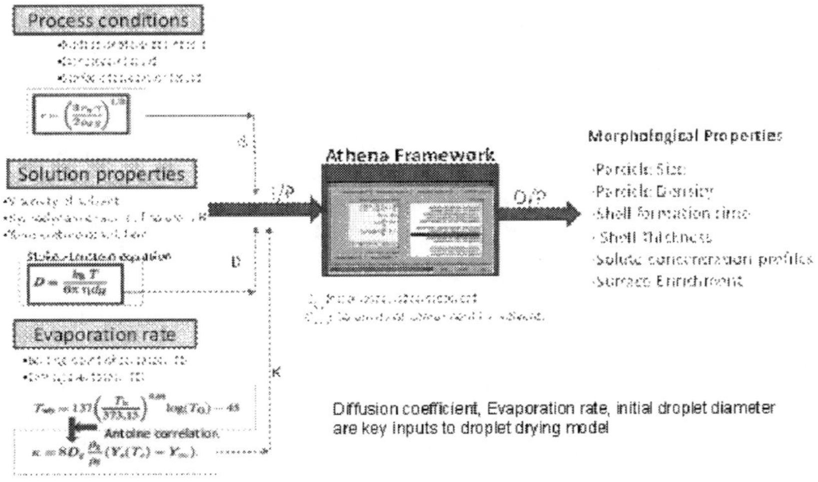

***Figure 1.* The model framework combines process conditions, solution properties, and evaporation rate in an Athena framework to generate predictions of particle morphological properties**

Results:

This particle formation model was shown to compare well against experimental results in the literature. Figure 2 illustrates a comparison of the model predictions of spray dried trehalose against results from Vehring et al.[ii] The samples evaluated cover a low Peclet number range of 0.44 to 1, with an initial droplet diameter of 20 µm resulting in a dry particle diameter of 1.7 to 1.8 µm. The model accurately predicts the measured dry particle density, which is near the density of amorphous trehalose indicating solid particles.

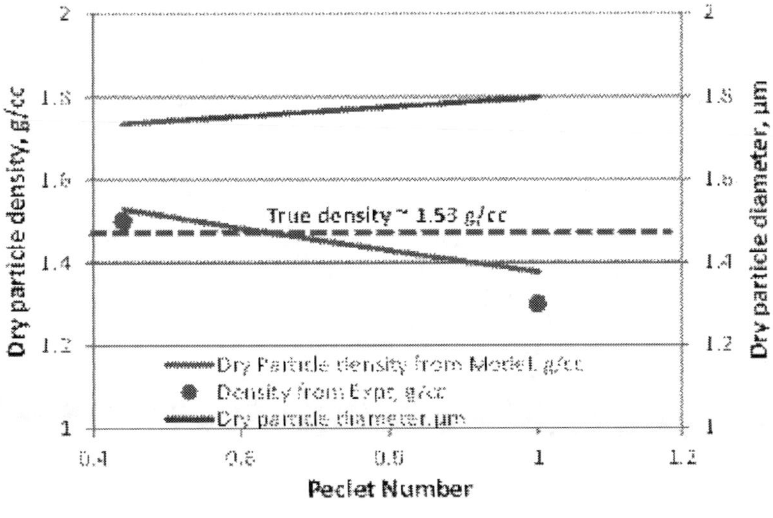

Figure 2. Comparison of experimental results from Vehring et al and model predictions of trehalose particles formed through spray drying.[ii]

It was also found to predict well the experimental results from Vehring et al.[ii] for spray drying of glycoprotein solution as shown in Figure 3. As the Peclet number increases, the particle diameter increases while the density decreases. Both the experimental and modeled particle densities (0.1 to 0.5 g/cc) are significantly lower than the density of solid glycoprotein. This indicates hollow particles with void space, a result confirmed through microscopy of the generated particles.

*** Formatting Issue - Best Available Paper/Graphic ***

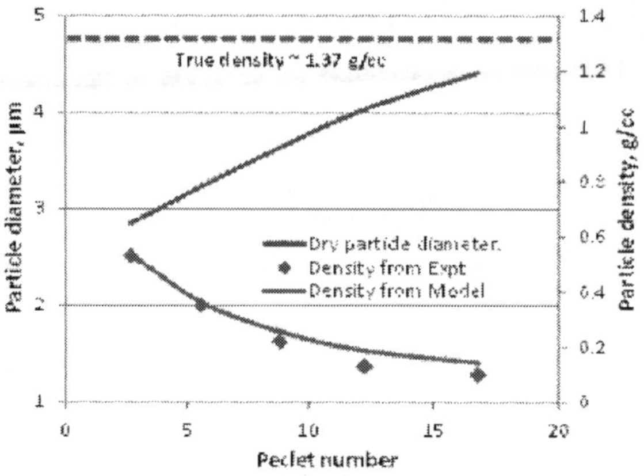

Figure 3. Comparison of experimental results from Vehring et al and model predictions of glycoprotein particles formed through spray drying.[ii]

To apply the droplet drying model to spray drying API-HPMCAS dispersions, fundamental solution properties were characterized. The viscosities of such solutions were found to vary strongly based on the specific HPMCAS polymer and spray drying solvent. As shown in Figure 4, at similar concentrations a wide range of viscosities were observed indicating very different polymer configurations in solution. Similar measurements were conducted to quantify the diffusivity of polymer and API solutes in solution and the surface tension of these solutions. Having quantified these properties, predictions around the droplet drying process and resulting particle properties will be shared. It will be shown that depending on the solvents selected for spray drying, diffusion controlled or surface recession controlled drying kinetics can occur resulting in very different particle morphologies and compositional uniformity.

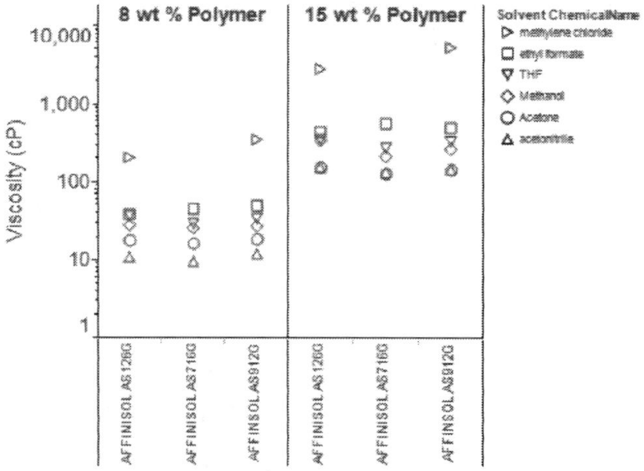

Figure 4: Viscosity of Polymer/Solvent Solutions that serve as inputs for particle formation modeling

Conclusions:

Experimental analysis of formulation solution properties and polymer chemistry provide a solid framework for building modeling capabilities to predict the formation of solid particles from liquid droplets in the spray dried dispersion process. The two examples presented highlight the dramatically different potential outcomes from a spray drying process, from solid particles of high density to low density hollow particles. The ability to predict the nature of the particles formed based on formulation and process properties provides the formulator with a valuable tool to optimize this process to hone in on a targeted outcome prior to conducting time consuming experiments. Particle morphology can play a critical role in downstream processing and ultimately the application performance. Combining polymer chemistry and formulation expertise with modeling capabilities provides formulators with a streamlined route to product development.

---

[i] Chen, Xiao Dong; Sidhu, Harvinder; Nelson, Mark; "Theoretical probing of the phenomenon of the formation of the outermost surface layer of a multi-component particle, and the surface chemical composition after the rapid removal of water in spray drying," Chem. Eng. Sci. 66 (2011) 6375-6384.

[ii] Vehring, Reinhard; Foss, Wiillard R.; Lechuga-Ballesteros, David; "Particle formation in spray drying," Aerosol Science 38 (2007) 728-746.

458912 Modeling Dissolution of Polymer Strip Films Loaded with Poorly Water-Soluble Drug Nanoparticles

Abstract:

Thanks to a wide variety of excipients, APIs, and their interactions, reliable prediction of the performance of any solid dosage form is a significant challenge. This is especially true for swellable polymeric strip films containing poorly water-soluble drug nanoparticles, as several different mechanisms may potentially control the rate of drug release. These limiting mechanisms include diffusion of the dissolved drug out of the swollen polymer gel layer, erosion of the polymer matrix, and dissolution of the poorly water-soluble drug particles. The objective of this work is to develop a predictive model for drug release from nanoparticle-laden polymer strip films based on various formulation parameters and inherent drug/excipient properties. This model utilizes first principles to estimate evolution of film thickness, dissolved drug concentration, drug particle size, and water concentration throughout the film over time during the dissolution process. Diffusion of dissolved drug and solvent, drug particle dissolution, polymer erosion, and film swelling/shrinkage are all taken into consideration. Results from this model will be used to better understand the fundamental driving forces behind drug release from poorly water-soluble drug particle-laden polymer strip films, allowing for more intelligent formulation design of targeted applications.

468415 A Novel Method for Grindability Assessment of Pharmaceutical Powders

Abstract:

In the early stages of development of a new Active Pharmaceutical Ingredient (API), insufficient material quantity is usually available for addressing process engineering issues, such as milling. Mechanical properties that are accountable for size reduction are hardness, toughness and stiffness (for semi-brittle failure), and these can be characterised using a small sample quantity by nanoindentation. However, the use of nano-indentation may not be appropriate for strain-rate sensitive solids, and hence impact testing, backed up by breakage analysis, can give better representative information on grindability. Unfortunately, there is no commercially-available single particle impact testing for fine particulate solids. In contrast, dry powder dispersers, such as the Scirocco powder disperser of Malvern Mastersizer 2000, are widely available. In this device, particles are dispersed by an air jet in a Venturi eductor configuration, impacted at a bend and presented to the laser light beam for size analysis by laser light diffraction. So this device has the potential to be used for testing the breakage propensity (grindability) of particles. However, particles of different sizes accelerate to different velocities, due to short acceleration length, and break to different extents; hence the results will be 'fuzzy', unless the relation between particle size and impact velocity and nozzle pressure is established.

Three dimensional multiphase CFD simulations of the Scirocco disperser have been carried out to analyse the air flow field, following which particle trajectories and impact velocities have been calculated. These calculations are then used in the estimation of the breakage propensity parameter. The extent of breakage and change in the surface area are then related to the breakage propensity parameter, for which a good correlation is shown for a range of particle sizes of several organic particles. The obtained information shows the ease of grindability of different materials relative to each other.

457458 Understanding the Role of Hpmcas in Phenytoin Solid Dispersion Formulation Using Coarse Grained Molecular Dynamics Simulations

Abstract:

Many orally administrated drug candidates currently being developed are highly lipophilic, which necessitates the use of an excipient to stabilize and increase their solubility in the gastrointestinal tract. Hydroxypropyl methylcellulose acetate succinate (HPMCAS) has been identified to be a very effective excipient; however, to date very little understanding has been established on how HPMCAS interacts with drug molecules at a molecular level, and how different structural variables such as molecular weight and substitution pattern affect the performance of HPMCAS as an excipient. In this study, we employed coarse-grained (CG) molecular dynamics (MD) simulations to model HPMCAS and a drug candidate, phenytoin, in an implicit water environment. We parameterized our CG model using the radial distribution functions (RDFs) from atomistic simulations of short HPMCAS oligomers and applied the model to simulate long chains. We studied the effect of individual substitution type (e.g. methyl, hydroxypropyl, acetate etc.) by simulating homogenous polymer chains with drug molecules. We found the conformation and the structural details of the polymer-drug complex are affected by monomer substitution type, the amount of polymer loading, and polymer length. Specifically, bulky hydroxypropyl group results in a more loosely packed polymer matrix, thus allowing drug molecules to diffuse through the matrix more easily. The use of high polymer concentration (>80 wt%) can significantly reduce the size of the drug aggregates in the complex. We determined the drug release rate for different polymer-drug complex by running simulations for over 50Î¼s. We also modeled the heterogeneous copolymer and investigated the effect of deprotonated succinate substitution on the release behavior. We found that the polymer-drug complex swells when succinate groups switch from protonated to deprotonated state, thus facilitates the release of the drug molecules. Our simulation box size is comparable to the size of polymer-drug complex identified from the experiment. This study provides insights to help optimizing the composition of HPMCAS (e.g. ratio of acetate to succinate group) for any given drug.

464851 Assessing the Efficacy of Thermo-Responsive Poly(N-isopropylacrylamide) in Phenytoin Solid Dispersion Formulation Using Atomistic Molecular Dynamics Simulations

Abstract:

Abstract for AICHE Annual Meeting 2016

Submission Deadline: May 9th, 2016

Session: 26004 Amorphous Solid Dispersion for Drug Product

Category: Pharmaceutical Discovery, Development, and Manufacturing Forum

Assessing the Efficacy of Thermo-Responsive Poly(N-isopropylacrylamide) in Phenytoin Solid Dispersion Formulation Using Atomistic Molecular Dynamics Simulations

Soroush Moghadam[1] and Ronald G. Larson[2]

[1]Department of Mechanical Engineering, University of Michigan, Ann Arbor, MI, 48109

[2]Department of Chemical Engineering, University of Michigan, Ann Arbor, MI, 48109

For an active pharmaceutical ingredient (API) to be absorbed from the gastro-intestinal tract into the bloodstream, it must be sufficiently soluble in the GI fluid. However, more than 40% of new chemical entities in the pharmaceutical industry are poorly soluble, leading to drug aggregation and crystallization, which necessitates the use of an excipient to stabilize and increase the solubility. Solid dispersions can maintain the API in an amorphous form and result in a supersaturated API solution upon its dissolution in the GI tract. As the hydrophilic component, thermo-sensitive poly(N-isopropylacrylamide), namely pNIPAAm, has been suggested to provide desired controlled release properties due to its lower critical solution temperature (LCST~32°C), biocompatibility and also large design space for its functional groups. Although many simulation studies analyzed the LCST behavior of pNIPAAm chains in water, to date there has not been a computational analysis of how pNIPAAm interacts with drug molecules at molecular levels as a function of temperature and how we can tune these properties to reach desired stability of polymer-drug complex. In this study, we performed all-atom molecular dynamic (AA-MD) simulations to model pNIPAAm and a drug candidate, phenytoin, in an explicit water environment. AMBER forcefield and solvated ensemble averaging of partial charges have been implemented. After validating the forcefield parameters using the well-known lower critical solution behavior of pNIPAAm, we simulated the polymer-drug complex in water and its behavior at temperatures below (295K) and above the LCST (310K). The effect of copolymerization with hydrophilic elements (e.g. Dimethylacylamide (DMA)) on the excipient behavior has been studied. We found that existence of hydrophilic comonomers in a pNIPAAm chain dramatically affects its temperature transition dependence. However, It affects affinity of the excipient functional groups to interact with drug molecules as well. Using radial distribution functions, we found that there is an optimum comonomer molar fraction of around 20-30% DMA at which interaction with phenytoin drug molecules is strongest. Furthermore, maintaining the weight percents of the components, using shorter polymer chains close to pNIPAAm reported persistence lengths can significantly influence the drug-polymer interaction upon heating. Although the length and time scales of our simulations are not high enough to be conclusive, this study provides insights to help optimizing the composition of pNIPAAm for any given drug.

KEYWORDS: Polymer Drug Aggregates, Molecular Dynamics, pNIPAAm

474782 Solid State Form Understanding Applied to the Design of a Spray Dried Dispersion Process: A Quality By Design Approach to Optimization of Feed Solution Loading

Abstract:

Spray drying is a common unit operation across many industries due to the ability to control bulk powder properties and isolate and dry product from a liquid feed in a single operation. For the pharmaceutical industry, spray drying offers the control of physical properties of crystalline or amorphous active pharmaceutical ingredients (APIs), as a neat or formulated drug product that can lead to improved bioavailability of the drug candidate. Drug candidate molecules are often formulated as a solid dispersion, incorporating both API and a polymer excipient for enhanced dissolution performance and improved physical stability. In order to reliably achieve solid form control to the amorphous state during the spray drying operation, the API must be fully solubilized in the spray dryer feed solution, eliminating crystalline API material. The crystalline form of the API may or may not be the most stable form in the selected spray drying solvent. To ensure robust solid state form control of the spray dried dispersion (SDD), not only should the form landscape of the API in the spray drying solvent be well understood the process control capabilities during feed solution preparation must be considered as well. Numerical modeling of the solubility of relevant API forms, in conjunction with feed solution loading and process control capability should be used to determine an optimal feed solution loading. The operating point can be selected through optimization based upon on selected risk-based criteria and control capability. Applying this methodology to Compound A, the throughput of the spray drying process was improved two-fold, while ensuring a robust control strategy to deliver the desired API in an amorphous form as a spray dried dispersion.

468548 Ternary Dispersions of Itraconazole: A Comparison of Hot Melt Extrusion and Spray Drying

Abstract:

Ternary dispersions of itraconazole: a comparison of hot melt extrusion and spray drying

Mark T. Davis[1]*, Catherine B. Kelly[1], & Gavin M. Walker[1, 2]

[1]Synthesis and Solid State Pharmaceutical Centre (SSPC), Bernal Institute,

University of Limerick, Limerick, Ireland

[2]Pharmaceutical Manufacturing Technology Centre, PMTC, University of Limerick, Limerick, Ireland

Tel.: +353-(0)-61- 213096, E-mail: mark.davis@ul.ie

ABSTRACT

BCS Class II drug itraconazole (ITR) represents a difficult formulation challenge due to its very low aqueous solubility and poor dissolution kinetics. It has been extensively studied in an attempt to achieve higher bioavailability and therefore better dissolution. The amorphous state represents a disordered structure relative to the crystalline form which has been shown to have an advantage of higher dissolution rates and better oral absorption. Much work has been carried out on the performance of single polymers in the formulation of the amorphous form as an amorphous solid dispersion however evidence exists that ternary amorphous solid dispersions (ASDs) containing two rationally chosen biocompatible polymers may exhibit better amorphous stability and dissolution profiles than their binary

equivalents.

In this study, ITR was formulated into a number of amorphous solid dispersions using hot melt extrusion (HME) and spray drying (SD). The systems comprised ternary formulations of itraconazole with Soluplus® and/or HPMCP. Initial studies within our group have indicated that although HPMCP is an enteric polymer it interacts more favourably with ITR than Soluplus does and so should offer amorphous stabilisation benefits. Soluplus, however, is an excellent aqueous phase solubiliser for ITR so contributes to the dissolution properties of the ASD.

A drug loading of 40% was chosen to produce tablets of 500 mg containing 200 mg of active pharmaceutical ingredient (API). Equivalent physical mixtures were prepared to measure the solubility benefit of the amorphous state. All formulations were shown to be completely free of either liquid-crystalline or crystalline ITR using both wide-angle XRD and hyper differential scanning calorimetry (DSC). Further analysis of physical characteristics was carried out using SEM, laser scattering and density measurements.

Dissolution testing was carried out on both the tablets and the physical mixture using USP dissolution apparatus 2 (paddle at 37 °C) in biorelevant media. 5ml aliquots were removed at regular intervals and, following production and validation of an absorbance - concentration calibration curve for ITR, dissolved drug was quantified using UV-Vis spectrophotometry. Differences in dissolution rate and extent were rationalised in terms of the preparation methods, hydrogen bonding and the material properties.

469561 Tunable Drug Loaded Stable and Fast Release Amorphous Composites Production Using Mesoporous Neusilin ®UFL2

Abstract:

Literature has revealed that there may exist possibility of greatly reduced mobility of drug molecules due to confinement, which could occur through several different manners. Impregnation of poorly water drug solutions in to mesoporous materials is one such attractive approach for nano-confinement that leads not only to creation of amorphous state but also its stability due to what may be the poor molecular mobility. Such as approach is examined here to increase the aqueous solubility and thus the bioavailability of poorly soluble drugs since it allows formation of drug amorphous state which generally exhibit higher apparent solubilities than their crystalline counterparts. Unfortunately, mesoporous materials such as silica are expensive and proprietary, and not easily available in bulk quantities. In this work, Neusilin® UFL2 as commercially available mesoporous amorphous materials was used to prepare amorphous drug composites of the model poorly soluble drugs Fenofibrate and Itraconazole via impregnation. Drug loading could be tuned as low as 1.0% (w/w) to as high as 30%. PXRD, DSC, and Raman spectroscopy were used to assess the amorphousness/crystallinity of the drug in the impregnated composites after formation as well as after storage. It was shown that the impregnated drug composites are stable for three months even stored at 40°C and 75% relative humidity. Re-crystallization of the amorphous drug was completely suppressed in the composites due to the confinement effect provided by the mesopores of Neusilin® UFL2. The amorphous drug composites result in higher apparent solubility and faster release/dissolution rate of the model drugs compared to crystalline counterparts. UV imaging technique, a real-time and sample sparing method, was used to qualitatively differentiate the amorphous and crystalline drug dissolution besides the conventional USP II dissolution. In conclusion, preparation of amorphous drug composites using Neusilin® UFL2 appears to be promising as a stable amorphous drug delivery vehicle providing a rapid drug release and subsequently bioavailability enhancement.

Keywords: Mesoporous Neusilin®, Impregnation, Amorphous Drug Composites, Stability, Dissolution Enhancement

458072 Hot Melt Extrusion of Ternary Solid Dispersions of Itraconazole: Amorphous *Vs* Liquid Crystal Intermediate

Abstract:

Abstract

It has been demonstrated that ternary solid dispersion compositions can be more efficient than binary compositions in terms of their excellent physical stability and ability to inhibit the recrystallization of API's in bio-relevant media. This study is mainly focused on investigating the preparation of ternary solid dispersions of itraconazole, a BSC Class (II) model drug, containing a water-soluble polymer and/or a water-insoluble polymer using hot melt extrusion. The formation, physio-chemical characterization, stability, and phase transformation in the context of drug formulation and development of amorphous/liquid crystals of itraconazole was investigated in detail.

Five combinations of ternary solid dispersions comprising itraconazole, hydroxy propyl methyl cellulose phthalate (HPMCP-HP50) and/or soluplus were prepared by hot melt extrusion (HME) using a 16 mm pharmalab co-rotating twin screw extruder and L/D ratio of 40: 1 (Thermo Fisher Scientific, Karlsruhe, Germany). Information regarding the miscibility of the drug with the excipient, physical state, phase transformation, possible interactions and proper storage conditions were obtained by employing small angle powder X-ray diffractometry (SAXS), differential scanning calorimetry (DSC), Raman spectroscopy and infrared spectroscopic (FT-IR) measurements for specific polymer-polymer and drug-polymer interactions.

The SAXS results showed that itraconazole in the solid dispersion was in the amorphous state and indicated the absence of liquid crystalline itraconazole. The Raman and FT-IR analysis indicated the formation of hydrogen bonding between itraconazole and HPMCP-HP50/soluplus molecules and that itraconazoleâ€'HPMCP-HP50 interactions were stronger than itraconazoleâ€'soluplus interactions. From the DSC scans melt extrudates of itraconazole and HPMCP-HP50/soluplus exhibited one glass transition (Tg) and an absence of an endotherm around the melting point of itraconazole confirming that the API is present only in its amorphous state and that each formulation consists of a single homogenous amorphous phase of Itraconazole and HPMCP-HP50 and/or soluplus.

AUTHOR INDEX

AUTHOR INDEX

AUTHOR INDEX

AUTHOR INDEX

AUTHOR INDEX

AUTHOR INDEX

AUTHOR INDEX

AUTHOR INDEX

AUTHOR INDEX

AUTHOR INDEX